Exocytosis: From Molecules to Cells

Online at: https://doi.org/10.1088/978-0-7503-3771-7

Biophysical Society–IOP Series

About the Series

The Biophysical Society and IOP Publishing have forged a new publishing partnership in biophysics, bringing the world-leading expertise and domain knowledge of the Biophysical Society into the rapidly developing IOP ebooks program.

The program publishes textbooks, monographs, reviews, and handbooks covering all areas of biophysics research, applications, education, methods, computational tools, and techniques. Subjects of the collection will include: bioenergetics, bioengineering, biological fluorescence, biopolymers *in vivo*, cryo-electron microscopy, exocytosis and endocytosis, intrinsically disordered proteins, mechanobiology, membrane biophysics, membrane structure and assembly, molecular biophysics, motility and cytoskeleton, nanoscale biophysics, and permeation and transport.

A full list of titles published in this series can be found here: https://iopscience.iop.org/bookListInfo/iop-series-in-biophysical-society.

Exocytosis: From Molecules to Cells

Edited by
Arun Anantharam
Department of Neuroscience, University of Toledo Medical School, Toledo, OH, USA

Jefferson Knight
Department of Chemistry, University of Colorado Denver, Denver, CO, USA

IOP Publishing, Bristol, UK

ISBN 978-0-7503-3771-7 (ebook)
ISBN 978-0-7503-3769-4 (print)
ISBN 978-0-7503-3772-4 (myPrint)
ISBN 978-0-7503-3770-0 (mobi)

DOI 10.1088/978-0-7503-3771-7

Version: 20221201

IOP ebooks

British Library Cataloguing-in-Publication Data: A catalogue record for this book is available from the British Library.

Published by IOP Publishing, wholly owned by The Institute of Physics, London

IOP Publishing, No.2 The Distillery, Glassfields, Avon Street, Bristol, BS2 0GR, UK

US Office: IOP Publishing, Inc., 190 North Independence Mall West, Suite 601, Philadelphia, PA 19106, USA

To our mentors.

Contents

Acknowledgements

The editors are deeply thankful to all the authors, anonymous peer reviewers, and artist who contributed to making this e-book a success. We have enjoyed working with you and getting to know you and your amazing contributions to science. Thank you for helping to make this e-book reflect the current state of the field. We also acknowledge Dr. John Heuser for contributing electron micrographs that helped to inspire the cover art, in addition to his many other contributions to the field over the years.

We thank the staff at IOP Publishing for their timely efforts to bring this book to publication. We also thank the Biophysical Society for sponsoring this book series and inviting us to edit a volume for it. We especially acknowledge the vibrant scientific community in the Membrane Fusion, Fission, and Traffic subgroup.

Editor biographies

Arun Anantharam

Arun Anantharam is an Associate Professor of Neurosciences at the University of Toledo. He earned his B.A. from Columbia University where he double-majored in English Literature and Neuroscience, and a Ph.D. from Cornell University where he was mentored by Dr. Lawrence Palmer in the Department of Physiology and Biophysics. He performed his postdoctoral studies with Drs. Daniel Axelrod (Physics) and Ronald W. Holz (Pharmacology) at the University of Michigan. Arun has been a National Academies Education Fellow in the Life Sciences and Chair of the Membrane Fusion, Fission, and Traffic Subgroup of the Biophysical Society. His research focuses on the mechanisms of calcium sensing and exocytosis in neuroendocrine cells.

Jefferson Knight

Jefferson Knight received a B.S. in Chemistry from the University of North Carolina, Chapel Hill in 2000 and a Ph.D. in Pharmacology from Yale University in 2006 under the mentorship of Dr. Andrew Miranker. After completing postdoctoral studies in the lab of Dr. Joseph Falke at the University of Colorado, Boulder, he joined the Department of Chemistry at the University of Colorado Denver in 2010. He was named a Henry Dreyfus Teacher-Scholar in 2018 and a Fulbright Scholar in 2019. Jeff is currently Professor of Chemistry and Director of the BS Biochemistry program at the University of Colorado Denver. His research focuses on the molecular biophysics and biochemistry of protein-membrane interactions involved in calcium-regulated membrane fusion and insulin secretion. He has been a member of the Biophysical Society since 2006.

List of contributors

Arun Anantharam
Department of Neuroscience, University of Toledo Medical School, Toledo, OH, USA

Daniel Axelrod
Department of Physics and department of LSA Biophysics, University of Michigan, Ann Arbor, MI, USA

Huan Bao
Department of Molecular Medicine, The Scripps Research Institute, Jupiter, FL, USA

Natasha Buwa
Department of Biochemistry and Molecular Biotechnology, University of Massachusetts Chan Medical School, Worcester, MA, USA

J David Castle
Department of Cell Biology and Center for Membrane and Cell Physiology, University of Virginia School of Medicine, VA, USA

Robert E Coffman
Neuroscience Center and Department of Cell Biology and Physiology, Brigham Young University, Provo, UT, USA

Sarah Dallo
Department of Molecular Medicine, The Scripps Research Institute, Jupiter, FL, USA

Yuki Hashimotodani
Graduate School of Brain Science, Doshisha University, Kyoto, Japan

Jenny E Hinshaw
Laboratory of Cell and Molecular Biology, National Institute of Diabetes and Digestive and Kidney Diseases, National Institutes of Health, Bethesda, MD, USA

Ronald W Holz
Department of Pharmacology, University of Michigan, Ann Arbor, MI, USA

Meyer B Jackson
Department of Neuroscience, University of Wisconsin–Madison, USA

Jophin Joseph
Department of Mechanical Engineering, University of Michigan, Ann Arbor, MI, USA

Jefferson D Knight
Department of Chemistry, University of Colorado Denver, Denver, CO, USA

Michelle K Knowles
Department of Chemistry and Biochemistry, University of Denver, Denver, CO, USA
and
Molecular and Cellular Biophysics Program, University of Denver, Denver, CO, USA

Nidhi Kundu
Laboratory of Cell and Molecular Biology, National Institute of Diabetes and Digestive and Kidney Diseases, National Institutes of Health, Bethesda, MD, USA

Shan-Shan Lin
Department of Mechanical Engineering, University of Michigan, Ann Arbor, MI, USA

Allen P Liu
Department of Mechanical Engineering, University of Michigan, Ann Arbor, MI, USA
and
Department of Biomedical Engineering, University of Michigan, Ann Arbor, MI, USA
and
Department of Biophysics, University of Michigan, Ann Arbor, MI, USA
and
Cellular and Molecular Biology Program, University of Michigan, Ann Arbor, MI, USA

Anarkali Mahmood
Department of Chemistry and Biochemistry, University of Denver, Denver, CO, USA

Lisa M Mehlmann
Department of Cell Biology, UConn Health, Farmington, CT, USA

Takafumi Miki
Graduate School of Brain Science, Doshisha University, Kyoto, Japan

Ira Milosevic
Wellcome Centre for Human Genetics, Nuffield Department of Medicine, NIHR Oxford Biomedical Research Centre, University of Oxford, Oxford, UK
and
Multidisciplinary Institute for Ageing, University of Coimbra, Coimbra, Portugal

Peter Mühlenbrock
Georg-August-Universität Göttingen, Institute of Organic and Biomolecular Chemistry, Göttingen, Germany

Mary Munson
Department of Biochemistry and Molecular Biotechnology, University of Massachusetts Chan Medical School, Worcester, MA, USA

Leonora Martínez-Núñez
Department of Biochemistry and Molecular Biotechnology, University of Massachusetts Chan Medical School, Worcester, MA, USA

Takeshi Sakaba
Graduate School of Brain Science, Doshisha University, Kyoto, Japan

Claudia Steinem
Max Planck Institute for Dynamics and Self-Organization, Göttingen, Germany

Ann M Wehman
Department of Biological Sciences, University of Denver, Denver, CO, USA
and
Molecular and Cellular Biophysics Program, University of Denver, Denver, CO, USA

Alan W Weisgerber
Department of Chemistry and Biochemistry, University of Denver, Denver, CO, USA

Dixon J Woodbury
Neuroscience Center and Department of Cell Biology and Physiology, Brigham Young University, Provo, UT, USA

IOP Publishing

Chapter 1

Introduction

Jefferson D Knight and Arun Anantharam

Exocytosis is the process by which cargo proteins or other compounds are packaged into membrane-bound secretory vesicles and released into the plasma membrane or extracellular environment. Pathways for inserting proteins into cellular membranes exist in all kingdoms of life; these pathways are thought to be the foundation for the evolution of organelle-based exocytosis. Major features of the secretory pathway are conserved from yeast to mammals, and key exocytotic proteins are expressed in most if not all mammalian cells. The controlled secretion of compounds allows a cell to influence its immediate environment, communicate chemically with other cells, and eliminate waste. In animals, a wide variety of cells use exocytosis for processes as diverse as controlling cell polarization, communicating with neighboring cells across a synapse, releasing hormones into circulation, and regulating sperm-egg fusion, among many other functions.

Biophysical and molecular insights into exocytosis now span more than 60 years. Christian de Duve suggested the term 'exocytosis' in 1963 to refer to the process of fusion at the cellular level that had been described a few years earlier by George Palade [1, 2]. Palade and de Duve were awarded the Nobel Prize for Physiology or Medicine in 1975 along with Albert Claude, whose centrifugation methods for isolating organelles enabled de Duve's studies of lysosomes and Palade's studies on ribosomes and organelle trafficking. These three pioneers joined the ranks of prior Nobel laureates Alan Hodgkin, Andrew Huxley, and Sir John Eccles (1963), who, along with Sir Bernard Katz (1970), had been honored for their studies on neurotransmission. Since then, several more Nobel Prizes in Physiology or Medicine have recognized advances in our understanding of cellular membrane trafficking and exocytosis components, including those awarded to Günter Blobel (1999); Randy Schekman, Thomas Südhof, and James Rothman (2013); and Yoshinori Ohsumi (2016). As our collective understanding has progressed from the level of cells to organelles to molecules, biophysical approaches have played key roles in many of the seminal findings. No doubt the biophysics community will continue to be essential to the study of exocytosis for many years to come.

doi:10.1088/978-0-7503-3771-7ch1

1.1 Purpose and scope of this e-book

This book updates the biophysics community on recent developments in our understanding of exocytosis pathways spanning from the molecular to the cellular and organism levels, including reconstituted systems and *in vivo* studies. Its broad scope provides readers with a global view that illuminates common threads among pathways in different organisms and cell types as well as revealing the broad questions that remain unanswered. Thus, we encourage readers to read beyond the chapter on their closest system of interest and learn more about the ideas, methods, and experimental systems employed by other arms of our research community.

This book is intended as a reference both for experienced researchers and newcomers to the fields of membrane trafficking and exocytosis. For students and others who are new to the field, each chapter in this book provides a current perspective that introduces overarching concepts and particular areas of interest. For experienced researchers, the breadth of topics provides a view beyond one's particular field, emphasizing connections between different subdisciplines. Diverse perspectives are included from authors with expertise spanning cell biology, chemistry, engineering, and medicine. We hope that reading about exocytosis in a broad variety of cell types and from researchers in different disciplines will help to overcome some of the silo effects that inevitably exist within such a broad and diverse field. We have witnessed many advances in the field in recent years. These advances have been documented in countless research articles as well as in a number of excellent reviews and more focused compendia. Apologies are due to the authors of these works that we, or other contributors to this book, have not been able to cite. Our goal as editors has been to prepare a collection of chapters that reflect a sampling of the breadth of molecular, methodological, and systemic advancements in the study of exocytosis pathways over the last 15–20 years.

The chapters of this book summarize these recent advances and how they have impacted our collective current understanding throughout many exocytotic systems that are of interest to the biophysics community. The first six chapters after this Introduction provide an in-depth review of the current understanding of exocytosis pathway stages and the major molecular players, from secretory vesicle biogenesis to fusion pore morphology and the role of membrane tension. The next section comprises three chapters on insights into *in vitro* reconstituted systems, including two on reconstituted systems of SNARE-catalyzed membrane fusion and one summarizing the known and postulated roles of anesthetics in influencing membrane fusion and exocytosis. The final section includes five chapters that provide an integrated view of exocytosis in individual cellular systems: yeast, chromaffin cells, sperm and egg, calyx of Held, and exosome production. This is far from an exhaustive list of cellular exocytotic systems, but these chapters provide excellent in-depth descriptions of these representative systems. Some of these systems are well-studied (e.g. yeast, chromaffin cells, etc) while, in others, our molecular and biophysical understanding of the system is still emerging. The reader should come away with a sense of the similarities that exist between these systems and an understanding of what is known or should be known.

Because such a book can never cover every aspect of a field, we direct the reader to additional recent reviews on relevant topics. These include reviews on important lipids [3] and proteins in the exocytosis pathway, such as Rab27 effectors [4], Munc13 and Munc18 [5, 6], and synaptotagmins [7] including synaptotagmin-1 [8] and synaptotagmin-7 [9]; the recent recognition of liquid–liquid phase separation [10, 11] and its possible connection to synaptic vesicle cycling [12, 13]; contributions from theoretical physics [14, 15] and molecular dynamics simulations [16] toward understanding membrane fusion; as well as summaries of recent advances in understanding of neurotransmitter release [17, 18], synaptic facilitation [19], and insulin secretion [20, 21]. A recent historical perspective on chromaffin cells is also available for further reading [22]. Many more excellent review articles are cited throughout the various chapters of this book.

References and recommended reading

[1] de Duve C 1963 *Ciba Foundation Symp. on Lysosomes* 126 (Boston: Little, Brown)

[2] Palade G E 1959 Subcellular Particles: A Symp. Held during the Meeting of the Society of General Physiologists at the Marine Biological *Laboratory, Woods Hole, Mass.* 1958 ed T Hayashi 64–83 (New York: Ronald Press)

[3] Binotti B, Jahn R and Perez-Lara A 2021 An overview of the synaptic vesicle lipid composition *Arch. Biochem. Biophys.* **709** 108966

[4] Izumi T 2021 *In vivo* roles of Rab27 and its effectors in exocytosis *Cell Struct. Funct.* **46** 79–94

[5] Zhang Y and Hughson F M 2021 Chaperoning SNARE folding and assembly *Annu. Rev. Biochem.* **90** 581–603

[6] Brunger A T, Choi U B, Lai Y, Leitz J, White K I and Zhou Q 2019 The pre-synaptic fusion machinery *Curr. Opin. Struct. Biol.* **54** 179–88

[7] Volynski K E and Krishnakumar S S 2018 Synergistic control of neurotransmitter release by different members of the synaptotagmin family *Curr. Opin. Neurobiol.* **51** 154–62

[8] Brunger A T, Leitz J, Zhou Q, Choi U B and Lai Y 2018 Ca^{2+}-triggered synaptic vesicle fusion initiated by release of inhibition *Trends Cell Biol.* **28** 631–45

[9] MacDougall D D, Lin Z, Chon N L, Jackman S L, Lin H, Knight J D and Anantharam A 2018 The high-affinity calcium sensor synaptotagmin-7 serves multiple roles in regulated exocytosis *J. Gen. Physiol.* **150** 783–807

[10] Milovanovic D and De Camilli P 2017 Synaptic vesicle clusters at synapses: a distinct liquid phase? *Neuron* **93** 995–1002

[11] Lautenschlager J 2022 Protein phase separation hotspots at the presynapse *Open Biol.* **12** 210334

[12] Wu X, Qiu H and Zhang M 2022 Interactions between membraneless condensates and membranous organelles at the presynapse: a phase separation view of synaptic vesicle cycle *J. Mol. Biol.* 167629

[13] Sauvola C W and Littleton J T 2021 SNARE regulatory proteins in synaptic vesicle fusion and recycling *Front. Mol. Neurosci.* **14** 733138

[14] Mion D, Bunel L, Heo P and Pincet F 2022 The beginning and the end of SNARE-induced membrane fusion *FEBS Open Bio.* https://doi.org/10.1002/2211-5463.13447

[15] Risselada H J and Mayer A 2020 SNAREs, tethers and SM proteins: how to overcome the final barriers to membrane fusion? *Biochem. J.* **477** 243–58

[16] Risselada H J and Grubmuller H 2021 How proteins open fusion pores: insights from molecular simulations *Eur. Biophys. J.* **50** 279–93

[17] Rizo J 2022 Molecular mechanisms underlying neurotransmitter release *Annu. Rev. Biophys.* **51** 377–408

[18] Dittman J S and Ryan T A 2019 The control of release probability at nerve terminals *Nat. Rev. Neurosci.* **20** 177–86

[19] Jackman S L and Regehr W G 2017 The mechanisms and functions of synaptic facilitation *Neuron* **94** 447–64

[20] Thurmond D C and Gaisano H Y 2020 Recent insights into beta-cell exocytosis in type 2 diabetes *J. Mol. Biol.* **432** 1310–25

[21] Omar-Hmeadi M and Idevall-Hagren O 2021 Insulin granule biogenesis and exocytosis *Cell. Mol. Life Sci.* **78** 1957–70

[22] Anantharam A and Kreutzberger A J B 2019 Unraveling the mechanisms of calcium-dependent secretion *J. Gen. Physiol.* **151** 417–34

Part I

Components of exocytotic pathways

IOP Publishing

Exocytosis: From Molecules to Cells

Arun Anantharam and Jefferson Knight

Chapter 2

Biogenesis of secretory granules

J David Castle

Secretory granules are specialized organelles for intracellular storage and stimulus-dependent exocytosis in regulated secretory cells. Granule formation begins in association with the *trans*-Golgi network (TGN). Here, the bulk of secretory cargoes is largely sorted and, along with a cohort of membrane proteins, becomes segregated within condensing vacuoles from proteins targeted at other destinations within the cell. Upon scission from the TGN, condensing vacuoles become immature granules and are progressively modified to become mature, stimulus-competent secretory granules. The purpose of this chapter is to review and update many of the earlier findings regarding the secretory sorting and granule maturation processes. Complementary to this goal are discussions of emerging realizations concerning the multiple and important roles that lipids play in regulating granule formation and the multiple membrane trafficking events that act to refine the composition of secretory granule membranes before their utilization for exocytosis. The latter encompass export and import pathways including recycling and reutilization routes that together act to fulfill the maturation process. Finally, one of the most exciting realizations in this field is that many if not most regulated secretory cells contain distinct subpopulations of secretory granules. Emerging data indicate that the subpopulations have differing compositions of secretory proteins and membrane proteins that regulate both small molecular import and responsiveness to exocytotic stimuli. Thus, important goals for future studies are to explore how and at what levels the biogenetic pathways for granule subpopulations are distinguished and to elucidate how the separate pathways contribute to the physiology of secretory responses.

2.1 Introduction

The secretory pathway of eukaryotic cells is the major intracellular highway supporting the production and export of macromolecules that serve a broad range of functions encompassing digestion, communication, construction of an extracellular supporting framework and defense. The pathway is membrane-bounded, originates in the endoplasmic reticulum, and leads to the cell surface. It is interconnected with import pathways that internalize and process materials from the extracellular environment. Alongside these activities, the secretory pathway engages in the construction and distribution of membrane components that make up the pathway's boundaries and various proteins that function within specific organelles of this network. While the core pathway operates continually, a broad range of unicellular organisms and numerous (perhaps nearly all) different cell types within multicellular organisms have developed the auxiliary capability to select and store a subset of secretory products and to mobilize and release them in response to external signaling. Following passage through the Golgi apparatus, these products are progressively segregated, often processed into active forms, and are concentrated to levels that considerably exceed immediate biosynthetic capacity. Here, they are stored within distinct limiting membranes until cellular signaling stimulates their mobilization for exocytosis. In certain cell types, e.g. peptide hormone-secreting endocrine cells, digestive enzyme- and mucus-secreting cells, and hematopoietic-derived cells that protect against pathogens, storage granule formation has been amplified to such an extent that it constitutes the major activity of the cell. Consequently, these kinds of cell are considered to be regulated secretory cells.

The main focus of this chapter is on the biogenesis of secretory (storage) granules (SGs), often designated dense core vesicles (DCVs) in endocrine/neuroendocrine cells. More than 20 years ago, I contributed to a review of earlier insights about SG formation [1]. Thus, the principal goals here are to briefly update many of the earlier findings and to discuss a wide range of more recent observations that reveal multiple steps of trafficking (including recycling) events. These encompass the process of initial formation in condensing vacuoles (CVs) associated with the TGN and progressive conversion to immature granules (IGs) and mature SGs. Moreover, while there have been a few previous reports that some regulated secretory cells produce more than one type of SG, there is a growing realization that the presence of distinct SG subpopulations may be a more general phenomenon. This will likely necessitate future refinement or reinterpretation of earlier findings. In addition, relationships to lysosome-related secretory organelles (LROs), identified in hematopoietic-derived cell types, melanocytes, keratinocytes, pneumocytes, and endothelial cells, have emerged. These realizations now drive consideration of whether LROs might serve as a model for a granule population that is maintained alongside classical SGs. Thus, I will present a current understanding of the distinct compositions and regulation of granule subpopulations and will also consider how they are formed as mainly learned from studies of pancreatic beta cells. These findings are considered in the context of physiological and pathological findings that provide interesting new insights about the potential roles of the subpopulations. Finally, an increased appreciation of granule biogenesis may serve as a basis for an expanded understanding that links cell biology to both physiology and disease.

2.2 Secretory protein sorting—beginning in the TGN

The biosynthetic highway proceeding from the endoplasmic reticulum (ER) through the Golgi complex reaches a point of divergence in the TGN. Here, anterograde pathways that lead directly to the plasma membrane, to endocytic compartments including lysosomes, and to forming SGs begin to branch from one another. For conceptual purposes, the TGN-associated branch points are typically drawn as well-defined abrupt exit routes along which proteins intended for distinct final destinations are selectively targeted (e.g. as depicted in widely used cell biology texts). In reality, the process of divergence is progressive and hierarchical such that it extends into the compartments (IGs and endosomes) that are distal to the TGN within the membrane trafficking network. As one consequence of this more delocalized arrangement, the membranes serving as carriers emanating from the TGN, especially those bounding CVs, are compositional hybrids that progressively become more refined as they continue various sorting processes including those that were initiated in the TGN. Further, while active receptor-based sorting events that determine the destination of individual proteins have been well characterized for certain classes of proteins, such as lysosomal prohydrolases and lysosomal membrane proteins [2, 3], some other proteins including various secretory and membrane proteins may advance along available route(s) without requiring individual protein selection. Interestingly, for the abundant quantities of secretory proteins that are destined to be concentrated and stored prior to release, entry into the CV mostly appears to be default-like, in that regulated secretory cells direct the majority of their secretory proteome in this direction. Further, such trafficking can be mimicked by exogenously expressed proteins that have no physiological relationship to what is being stored (e.g. [4–6]). Rather, most of the proteins for regulated secretion accumulate and are retained selectively within CVs and IGs by a passive sorting process involving intermolecular interactions that are collectively termed condensation. Here, a range of protein–protein interactions (both homotypic and heterotypic), binding of divalent cations (e.g. Ca^{2+} and Zn^{2+}) and other small molecules (e.g. adenosine triphosphate (ATP), biogenic amines), and association with proteins that facilitate adsorption to surrounding SG membranes [7–11] collaborate to create highly concentrated quanta of stored proteins (>100–200 mg ml^{-1}) with low osmotic activity. In most but not all regulated secretory cell types, the condensation process is facilitated by mildly acidic pH values [12–18] driven by a proton pump (H^+-ATPase) and complementary chloride channel (ClC3) associated with the TGN and evolving granule membranes [19–22].

Notably, various integral membrane and membrane-associated proteins that are destined to reside in SGs, e.g. granins, prohormone convertases, and exocrine pancreatic GP-2, participate in the selective condensation process [10, 14, 23–26]. Recent evidence indicates that such condensation interactions may begin in the TGN and may significantly facilitate transit of prohormones such as proinsulin into CVs [27, 28]. Moreover, the granins in neuroendocrine cells appear to aid condensation through a combination of liquid–liquid phase separation and prohormone binding [28]. Electron microscopic studies conducted over many years have shown that condensation also entails structural remodeling as mature hormones in endocrine cells and exocrine proteins, particularly in salivary glands, are progressively packaged for storage [29–31]. In contrast, proteins that fail to participate in condensation may be excluded; they have an opportunity to exit maturing SGs [5, 32–34] and be sorted to other destinations. With this overall arrangement, one can appreciate that the clathrin-coated buds that frequently stud condensing vacuoles and IGs generate vesicular carriers that may be engaged in any of a combination of activities. These include receptor-mediated sorting of lysosomal prohydrolases [35], constitutive-like secretion directed through endosomes [32, 36, 37], and trafficking of proteins that limit maturing SG membrane function and/or facilitate endosomal function (e.g. syntaxin6, VAMP4, synaptotagmin4, carboxypeptidase D (CPD)) [35, 38–43].

A variety of proteins, particularly in endocrine and neuroendocrine cells, are synthesized as inactive precursors. They undergo proteolytic processing to their bioactive forms by the successive action of proprotein convertases, carboxypeptidases, and, in some cases, N-terminal acetylation or C-terminal amidation (via the peptidylglycine α-amidating mono-oxygenase/lyase (PAM)) prior to or during their condensation in SGs. While processing may begin in the TGN and its associated CV, it is clear, as is well illustrated for both proinsulin and proopiomelanocortin, that processing can facilitate condensation [44] and that the bulk of it occurs for these prohormones within the acidic interior of IGs [19, 45–47]. Notably, processing machinery is similarly co-packaged with the bioactive proteins, initially as pro-enzymes that are cleaved and activated in IGs [7, 48–50]. Moreover, the membrane-associated components of the processing machinery have had additional value as a result of efforts to clarify their own interactions that contribute to secretory sorting (reviewed in [10]), to track the fate of granule membranes post-exocytosis [34, 51], and to understand how they might aid in seeding granule formation [52].

Collectively, the preceding strategies evolved as a useful model for the sorting and packaging of the contents of several kinds of classical SG (1). However, consideration of the formation of LROs has provided compelling examples in which active sorting is used for delivering various non-aggregated secretory components and selected membrane proteins secondarily to immature organelles to facilitate the ongoing maturation process. Here, transport vesicles intersect and fuse with immature LROs. Beautiful examples of this process have been provided in studies of the biogenesis of mucocysts in *Tetrahymena* [53] and Weibel–Palade bodies in endothelial cells [54]. In the case of mucocysts, a sortilin/VPS10 family receptor, Sor4p, binds an array of proteins including *Gr*anule *t*ip (Grt) proteins, *I*nduced upon *g*ranule *r*egeneration (Igr) proteins, and a cathepsin family (aspartyl) processing protease (Cth) in the TGN. Thus, Sor4p's action is similar to mannose-6-phosphate receptor (MPR) mediated trafficking of lysosomal prohydrolases. It then delivers its cargoes to immature LROs via a Rab7-containing endosomal intermediate and an ensuing fusion event mediated by the syntaxin7-like SNAp REceptor (SNARE) Stx7l1p and a specialized Class C Core Vacuole/Endosome Transport (CORVET) scaffolding complex [55, 56]. As a consequence, partially condensed Granule lattice (Grl) protein precursors are proteolytically processed by Cth to enable completion of Grl lattice assembly [53, 56]. Interestingly, examples of sortilin-family receptors being involved in the formation of mammalian SGs are also increasingly recognized—in the case of DCVs in neurons [57] and insulin granules in endocrine pancreatic beta cells [58]. The latter will be discussed in more detail below while considering the presence and roles of SG subpopulations. Thus, while passive sorting of secretory cargoes via condensation may be the major contributor to SG and LRO formation, active receptor-based sorting may also contribute to selective protein delivery (10). Delivery may occur at different levels, depending on the cell type, protein, and (possibly) granule subtype. Figure 2.1 depicts the post-Golgi pathways followed by secretory proteins in regulated secretory cells and the interactions that contribute to the condensation process in forming SGs.

2.3 The roles of lipids in the early events of SG formation

Within the secretory pathway, the composition and transbilayer distribution of selected lipids becomes noticeably shifted in the TGN and its derivative membrane-bounded carriers, as compared to earlier compartments that are specialized for biosynthesis and posttranslational modifications. Phospholipids become enriched in fatty acids with longer chains and a higher overall degree of saturation [59]. There is a marked increase in the content of sphingomyelin and complex glycosphingolipids, whose synthesis is completed within the trans-Golgi/TGN [60–62]. Cholesterol content is dramatically increased [63], largely driven by its non-vesicular transport at ER–TGN contacts [64, 65]. Finally, there are striking transbilayer asymmetries of sphingolipids, which are synthesized on the luminal leaflet, and of phosphatidylserine, which becomes concentrated on the cytoplasmic leaflet via transbilayer flipping [66, 67]. These changes distinguish the composition and support the functions of downstream organelles including the cell surface. In addition, concentration of phosphatidylserine enables electrostatic interactions that are needed by post-TGN membranes to bind regulatory and membrane-shaping proteins and components of the cytoskeleton [68–71]. Accompanying these changes, phosphatidylinositol-4-phosphate (PI4P), which is synthesized in part in the TGN and is highly concentrated there [64, 72, 73], has been identified as an organelle identity marker. PI4P is a key regulator alongside the Rab and ADP-ribosylation factor (ARF) guanine triphosphate (GTP)-ases that functions in numerous TGN-associated processes, including import of cholesterol and other lipids from the ER and formation of TGN-derived vesicular pathways including SGs [64, 67, 74–77]. Interestingly, the role of PI4P in these events may be governed not only by its synthesis and turnover but also by its protonation/deprotonation as a reflection of cytoplasmic pH in the physiological range. While this has been illustrated in *Saccharomyces cerevisiae*, there are hints that it likely applies in metazoan cells as well [78].

Relatively little is known about how the exchange to more saturated fatty acyl chains impacts the formation of vesicular carriers emanating from the TGN [79]. In contrast, exciting recent findings using a fluorescent sphingomyelin biosensor have shown that sphingomyelin is highly enriched in carriers engaged in TGN-to-cell surface transport of GPI-anchored proteins and other selected cargoes and much less so in carriers bearing other plasma membrane-targeted proteins [65, 80]. Further, because little of the biosensor is detected in endocytic compartments, TGN-derived carriers targeted at endosomes are likely sphingomyelin-poor [81]. Further insights have established a collaboration between sphingomyelin synthesis by sphingomyelin synthase SMS1, import of Ca^{2+} into the TGN by the secretory pathway calcium ATPase SPCA1, and oligomerization of the Golgi resident calcium

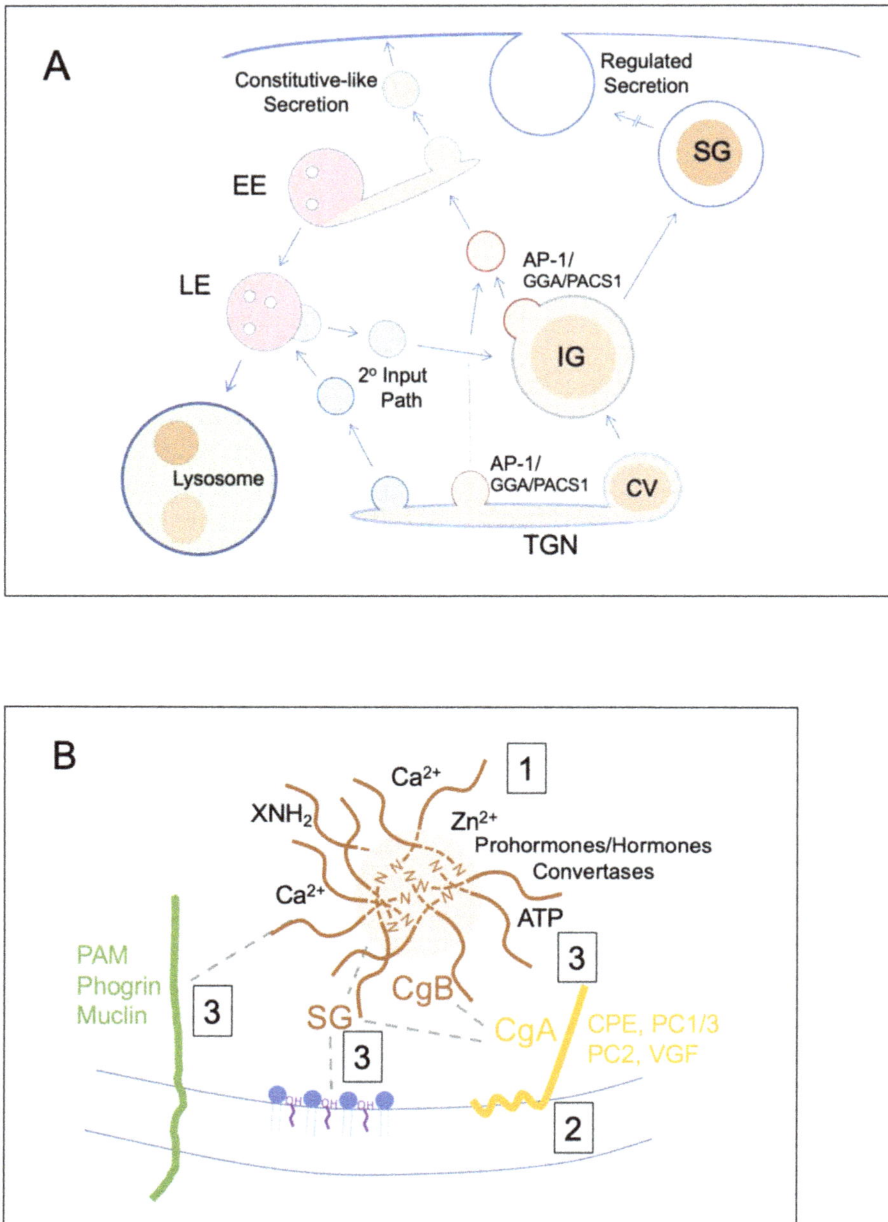

Figure 2.1. (a) Sorting of cargoes to the regulated secretory pathway. Newly synthesized secretory proteins exit the endoplasmic reticulum, traverse the Golgi, and within the TGN, they mostly enter associated CVs. Here, they are passively sorted and retained by forming intermolecular associations with other secretory proteins and with components of the nascent granule membranes. This condensation process is aided by Ca^{2+} and mildly acidic pH. Upon scission from the TGN, the CV becomes an IG. As the IG matures to form an SG, it continues active receptor-mediated sorting of lysosomal prohydrolases that began in the TGN via exiting clathrin-coated vesicles formed with the aid of the AP-1 adaptor, *G*olgi-localizing, *g*amma-adaptin ear homology domain, *A*RF-binding (GGA), and *p*hosphofurin *a*cidic *c*luster *s*orting protein (PACS-1)

complexes. This pathway also serves as a source of constitutive-like secretion of proteins that fail to condense and for the removal of proteins that function either transiently in IGs or act elsewhere. The IGs, especially immature lysosome-related secretory organelles (LROs), receive additional secretory cargoes that are actively sorted by sortilin-family receptors. They are transported via a secondary import pathway. The one shown is routed through late endosomes (LEs), but other routes (e.g. through early endosomes (EEs)) may apply depending on the receptor, cargo, and/or cell type. (b) Condensation and membrane association within forming neuroendocrine SGs: (1) intermolecular interactions among secretory proteins and with divalent ions, especially Ca^{2+}, Zn^{2+} and small molecules such as ATP and biogenic amines (XNH_2); (2) anchoring of membrane-associated proteins by a charged α-helical domain that has a hydrophobic patch; (3) interactions of secretory proteins with transmembrane proteins, membrane-associated proteins, and cholesterol-enriched lipid domains. The specific proteins identified have been highlighted previously (see the review by Dikeakos and Reudelhuber [10]). Notably, while these different types of interaction broadly contribute to condensation processes in most types of forming SGs, the extent to which they individually contribute to condensation is known to differ depending on the cell type and granule type [10]. Abbreviations in (b): XNH_2, biogenic amines; CPE, carboxypeptidase E; PC1/3 and PC2, prohormone convertases; CgA, CgB, SG, and VGF are granins or granin-like proteins. For CgB, recent evidence has highlighted a contribution of its intrinsically disordered N-terminal domain to condensation through liquid–liquid phase separation (dashed segments coalesced within a lightly shaded sphere). CgB, in turn, facilitates the sorting and packaging of prohormones, hormones, and prohormone convertases [28].

binding protein Cab45 in linking glycosylphosphatidylinositol (GPI)-protein trafficking to the sphingomyelin-enriched carriers [80]. Together with earlier findings [63, 82, 83], these observations imply that the membranes forming the TGN are heterogeneously populated by lipid microdomains that serve as distinct origins for post-Golgi carriers that differ in their final destinations. To date, the role of sphingomyelin-enriched domains in the formation of SGs has not been examined directly. However, several indicators argue that similar lipid-based sorting makes important contributions to SG biogenesis. These include cholesterol enrichment in SGs [84], the inhibitory effects of blocking sphingomyelin synthesis on stimulated insulin secretion [85], and the recent demonstration that pancreatic beta cells produce subpopulations of SGs with very different sphingomyelin contents ([86]; highlighted below in the section on granule subpopulations).

It has been known for many years that the lipid content of SG membranes generally is quite cholesterol-rich. The presence of cholesterol at ~35 mole% of total lipid [87] is comparable to the content observed in plasma membranes [59, 68] and well above the ~5 mole% in the ER where cholesterol is synthesized and regulated [88, 89]. In aggregate, SG membranes make up a very sizable fraction of cellular cholesterol content, and the level is tightly regulated. Numerous studies have shown that SG formation, efficient content storage, and SG maturation all require cholesterol [84, 90–93]. Moreover, SG stability and competence for exocytosis are impaired by either cholesterol deficiency or excess [84, 94–98]. It has generally been assumed that cholesterol enrichment in TGN-derived secretory carriers occurs concurrently with sphingolipid synthesis and entails passive assembly of cholesterol and sphingo- (and phospho-) lipid condensed complexes by thermodynamic trapping [71]. However, active concentration mechanisms that include the collaborative action of oxysterol binding protein (OSBP) and two ATP binding cassette (ABC) transporters ABCG1 and ABCA1 play essential and cholesterol-dependent roles during SG biogenesis, as illustrated for insulin granules in pancreatic beta cells [84]. OSBP acts proximally at ER–TGN contact sites to drive the concentration of cholesterol in the TGN and likely also enhances non-vesicular import of ceramide for sphingomyelin synthesis [72, 99]. Subsequent actions of ABCs G1 and A1 at the TGN [100] are envisioned to catalyze phospholipid-mediated translocation of cholesterol from cytoplasmic-to-luminal leaflet [101–103] to drive the formation of sphingolipid/cholesterol-enriched domains in quantities that are large enough to support SG formation [84]. Indeed, this machinery is thought to enable the net biogenesis of new SG membranes, which is needed to replace granules that are turned over both continually and conditionally by autophagy/crinophagy [104–108]. In addition, new lipid input may be required to readjust the lipid content of granule membranes that are reutilized post-exocytosis but have been compositionally edited during endocytic recycling. Overall, non-vesicular transport from the ER and concentration of cholesterol in forming SGs appears to be closely coupled to the production of secretory proteins, thereby enabling matching of the levels of cargo with stable storage containers [84].

Quite interestingly, an additional and apparently distinct contribution to cholesterol regulation during the formation of insulin granules has been revealed by recent studies of another ABC transporter, ABCA12 [109]. Although subcellular localization of ABCA12 has not been addressed, transporter deficiency alters SG structure, impairs insulin secretion, and results in enhanced cell surface cholesterol similar to that characterized for ABCG1 and ABCA1 transporter deficiencies [98, 110]; in addition, it may affect SG interactions with the actin cytoskeleton [109]. These pleiotropic effects are suggestive that ABCA12 may act at sites that are similar to those characterized for the ABCs G1 and A1 [100]. Because loss of ABCA12 has previously been shown to impair the loading of glucosylceramides into cutaneous (keratinocyte) lamellar bodies [111], the primary action of the transporter might be to similarly regulate the glycolipid content of insulin SG membranes. These mechanisms for ABC transporters, particularly ABCs G1 and A1, in SG biogenesis may involve coordinated actions with P4-ATPases (flippases) that catalyze luminal-to-cytosolic leaflet translocation of selected phospholipids. Indeed, relevant flippases are well-known residents of the TGN and SGs [67, 112, 113] and are required for stimulated insulin secretion [113]. In this regard, combined ABC transporter-catalyzed inward/P4-ATPase-catalyzed outward phospholipid translocations could serve to couple cholesterol translocation to amplified formation of cholesterol-enriched granule membranes. For the present, a tentative/speculative picture of how cholesterol- and sphingomyelin regulatory machinery in the TGN might contribute to the formation of condensing vacuoles and the ensuing scission to generate IGs is illustrated in figure 2.2. Finally, yet another cholesterol-related and phosphoinositide-binding protein, STARD10, may support insulin granule formation, as its deficiency alters granule morphology and increases unstimulated secretion of proinsulin. However, at present there is limited insight regarding the way in which it might contribute to granule biogenesis [114].

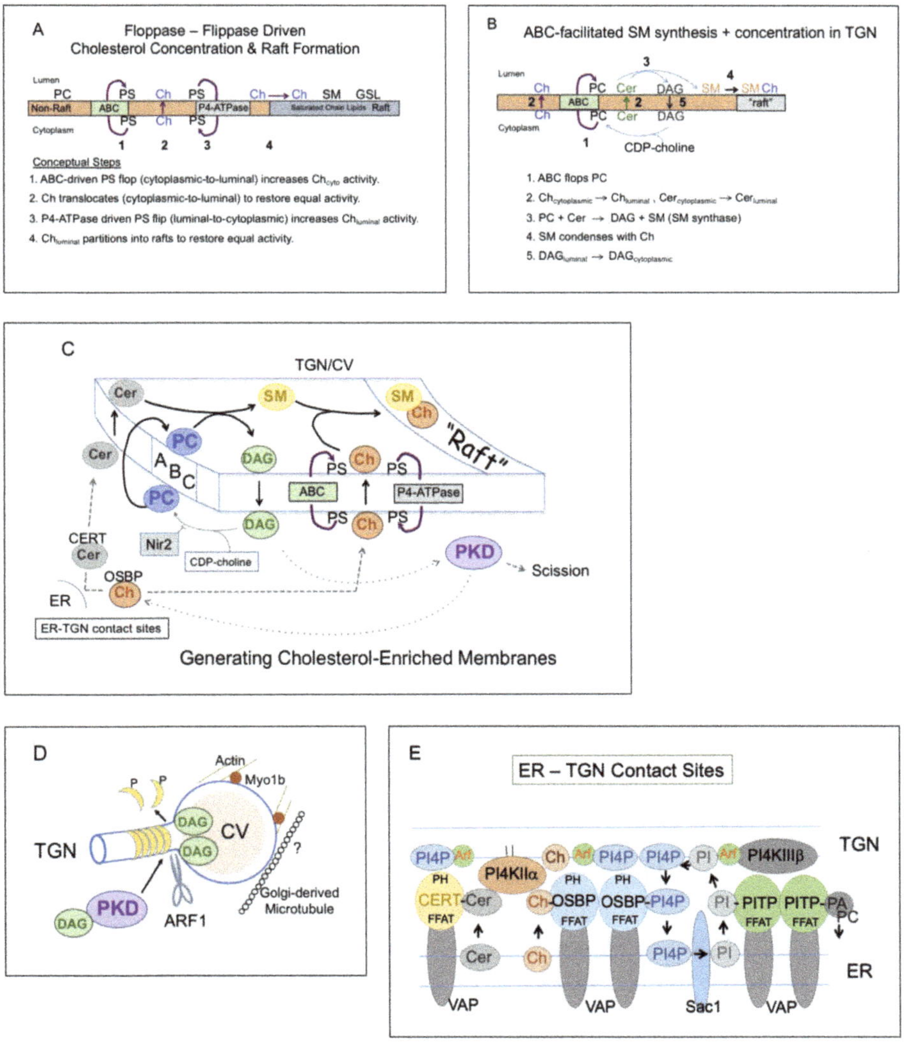

Figure 2.2. Model envisioned for the formation of cholesterol- and sphingomyelin-enriched secretory granule membranes in the TGN, a process aided by ER–TGN contact sites and ultimately coupled to the scission of CVs to form IGs. (a and b) The prospective role of ABC transporters in enabling the biogenesis of cholesterol- and sphingomyelin-enriched 'raft-like' domains that seed CV formation. The ABCs G1 and A1 are both known to catalyze cytoplasmic-to-luminal leaflet translocation of phosphatidylserine (PS) and phosphatidylcholine (PC). Translocation in turn increases the activity (higher chemical potential) of non-polar lipids such as cholesterol and ceramide and promotes their relocation to the luminal leaflet. To facilitate concentration of cholesterol on the luminal leaflet, one ABC transporter is predicted to act as part of a cycle with a P4-ATPase. Thus, ABC-stimulated inward translocation of PS stimulates cholesterol internalization, and the outward translocation of PS by P4-ATPase drives the formation of cholesterol-sphingolipid condensed complexes. The other ABC is suggested to translocate PC to the inner leaflet to stimulate ceramide concentration on the luminal leaflet. Within the TGN, PC and ceramide are substrates for sphingomyelin synthase 1 (SMS1). The sphingomyelin that is produced condenses in complexes with cholesterol in the 'raft-like' domains. (c) Cartoon envisioning the placement of the TGN-associated activities described in panels A and B in the context of the ER–TGN contact sites where cholesterol and ceramide are imported via OSBP and CERT, respectively.

Note further that sphingomyelin synthesis generates diacylglycerol (DAG), which on an as-needed basis either enables further local synthesis of PC [279] for increased membrane biogenesis or activates protein kinase D (PKD). PKD in turn serves as a regulator of cholesterol and ceramide import and importantly also activates scission of the CV from the TGN. Phosphatidylinositol (PI) transfer proteins (PITPs), also localized at ER–TGN contact sites, regulate PC synthesis [280, 281] and supply PI as substrates for PI4Ks [282, 283]. (d) PKD activated by DAG mediates scission of the CV from the TGN by phosphorylating the crescent Bin/amphiphysin/Rvs (BAR) protein arfaptin. Dissociation of phosphorylated arfaptin enables ARF1-mediated scission. Scission may be aided by forces applied by actin/myosin-1b and microtubular components of the cytoskeleton. (e) Cartoon of an ER–TGN contact site showing the transporters for cholesterol (Ch), ceramide (Cer), and PI that span the contact through interaction with VAMP-associated protein (VAP) in the ER and PI4P and ARF1 in the TGN. The transporters facilitate import of these lipids into the TGN. Also shown are TGN-associated PI 4-kinases—PI4KIIα, which is palmitoylated and associated with lipid-ordered domains and PI4KIIIβ, which associates with lipid-disordered domains via its amphipathic lipid packing sensor (ALPS) motif. Sac1 localized to the ER membrane is a PI4P phosphatase that generates PI as well as a TGN:ER concentration gradient of PI4P that facilitates cholesterol concentration in the TGN [74].

As highlighted above, PI4P that is associated with the TGN is a key organizer of the trafficking of TGN-derived transport vesicles at many levels, including the scission of nascent IGs discussed in the following section. In fulfilling these roles, PI4P, often in coincidence with ARF-GTP, has numerous effector actions. These include: (1) promoting non-vesicular lipid transport at ER–TGN contact sites [76, 77]; (2) activating nucleotide exchange factors for small GTPases that promote vesicular budding via both transbilayer phospholipid flipping [115] and association of clathrin adaptors (GGA and AP-1) (2,3); (3) facilitating the binding of proteins, e.g. golgi phosphoprotein GOLPH3 that function in Golgi shaping/organization, regulate glycolipid synthesis, and promote scission of TGN-derived transport vesicles [62, 75]; and (4) supporting PKD, which is a key regulator of vesicle scission including nascent IGs [116, 117]. PI4P constitutes only a few percent of total cell lipid. Its levels are coupled to sphingolipid flow [118] and are regulated by a combination of robust synthesis (via PI 4-kinases), additional phosphorylation (especially to $PI4,5P_2$), and dephosphorylation (principally by the phosphatase Sac1) [74, 119, 120]. Notably, two PI 4-kinases, PI4KIIα and PI4KIIIβ, are associated with the TGN. PI4KIIα is mostly palmitoylated, leading to its enrichment in lipid-ordered domains. In contrast, PI4KIIIβ is concentrated in lipid-disordered domains of the TGN by membrane insertion of its ALPS amphiphilic helix and by binding the ACBD3 (Acyl CoA binding domain) adaptor protein [64, 77, 121]. Thus, as previously noted [64], the two kinases likely generate different pools of PI4P. These may support distinct (or complementary) processes, possibly including separate populations of nascent vesicles that are distinguished by their lipid compositions.

Beyond facilitating the scission of nascent IGs, PI 4-kinases may also contribute to protein sorting during granule formation [122]. In addition, they have later roles in granule maturation [123, 124]. Indeed, the functions of PI4K and PI4P necessarily extend into IGs to support ongoing export pathways such as AP-1-mediated vesicular export that were initiated in the TGN [35, 39, 125]. However, their post-TGN roles are likely limited by the lifetimes of the phosphoinositide and kinase on the nascent granule. In yeast secretory vesicles, it is well established that reduced PI4P content is essential for an ensuing Rab-mediated cascade of events that facilitate exocytosis [126, 127]. Related recent findings indicate that the PI4P phosphatase Sac2 resides on insulin granules (in addition to its presence on endosomes). Moreover, Sac2 is required to hydrolyze PI4P to enable the SGs to dock at the plasma membrane [128].

2.4 Scission of nascent IGs from the TGN

As secretory protein sorting proceeds in the dilating CV, its connection to the TGN is eventually severed to generate an IG. Many issues regarding how scission is timed and brought about in relation to the influx of protein destined for storage and the genesis of the surrounding granule membrane are now partially understood. Because storage granules normally tend to have characteristic and quite similar sizes in a given cell type, the contributing events must be closely coordinated. Insights substantially derived from the study of dense core vesicles in endocrine cells and especially of insulin granule formation in pancreatic beta cells suggest that the small GTPase ARF1, PI4P, PKD, and the machinery that regulates lipid distribution collaborate. Together, they reliably link the flux and packaging of SG components to the severance of the nascent granule's connection to the TGN. ARF, PI4P, and PKD all regulate the import of cholesterol and sphingomyelin precursors from the ER [74, 77, 129, 130]; ARF and PI4P regulate cargo protein flux through the Golgi [131]; cholesterol and sphingomyelin (the latter synthesized in the TGN) facilitate the biogenesis of SG membranes and membrane–cargo interactions in the CV [7, 84, 91, 109]; and DAG (diacylglycerol) produced either during sphingomyelin synthesis or by phospholipase C (PLCβ) activates PKD, which together with ARF and PI4P enables scission [117, 132, 133].

Regarding the scission process itself, arfaptins have been identified as negative regulatory proteins. They contain membrane-shaping crescent BAR domains and likely enwrap and stabilize the TGN and contiguous CV junctions by binding to PI4P and the ARF-family GTPase Arl1 [133, 134]. This action blocks presumptive scission sites [117]. In the case of insulin granules, phosphorylation of arfaptin-1 by PKD stimulates its dissociation from the TGN, giving access to the guanine nucleotide exchange factor BIG2, which activates ARF. In turn, ARF triggers scission and is thought to be aided by DAG, which promotes tight membrane curvature conducive to severing the CV–TGN tubule connection [117, 135] (figure 2.2(d)). Alternatively, phosphatidic acid produced by phosphorylation of DAG or by phospholipase D action may serve as a fusogenic lipid [136, 137]. Because tubules and vesicles in other contexts are well known to be differentially enriched in lipids that favor lipid-disordered versus lipid-ordered domains, respectively [138–141], the scission process likely also contributes to selective lipid enrichment during IG formation.

Other studies involving the perturbation of early events in granule formation and the coupled scission process have added to our insights into the regulatory roles played by granule membrane lipids, arfaptins, and PKD. Pharmacological inhibition of sphingomyelin synthesis and inhibition of cholesterol import to the TGN by RNA interference (RNAi)-mediated knockdown of OSBP both inhibit insulin granule formation and ultimately the stimulated secretion of insulin [84, 85]. Similarly, pharmacological inhibition of PKD blocks granule scission and, as a consequence, blocks secretion and increases insulin degradation [108, 117]. Interestingly, knockdown of arfaptin-1 causes a marked reduction in the size of insulin granules and curiously also leads to their dispersion in the cytoplasm rather than accumulation beneath the plasma membrane [117]. Potentially, premature scission in the absence of arfaptin-1 may account for the size reduction and may also perturb the ability of the granules to undergo productive cytoskeletally mediated transport to the cell periphery. Similarly to PKD inhibition, the expression of mutant arfaptin-1 S132A, which cannot be phosphorylated by PKD, blocks the scission process and increases the degradation of nascent granules containing both insulin and proinsulin [108]. In contrast, PKD-mediated phosphorylation of its substrates can be stimulated in beta cells by carbachol [117], suggesting that signaling may be able to accelerate scission in relation to the other events that are coupled in granule formation. Consequently, the timing of the onset and extent of PKD activation are prospective key determinants of the characteristic granule sizes that are observed in different cell types or in the same cells under different conditions [142].

The preceding discussion of scission has focused on the membrane connection between the CV and the TGN. However, the precedence of other TGN-associated carrier budding events indicates that scission is likely to be aided by forces generated by cytoskeletal interactions involving myosin-1b and F-actin [143]. In regulated secretory cells, findings have localized myosin-1b to dense core vesicles of PC12 cells and have further suggested through the use of RNAi-mediated depletion that it participates along with F-actin and the Arp2/3 complex in IG formation [144]. Importantly, these observations have gained stronger support from recent studies of insulin-secreting cells, in which close association of myosin-1b with nascent insulin granules near the TGN was demonstrated by super-resolution microscopy. In addition, RNAi-mediated loss of this myosin impaired the egress of SGs to the peripheral cytoplasm and significantly decreased glucose-stimulated insulin secretion [145]. Other quite novel studies have uncovered a role for a special class of Golgi-derived microtubules in facilitating the budding of nascent insulin granules from the TGN. Formation of these microtubules is stimulated by elevated glucose concentration through a mechanism that involves cyclic-AMP and Epac2A, a guanine nucleotide exchange factor for the small GTPase Rap1. In contrast, selectively blocking microtubule formation led to insulin accumulation within the Golgi and inhibited ensuing granule formation [146]. Notably, a recent high-resolution reconstruction of microtubule organization in pancreatic beta cells has presented potentially contrasting evidence that Golgi-associated microtubules are actually quite sparse and that the number may change at most modestly with increasing glucose concentration [147]. However, given that this extensive and impressive analysis was necessarily restricted to a limited number of cells, it may be too early to draw strong conclusions about the active role of Golgi-associated microtubules in SG formation. Thus, whether and how actin/myosin and microtubule interactions collaborate or act sequentially in this context still needs to be addressed. Figure 2.2 presents proposed models for the processes that collectively coordinate granule membrane biogenesis and scission from the TGN, as revealed primarily by the study of insulin granules.

2.5 Granule maturation—a multi-step process

Granule maturation is discussed here as if it begins following scission of the newly formed IG from the TGN. However, many of the early events following scission are continuations of processes that began in TGN-attached CVs. This is consistent with the progressive nature of secretory sorting activities and the localizations of the machinery and trafficking events that refine IG composition. Because sorting of secretory cargoes has already been discussed, the focus here is on membrane interaction and trafficking pathways that are coupled to granule maturation. Accordingly, a succession of topics will be considered. These include: possible homotypic IG fusion, exit from IGs, trafficking related to retention in IGs, entry of SG-resident proteins, recycling routes post-exocytosis, the contribution of cholesterol to gaining competence for exocytosis, and notable membrane-associated proteins that assist SG formation. While these topics are treated individually, there is still much to be learned concerning how they are coupled or overlap in the overall maturation process.

Homotypic fusion of IGs—a possible early step in maturation: quite early on, classic studies in pituitary lactotrophs presented morphological evidence for the presence of irregularly shaped CVs or early IGs located in proximity to the Golgi and containing several foci of condensed secretory content within a single membrane-bounded container. In contrast, more distantly located mature SGs had large, nearly round profiles surrounding a single internal content mass [148]. These observations spawned an interest in the possibility that IG–IG homotypic fusion is an early step in the SG maturation process, although distinguishing it from events that occur prior to scission was challenging. Subsequent investigations arguing that IG maturation contributes to a volume increase in the SGs of endocrine and exocrine cells [124, 149, 150] and that SNARE (syntaxin6)-mediated IG–IG fusion can be reproduced *ex vivo* with isolated IGs [40, 150–153] have led to acceptance that this process can happen. However, whether homotypic fusion is an essential step in granule maturation *in situ* remains open to question at present and still needs further evaluation. It might occur conditionally as a reflection of naturally occurring variations in the timing of content condensation in relation to scission from the TGN among different types of regulated secretory cells. In addition, it may apply where there are demonstrable wholesale differences in IG and SG size. Importantly, in *Drosophila* larval salivary glands, where maturation of one-time-use SGs with enormous storage capacity is hormone-regulated and has been imaged in real time, homotypic fusion of IGs has been shown convincingly [154]. However, in another case—the maturation of Weibel–Palade bodies in endothelial cells—in which one might suspect homotypic fusion is necessary to create their highly elongated shapes, it has been shown that it does not occur [155]. Furthermore, other processes increase the complexity of deciphering whether and how IG–IG fusion might necessarily contribute to maturation. These include: the presence of Golgi-to-IG or endosome-to-IG vesicular trafficking during granule maturation that requires heterotypic fusion of vesicles and IGs [156, 157]; the presence of distinct subpopulations of SGs in endocrine cells that have different contents, phospholipid compositions, and sizes [15, 86, 158]; and the presence of 'multi-granules' associated with granule turnover by crinophagy [105].

Exit from IGs: it is well established in various endocrine and exocrine cell types that newly formed IGs continue the process of MPR-mediated sorting of lysosomal prohydrolases into AP-1/clathrin-coated vesicles that is initiated in the TGN. Indeed, ligand and receptor have both been immunocytochemically localized to budding vesicles on IGs, stimulus-dependent secretion of newly synthesized prohydrolases has been demonstrated during the time window that they are transiting through IGs, and prohydrolases are retained in mature SGs in the absence of MPR [35, 159]. Additionally, AP-1/clathrin-mediated exit is followed by membrane proteins that function either transiently during IG biogenesis or subsequently at distal sites following exit from IGs, but not in mature SGs. Such proteins include the protease furin [160], the SNAREs VAMP4 and syntaxin6, the SNARE-regulator synaptotagmin4 [35, 38, 39], and vti1a, another partnering SNARE for syntaxin6 [161]. Egress of these proteins is variously assisted by GGA proteins or by PACS-1, and specific details of their actions are reviewed in [162]. Perturbation of this exit process not only leads to retention of the proteins in mature SGs but also has been shown to inhibit prohormone convertase activation and associated maturation of stored granins as well as regulated secretion [38, 40]. Exit also appears to be regulated by myosin Va, although in this case, perturbation of its function does not have an effect on IG-associated proprotein processing [163]. Interestingly, while AP-1/clathrin-containing complexes have been the major focus as mediators of IG exit traffic, this may not be their only role during maturation. Nor is the AP-1/clathrin-mediated pathway the only exit route. Studies of Weibel–Palade bodies in endothelial cells have shown that AP-1 and clathrin form a stabilizing coat that fully surrounds the evolving IG and appears unrelated to vesicular exit trafficking [164], while CPD, a proprotein-processing enzyme [41], appears to use a distinct IG exit route. The pathway followed by CPD differs from that followed by MPR and does not involve direct interactions with AP-1. Although AP-1 deficiency affects CPD egress [42], emerging evidence implicates departure by a Rab2-mediated pathway discussed below. Perturbation of the latter pathway leads to retention of CPD in mature SGs [43].

Retention-associated trafficking: elegant studies of DCV formation in *C. elegans* neurons have uncovered a fascinating process by which certain secretory cargoes and membrane proteins are conditionally retained in forming and maturing IGs by a pathway that is mediated by the GTPase Rab2 and its various effectors [165–168]. Moreover, many of the same events have been documented more recently in analyzing the biogenesis of insulin storage granules in pancreatic beta cells [43, 156, 169]. Most of the components of this pathway—Rab2, RIC-19 (and its mammalian ortholog ICA69), TBC-8 (Rab2 GTPase activating protein (GAP)), CCCP-1 (and its ortholog CDCC186), RUND-1, and the *e*ndosome-*a*ssociated *r*ecycling *p*rotein (EARP) trafficking complex—are localized to the TGN and nearby IGs [156, 165–169]. The membrane interaction of the EARP complex, which also distributes in part to endosomes, is enabled by an interacting protein, EIPR-1 [169]. Normally, both aggregated and non-aggregated secretory cargoes are efficiently retained during IG maturation. However, mutations or deficiencies in Rab2 pathway components lead to substantial and selective loss of non-aggregated cargoes to RAB-5-positive EEs as well as partial loss of the IA-2/phogrin ortholog IDA-1. Pathway disruptions do not interfere with proprotein processing and with eventual exocytosis of the content-poor mature SGs in *C. elegans* neurons [165, 166]. These features distinguish this pathway from those involved in entry traffic to immature mucocysts (described under secretory protein sorting above) and in acquisition of competence for regulated exocytosis (described below).

Interestingly, cargo loss incurred by Rab2 pathway deficiency can be prevented by dominant active RAB-5 (Q78L), which inhibits endosomal fusion, or by agents that interfere with endosomal PI3P [165, 166]. These perturbations argue that cargo retention during IG formation and maturation necessarily involves an EE-to-IG trafficking interaction. Several factors support this deduction. First, there is a strong resemblance between the EARP complex and the Golgi-associated retrograde protein (GARP) complex that supports endosome-to-TGN trafficking [156]. Second, the Rab2 effector CCCP-1 has vesicle-like tethering properties [170]. Finally, candidate IG-associated SNAREs (SNAP29, vti1, and syntaxin6) that could mediate endosome-IG fusion have been identified [156, 161]. However, at the same time, IG-associated components of the Rab2 pathway normally ensure efficient retention of soluble cargoes and of IDA-1. Thus, traffic mediated by the Rab2 pathway appears to include a filter that stabilizes IG composition with deficiencies compromising its fidelity in retention. While speculative at present, it is possible that such a filter may be created by an IG-associated scaffolding and/or a local environment that is conducive to specific cargo condensation and retention. Potential candidates that may contribute to the filter are RIC-19, HID-1, and their mammalian counterparts, which are discussed in the following paragraphs. Such restricted interactions between IGs and EEs may resemble the recently reported transient fusion/restricted cargo transfer between early and recycling endosomes [171] and between the TGN and recycling endosomes [172].

Additional findings regarding the function of the Rab2 pathway in retention-associated trafficking have suggested that it may aid in maintaining early acidification during maturation and thereby promote content protein associations during condensation. Acidification by V-ATPase is facilitated by interaction with Vps50, a component of the EARP complex [173] and by the membrane-associated protein HID-1 [150, 174], which has been implicated as a Rab2 pathway component in *C. elegans*. Further, mammalian HID-1 loss results in profound reduction of secretogranin II and exogenous ANF-GFP (atrial natriuretic peptide-green fluorescent protein) in neuroendocrine DCVs in PC12 cells [174]. Notably, while proprotein processing is unaffected by Rab2 pathway disruption in *C. elegans* neurons, loss of HID-1 in pancreatic beta cells results in substantial accumulation of proinsulin in poorly condensed IGs in the perinuclear cytoplasm. The authors attribute the changes to a combination of defective acidification, slowed proteolytic activation of prohormone convertases, and consequent inefficiency in proinsulin processing [150].

Separate and quite interesting insights have come from the study of RIC-19-related proteins ICA69 (the ortholog of RIC-19) and PICK1 in both mammalian cells and *Drosophila* growth-hormone-secreting cells [175, 176]. As for RIC-19, genetic studies place the actions of ICA69 and PICK1 downstream of Rab2 [177] and thus make them potential contributors to retention-associated trafficking. In the context of SG formation, ICA69 and PICK1 together are associated (likely as heterodimers) with proinsulin-containing CVs and IGs near the TGN [175, 176]. In contrast, PICK1 alone (presumably as homodimers) associates with SGs in the peripheral cytoplasm, which may increase the potential for interactions mediated by its PDZ (PSD-95, ZO-1, Disheveled) domain [175, 176]. However, as for RIC-19 bound to forming DCVs in *C. elegans* neurons, the association is lost in fully mature SGs that are in proximity to their sites of exocytosis [166, 175, 178]. In further agreement, deficiency in PICK1 does not affect the competence of individual SGs to undergo exocytosis in adrenal chromaffin or pancreatic beta cells. However, reductions in SG number, size, docking at the cell surface, and selected cargo loss (as for the HID-1 deficiency discussed above) have been identified [179, 180]. Interestingly, a pancreatic beta-cell-specific knockout of PICK1 strongly increased both the cellular level and secretion of proinsulin. This reflected not only reduced prohormone convertase action, similar to that for HID-1 loss, but also increased prohormone biosynthesis triggered by insulin insufficiency in the peripheral circulation [180]. RIC-19, ICA69, and PICK1 are all members of the crescent BAR family of membrane-shaping proteins alongside arfaptin, which has been implicated in the scission of CVs from the TGN, as discussed above. An amphipathic helix adjacent to the BAR domain is key to curvature sensing on sphere-like surfaces as condensation proceeds [181], while the PDZ domain of PICK1 has been shown to be essential for secretory function [179]. Intriguing recent studies of PICK1 missense mutations identified in diabetic patients show that interference with BAR domain interactions decreases insulin retention and enhances vesicular egress (including loss of syntaxin6) associated with IG maturation in beta cells [178]. Thus, the roles of RIC-19, ICA69, and PICK1 in SG maturation likely involve a combination of stabilizing the shape and composition of IGs throughout biogenesis and mediating the entry and exit of selected proteins [175, 176, 178]. While the link between these proteins and Rab2 makes them candidates for assisting retention-associated trafficking events, their actions probably extend to other entry/exit pathways that contribute to the granule formation process.

Entry into IGs: the preceding sections highlighting the normal exit of itinerant proteins and retention-associated protection against the aberrant exit of SG residents are complementary to considering the pathways by which resident membrane proteins become SG-associated in the first place. Because SGs initially develop from the TGN, direct passage and sorting of membrane proteins into CVs is an obvious candidate pathway. Findings to date show that some of the proteins are indeed directly incorporated into forming granules. These proteins include furin, CPD, CPE, and PAM in neuroendocrine cells [34, 42, 51, 160, 182, 183] and

P-selectin in endothelial cells [54]. However, other proteins appear to bypass direct inclusion and are instead incorporated into forming granules later. In this case, their possible indirect routes leading to entry into IGs include a TGN-derived vesicular pathway that may directly fuse with the maturing granule or transit through endosomes (or even through the plasma membrane and then endosomes) to reach IGs [156, 157, 161, 169, 184, 185]. Proteins that appear to be indirectly incorporated include synaptotagmin1 (syt1), the VAMP family R-SNAREs such as VAMPs 2 and 3, and IA-2 and phogrin in neuroendocrine cells [161, 186, 187]. The vesicular monoamine transporter VMAT2 may also be indirectly incorporated [185], although recent observations raise the possibility of incorporation at the TGN ([188], discussed further below). In endothelial cells, the tetraspanin CD63 is indirectly incorporated [54], while in *Drosophila* larval salivary glands, CD63, syt1, SNAP24, Past1, and syntaxin16 are incorporated post-TGN into already formed IGs that are destined to become single-use exocrine granules [124, 189, 190]. For syt1 and SNAP24, it is quite interesting that an ER-anchored protein, Hobbit, is required for IG entry, raising the possibility that interorganellar interactions involving the ER may regulate import [189].

The membrane proteins identified above that are directly incorporated into forming granules share the common feature that they all function either within the granule interior in activities related to protein processing and condensation or extracellularly post-exocytosis. Sorting determinants in their luminal domains may aid in direct targeting through interactions related to condensation (reviewed in [162]). Moreover, the directly incorporated proteins have cytoplasmic sorting motifs that may guide incorporation (or the reincorporation of recycled proteins, as discussed in the recycling section below). The resident SG membrane proteins identified above that follow indirect routes to IG entry share the common feature that they all have functions on the cytoplasm-facing granule surface. Their roles include regulating exocytosis [186, 191], small molecular neurotransmitter import [185], promoting secretory competence [54], and lipid phosphatase activity [192]. Normally, maturing IGs experience a delay before they are responsive to secretory stimulation, which may, in part, reflect the time lag associated with the indirect trafficking process. Early pulse-chase biosynthetic labeling studies showed that IGs in pancreatic beta cells and PC12 cells require ~30 min beyond exit from the TGN to become responsive for stimulated exocytosis [32, 151, 182], while even longer lags have been implicated in other cell types [161, 186]. In addition, perturbations affecting proteins that enter IGs indirectly can inhibit full maturation and can ultimately compromise regulated exocytosis [124, 184, 189].

The recent attention focused on the presence of indirect TGN-derived delivery to IGs has been inspired in part by the analysis of the role of the Q-SNAREs vti1a/b in adrenal chromaffin cells and hippocampal neurons [157, 161]. Vti1 isoforms are concentrated in and around the Golgi/TGN and have been implicated in fusion events related to Golgi and endosome functions in other cell types (see the references in [161]). Vti1a, which affects chromaffin DCV maturation, shows strong perinuclear colocalization with syntaxin6 (one of its prospective SNARE partners) but, notably, not with chromogranin B that marks TGN-associated CVs and newly formed IGs. Moreover, there is little colocalization of vti1a with VAMP2, which marks mature DCVs and is exclusively concentrated near the cell periphery. Abrogation of vti1a decreases the VAMP2 content, number, and size distribution of DCVs but does not alter release kinetics during exocytosis [161]. In neurons, vti1a/b isoforms are concentrated in soma, and they appear to have redundant functions in regulated secretion. Using double knockouts of both isoforms, investigators showed that loss of Vti1 did not alter the formation of nascent, secretory cargo-containing DCVs. However, the DCVs were ultimately compromised in terms of their ability to undergo regulated exocytosis. As in the chromaffin cell study, this correlated with reduced VAMP2 content at sites of exocytosis. Parallel findings identified perturbations of synaptic vesicle formation and delivery of retrograde trafficking to the Golgi/TGN [157]. Thus, the upshot of both studies is that vti1 acts at the TGN on a vesicular pathway that is distinct from CVs but is necessary for their maturation. Its role may be in the fusion of TGN-derived vesicles with IGs or in delivering proteins to the TGN for incorporation into the TGN-derived indirect delivery pathway.

Additional evidence for the indirect TGN-derived pathway for delivering granule membrane proteins has been provided by tracking newly synthesized VAMP2, VAMP3, and syt1 in chromaffin cells in which an extended window of maturation provided clear temporal and spatial resolution [186]. Here, all three newly synthesized proteins did not colocalize with nascent or newly formed IGs located near the Golgi and required hours to appear in SGs that were located at the cell periphery. Moreover, newly synthesized syt1 in transit beyond the TGN was shown to intersect with syt1 internalized from the cell surface post-endocytosis before colocalizing with maturing IGs. Because intersection occurred toward the cell periphery, this observation supports the likely fusion of TGN-derived carriers containing newly synthesized syt1 with endosomes prior to granule delivery. Consequently, the indirect route traced by syt1 very likely links to pathways that are regulated by AP-3 adaptors, especially as previous studies (discussed below) establish a role for AP-3 in syt1 delivery to DCVs [184, 191].

Consideration of the localization and trafficking of phogrin and VMAT2 provides further insight and highlights prospective complexities at the origin of what may be more than one indirect TGN-derived pathway. Earlier studies showed that phogrin strongly colocalizes with γ-adaptin in the perinuclear region and that its cytoplasmic acidic dileucine-based sorting motif binds to AP-1 but apparently not to AP-3. Moreover, mutation of the acidic residues to alanine leads to profound accumulation in the TGN [193]. Notably, VMAT2 has a nearly identical motif, and the corresponding mutation similarly leads to perinuclear accumulation [191, 194]. However, in this case strong evidence has been presented that VMAT2 binds to AP-3 (discussed further below) and colocalizes with TGN-associated vesicular buds labeled by AP-3 and associated VPS41 [188, 195], suggesting a distinct route. In the recent study, the authors argued in favor of direct delivery of VMAT2 to TGN-associated nascent DCVs based on some colocalization with the secretory protein BDNF (bone-derived neurotropic factor) at AP-3-labeled foci, which can be amplified by low temperature-induced retention in the TGN [188]. Because BDNF undergoes receptor-mediated sorting by sortillin [57], a more likely scenario may be that colocalization reflects association with a budding vesicle that indirectly mediates delivery to maturing DCVs. This would be analogous to sortillin-based delivery to LROs [53]. Moreover, association of VMAT2 with a transport vesicle rather than a TGN-associated DCV seems more compatible with the absence of condensed secretory content in VPS41-coated TGN-associated vesicular buds observed by electron microscopy [195].

A second recent study has argued against the role of adaptors in mediating the sorting of phogrin and VMAT2 at the TGN. Here, newly synthesized phogrin and VMAT2 were reported to enter by default into a TGN-derived constitutive secretory pathway leading to the cell surface before endocytosis and delivery to maturing DCVs [185]. While the data presented support this possibility, a potential caveat stems from the experimental strategy of creating artificial retention and buildup of the newly synthesized proteins in the ER, followed by wholesale release for transport [196]. Thus, much of the trafficking observed, including passage through the cell surface, may reflect overflow beyond the capacity of the normal sorting and delivery routes. In the case of VAMP2 and 3 and syt1 (noted above for their indirect delivery and late entry during DCV maturation) all three proteins contain similar acidic dileucine (EExxxLL)-like motifs implicated in adaptor-mediated sorting of VMAT2 and phogrin [197, 198]. Compelling evidence supports a role for AP-3 in the sorting of syt1 [184, 191]. However, the sorting of VAMPs 2 and 3 appears to be independent of AP-3, as subcellular fractionation of PC12 cells shows that association of both VAMPs with DCVs is unaffected by the loss of AP-3 [184, 191]. Altogether, these observations make a case for indirect delivery of a cohort of membrane proteins that function on the cytoplasmic surface of SGs. Evidently, however, further studies are needed to clarify how entry into indirect TGN-derived pathways operates and which proteins follow a common route.

With regard to tracking and comparing the indirect pathways directed to IGs, little or nothing is known about the possible direct fusion of TGN-derived carriers with maturing IGs. However, a route that includes delivery through endosomal intermediates is readily related to numerous earlier investigations of endosomal pathways that function in the targeting of regulated secretory pathways and other post-TGN locations. At the endosomal level, there is broad support for a role of AP-3-mediated trafficking in the biogenesis of multiple organelles (including lysosomes) and a range of secretory organelles that spans synaptic vesicles, LROs, and neuroendocrine IGs (iDCVs) [54, 55, 184, 185, 199–202]; reviewed in [200, 203, 204]. An apparent exception appears to be the genesis of exocrine glue granules in *Drosophila* larval salivary glands, in which AP-1 has been implicated instead [122, 190, 205]. Importantly, AP-3 actions at endosomes, which in many cases involve interactions with *B*iogenesis of *L*ysosome-Related *O*rganelles *C*omplex (BLOC) complexes (discussed below) appear to be complementary to the actions at the TGN described above, which appear to be BLOC-independent [188]. In contrast, VPS41 actions as an AP-3 vesicle coat appear to be mainly restricted to the TGN (aside from associations with the late endosomal HOPS (homotypic fusion and protein sorting) complex) [188]. Thus, an ongoing challenge will be to evaluate the extent to which granule maturation is supported by actions likely at both sites.

Two distinct versions of the AP-3 complex have been characterized: Ap3b1, which is ubiquitously expressed, and Ap3b2, which is specifically expressed alongside Ap3b1 in neuronal and neuroendocrine cells. They each have four subunits δ, β3, μ3, and σ3. Ap3b1 and Ap3b2 share common δ and σ3 subunits but have differing β3 and μ3 subunits: β3A, μ3A, and β3B, μ3B. Notably, μ3A and μ3B bind tyrosine-based sorting motifs and the interface between σ3 and δ binds dileucine motifs, thus facilitating distinct trafficking of membrane proteins [204]. Mice that have mutations in δ and β3A (*mocha* and *pearl*, respectively) and genetic or RNAi-generated deficiencies of β3B have been very useful in clarifying the multiple systemic phenotypes of δ and β3A and the neurological deficiencies of δ and β3B. Particularly revealing studies related to IG entry that place Ap3b1-mediated trafficking distal to the formation of immature LROs have come from the examination of mucocysts in *Tetrahymena* and of Weibel–Palade bodies in endothelial cells. Following heterotypic fusion of the TGN- and endosome-derived pathways that deliver separate cargoes to create *Tetrahymena* immature mucocysts, it is quite clear that genetically-induced loss of μ3 (the single isoform in this system) specifically blocks the ensuing conversion to mature mucocysts [55]. For Weibel–Palade bodies, the immature LROs containing von Willebrand factor protomers and P-selectin require Ap3b1 to deliver the tetraspanin CD63 and complete maturation. Only then is the LRO able to export fully functional content during exocytosis [54]. In mice, mutation or genetically-targeted loss of either or both Ap3b1 and Ap3b2 dysregulates secretory cargo release in adrenal chromaffin and pancreatic beta and alpha cells. In all three regulated secretory cell types, basal release of SGs was elevated, while stimulated exocytosis was either decreased or unaffected, depending how it was measured [184]. Of further note, combined silencing of Ap3b1 and Ap3b2 was required to reach the same level of perturbation caused by coordinate loss of both complexes via mutation of the common δ subunit (*mocha* mouse) [184]. Thus, Ap3b1 and Ap3b2 appear to have redundant roles in neuroendocrine IG maturation. Interestingly, a further examination of induced deficiency in the adaptor complex AP-1 in combination with AP-3 deficiency potentiated inhibitory effects on secretion [184]. The investigators have suggested that loss of AP-1 may prevent exit from IGs of components, e.g., synaptotagmin4, that are known to be inhibitory to stimulus-dependent exocytosis [38]. However, as described above, there is also the possibility of redundant roles of AP-1 and AP-3 in delivering fusion regulatory machinery into or along the TGN-derived indirect pathway.

As highlighted in the characterization of mutations that lead to Hermansky–Pudlak Syndrome and their analysis in mouse models, AP-3 complexes have a strong association with components of the BLOC complexes 1–3. Indeed, BLOC complexes collaborate with AP-3 complexes in the biogenesis of lysosomes, LROs, and subpopulations of synaptic vesicles [203, 204]. Moreover, phenotypes of AP-3 and BLOC deficiencies are often quite similar, implicating action in a common pathway. Of particular interest are findings that point to their interactions with membrane proteins that function in the uptake and sequestration of a range of signaling molecules, e.g. VGLUT (glutamate), VGAT (gaba), VMAT2 (catecholamines), and of machinery for intercompartmental fusion, e.g. VAMP7, syntaxin 13, and SNAP25 [200, 206]. Along with the role of AP-3 in syt1 targeting [184, 191] discussed above, BLOC support is likely to be substantially involved in supplying immature IGs with newly synthesized or recycled machinery that functions in signaling and regulated exocytosis. Notably, other studies of adrenal chromaffin (and PC12 cell) DCVs have called attention to the possibility that BLOC actions in promoting IG maturation may be more varied, in that disruption of different subunits of BLOC-1 elicits contrasting changes. Mutation of the *muted* subunit results in increased DCV size characteristics accompanied by increased secretory content of chromogranin A and reduced docking and filling of the readily releasable pool of DCVs at the cell surface. However, disruption of *dysbindin* results in decreased chromogranin A and is more similar to mutations in the β3A subunit of Ap3b1 (*pearl* mouse) [207]. While the underlying basis for these different phenotypes is not understood, they point to potential heterogeneity of the BLOC-1 function in granule maturation pathway(s), suggesting differential effects on the maturation of distinct subpopulations of granules (further discussed below).

As final notes with respect to the roles of AP-3 and VPS41 and the multiple import pathways that support granule maturation, it will be especially important for ongoing studies to evaluate: (a) whether or not the actions of both Ap3b1 and Ap3b2 occur at the TGN as well as at endosomes; (b) whether and how VPS41 engages each AP-3 isoform, and (c) whether VPS41 and BLOC interactions with AP-3 are mutually exclusive. Finally, VPS41, which broadly affects neuroendocrine DCV biogenesis and secretion [195, 208] has also been identified as a Rab2 interacting protein [209], while the dysbindin subunit of BLOC-1 interacts with EIPR-1 and selected EARP subunits [210]. Thus, while entry into SGs, retention-associated trafficking, and SG membrane maturation have been treated (with some justification) as distinct processes in SG maturation, their close interaction or even potential overlap will need further evaluation. Moreover, pursuit of these interesting interactions might also clarify whether the Rab2 pathway not only functions in retention trafficking but also delivers proteins for IG entry.

Recycling routes of SG membranes and their proteins: in contrast to secretory cargoes, which are all newly synthesized, only a fraction of granule membrane proteins is newly synthesized. The rest are long-lived and are extensively recycled post-exocytosis for reutilization. During compensatory endocytosis following exocytosis, SG membrane proteins can follow different internalization routes. Some diffuse from the SG into the plasma membrane and are recovered by clathrin-mediated endocytosis, as exemplified by fluorescent-tagged VAMP2. Others, however, are recovered directly at the fusion site in association with residual granule membrane components, as exemplified by phogrin, VMAT2, and dopamine beta hydroxylase (DBH) [211–216]. In this case, recovery also includes clathrin-mediated endocytosis [217–219], but its extent may vary depending on whether granule discharge has proceeded by full fusion and flattening into the plasma membrane or transient fusion pore opening and closure. Regardless, SG membrane components mainly converge on EEA1-positive EEs [215] that have been marked by tracers of constitutive endocytosis [220]. This is also a likely point of convergence for proteins such as furin and CPD that, following exit from maturing IGs, cycle through the plasma membrane en route to reutilization [41, 160].

From EEs, the pathways to reutilization vary for different proteins. Two proteins, P-selectin and PAM, have been shown to pass through LEs before reaching the Golgi/TGN for reincorporation alongside their newly synthesized counterparts [51, 54, 221]. In the case of P-selectin, two short linear sorting motifs act in succession to mediate the steps from early-to-LE and LE-to-TGN [54]. For PAM, recycling through LEs appears to be regulated by phosphorylation within a motif enriched in acidic amino acid residues [222]. In contrast, CPD and phogrin both follow a pathway that leads from EEs to a syntaxin6-positive portion of the TGN that is distinct from the site where TGN38 is concentrated [182, 187]. Interestingly, the recipient compartment has been implicated as a site where syntaxin6 and vti1 may mediate retrograde carrier fusion [182]. Thus, as with the pathway through LEs, this second return route brings recycled proteins back to where they can be reincorporated alongside newly synthesized counterparts. A third route followed by recycled syt1 is a short circuit, which appears to bypass the TGN-associated syntaxin6-containing compartment. It is restricted to the early portion of the endosomal system but excluded from Rab11-positive recycling endosomes [186]. As noted in the preceding section, newly synthesized and recycled syt1 appear to meet in an endosomal compartment before proceeding to enter IGs together [186]. Given the well-established dependence of syt1 delivery on AP-3 [184, 191], AP-3 may be involved in mediating the common delivery of newly synthesized and recycled syt1 from EEs to IGs. For VMAT2, which is also regulated by AP-3 sorting, the recycling route appears distinct from that followed by syt1. Very interesting recent findings show that VMAT2 traffics from EEA1-marked EEs back to the TGN along a pathway marked by SNX5 [188], a regulator of retrograde sorting [223]. SNX5 and AP-3 both directly bind VMAT2, exhibiting partially overlapping interactions at VMAT2's acidic dileucine sorting motif. Consequently, retrograde transport of VMAT2 via SNX5 and targeting to DCVs mediated by TGN-associated AP-3 act serially. Moreover, because the interactions compete with one another, the SNX5 association apparently limits binding between VMAT2 and the portion of endosomal AP-3 located in a neighboring early endosomal domain [188]. These intriguing findings should stimulate follow-up studies to evaluate whether phogrin, VAMPs 2 and 3, and other recycling DCV proteins interact with SNX5 and thus traffic similarly or whether, like syt1, they are are unaffected by SNX5 [188] and follow short-circuit recycling. Further, do interactions with BLOC complexes or other early endosomal sorting machinery preclude SNX5 interaction and thus promote bypassing the return to the TGN?

Finally, the proposed role of the recently discovered endosome-associated, C2 domain-containing protein BAIAP3 in SG membrane recycling [224] will need to be directly tested using candidate SG membrane proteins to identify which recycling proteins and pathways are involved. Figure 2.3 attempts to summarize current knowledge regarding the trafficking of newly synthesized and recycled granule membrane proteins.

Impact of cholesterol on IG maturation: thus far, most of the discussion of events affecting SG maturation distal to the TGN has focused on pathways and their cargoes. Little is known about how lipid composition, which is a distinguishing

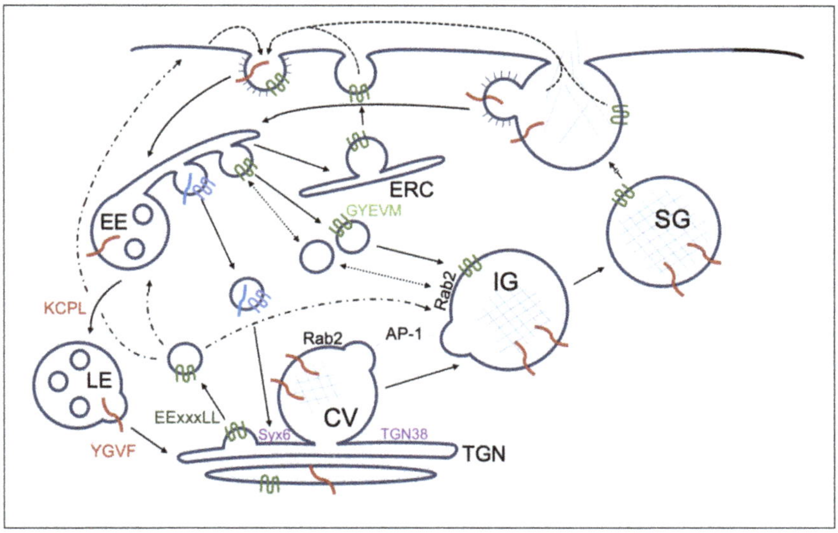

Figure 2.3. Trafficking pathways identified for newly synthesized and recycled SG integral membrane proteins during SG biogenesis. The red protein characterizes the trafficking of P-selectin and PAM and the early trafficking of CPD. All have functions on the luminal surface. The green protein characterizes the trafficking of CD63, phogrin, and (likely) VMAT2, which all function on the cytoplasmic surface. Newly synthesized red and green proteins enter the TGN from the Golgi. Here, red proteins enter CVs where they are retained by interactions with secretory cargoes. CPD, which is transient in forming and maturing IGs, exits via a Rab2-mediated endosomal interaction (dotted double-headed arrows) alongside AP-1-mediated exit. In contrast, green proteins are routed into TGN-derived parallel pathways. An acidic dileucine motif in the cytoplasmic tails of phogrin and VMAT2 may facilitate protein sorting. Newly synthesized syt1 and VAMPs 2 and 3, which share closely related sorting motifs, may traffic similarly. Targeting to reach IGs may proceed along alternative routes that are identified by dashed–dotted arrows—either directly to IGs or indirectly through endosomes, potentially including trafficking to the plasma membrane. Note that another potential route leading from TGN through Rab7-containing LEs to IGs has been characterized for *Tetrahymena* mucocyst trafficking (see figure 1). During post-exocytosis recycling (indicated by dotted arrows), internalization mainly involves clathrin-mediated endocytosis, either at the site of exocytosis or at displaced sites for proteins that have diffused into the plasma membrane. Internalized proteins converge on EEs. From here, P-selectin and PAM return to the TGN through LEs; the relevant motifs that guide P-selectin's recycling steps [54] are identified in red. In contrast, recycling phogrin, VMAT2, and CPD (blue-labeled proteins) return to a syntaxin6-containing subdomain of the TGN for reincorporation into CVs (CPD) or the indirect TGN-derived pathways (phogrin, VMAT2). Retrograde recycling of VMAT2 has been shown to be regulated by SNX5 [188]. Syt1 uses a short circuit for recycling involving AP-3-mediated sorting by EEs. This same route is followed by CD63, which together with its sorting motif [54] are shown in green. It is possible that the Rab2 pathway operating between EEs and IGs functions in the input of newly synthesized and/or recycled proteins into IGs; however, cargoes have not yet been identified. The endosomal recycling compartment (ERC) enables repeated surface recycling/reinternalization until sorting occurs.

feature among post-Golgi pathways, impacts the maturation process. Interestingly, a distinct late (post-TGN) step has been implicated as a result of examining the roles of the ABC transporters ABCG1 and ABCA1 and cholesterol in supporting insulin granule exocytosis. As discussed in relation to the early role of lipids in SG formation above, these transporters acting in the TGN are thought to aid in driving the formation of cholesterol and sphingolipid-enriched SG membranes. In particular, ABCG1 deficiency results in the formation of unstable IGs that have been deduced to undergo lysosomal degradation, likely by crinophagy [84]. Importantly, transporter-deficient cells (in this case either ABCG1 or ABCA1) exhibited a second-level defect that became apparent after accounting for the effects of lysosomal degradation. The cells were deficient in the ability to undergo glucose- (and also K^+ depolarization) stimulated exocytosis. Addition to transporter-deficient cells of a low level of exogenous cholesterol (that did not alter the total cellular content) fully restored exocytotic competence to the level normally observed for mature SGs [84]. Strikingly, using a previously characterized assay [86], follow-up studies have shown that SGs purified from the transporter-deficient cells are defective when tested *in vitro* for their ability to fuse to SNARE-containing supported planar bilayers. Moreover, when cholesterol was added to the cells prior to granule purification, the fusion competence of the granules *in vitro* was fully restored (Kreutzberger *et al*, unpublished observations). The implication of these findings is that the added cholesterol corrects or bypasses an ABC transporter-dependent trafficking step that is required for granule maturation. While not yet addressed, this outcome is consistent with the actions of ABCs G1 and A1 (and thus cholesterol enrichment) on processes that either relate to the origin of the indirect pathway in the TGN or to delivery from endosomes to IGs. Both the TGN and endosomes are sites where ABCs G1 and A1 are known to accumulate during their trafficking itinerary [100].

Other SG formation-assisting proteins of note: among membrane-associated proteins that contribute not only to the maturation of IGs but also to ultimately regulating the exocytosis of SGs are selected Rab family GTPases, *do*uble *C2* domain-containing protein (Doc2B) and *c*alcium-dependent *a*ctivator of *p*rotein *s*ecretion (CAPS). Rab3, long regarded as an identity marker for classical SGs [225], has four isoforms (A–D) that typically are all expressed in exocrine- and endocrine-regulated secretory cells and are granule-associated via C-terminal prenylation. They have partially distinct but also some overlapping functions with respect to proprotein processing, IG exit traffic, granule size, and longevity [105, 163, 226–229]. Rab27A (and in certain cell types isoform B) associates with both LROs and SGs [204, 230–234]. Both Rabs 3 and 27A localize to the same vesicles, where they have complementary roles that in part reflect distinct interaction partners [235]. Moreover, both Rabs share the same activating GDP/GTP exchange protein, guanine nucleotide exchange factor (GEF), promoting their common recruitment to IGs. However, they have different GAPs, which are thought to promote repeated cycling of Rab3 on/off membranes but more sustained binding of Rab27A (reviewed in [235]). Importantly, with respect to contributions to maturation, the association of these Rabs with IGs appears to occur ~20 min after scission from the TGN, as timed following release from a 20 °C trafficking block [236]. The timing is well suited for the proposed roles of the two Rabs in relocation of SGs to the cell periphery and navigating the cortical cytoskeleton prior to their regulatory roles in exocytosis [237, 238]. Interestingly, the rabphilin homolog Noc2 is an effector for both Rab3A and Rab27A separately. In the case of Rab27A, a portion of Noc2 forms a tripartite complex with Rab2 that specifically localizes to perinuclear maturing IGs in pancreatic beta cells, while the binary Rab27A-Noc2 complex localizes to mature SGs at the cell periphery. Moreover, the tripartite complex, which facilitates proinsulin/insulin processing, is separate from Rab2's maturation-supporting interactions with ICA69 and PICK1 on IGs described above [239]. Collectively, the range of sequential Rab interactions points to critical roles that span the entire biogenesis and ultimate function of SGs.

A striking variation on the timing of Rab27A association with IGs is illustrated for the LRO-like cytolytic granules in cytotoxic T cells and NK (Natural Killer) cells, where both the association and role of Rab27A occur late and just prior to exocytosis. Here, upon target cell recognition and activation, there is a coalescence at the immune synapse of the immature cytolytic granules with both Rab11-positive recycling endosomes containing Munc13-4 and Rab7-positive LEs containing Rab27A, thereby uniting the cytolytic cargo with the necessary machinery to enable immediate release [240]. Finally, in addition to Rabs 3 and 27A, Rab1 and Rab11 have recently been implicated in the maturation of IGs and secretory discharge of the large glue granules in *Drosophila* larval salivary glands. Interestingly, Rab1 prevents premature exocytosis during maturation, while Rab11 is required for Rab1 association with IGs and also for mature SG exocytosis [241].

Doc2B and calcium-dependent activator proteins for secretion (CAPS) have both been extensively studied with respect to their regulatory roles in SG and synaptic vesicle exocytosis [242–244], which are not discussed here. However, a variety of studies point either directly or indirectly to earlier roles during IG formation and maturation. Moreover, increased interest in understanding the range of their functions has stemmed from the recent discovery that Doc2B and CAPS associate with distinct subpopulations of SGs in insulin-secreting cells [86]. While both Doc2B and CAPS have been identified as granule-associated calcium-binding proteins, there is in each case a sizable cytosolic pool as well as a portion associated with the plasma membrane [242, 245, 246], which in the case of Doc2B is enhanced by calcium [247]. Attention to granule-associated Doc2B has been amplified by a series of discoveries identifying Doc2B deficiency as a biomarker for early onset type one diabetes, which can be mimicked in mice treated with proinflammatory cytokines TNFα, interferon-γ, and IL-1β. Remarkably, the same cytokine treatment causes selective loss of a subset of insulin granules, and both diabetic phenotype and granule loss can be prevented by overexpression of Doc2B [86, 248, 249]. While the mechanism of granule loss is not known, the results suggest that Doc2B may aid in either stabilizing IGs against depletion (lysosomal degradation or unstimulated secretion) or facilitating their repletion (formation or maturation).

Examination of CAPS, which has two isoforms (1 and 2) has uncovered roles that include not only function in exocytotic priming (the basis of CAPS discovery [243]) but also span IG biogenesis and maturation. Indeed, a deficiency of CAPS, which binds class II ARFs 4/5 at the TGN, inhibits IG formation and leads to chromogranin accumulation in the TGN in PC12 cells [250]. CAPS deficiency also dissipates the pH gradient in maturing DCVs [251] and thereby interferes with vesicular monoamine uptake [252, 253]. Finally, the analysis of CAPS loss in pancreatic beta cells has begun to distinguish distinct roles for CAPS1 and 2. CAPS1 appears to stabilize newly formed IGs (as its loss decreases cellular insulin content with a negligible effect on proinsulin), decreases granule number, and appears to increase lysosomal degradation. In contrast, CAPS2 appears to be largely responsible for facilitating insulin secretion, as its loss does not affect insulin content or granule number but decreases the incidence of SGs docked at the plasma membrane and inhibits phasic insulin secretion, particularly the second phase [254]. Coincidentally, cell fractionation studies show enrichment of CAPS2 in purified insulin SGs, while CAPS1 is restricted to lower-density fractions [86].

2.6 Granule subpopulations in regulated secretory cells

It has been known for quite some time that certain types of endocrine/neuroendocrine as well as hematopoietically-derived regulated secretory cell produce subpopulations of SGs with distinct secretory cargoes. In cells such as neutrophils, production can be attributed to asynchronous synthesis during development [255]. However, in cells such as pituitary sommatomammotrophs, the production and packaging of segregated cargoes occurs simultaneously [256, 257] and in *Aplysia* bag cells as well as mammalian neuroendocrine cells, even following the processing of a common precursor polypeptide [15, 158]. In these cases, sorting has been attributed to homotypic cargo associations that occur during condensation, which begins in CVs associated with the TGN. The *Aplysia* results were especially insightful in that they clarified the origin of the two types of regulated carrier from different sites within the TGN and also highlighted their distinct functions in autocrine and paracrine/endocrine regulation [15]. In the following years, other studies have uncovered differential sorting of granins (chromogranin A/secretogranin III vs. secretogranin II) among endocrine granule subpopulations [8] and also functional heterogeneity in the response to exocytotic signaling [8, 158, 258, 259]. Similar heterogeneity identified in exocrine cells [260, 261] and endothelial cells [262] raised the possibility that the presence of granule subpopulations might be a general feature of regulated secretory cells. Notably, where sensitivity to exocytotic signaling has been used as a distinguishing characteristic among subpopulations, analyses of a particular cell type have often focused on tracking the export of major cargoes that are shared by the granule subpopulations, and less attention has been paid to differences in secretory composition that might relate to sorting during granule biogenesis and distinct pathway function.

Recent studies have identified key differences among membrane polypeptides in granule subpopulations, especially synaptotagmin (syt) isoforms that serve as calcium triggers of exocytosis; thus likely explaining one way in which regulated secretory cells are able to respond to stronger and weaker stimulatory signals. Indeed, in both adrenal chromaffin cells and a pancreatic beta cell line, syt7 (which responds to intracellular calcium in the low micromolar or submicromolar range), and syt1 in chromaffin cells or syt9 in beta cells (which each require higher calcium, i.e. $>10\ \mu M$) [263] mainly localize into two separate granule subpopulations [86, 258, 264, 265]. Moreover, one study of neuroendocrine PC12 cells has provided evidence that three syts capable of stimulating exocytosis (isoforms 1, 7, and 9) may each reside in distinct SG subpopulations [263]. Among these studies, the one focusing on insulin granules combined the characterization of calcium-triggered SNARE-mediated fusion of purified SGs to planar-supported plasma membrane bilayers with compositional analyses of the two major SG subpopulations that were separated from one another by immunoadsorption with isoform-specific anti-syt antibodies [86]. Together, these approaches led to quite a range of insights that bear on both the formation and function of the SG subtypes. In particular, planar bilayer studies readily distinguished the syt7- and syt9-enriched insulin SG subpopulations based on their distinct calcium sensitivities for fusion and further identified their differing fusion pore kinetics, which were confirmed *in situ*. The compositional analyses of syt7- and syt9-enriched SGs (hereafter called syt7 and syt9 granules) were especially striking in that they revealed the presence of different secretory cargoes that were packaged alongside insulin. These included not only polypeptides but also small signaling molecules that are sequestered by membrane transporters during IG maturation. The same studies showed that the membranes of the syt7 granules are enriched in cholesterol and sphingomyelin, while the syt9 granule membranes contain significantly less cholesterol and almost no sphingomyelin [86]. Notably, these contrasting cholesterol/sphingomyelin-rich and -poor lipid compositions are reminiscent of those that have been characterized for plasma membrane- and endosome-targeted pathways emanating from the TGN in cells that are not specialized for regulated secretion [63, 80, 81]. This raises interest in possible biogenetic interrelationships to consider in future studies.

At present, it is not known whether comparable lipid compositional differences exist for granule subpopulations in other cell types. However, examining those observed for syt7 and syt9 insulin SGs in the light of other findings regarding the organization of the TGN and early events in granule formation leads to interesting questions for further investigation. First, based on the precedent that distinct SG subpopulations originate separately within the TGN [15], will comparable results be observed for syt7 and syt9 SGs using fluorescent biosensors for sphingomyelin [80, 81]? And are the origins within the same or possibly distinct TGN cisternae [266]? Second, TGN-associated PI 4-kinase activity is well known to support granule formation (discussed above). Consequently, do PI4KIIα, which preferentially associates with lipid-ordered domains, and PI4KIIIβ, which associates with lipid-disordered domains [64], respectively, support the formation of cholesterol/sphingomyelin-rich syt7 and cholesterol/sphingomyelin-poor syt9 IGs? As a secondary upshot of this possibility, distinct scission mechanisms drawing on different pools of PI4P may sever the two types of IG from the TGN. Accordingly, does DAG derived from sphingomyelin synthesis facilitate selective severing of cholesterol/sphingomyelin-enriched syt7 granule precursors? And is phosphatidic acid used to promote severing of cholesterol/sphingomyelin-poor syt9 granule precursors? The latter possibility may potentially be analogous to the selective formation of basolateral transport vesicles by the coupled action of PI4KIIIβ-generated PI4P and LPAATδ (lysophosphatidic acid acyltransferase) in other cell types [267]. Third, granin family members, which collectively are ubiquitous packaging agents in endocrine SGs [268] exhibit isoform-specific partitioning among SG subpopulations in pituitary gonadotrophs and corticotrophs [8, 269]. The isoforms differ further in their ability to bind cholesterol-enriched membranes [8]. Thus, are specific granins active determinants for the selective formation of granule subtypes and of syt7 and syt9 insulin SGs in particular? Fourth, now that cholesterol/sphingomyelin-rich and -poor SG subpopulations can be recognized, will future studies show whether one or both types serve as sites for exit- and retention-associated trafficking mediated by the AP-1 and Rab2 pathways during IG maturation? Finally, lipid composition may be a determinant that distinguishes the association and prospective subpopulation-specific actions of Doc2B and CAPS on syt7 and syt9 granules. Is this the case?

The compositional differences between syt7 and syt9 insulin SGs also affect the thinking about what may distinguish the two maturing subpopulations with respect to acquiring competence for regulated exocytosis and their differing contents of small molecular signaling molecules. First, as described above, deficiencies in the ABC transporters G1 and A1 affected insulin IG formation at two levels—initially during the formation of stable IGs that were not susceptible to lysosomal degradation and subsequently during acquisition of the ability to undergo stimulus-dependent exocytosis. Moreover, both these consequences could be bypassed by exogenous cholesterol addition [84]. With the necessity to consider two granule subpopulations, it is of significant interest to discover whether each of the transporters supports one of the two granule subtypes and also whether support is provided at both levels. This consideration further begs the question: how and at what point during maturation does exogenous cholesterol ameliorate the insufficiencies of both syt7 granules, which are cholesterol-enriched and especially syt9 granules, which are not cholesterol-enriched? Because both cholesterol and the two ABCs accumulate in the TGN and endosomes [100], is the cholesterol bypass occurring at one or both sites? Second, because it is known that both AP3b1 and AP3b2 engage along with BLOC complexes in endosomal sorting in other contexts [203, 204], the apparently redundant roles of Ap3b1 and Ap3b2 in SG formation [184] may actually reflect their actions in the maturation of two separate granule subpopulations, similar to what has been proposed for trafficking to lysosomes and synaptic vesicles in neurons [200]. In this case, exocytotic machinery and small molecule transporters would be ligands for targeting to their respective granule types. While the differing ligand preferences of Ap3b1 and Ap3b2 may distinguish at least some of the destinations, it also seems likely that membrane cholesterol level may be a contributing factor, as it governs the delivery of a cohort of proteins from endosomes to synaptic vesicles [270] and may also contribute to distinguishing the sorting of syt7, which is palmitoylated, from syt9, which is not [86]. Moreover, since the distribution of syt7 is altered by SNX5 deficiency but unaffected by AP-3 deficiency [188], its SG destination may be determined at the TGN rather than at endosomes.

It has been an especially surprising discovery in the study of insulin granule subpopulations that beta cells treated with proinflammatory cytokines to mimic the onset of type-1 diabetes [248] lose their syt7 SGs selectively. In contrast, cells receiving extended palmitate treatment to mimic the lipotoxicity associated with type-2 diabetes [271, 272] lose their syt9 granules selectively [86]. While pathological correlations relating to the differing compositions and susceptibilities of syt7 and syt9 granules are highly interesting and summarized elsewhere [86], these new insights are also instructive regarding SG biogenesis when considered from the standpoint of existing findings in the literature. Previously, studies carried out on pancreatic beta cells of mice that lacked one of the sortilin sorting receptors, Sorcs1, identified the essential role of the receptor in preventing diabetes in animals that either carried the *ob/ob* mutation or were given a diet enriched in glucose and palmitate. In contrast, Sorcs1-deficient animals given a normal diet were unaffected [58]. In view of the finding that syt9-containing SGs are selectively lost as a consequence of extended palmitate treatment, these intriguing results would be explained if the formation of syt7 granules were specifically dependent on Sorcs1. Thus, in the absence of syt7 SGs, syt9 SGs alone would support insulin requirements in Sorcs1-deficient mice on a normal lean diet, but the simultaneous loss of both SG subpopulations on an *ob/ob* background or palmitate-enriched diet would lead to diabetes. Moreover, the realization that syt7 granules may be Sorcs1-dependent raises the possibility that they may strongly resemble LROs, especially as the formation of LROs in *Tetrahymena* has been shown to be dependent on the sortilin receptor Sor4p [53] and syt7 has been implicated in the exocytosis of cytolytic LROs in cytotoxic T cells [273]. In contrast, syt9 granules are likely to more strongly resemble classical SGs. If these predictions are valid, then it will be fascinating to determine the underlying basis of Sorcs1 dependence in syt7 insulin SGs. Further, Sorcs1-based sorting and targeting could potentially explain the selective presence of insulin-like growth factor 2 (IGF2), islet amyloid polypeptide (IAPP), and macrophage inhibitory factor (MIF) in syt7 granules [86], which is analogous to the sortilin-dependent sorting of BDNF to neuronal and neuroendocrine dense core vesicles [57]. Figure 2.4 illustrates the current understanding of how dual regulated secretory pathways may be formed in tandem and highlights compositional differences that have been discovered among subpopulations of insulin SGs.

2.7 Exploring SG biogenesis—where we are and going forward

This review of secretory granule biogenesis has emphasized the collaboration of the TGN and the endosomal network in setting the origin and completing the maturation of storage compartments that are mobilized for stimulus-dependent exocytosis. The overall process is quite dynamic; multiple import and export events shape the composition of both secretory cargoes and the surrounding membranes. Since the time of earlier reviews, it has become quite clear that the bulk of exportable proteins enter the nascent granule at the TGN without the use of structurally-encoded sorting signals and are retained by passive interactions between protein cargoes and with ions and small molecules. However, there has also been an emerging realization that active receptor-mediated delivery may complement this process as a secondary input more broadly than originally appreciated from the definitive demonstration of these events in the context of LRO biogenesis and the

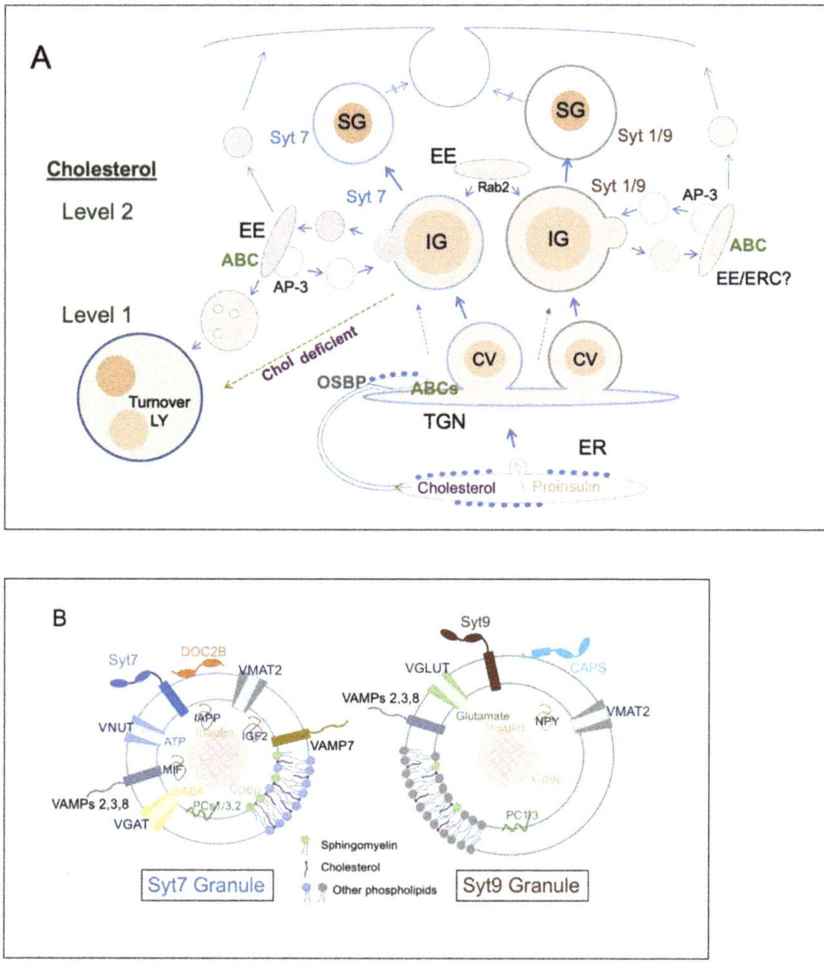

Figure 2.4. (a) Depiction of the formation of two granule subpopulations in endocrine pancreatic beta cells and their dependence on cholesterol and AP-3 mediated sorting. Distinct CVs arise from either cholesterol-/sphingomyelin-enriched or lower-cholesterol/sphingomyelin-poor domains of the TGN (possibly different cisternae). The former domain is depicted in proximity to ER–TGN contact sites where OSBP, CERT, and ABC transporters collaborate to generate CVs with higher cholesterol/sphingomyelin content. Failure to acquire sufficient cholesterol (indicated by the Level 1 cholesterol effect) leads to partial loss of the deficient IGs by lysosomal degradation. Maturation of these IGs is projected to involve AP-3 (likely Ap3b1) mediated import trafficking. IG acquisition of syt7 appears AP-3 independent and may be facilitated by palmitoylation. Acquisition of exocytosis competence is contingent on ABC transporter-mediated cholesterol accumulation (indicated by the Level 2 cholesterol effect). Syt7-containing SGs resemble LROs and are highly sensitive to Ca^{2+}-triggered release. CVs that are generated in lower-cholesterol/sphingomyelin-poor membranes of the TGN have little sensitivity to cholesterol deficiency that results from ABC transporter depletion (the Level 1 cholesterol effect). Thus, the formation of these IGs results in minimal loss to lysosomal degradation. Here, AP-3 (likely Ap3b2) mediated import is the prospective source of Syt9. As in the case of syt7 granules, however, ABC transporter-mediated cholesterol accumulation is essential (Level 2) for acquisition of exocytosis competence by syt9 granules even though syt9 granules have intrinsically lower cholesterol than syt7 granules. Thus, the Level 2 cholesterol dependence for achieving exocytosis competence is likely linked to

cholesterol requirements for IG import pathways that originate in cholesterol-enriched domains of either the TGN or endosomes. Both these donor compartments are sites where the transporters are enriched. Syt1 is also included alongside syt9, corresponding to the analogous lower calcium sensitivity subpopulation of SGs in chromaffin cells. Note that incomplete fidelity of AP-3 mediated maturation may be the source of the limited colocalization of syt isoforms in SGs, as has been observed for syt7 and syt1 in chromaffin cells [258, 264, 265]. Dotted blue and gray arrows represent TGN-derived pathways supporting granule formation. While AP-1-mediated exit trafficking and Rab2-mediated retention trafficking are depicted as being linked to the formation of both types of granule, this will need to be addressed, along with many other aspects of this model, in future studies. (b) Visual summary of the distinct sizes and protein, lipid, and small molecule compositions of syt7 and syt9 granules comprising the two subpopulations of insulin SGs. Panel B is adapted with permission from [86]. Organelle abbreviations: ER, endoplasmic reticulum; TGN, trans-Golgi network; CV, condensing vacuole; IG, immature granule; SG, mature secretory granule; EE, early endosome; ERC, endosomal recycling compartment; LY, lysosome.

targeting of selected neuropeptides to neuronal dense core vesicles. Moreover, while a number of findings point to ways that secretory cargoes interact with the surrounding specialized granule membranes during condensation, new challenges have emerged, namely, to decipher how both common and especially distinct secretory cargoes are distributed between two SG subpopulations and where retention involves membranes that differ compositionally.

Regarding the composition of SG membranes, aggregate findings so far suggest that the incorporation of proteins that have intracompartmental or extracellular functions into nascent granules occurs directly into CVs, where they are retained by interaction with soluble cargoes. In contrast, the proteins with 'cytoplasmic' actions that transport small molecules into SGs or that control exocytosis so far seem to follow indirect routes that may fuse with maturing IGs or intersect with endosomes on their way to granule entry. Will further studies support this distinction? As with secretory content, there is now a premium to understanding how these latter proteins are targeted and selectively distributed between the two (or possibly even more) granule subpopulations, especially as the subpopulations prospectively differ in contributing to the physiology of their host organisms [15, 86].

As more has been learned about the branch points and intersections of granule formation pathways, similarities, including overlapping machinery, have been identified with other pathways that operate in the TGN-endosomal network. For example, cholesterol- and sphingomyelin-enriched membranes serve as constitutive carriers to the cell surface of selected soluble and membrane proteins, while carriers with lower cholesterol and sphingomyelin contents service different cell surface domains and endosomes [65, 80, 81, 274]. How are these configured and distinguished from forming SGs? Similarly, AP-3 mediated endocytic trafficking to synaptic-like vesicles and lysosomes and to SGs operate in the same cells [200]. To what extent do differing destinations involve binary decisions based on membrane lipid composition, on specific sorting signals, and on the distinct localizations of specific AP-3 complexes? As more is learned about mechanistic distinctions in selective targeting to granule subpopulations, the foundation will be set for exploring these larger issues of interface and divergence from other pathways.

One of the more profound insights from research over the past two decades has been the realization that lipids play central and intertwined roles in coordinating the range of processes involved in SG biogenesis. Indeed, PI4P has emerged as a master regulator of cholesterol dynamics, sphingomyelin synthesis, and the mechanics of SG formation and maturation. These three lipids, together with PKD, link and orchestrate events all the way from ER export to the maturation of SGs within a tightly controlled assembly line to match the influx of cargoes to the supply of membrane containers and to ensure that cargo–carrier interactions and export/import trafficking take place while producing the final product. Indeed, experimental or dietary manipulations that perturb the levels of any of these lipids generally compromise storage capacity and/or stimulus-dependent exocytosis and often lead to increased SG degradation by crinophagy. The insights that have been gained now enable a new range of refined studies to take place that aim to determine how specialized pools of PI4P, cholesterol, and possibly sphingolipids distribute and enable the formation of two types of SG with distinct roles and whether lipid perturbations disproportionately affect one subpopulation or the other or lead to compensatory changes between them.

Finally, with the knowledge that the existence of two subpopulations of SGs in regulated secretory cells may be a general phenomenon, an added level of evaluation is now needed when exploring the effects of mutations or deficiencies, especially those that either conditionally or only partially perturb a secretory response. It is already possible to appreciate that different SG subpopulations likely contribute differentially to the two temporal phases of experimental glucose-stimulated insulin secretion [86]. Thus, treatments that selectively alter or perturb one phase or the other (e.g. [180, 275]) or involve specific granule proteins such as syt7 [276, 277] may be assignable to a particular subset of granules. In general, the discovery that regulated secretory cells form two (or possibly more) subpopulations of SGs that differ in sensitivity in responding to secretory stimuli and have at least partially distinct compositions of secretory cargoes will motivate reconsideration of a range of previous findings. Moreover, this realization will hopefully stimulate future investigations that clarify the mechanisms that distinguish the formation of individual granule subtypes in parallel, refine understanding of their relative contributions to secretory homeostasis in beta cells [272, 278] as well as other cell types, and delineate their contribution to physiological responses that reflect the timing as well as the strength of secretory signals.

Acknowledgments

I am very grateful to Michael Ailion (University of Washington) for his helpful comments and insightful discussion during the review process. I am further very grateful to Peter Arvan (University of Michigan) for numerous discussions over many years and for helpful comments on the manuscript, and to Volker Kiessling (University of Virginia), and Alex Kreutzberger (Harvard University) for helpful comments on the manuscript.

References

[1] Arvan P and Castle D 1998 Sorting and storage during secretory granule biogenesis: looking backward and looking forward *Biochem. J.* **332** 593–610

[2] Kornfeld S and Mellman I 1989 The biogenesis of lysosomes *Annu. Rev. Cell. Biol.* **5** 483–525

[3] Braulke T and Bonifacino J S 2009 Sorting of lysosomal proteins *Biochim. Biophys. Acta* **1793** 605–14

[4] Meskini E R, Galano G J, Marx R, Mains R E and Eipper B A 2001 Targeting of membrane proteins to the regulated secretory pathway in anterior pituitary endocrine cells *J. Biol. Chem.* **276** 3384–93

[5] Castle A M, Schwarzbauer J E, Wright R L and Castle J D 1995 Differential targeting of recombinant fibronectins in AtT-20 cells based on their efficiency of aggregation *J. Cell Sci.* **108** 3827–37

[6] Castle A M, Huang A Y and Castle J D 1998 Immunoglobulin-derived polypeptides enter the regulated secretory pathway in AtT-20 cells *FEBS Lett.* **439** 341–5

[7] Dhanvantari S and Loh Y P 2000 Lipid raft association of carboxypeptidase E is necessary for its function as a regulated secretory pathway sorting receptor *J. Biol. Chem.* **275** 29887–93

[8] Sun M, Watanabe T, Bochimoto H, Sakai Y, Torii S and Takeuchi T *et al* 2013 Multiple sorting systems for secretory granules ensure the regulated secretion of peptide hormones *Traffic* **14** 205–18

[9] De Lisle R C 1994 Characterization of the major sulfated protein of mouse pancreatic acinar cells: a high molecular weight peripheral membrane glycoprotein of zymogen granules *J. Cell. Biochem.* **56** 385–96

[10] Dikeakos J D and Reudelhuber T L 2007 Sending proteins to dense core secretory granules: still a lot to sort out *J. Cell Biol.* **177** 191–6

[11] Reck J, Beuret N, Demirci E, Prescianotto-Baschong C and Spiess M 2022 Small disulfide loops in peptide hormones mediate self-aggregation and secretory granule sorting *Life Science Alliance* **5** e202101279

[12] Chanat E and Huttner W B 1991 Milieu-induced, selective aggregation of regulated secretory proteins in the trans-Golgi network *J. Cell Biol.* **115** 1505–19

[13] Gerdes H H, Rosa P, Phillips E, Baeuerle P A, Frank R and Argos P *et al* 1989 The primary structure of human secretogranin II, a widespread tyrosine-sulfated secretory granule protein that exhibits low pH- and calcium-induced aggregation *J. Biol. Chem.* **264** 12009–15

[14] Colomer V, Kicska G A and Rindler M J 1996 Secretory granule content proteins and the luminal domains of granule membrane proteins aggregate *in vitro* at mildly acidic pH *J. Biol. Chem.* **271** 48–55

[15] Sossin W S, Fisher J M and Scheller R H 1990 Sorting within the regulated secretory pathway occurs in the trans-Golgi network *J. Cell Biol.* **110** 1–12

[16] Arvan P, Rudnick G and Castle J D 1985 Relative lack of ATP-driven H^+ translocase activity in isolated parotid secretory granules *J. Biol. Chem.* **260** 14945–52

[17] Arvan P and Castle J D 1986 Isolated secretion granules from parotid glands of chronically stimulated rats possess an alkaline internal pH and inward-directed H^+ pump activity *J. Cell Biol.* **103** 1257–67

[18] Sobota J A, Ferraro F, Bäck N, Eipper B A and Mains R E 2006 Not all secretory granules are created equal: partitioning of soluble content proteins *Mol. Biol. Cell* **17** 5038–52

[19] Orci L, Ravazzola M, Amherdt M, Madsen O, Perrelet A and Vassali J-D *et al* 1986 Conversion of proinsulin to insulin occurs coordinately with acidification of maturing secretory vesicles *J. Cell Biol.* **103** 2273–81
[20] Wu M M, Grabe M, Adams S, Tsien R Y, Moore H P and Machen T E 2001 Mechanisms of pH regulation in the regulated secretory pathway *J. Biol. Chem.* **276** 33027–35
[21] Barg S, Huang P, Eliasson L, Nelson D J, Obermüller S and Rorsman P *et al* 2001 Priming of insulin granules for exocytosis by granular Cl(−) uptake and acidification *J. Cell Sci.* **114** 2145–54
[22] Deriy L V, Gomez E A, Jacobson D A, Wang X, Hopson J A and Liu X Y *et al* 2009 The granular chloride channel ClC-3 is permissive for insulin secretion *Cell Metabolism* **10** 316–23
[23] Leblond F A, Viau G, Lainé J and Lebel D 1993 Reconstitution *in vitro* of the pH-dependent aggregation of pancreatic zymogens en route to the secretory granule: implication of GP-2 *Biochem.* J **291** 289–96
[24] Milgram S L, Mains R E and Eipper B A 1996 Identification of routing determinants in the cytosolic domain of a secretory granule-associated integral membrane protein *J. Biol. Chem.* **271** 17526–35
[25] Michaux G, Pullen T J, Haberichter S L and Cutler D F 2006 P-selectin binds to the D'-D3 domains of von Willebrand factor in Weibel–Palade bodies *Blood.* **107** 3922–4
[26] Dikeakos J D, Lacombe M-J, Mercure C, Mireuta M and Reudelhuber T L 2007 A hydrophobic patch in a charged alpha-helix is sufficient to target proteins to dense core secretory granules *J. Biol. Chem.* **282** 1136–43
[27] Bearrows S C, Bauchle C J, Becker M, Haldeman J M, Swaminathan S and Stephens S B 2019 Jul 1 Chromogranin B regulates early-stage insulin granule trafficking from the Golgi in pancreatic islet β-cells *J. Cell Sci.* **132**
[28] Parchure A, Tian M, Boyer C K, Bearrows S C, Rohli K E and Zhang J *et al* 2021 Liquid–liquid phase separation facilitates the biogenesis of secretory storage granules *bioRxiv*:2021.12.22.472607
[29] Orci L, Ravazzola M, Amherdt M, Madsen O, Vassali J-D and Perrelet A 1985 Direct identification of prohormone conversion site in insulin-secreting cells *Cell.* **42** 671–81
[30] Tandler B and Phillips C J 1993 Structure of serous cells in salivary glands *Microsc. Res. Tech.* **26** 32–48
[31] Syed Z A, Zhang L, Tran D T, Bleck C K E and Hagen Ten K G 2021 Orchestrated restructuring events during secretory granule maturation mediate intragranular cargo segregation *bioRxiv*:2021.08.16.456250
[32] Arvan P, Kuliawat R, Prabakaran D, Zavacki A M, Elahi D and Wang S *et al* 1991 Protein discharge from immature secretory granules displays both regulated and constitutive characteristics *J. Biol. Chem.* **266** 14171–4
[33] Castle A M, Huang A Y and Castle J D 1997 Passive sorting in maturing granules of AtT-20 cells: the entry and exit of salivary amylase and proline-rich protein *J. Cell Biol.* **138** 45–54
[34] Milgram S L, Eipper B A and Mains R E 1994 Differential trafficking of soluble and integral membrane secretory granule-associated proteins *J. Cell Biol.* **124** 33–41
[35] Klumperman J, Kuliawat R, Griffith J M, Geuze H J and Arvan P 1998 Mannose 6-phosphate receptors are sorted from immature secretory granules via adaptor protein AP-1, clathrin, and syntaxin 6-positive vesicles *J. Cell Biol.* **141** 359–71

[36] Kuliawat R and Arvan P 1992 Protein targeting via the 'constitutive-like' secretory pathway in isolated pancreatic islets: passive sorting in the immature granule compartment *J. Cell Biol.* **118** 521–9

[37] Turner M D and Arvan P 2000 Protein traffic from the secretory pathway to the endosomal system in pancreatic β-cells *J. Biol. Chem.* **275** 14225–30

[38] Eaton B A, Haugwitz M, Lau D and Moore H P 2000 Biogenesis of regulated exocytotic carriers in neuroendocrine cells *J. Neurosci.* **20** 7334–44

[39] Ahras M, Otto G P and Tooze S A 2006 Synaptotagmin IV is necessary for the maturation of secretory granules in PC12 cells *J. Cell Biol.* **173** 241–51

[40] Kakhlon O, Sakya P, Larijani B, Watson R and Tooze S A 2006 GGA function is required for maturation of neuroendocrine secretory granules *EMBO J.* **25** 1590–602

[41] Varlamov O, Wu F, Shields D and Fricker L D 1999 Biosynthesis and packaging of carboxypeptidase D into nascent secretory vesicles in pituitary cell lines *J. Biol. Chem.* **274** 14040–5

[42] Bonnemaison M, Bäck N, Lin Y, Bonifacino J S, Mains R and Eipper B 2014 AP-1A controls secretory granule biogenesis and trafficking of membrane secretory granule proteins *Traffic* **15** 1099–121

[43] Cattin-Ortolá J, Topalidou I, Lau H-T, Hummer B, Asensio C S and Ong S-E *et al* 2019 CCDC186 controls dense-core vesicle cargo sorting by exit *bioRxiv*:616458

[44] Huang X F and Arvan P 1995 Intracellular transport of proinsulin in pancreatic beta-cells. Structural maturation probed by disulfide accessibility *J. Biol. Chem.* **270** 20417–23

[45] Orci L, Ravazzola M, Storch M J, Anderson R G, Vassalli J D and Perrelet A 1987 Proteolytic maturation of insulin is a post-Golgi event which occurs in acidifying clathrin-coated secretory vesicles *Cell* **49** 865–8

[46] Tooze J, Hollinshead M, Frank R and Burke B 1987 An antibody specific for an endoproteolytic cleavage site provides evidence that pro-opiomelanocortin is packaged into secretory granules in AtT20 cells before its cleavage *J. Cell Biol.* **105** 155–62

[47] Schnabel E, Mains R E and Farquhar M G 1989 Proteolytic processing of pro-ACTH/endorphin begins in the Golgi complex of pituitary corticotropes and AtT-20 cells *Mol. Endocrinol.* **3** 1223–35

[48] Fricker L D and Devi L 1993 Posttranslational processing of carboxypeptidase E, a neuropeptide-processing enzyme, in AtT-20 cells and bovine pituitary secretory granules *J. Neurochem.* **61** 1404–15

[49] Vindrola O and Lindberg I 1992 Jul Biosynthesis of the prohormone convertase mPC1 in AtT-20 cells *Mol. Endocrinol.* **6** 1088–94

[50] Muller L and Lindberg I 1999 The cell biology of the prohormone convertases PC1 and PC2 *Prog. Nucleic Acid Res. Mol. Biol.* **63** 69–108

[51] Bäck N, Rajagopal C, Mains R E and Eipper B A 2010 Secretory granule membrane protein recycles through multivesicular bodies *Traffic* **11** 972–86

[52] Bäck N, Luxmi R, Powers K G, Mains R E and Eipper B A 2020 Peptidylglycine α-amidating monooxygenase is required for atrial secretory granule formation *Proc. Natl Acad. Sci. USA* **117** 17820–31

[53] Briguglio J S, Kumar S and Turkewitz A P 2013 Lysosomal sorting receptors are essential for secretory granule biogenesis in Tetrahymena *J. Cell Biol.* **203** 537–50

[54] Harrison-Lavoie K J, Michaux G, Hewlett L, Kaur J, Hannah M J and Lui-Roberts W W Y *et al* 2006 Jun P-selectin and CD63 use different mechanisms for delivery to Weibel–Palade bodies *Traffic* **7** 647–62

[55] Kaur H, Sparvoli D, Osakada H, Iwamoto M, Haraguchi T and Turkewitz A P 2017 An endosomal syntaxin and the AP-3 complex are required for formation and maturation of candidate lysosome-related secretory organelles (mucocysts) in *Tetrahymena thermophila* *Mol. Biol. Cell* **28** 1551–64

[56] Sparvoli D, Richardson E, Osakada H, Lan X, Iwamoto M and Bowman G R *et al* 2018 Remodeling the specificity of an endosomal CORVET tether underlies formation of regulated secretory vesicles in the ciliate *Tetrahymena thermophila* *Curr. Biol.* **28** 697–710.e13

[57] Chen Z-Y, Ieraci A, Teng H, Dall H, Meng C-X and Herrera D G *et al* 2005 Sortilin controls intracellular sorting of brain-derived neurotrophic factor to the regulated secretory pathway *J. Neurosci.* **25** 6156–66

[58] Kebede M A, Oler A T, Gregg T, Balloon A J, Johnson A and Mitok K *et al* 2014 SORCS1 is necessary for normal insulin secretory granule biogenesis in metabolically stressed β cells *J. Clin. Invest.* **124** 4240–56

[59] Mesmin B and Maxfield F R 2009 Intracellular sterol dynamics *BBA – Mol. Cell Biol. Lipids* **1791** 636–45

[60] Futerman A H and Riezman H 2005 The ins and outs of sphingolipid synthesis *Trends Cell Biol.* **15** 312–8

[61] Hayashi Y, Nemoto-Sasaki Y, Matsumoto N, Hama K, Tanikawa T and Oka S *et al* 2018 Complex formation of sphingomyelin synthase 1 with glucosylceramide synthase increases sphingomyelin and decreases glucosylceramide levels *J. Biol. Chem.* **293** 17505–22

[62] Rizzo R, Russo D, Kurokawa K, Sahu P, Lombardi B and Supino D *et al* 2021 Golgi maturation-dependent glycoenzyme recycling controls glycosphingolipid biosynthesis and cell growth via GOLPH3 *EMBO J.* e107238

[63] Klemm R W, Ejsing C S, Surma M A, Kaiser H-J, Gerl M J and Sampaio J L *et al* 2009 Segregation of sphingolipids and sterols during formation of secretory vesicles at the trans-Golgi network *J. Cell Biol.* **185** 601–12

[64] Mesmin B, Bigay J, Polidori J, Jamecna D, Lacas-Gervais S and Antonny B 2017 Sterol transfer, PI4P consumption, and control of membrane lipid order by endogenous OSBP *EMBO J.* **36** 3156–74

[65] Wakana Y, Hayashi K, Nemoto T, Watanabe C, Taoka M and Angulo-Capel J *et al* 2021 The ER cholesterol sensor SCAP promotes CARTS biogenesis at ER-Golgi membrane contact sites *J. Cell Biol.* **220** 10647

[66] Fairn G D, Schieber N L, Ariotti N, Murphy S, Kuerschner L and Webb R I *et al* 2011 High-resolution mapping reveals topologically distinct cellular pools of phosphatidylserine *J. Cell Biol.* **194** 257–75

[67] Hankins H M, Baldridge R D, Xu P and Graham T R 2015 Role of flippases, scramblases and transfer proteins in phosphatidylserine subcellular distribution *Traffic* **16** 35–47

[68] van Meer G, Voelker D R and Feigenson G W 2008 Membrane lipids: where they are and how they behave *Nat. Rev. Mol. Cell Biol.* **9** 112–24

[69] Raghupathy R, Anilkumar A A, Polley A, Singh P P, Yadav M and Johnson C *et al* 2015 Transbilayer lipid interactions mediate nanoclustering of lipid-anchored proteins *Cell* **161** 581–94

[70] Lee S, Uchida Y, Wang J, Matsudaira T, Nakagawa T and Kishimoto T *et al* 2015 Transport through recycling endosomes requires EHD1 recruitment by a phosphatidylserine translocase *EMBO J.* **34** 669–88
[71] Holthuis J C M and Menon A K 2014 Lipid landscapes and pipelines in membrane homeostasis *Nature* **510** 48–57
[72] Banerji S, Ngo M, Lane C F, Robinson C-A, Minogue S and Ridgway N D 2010 Oxysterol binding protein-dependent activation of sphingomyelin synthesis in the golgi apparatus requires phosphatidylinositol 4-kinase IIα *Mol. Biol. Cell* **21** 4141–50
[73] Balla T 2013 Phosphoinositides: tiny lipids with giant impact on cell regulation *Physiol. Rev.* **93** 1019–137
[74] Mesmin B, Bigay J, Moser von Filseck J, Lacas-Gervais S, Drin G and Antonny B 2013 A four-step cycle driven by PI(4)P hydrolysis directs sterol/PI(4)P exchange by the ER-Golgi tether OSBP *Cell* **155** 830–43
[75] Dippold H C, Ng M M, Farber-Katz S E, Lee S-K, Kerr M L and Peterman M C *et al* 2009 GOLPH3 bridges phosphatidylinositol-4-phosphate and actomyosin to stretch and shape the Golgi to promote budding *Cell* **139** 337–51
[76] Wang Y, Mousley C J, Lete M G and Bankaitis V A 2019 An equal opportunity collaboration between lipid metabolism and proteins in the control of membrane trafficking in the trans-Golgi and endosomal systems *Curr. Opin. Cell Biol.* **59** 58–72
[77] De Matteis M A, Wilson C and D'Angelo G 2013 Phosphatidylinositol-4-phosphate: the Golgi and beyond *Bioessays* **35** 612–22
[78] Shin J J H, Liu P, Chan L J, Ullah A, Pan J and Borchers C H *et al* 2020 pH biosensing by PI4P regulates cargo sorting at the TGN *Dev. Cell* **52** 461–4
[79] Harayama T, Eto M, Shindou H, Kita Y, Otsubo E and Hishikawa D *et al* 2014 Lysophospholipid acyltransferases mediate phosphatidylcholine diversification to achieve the physical properties required *in vivo Cell Metab.* **20** 295–305
[80] Deng Y, Pakdel M, Blank B, Sundberg E L, Burd C G, Blume and von J 2018 Activity of the SPCA1 calcium pump couples sphingomyelin synthesis to sorting of secretory proteins in the trans-Golgi network *Dev. Cell* **47** 464–8
[81] Deng Y, Rivera-Molina F E, Toomre D K and Burd C G 2016 Sphingomyelin is sorted at the trans Golgi network into a distinct class of secretory vesicle *Proc. Natl. Acad. Sci. USA* **113** 6677–82
[82] van Meer G, Stelzer E H, Wijnaendts-van-Resandt R W and Simons K 1987 Sorting of sphingolipids in epithelial (Madin–Darby canine kidney) cells *J. Cell Biol.* **105** 1623–35
[83] Kiyokawa E, Baba T, Otsuka N, Makino A, Ohno S and Kobayashi T 2005 Spatial and functional heterogeneity of sphingolipid-rich membrane domains *J. Biol. Chem.* **280** 24072–84
[84] Hussain S S, Harris M T, Kreutzberger A J B, Inouye C M, Doyle C A and Castle A M *et al* 2018 Control of insulin granule formation and function by the ABC transporters ABCG1 and ABCA1 and by oxysterol binding protein OSBP *Mol. Biol. Cell* **29** 1238–57
[85] Subathra M, Qureshi A and Luberto C 2011 Sphingomyelin synthases regulate protein trafficking and secretion *PLoS One* **6** e23644
[86] Kreutzberger A J B, Kiessling V, Doyle C A, Schenk N, Upchurch C M and Elmer-Dixon M *et al* 2020 Distinct insulin granule subpopulations implicated in the secretory pathology of diabetes types 1 and 2 *eLife* **9** 2145

[87] Westhead E W 1987 Lipid composition and orientation in secretory vesicles *Ann. NY Acad. Sci.* **493** 92–100

[88] Radhakrishnan A, Sun L-P, Kwon H J, Brown M S and Goldstein J L 2004 Direct binding of cholesterol to the purified membrane region of SCAP: mechanism for a sterol-sensing domain *Mol. Cell* **15** 259–68

[89] Goldstein J L and Brown M S 2015 A century of cholesterol and coronaries: from plaques to genes to statins *Cell* **161** 161–72

[90] Wang Y, Thiele C and Huttner W B 2000 Cholesterol is required for the formation of regulated and constitutive secretory vesicles from the trans-Golgi network *Traffic* **1** 952–62

[91] Hosaka M 2003 Secretogranin III binds to cholesterol in the secretory granule membrane as an adapter for chromogranin A *J. Biol. Chem.* **279** 3627–34

[92] Gondre-Lewis M C 2006 Abnormal sterols in cholesterol-deficiency diseases cause secretory granule malformation and decreased membrane curvature *J. Cell Sci.* **119** 1876–85

[93] Tsuchiya M, Hosaka M, Moriguchi T, Zhang S, Suda M and Yokota-Hashimoto H *et al* 2010 Cholesterol biosynthesis pathway intermediates and inhibitors regulate glucose-stimulated insulin secretion and secretory granule formation in pancreatic β-cells *Endocrinology* **151** 4705–16

[94] Brunham L R, Kruit J K, Pape T D, Timmins J M, Reuwer A Q and Vasanji Z *et al* 2007 β-cell ABCA1 influences insulin secretion, glucose homeostasis and response to thiazolidinedione treatment *Nat. Med.* **13** 340–7

[95] Hao M, Head W S, Gunawardana S C, Hasty A H and Piston D W 2007 Direct effect of cholesterol on insulin secretion: a novel mechanism for pancreatic-cell dysfunction *Diabetes* **56** 2328–38

[96] Xia F, Xie L, Mihic A, Gao X, Chen Y and Gaisano H Y *et al* 2008 Inhibition of cholesterol biosynthesis impairs insulin secretion and voltage-gated calcium channel function in pancreatic β-cells *Endocrinology* **149** 5136–45

[97] Hao M and Bogan J S 2009 Cholesterol regulates glucose-stimulated insulin secretion through phosphatidylinositol 4,5-bisphosphate *J. Biol. Chem.* **284** 29489–98

[98] Sturek J M, Castle J D, Trace A P, Page L C, Castle A M and Evans-Molina C *et al* 2010 An intracellular role for ABCG1-mediated cholesterol transport in the regulated secretory pathway of mouse pancreatic β cells *J. Clin. Invest.* **120** 2575–89

[99] Perry R J and Ridgway N D 2006 Oxysterol-binding protein and vesicle-associated membrane protein-associated protein are required for sterol-dependent activation of the ceramide transport protein *Mol. Biol. Cell* **17** 2604–16

[100] Harris M T, Hussain S S, Inouye C M, Castle A M and Castle J D 2018 Reinterpretation of the localization of the ATP binding cassette transporter ABCG1 in insulin-secreting cells and insights regarding its trafficking and function *PLoS One* 13 e0198383

[101] Kobayashi A 2006 Efflux of sphingomyelin, cholesterol, and phosphatidylcholine by ABCG1 *J. Lipid Res.* **47** 1791–802

[102] Sano O, Kobayashi A, Nagao K, Kumagai K, Kioka N and Hanada K *et al* 2007 Sphingomyelin-dependence of cholesterol efflux mediated by ABCG1 *J. Lipid Res.* **48** 2377–84

[103] Quazi F and Molday R S 2013 Differential phospholipid substrates and directional transport by ATP-binding cassette proteins ABCA1, ABCA7, and ABCA4 and disease-causing mutants *J. Biol. Chem.* **288** 34414–26

[104] Halban P A and Wolheim C B 1980 Intracellular degradation of insulin stores by rat pancreatic islets *In Vitro J. Biol. Chem.* **255** 6003–6
[105] Marsh B J, Soden C, Alarcón C, Wicksteed B L, Yaekura K and Costin A J *et al* 2007 Regulated autophagy controls hormone content in secretory-deficient pancreatic endocrine β-cells *Molecular Endocrinol.* **21** 2255–69
[106] Riahi Y, Wikstrom J D, Bachar-Wikstrom E, Polin N, Zucker H and Lee M-S *et al* 2016 Autophagy is a major regulator of beta cell insulin homeostasis *Diabetologia* **59** 1480–91
[107] Goginashvili A, Zhang Z, Erbs E, Spiegelhalter C, Kessler P and Mihlan M *et al* 2015 Insulin granules. insulin secretory granules control autophagy in pancreatic β cells *Science* **347** 878–82
[108] Pasquier A, Vivot K, Erbs E, Spiegelhalter C, Zhang Z and Aubert V *et al* 2019 Lysosomal degradation of newly formed insulin granules contributes to β cell failure in diabetes *Nat. Commun.* **10** 3312–4
[109] Ursino G M, Fu Y, Cottle D L, Mukhamedova N, Jones L K and Low H *et al* 2020 ABCA12 regulates insulin secretion from β-cells *EMBO Rep.* **21** e48692
[110] Kruit J K, Wijesekara N, Fox J E M, Dai X Q, Brunham L R and Searle G J *et al* 2011 Islet cholesterol accumulation due to loss of ABCA1 leads to impaired exocytosis of insulin granules *Diabetes* **60** 3186–96
[111] Akiyama M, Sugiyama-Nakagiri Y, Sakai K, McMillan J R, Goto M and Arita K *et al* 2005 Mutations in lipid transporter ABCA12 in harlequin ichthyosis and functional recovery by corrective gene transfer. *J. Clin. Invest.* **115** 1777–84
[112] Zachowski A, Henry J P and Devaux P F 1989 Control of transmembrane lipid asymmetry in chromaffin granules by an ATP-dependent protein *Nature* **340** 75–6
[113] Ansari I-U H, Longacre M J, Paulusma C C, Stoker S W, Kendrick M A and MacDonald M J 2015 Sep 18 Characterization of P4 ATPase phospholipid translocases (Flippases) in human and rat pancreatic beta cells: their gene silencing inhibits insulin secretion *J. Biol. Chem.* **290** 23110–23
[114] Carrat G R, Haythorne E, Tomas A, Haataja L, Müller A and Arvan P *et al* 2020 The type 2 diabetes gene product STARD10 is a phosphoinositide-binding protein that controls insulin secretory granule biogenesis *Mol. Metab.* **40** 101015
[115] Natarajan P, Liu K, Patil D V, Sciorra V A, Jackson C L and Graham T R 2009 Regulation of a Golgi flippase by phosphoinositides and an ArfGEF *Nat. Cell Biol.* **11** 1421–6
[116] Liljedahl M, Maeda Y, Colanzi A, Ayala I, Van Lint J and Malhotra V 2001 Protein kinase D regulates the fission of cell surface destined transport carriers from the trans-Golgi network *Cell* **104** 409–20
[117] Gehart H, Goginashvili A, Beck R, Morvan J, Erbs E and Formentini I *et al* 2012 The BAR domain protein Arfaptin-1 controls secretory granule biogenesis at the trans-Golgi network *Dev. Cell* **23** 756–68
[118] Capasso S, Sticco L, Rizzo R, Pirozzi M, Russo D and Dathan N A *et al* 2017 Sphingolipid metabolic flow controls phosphoinositide turnover at the trans-Golgi network *EMBO J.* **36** 1736–54
[119] Di Paolo G and De Camilli P 2006 Phosphoinositides in cell regulation and membrane dynamics *Nature* **443** 651–7
[120] Mesmin B, Kovacs D and D'Angelo G 2019 Lipid exchange and signaling at ER-Golgi contact sites *Curr. Opin. Cell Biol.* **57** 8–15

[121] Kutchukian C, Vivas O, Casas M, Jones J G, Tiscione S A and Simó S *et al* 2021 NPC1 regulates the distribution of phosphatidylinositol 4-kinases at Golgi and lysosomal membranes *EMBO J.* **40** e105990

[122] Burgess J, Del Bel L M, Ma C-I J, Barylko B, Polevoy G and Rollins J *et al* 2012 Type II phosphatidylinositol 4-kinase regulates trafficking of secretory granule proteins in *Drosophila Development* **139** 3040-50

[123] Lopes da Silva M, O'Connor M N, Kriston-Vizi J, White I J, Al-Shawi R and Simons J P *et al* 2016 Type II PI4-kinases control Weibel–Palade body biogenesis and von Willebrand factor structure in human endothelial cells *J. Cell Sci.* **129** 2096–105

[124] Ma C-I J, Yang Y, Kim T, Chen C H, Polevoy G and Vissa M *et al* 2020 An early endosome-derived retrograde trafficking pathway promotes secretory granule maturation *J. Cell Biol.* **219** 167

[125] Panaretou C and Tooze S A 2002 Regulation and recruitment of phosphatidylinositol 4-kinase on immature secretory granules is independent of ADP-ribosylation factor 1 *Biochem. J.* **3632** 289–95

[126] Mizuno-Yamasaki E, Medkova M, Coleman J and Novick P 2010 Phosphatidylinositol 4-phosphate controls both membrane recruitment and a regulatory switch of the Rab GEF Sec2p *Dev. Cell* **18** 828–40

[127] Ling Y, Hayano S and Novick P 2014 Osh4p is needed to reduce the level of phosphatidylinositol-4-phosphate on secretory vesicles as they mature *Mol. Biol. Cell* **25** 3389–400

[128] Nguyen P M, Gandasi N R, Xie B, Sugahara S, Xu Y and Idevall-Hagren O 2019 The PI(4)P phosphatase Sac2 controls insulin granule docking and release *J. Cell Biol.* **218** 3714–29

[129] Fugmann T, Hausser A, Schöffler P, Schmid S, Pfizenmaier K and Olayioye M A 2007 Regulation of secretory transport by protein kinase D-mediated phosphorylation of the ceramide transfer protein *J. Cell Biol.* **178** 15–22

[130] Ngo M and Ridgway N D 2009 Oxysterol binding protein-related protein 9 (ORP9) is a cholesterol transfer protein that regulates Golgi structure and function *Mol. Biol. Cell* **20** 1388–99

[131] Chang W-L, Chang C-W, Chang Y-Y, Sung H-H, Lin M-D and Chang S-C *et al* 2013 The Drosophila GOLPH3 homolog regulates the biosynthesis of heparan sulfate proteoglycans by modulating the retrograde trafficking of exostosins *Development* **140** 2798–807

[132] Baron C L and Malhotra V 2002 Role of diacylglycerol in PKD recruitment to the TGN and protein transport to the plasma membrane *Science* **295** 325–8

[133] Cruz-Garcia D, Ortega-Bellido M, Scarpa M, Villeneuve J, Jovic M and Porzner M *et al* 2013 Recruitment of arfaptins to the trans-Golgi network by PI(4)P and their involvement in cargo export *EMBO J.* **32** 1717–29

[134] Man Z, Kondo Y, Koga H, Umino H, Nakayama K and Shin H-W 2011 Arfaptins are localized to the trans-Golgi by interaction with Arl1, but not Arfs *J. Biol. Chem.* **286** 11569–78

[135] Szule J A, Fuller N L and Rand R P 2002 The effects of acyl chain length and saturation of diacylglycerols and phosphatidylcholines on membrane monolayer curvature *Biophys. J.* **83** 977–84

[136] Siddhanta A and Shields D 1998 Secretory vesicle budding from the trans-Golgi network is mediated by phosphatidic acid levels *J. Biol. Chem.* **273** 17995–8

[137] Tanguy E, Costé de Bagneaux P, Kassas N, Ammar M-R, Wang Q and Haeberlé A-M *et al* 2020 Mono- and poly-unsaturated phosphatidic acid regulate distinct steps of regulated exocytosis in neuroendocrine cells *Cell Rep.* **32** 108026

[138] Mukherjee S, Soe T T and Maxfield F R 1999 Endocytic sorting of lipid analogues differing solely in the chemistry of their hydrophobic tails *J. Cell Biol.* **144** 1271–84

[139] Maxfield F R and McGraw T E 2004 Endocytic recycling *Nat. Rev. Mol. Cell Biol.* **5** 121–32

[140] Sorre B, Callan-Jones A, Manneville J-B, Nassoy P, Joanny J-F and Prost J *et al* 2009 Curvature-driven lipid sorting needs proximity to a demixing point and is aided by proteins *Proc. Natl Acad. Sci. USA* **106** 5622–6

[141] Antonny B, Vanni S, Shindou H and Ferreira T 2015 From zero to six double bonds: phospholipid unsaturation and organelle function *Trends Cell Biol.* **25** 427–36

[142] Jamieson J D and Palade G E 1971 Synthesis, intracellular transport, and discharge of secretory proteins in stimulated pancreatic exocrine cells *J. Cell Biol.* **50** 135–58

[143] Almeida C G, Yamada A, Tenza D, Louvard D, Raposo G and Coudrier E 2011 Myosin 1b promotes the formation of post-Golgi carriers by regulating actin assembly and membrane remodelling at the trans-Golgi network *Nat. Cell Biol.* **13** 779–89

[144] Delestre-Delacour C, Carmon O, Laguerre F, Estay-Ahumada C, Courel M and Elias S *et al* 2017 Myosin 1b and F-actin are involved in the control of secretory granule biogenesis *Sci Rep.* **7** 5172

[145] Tokuo H, Komaba S and Coluccio L M 2021 In pancreatic β-cells myosin 1b regulates glucose-stimulated insulin secretion by modulating an early step in insulin granule trafficking from the Golgi *Mol. Biol. Cell* mbcE21030094

[146] Trogden K P, Zhu X, Lee J S, Wright C V E, Gu G and Kaverina I 2019 Regulation of glucose-dependent Golgi-derived microtubules by cAMP/EPAC2 promotes secretory vesicle biogenesis in pancreatic β cells *Curr. Biol.* **29** 2339–2350.e5

[147] Müller A, Schmidt D, Xu C S, Pang S, D'Costa J V and Kretschmar S *et al* 2021 3D FIB-SEM reconstruction of microtubule-organelle interaction in whole primary mouse β cells *J. Cell Biol.* **220** 1226

[148] Smith R E and Farquhar M G 1966 Lysosome function in the regulation of the secretory process in cells of the anterior pituitary gland *J. Cell Biol.* **31** 319–47

[149] Salpeter M M and Farquhar M G 1981 High resolution analysis of the secretory pathway in mammotrophs of the rat anterior pituitary *J. Cell Biol.* **91** 240–6

[150] Du W, Zhou M, Zhao W, Cheng D, Wang L and Lu J *et al* 2016 HID-1 is required for homotypic fusion of immature secretory granules during maturation *eLife.* **5** e18134

[151] Tooze S A, Flatmark T, Tooze J and Huttner W B 1991 Characterization of the immature secretory granule, an intermediate in granule biogenesis *J. Cell Biol.* **115** 1491–503

[152] Urbé S, Page L J and Tooze S A 1998 Homotypic fusion of immature secretory granules during maturation in a cell-free assay *J. Cell Biol.* **143** 1831–44

[153] Wendler F, Page L, Urbé S and Tooze S A 2001 Homotypic fusion of immature secretory granules during maturation requires syntaxin 6 *Mol. Biol. Cell* **12** 1699–709

[154] Tran D T and Hagen Ten K G 2017 Real-time insights into regulated exocytosis *J. Cell Sci.* **130** 1355–63

[155] Ferraro F, Kriston-Vizi J, Metcalf D J, Martin-Martin B, Freeman J and Burden J J *et al* 2014 A two-tier Golgi-based control of organelle size underpins the functional plasticity of endothelial cells *Dev. Cell* **29** 292–304

[156] Topalidou I, Cattin-Ortolá J, Pappas A L, Cooper K, Merrihew G E and MacCoss M J *et al* 2016 The EARP complex and its interactor EIPR-1 are required for cargo sorting to dense-core vesicles *PLoS Genet.* **12** e1006074

[157] Emperador-Melero J, Huson V, van Weering J, Bollmann C, Fischer von Mollard G and Toonen R F *et al* 2018 Vti1a/b regulate synaptic vesicle and dense core vesicle secretion via protein sorting at the Golgi *Nat. Commun.* **9** 3421

[158] Perello M, Stuart R and Nillni E A 2008 Prothyrotropin-releasing hormone targets its processing products to different vesicles of the secretory pathway *J. Biol. Chem.* **283** 19936–47

[159] Kuliawat R, Klumperman J, Ludwig T and Arvan P 1997 Differential sorting of lysosomal enzymes out of the regulated secretory pathway in pancreatic beta-cells *J. Cell Biol.* **137** 595–608

[160] Dittié A S, Thomas L, Thomas G and Tooze S A 1997 Interaction of furin in immature secretory granules from neuroendocrine cells with the AP-1 adaptor complex is modulated by casein kinase II phosphorylation *EMBO J.* **16** 4859–70

[161] Walter A M, Kurps J, de Wit H, Schöning S, Toft-Bertelsen T L and Lauks J *et al* 2014 The SNARE protein vti1a functions in dense-core vesicle biogenesis *EMBO J.* **33** 1681–97

[162] Bonnemaison M L, Eipper B A and Mains R E 2013 Role of adaptor proteins in secretory granule biogenesis and maturation *Front Endocrinol. (Lausanne)* **4** 101

[163] Kögel T and Gerdes H-H 2010 Roles of myosin Va and Rab3D in membrane remodeling of immature secretory granules *Cell Mol. Neurobiol.* **30** 1303–8

[164] Lui-Roberts W W Y, Collinson L M, Hewlett L J, Michaux G and Cutler D F 2005 An AP-1/clathrin coat plays a novel and essential role in forming the Weibel–Palade bodies of endothelial cells *J. Cell Biol.* **170** 627–36

[165] Edwards S L, Charlie N K, Richmond J E, Hegermann J, Eimer S and Miller K G 2009 Impaired dense core vesicle maturation in caenorhabditis elegansmutants lacking Rab2 *J. Cell Biol.* **186** 881–95

[166] Sumakovic M, Hegermann J, Luo L, Husson S J, Schwarze K and Olendrowitz C *et al* 2009 UNC-108/RAB-2 and its effector RIC-19 are involved in dense core vesicle maturation in caenorhabditis elegans *J. Cell Biol.* **186** 897–914

[167] Hannemann M, Sasidharan N, Hegermann J, Kutscher L M, Koenig S and Eimer S 2012 TBC-8, a putative RAB-2 GAP, regulates dense core vesicle maturation in Caenorhabditis elegans *PLoS Genet.* **8** e1002722

[168] Ailion M, Hannemann M, Dalton S, Pappas A, Watanabe S and Hegermann J *et al* 2014 Two Rab2 interactors regulate dense-core vesicle maturation *Neuron* **82** 167–80

[169] Topalidou I, Cattin-Ortolá J, Hummer B, Asensio C S and Ailion M 2020 EIPR1 controls dense-core vesicle cargo retention and EARP complex localization in insulin-secreting cells *Mol. Biol. Cell* **31** 59–79

[170] Cattin-Ortolá J, Topalidou I, Dosey A, Merz A J and Ailion M 2017 The dense-core vesicle maturation protein CCCP-1 binds RAB-2 and membranes through its C-terminal domain *Traffic* **18** 720–32

[171] Solinger J A, Rashid H-O, Prescianotto-Baschong C and Spang A 2020 FERARI is required for Rab11-dependent endocytic recycling *Nat. Cell Biol.* **22** 213–24

[172] Fujii S, Kurokawa K, Inaba R, Hiramatsu N, Tago T and Nakamura Y *et al* 2020 Recycling endosomes attach to the trans-side of Golgi stacks in *Drosophila* and mammalian cells *J. Cell Sci.* **133** jcs236935

[173] Paquin N, Murata Y, Froehlich A, Omura D T, Ailion M and Pender C L *et al* 2016 The conserved VPS-50 protein functions in dense-core vesicle maturation and acidification and controls animal behavior *Curr. Biol.* **26** 862–71

[174] Hummer B H, de Leeuw N F, Burns C, Chen L, Joens M S and Hosford B *et al* 2017 HID-1 controls formation of large dense core vesicles by influencing cargo sorting and trans-Golgi network acidification *Mol. Biol. Cell* **28** 3870–80
[175] Cao M, Mao Z, Kam C, Xiao N, Cao X and Shen C *et al* 2013 PICK1 and ICA69 control insulin granule trafficking and their deficiencies lead to impaired glucose tolerance *PLoS Biol.* **11** e1001541
[176] Holst B, Madsen K L, Jansen A M, Jin C, Rickhag M and Lund V K *et al* 2013 PICK1 deficiency impairs secretory vesicle biogenesis and leads to growth retardation and decreased glucose tolerance *PLoS Biol.* **11** e1001542
[177] Mallik B, Dwivedi M K, Mushtaq Z, Kumari M, Verma P K and Kumar V 2017 Regulation of neuromuscular junction organization by Rab2 and its effector ICA69 in *Drosophila Development* **144** 2032–44
[178] Andersen R C, Lycas M D, Schmidt J H, Christensen N R, Lund V K and Rombach J *et al* 2020 Coding variants identified in diabetic patients alter PICK1 BAR domain function in insulin granule biogenesis *bioRxiv*:2020.10.05.325951
[179] Pinheiro P S, Jansen A M, de Wit H, Tawfik B, Madsen K L and Verhage M *et al* 2014 The BAR domain protein PICK1 controls vesicle number and size in adrenal chromaffin cells *J. Neurosci.* **34** 10688–700
[180] Li J, Mao Z, Huang J and Xia J 2018 PICK1 is essential for insulin production and the maintenance of glucose homeostasis *Mol. Biol. Cell* **29** 587–96
[181] Herlo R, Lund V K, Lycas M D, Jansen A M, Khelashvili G and Andersen R C *et al* 2018 An amphipathic helix directs cellular membrane curvature sensing and function of the BAR domain protein PICK1 *Cell Rep.* **23** 2056–69
[182] Kuliawat R, Kalina E, Bock J, Fricker L, McGraw T E and Kim S R *et al* 2004 Syntaxin-6 SNARE involvement in secretory and endocytic pathways of cultured pancreatic *Mol. Biol. Cell* **15** 1690–701
[183] Arnaoutova I, Jackson C L, Al-Awar O S, Donaldson J G and Loh Y P 2003 Recycling of raft-associated prohormone sorting receptor carboxypeptidase E requires interaction with ARF6 *Mol. Biol. Cell* **14** 4448–57
[184] Sirkis D W, Edwards R H and Asensio C S 2013 Widespread dysregulation of peptide hormone release in mice lacking adaptor protein AP-3 *PLoS Genet.* **9** e1003812
[185] Hummer B H, Maslar D, Soltero-Gutierrez M, de Leeuw N F and Asensio C S 2020 Differential sorting behavior for soluble and transmembrane cargoes at the trans-Golgi network in endocrine cells *Mol. Biol. Cell* **31** 157–66
[186] Dembla E and Becherer U 2021 Biogenesis of large dense core vesicles in mouse chromaffin cells *Traffic* **22** 78–93
[187] Vo Y P, Hutton J C and Angleson J K 2004 Recycling of the dense-core vesicle membrane protein phogrin in min6 β-cells *Biochem. Biophys. Res. Commun.* **324** 1004–10
[188] Xu H, Chang F, Jain S, Heller B A, Han X and Liu Y *et al* 2022 SNX5 targets a monoamine transporter to the TGN for assembly into dense core vesicles by AP-3 *J. Cell Biol.* **221** e202106083
[189] Neuman S D and Bashirullah A 2018 Hobbit regulates intracellular trafficking to drive insulin-dependent growth during *Drosophila* development *Development* **145** dev161356
[190] Neuman S D, Terry E L, Selegue J E, Cavanagh A T and Bashirullah A 2021 Mistargeting of secretory cargo in retromer-deficient cells *Dis. Model Mech.* **14** dmm046417

[191] Asensio C S, Sirkis D W and Edwards R H 2010 RNAi screen identifies a role for adaptor protein AP-3 in sorting to the regulated secretory pathway *J. Cell Biol.* **191** 1173–87

[192] Caromile L A, Oganesian A, Coats S A, Seifert R A and Bowen-Pope D F 2010 The neurosecretory vesicle protein phogrin functions as a phosphatidylinositol phosphatase to regulate insulin secretion *J. Biol. Chem.* **285** 10487–96

[193] Torii S, Saito N, Kawano A, Zhao S, Izumi T and Takeuchi T 2005 Cytoplasmic transport signal is involved in phogrin targeting and localization to secretory granules *Traffic* **6** 1213–24

[194] Li H, Waites C L, Staal R G, Dobryy Y, Park J and Sulzer D L *et al* 2005 Sorting of vesicular monoamine transporter 2 to the regulated secretory pathway confers the somatodendritic exocytosis of monoamines *Neuron* **48** 619–33

[195] Asensio C S, Sirkis D W, Maas J W, Egami K, To T-L and Brodsky F M *et al* 2013 Self-assembly of VPS41 promotes sorting required for biogenesis of the regulated secretory pathway *Dev. Cell* **27** 425–37

[196] Boncompain G, Divoux S, Gareil N, de Forges H, Lescure A and Latreche L *et al* 2012 Synchronization of secretory protein traffic in populations of cells *Nat. Methods* **9** 493–8

[197] Grote E, Hao J C, Bennett M K and Kelly R B 1995 A targeting signal in VAMP regulating transport to synaptic vesicles *Cell* **81** 581–9

[198] Blagoveshchenskaya A D, Hewitt E W and Cutler D F 1999 Di-leucine signals mediate targeting of tyrosinase and synaptotagmin to synaptic-like microvesicles within PC12 cells *Mol. Biol. Cell* **10** 3979–90

[199] Peden A A, Oorschot V, Hesser B A, Austin C D, Scheller R H and Klumperman J 2004 Localization of the AP-3 adaptor complex defines a novel endosomal exit site for lysosomal membrane proteins *J. Cell Biol.* **164** 1065–76

[200] Newell-Litwa K, Salazar G, Smith Y and Faundez V 2009 Roles of BLOC-1 and adaptor protein-3 complexes in cargo sorting to synaptic vesicles *Mol. Biol. Cell* **20** 1441–53

[201] Grabner C P, Price S D, Lysakowski A, Cahill A L and Fox A P 2006 Regulation of large dense-core vesicle volume and neurotransmitter content mediated by adaptor protein 3 *Proc. Natl Acad. Sci. USA* **103** 10035–40

[202] Silm K, Yang J, Marcott P F, Asensio C S, Eriksen J and Guthrie D A *et al* 2019 Synaptic vesicle recycling pathway determines neurotransmitter content and release properties *Neuron* **102** 786–800.e5

[203] Newell-Litwa K, Seong E, Burmeister M and Faundez V 2007 Neuronal and non-neuronal functions of the AP-3 sorting machinery *J. Cell Sci.* **120** 531–41

[204] Bowman S L, Bi-Karchin J, Le L and Marks M S 2019 The road to lysosome-related organelles: Insights from Hermansky-Pudlak syndrome and other rare diseases *Traffic* **20** 404–35

[205] Burgess J, Jauregui M, Tan J, Rollins J, Lallet S and Leventis P A *et al* 2011 AP-1 and clathrin are essential for secretory granule biogenesis in *Drosophila Mol. Biol. Cell* **22** 2094–105

[206] Salazar G, Craige B, Styers M L, Newell-Litwa K A, Doucette M M and Wainer B H *et al* 2006 BLOC-1 complex deficiency alters the targeting of adaptor protein complex-3 cargoes *Mol. Biol. Cell* **17** 4014–26

[207] Hao Z, Wei L, Feng Y, Chen X, Du W and Ma J *et al* 2015 Impaired maturation of large dense-core vesicles in muted-deficient adrenal chromaffin cells *J. Cell Sci.* **128** 1365–74

[208] Burns C H, Yau B, Rodriguez A, Triplett J, Maslar D and An Y S *et al* 2021 Pancreatic β-Cell-specific deletion of VPS41 causes diabetes due to defects in insulin secretion *Diabetes* **70** 436–48

[209] Gillingham A K, Sinka R, Torres I L, Lilley K S and Munro S 2014 Toward a comprehensive map of the effectors of rab GTPases *Dev. Cell* **31** 358–73
[210] Huttlin E L, Ting L, Bruckner R J, Gebreab F, Gygi M P and Szpyt J *et al* 2015 The bioplex network: a systematic exploration of the human interactome *Cell* **162** 425–40
[211] Taraska J W, Perrais D, Ohara-Imaizumi M, Nagamatsu S and Almers W 2003 Secretory granules are recaptured largely intact after stimulated exocytosis in cultured endocrine cells *Proc. Natl Acad. Sci. USA* **100** 2070–5
[212] Taraska J W and Almers W 2004 Bilayers merge even when exocytosis is transient *Proc. Natl Acad. Sci. USA* **101** 8780–5
[213] Tsuboi T and Rutter G A 2003 Multiple forms of 'kiss-and-run' exocytosis revealed by evanescent wave microscopy *Curr. Biol.* **13** 563–7
[214] Perrais D, Kleppe I C, Taraska J W and Almers W 2004 Recapture after exocytosis causes differential retention of protein in granules of bovine chromaffin cells *J. Physiol.* **560** 413–28
[215] Ceridono M, Ory S, Momboisse F, Chasserot-Golaz S, Houy S and Calco V *et al* 2011 Selective recapture of secretory granule components after full collapse exocytosis in neuroendocrine chromaffin cells *Traffic* **12** 72–88
[216] Ory S, Ceridono M, Momboisse F, Houy S, Chasserot-Golaz S and Heintz D *et al* 2013 Phospholipid scramblase-1-induced lipid reorganization regulates compensatory endocytosis in neuroendocrine cells *J. Neurosci.* **33** 3545–56
[217] Bittner M A, Aikman R L and Holz R W 2013 A nibbling mechanism for clathrin-mediated retrieval of secretory granule membrane after exocytosis *J. Biol. Chem.* **288** 9177–88
[218] Patzak A and Winkler H 1986 Exocytotic exposure and recycling of membrane antigens of chromaffin granules: ultrastructural evaluation after immunolabeling *J. Cell Biol.* **102** 510–5
[219] Annaert W G, Partoens P, Slembrouck D, Bakker A, Jacob W and De Potter W P 1997 Rab3 dissociation and clathrin-mediated endocytosis, two key steps in the exo-endocytotic pathway of large dense-cored vesicles in primary cultures of superior cervical ganglia *Eur. J. Cell Biol.* **74** 217–29
[220] Bauer R A, Overlease R L, Lieber J L and Angleson J K 2004 Retention and stimulus-dependent recycling of dense core vesicle content in neuroendocrine cells *J. Cell Sci.* **117** 2193–202
[221] Bäck N, Kanerva K, Kurutihalli V, Yanik A, Ikonen E and Mains R E *et al* 2017 The endocytic pathways of a secretory granule membrane protein in HEK293 cells: PAM and EGF traverse a dynamic multivesicular body network together *Eur. J. Cell Biol.* **96** 407–17
[222] Steveson T C, Zhao G C, Keutmann H T, Mains R E and Eipper B A 2001 Access of a membrane protein to secretory granules is facilitated by phosphorylation *J. Biol. Chem.* **276** 40326–37
[223] Simonetti B, Paul B, Chaudhari K, Weeratunga S, Steinberg F and Gorla M *et al* 2019 Molecular identification of a BAR domain-containing coat complex for endosomal recycling of transmembrane proteins *Nat. Cell Biol.* **21** 1219–33
[224] Zhang X, Jiang S, Mitok K A, Li L, Attie A D and Martin T F J 2017 BAIAP3, a C2 domain-containing Munc13 protein, controls the fate of dense-core vesicles in neuroendocrine cells *J. Cell Biol.* **4** 2151–66
[225] Regazzi R, Vallar L, Ullrich S, Ravazzola M, Kikuchi A and Takai Y *et al* 1992 Characterization of small-molecular-mass guanine-nucleotide-binding regulatory proteins in insulin-secreting cells and PC12 cells *Eur. J. Biochem.* **208** 729–37

[226] Kögel T, Rudolf R, Hodneland E, Copier J, Regazzi R and Tooze S A *et al* 2013 Rab3D is critical for secretory granule maturation in PC12 cells *PLoS One* **8** e57321
[227] Riedel D, Antonin W, Fernandez-Chacon R, Alvarez de Toledo G, Jo T and Geppert M *et al* 2002 Rab3D is not required for exocrine exocytosis but for maintenance of normally sized secretory granules *Mol. Cell Biol.* **22** 6487–97
[228] Iezzi M, Escher G, Meda P, Charollais A, Baldini G and Darchen F *et al* 1999 Subcellular distribution and function of Rab3A, B, C, and D isoforms in insulin-secreting cells *Mol. Endocrinol.* **13** 202–12
[229] Schonn J-S, van Weering J R T, Mohrmann R, Schlüter O M, Südhof T C and de Wit H *et al* 2010 Rab3 proteins involved in vesicle biogenesis and priming in embryonic mouse chromaffin cells *Traffic* **11** 1415–28
[230] Fukuda M and Kuroda T S 2002 Slac2-c (synaptotagmin-like protein homologue lacking C2 domains-c), a novel linker protein that interacts with Rab27, myosin Va/VIIa, and actin *J. Biol. Chem.* **277** 43096–103
[231] Yi Z, Yokota H, Torii S, Aoki T, Hosaka M and Zhao S *et al* 2002 The Rab27a/granuphilin complex regulates the exocytosis of insulin-containing dense-core granules *Mol. Cell Biol.* **22** 1858–67
[232] Torii S, Zhao S, Yi Z, Takeuchi T and Izumi T 2002 Granuphilin modulates the exocytosis of secretory granules through interaction with syntaxin 1a *Mol. Cell Biol.* **22** 5518–26
[233] Zhao S, Torii S, Yokota-Hashimoto H, Takeuchi T and Izumi T 2002 Involvement of Rab27b in the regulated secretion of pituitary hormones *Endocrinology* **143** 1817–24
[234] Hannah M J, Hume A N, Arribas M, Williams R, Hewlett L J and Seabra M C *et al* 2003 Weibel–Palade bodies recruit Rab27 by a content-driven, maturation-dependent mechanism that is independent of cell type *J. Cell Sci.* **116** 3939–48
[235] Cazares V A, Subramani A, Saldate J J, Hoerauf W and Stuenkel E L 2014 Distinct actions of Rab3 and Rab27 GTPases on late stages of exocytosis of insulin *Traffic* **15** 997–1015
[236] Handley M T W, Haynes L P and Burgoyne R D 2007 Differential dynamics of Rab3A and Rab27A on secretory granules *J. Cell Sci.* **120** 973–84
[237] Izumi T 2007 Physiological roles of Rab27 effectors in regulated exocytosis *Endocrinol. J.* **54** 649–57
[238] Fukuda M 2006 Rab27 and its effectors in secretory granule exocytosis: a novel docking machinery composed of a Rab27.effector complex *Biochem. Soc. Trans.* **34** 691–5
[239] Matsunaga K, Taoka M, Isobe T and Izumi T 2017 Rab2a and Rab27a cooperatively regulate the transition from granule maturation to exocytosis through the dual effector Noc2 *J. Cell Sci.* **130** 541–50
[240] Ménager M M, Ménasché G, Romao M, Knapnougel P, Ho C-H and Garfa M *et al* 2007 Secretory cytotoxic granule maturation and exocytosis require the effector protein hMunc13-4 *Nat. Immunol.* **8** 257–67
[241] Neuman S D, Lee A R, Selegue J E, Cavanagh A T and Bashirullah A 2021 A novel function for Rab proteins during maturation of secretory granules *bioRxiv*:2021.03.01.433309
[242] Houy S, Groffen A J, Ziomkiewicz I, Verhage M, Pinheiro P S and Sørensen J B 2017 Doc2B acts as a calcium sensor for vesicle priming requiring synaptotagmin-1, Munc13-2 and SNAREs *eLife* **6** 3586
[243] Grishanin R N, Kowalchyk J A, Klenchin V A, Ann K, Earles C A and Chapman E R *et al* 2004 CAPS acts at a prefusion step in dense-core vesicle exocytosis as a PIP2 binding protein *Neuron* **43** 551–62

[244] Kabachinski G, Kielar-Grevstad D M, Zhang X, James D J and Martin T F J 2016 Resident CAPS on dense-core vesicles docks and primes vesicles for fusion *Mol. Biol. Cell* **27** 654–68
[245] Ke B, Oh E and Thurmond D C 2007 Doc2beta is a novel Munc18c-interacting partner and positive effector of syntaxin 4-mediated exocytosis *J. Biol. Chem.* **282** 21786–97
[246] Kabachinski G, Yamaga M, Kielar-Grevstad D M, Bruinsma S and Martin T F J 2014 CAPS and Munc13 utilize distinct PIP2-linked mechanisms to promote vesicle exocytosis *Mol. Biol. Cell* **25** 508–21
[247] Groffen A J A, Brian E C, Dudok J J, Kampmeijer J, Toonen R F and Verhage M 2004 Ca (2+)-induced recruitment of the secretory vesicle protein DOC2B to the target membrane *J. Biol. Chem.* **279** 23740–7
[248] Aslamy A, Oh E, Olson E M, Zhang J, Ahn M and Moin A S M *et al* 2018 Doc2b protects β-Cells against inflammatory damage and enhances function *Diabetes* **67** 1332–44
[249] Aslamy A, Oh E, Ahn M, Moin A S M, Chang M and Duncan M *et al* 2018 Exocytosis protein DOC2B as a biomarker of type 1 diabetes *J. Clin. Endocrinol. Metab.* **103** 1966–76
[250] Sadakata T, Shinoda Y, Sekine Y, Saruta C, Itakura M and Takahashi M *et al* 2010 Interaction of calcium-dependent activator protein for secretion 1 (CAPS1) with the class II ADP-ribosylation factor small GTPases is required for dense-core vesicle trafficking in the trans-Golgi network *J. Biol. Chem.* **285** 38710–9
[251] Crummy E, Mani M, Thellman J C and Martin T F J 2019 The priming factor CAPS1 regulates dense-core vesicle acidification by interacting with rabconnectin3β/WDR7 in neuroendocrine cells *J. Biol. Chem.* **294** 9402–15
[252] Speidel D, Bruederle C E, Enk C, Voets T, Varoqueaux F and Reim K *et al* 2005 CAPS1 regulates catecholamine loading of large dense-core vesicles *Neuron* **46** 75–88
[253] Brunk I, Blex C, Speidel D, Brose N and Ahnert-Hilger G 2009 Ca2+-dependent activator proteins of secretion promote vesicular monoamine uptake *J. Biol. Chem.* **284** 1050–6
[254] Speidel D, Salehi A, Obermueller S, Lundquist I, Brose N and Renström E *et al* 2008 CAPS1 and CAPS2 regulate stability and recruitment of insulin granules in mouse pancreatic β cells *Cell Metab.* **7** 57–67
[255] Bainton D F, Ullyot J L and Farquhar M G 1971 The development of neutrophilic polymorphonuclear leukocytes in human bone marrow *J. Exp. Med.* **134** 907–34
[256] Fumagalli G and Zanini A 1985 In cow anterior pituitary, growth hormone and prolactin can be packed in separate granules of the same cell *J. Cell Biol.* **100** 2019–24
[257] Hashimoto S, Fumagalli G, Zanini A and Meldolesi J 1987 Sorting of three secretory proteins to distinct secretory granules in acidophilic cells of cow anterior pituitary *J. Cell Biol.* **105** 1579–86
[258] Rao T C, Passmore D R, Peleman A R, Das M, Chapman E R and Anantharam A 2014 Distinct fusion properties of synaptotagmin-1 and synaptotagmin-7 bearing dense core granules *Mol. Biol. Cell* **25** 2416–27
[259] Rao T C, Santana Rodriguez Z, Bradberry M M, Ranski A H, Dahl P J and Schmidtke M W *et al* 2017 Synaptotagmin isoforms confer distinct activation kinetics and dynamics to chromaffin cell granules *J. Gen. Physiol.* **149** 763–80
[260] Huang A Y, Castle A M, Hinton B T and Castle J D 2001 Resting (basal) secretion of proteins is provided by the minor regulated and constitutive-like pathways and not granule exocytosis in parotid acinar cells *J. Biol. Chem.* **276** 22296–306

[261] Castle A M, Huang A Y and Castle J D 2002 The minor regulated pathway, a rapid component of salivary secretion, may provide docking/fusion sites for granule exocytosis at the apical surface of acinar cells *J. Cell Sci.* **115** 2963–73

[262] Emeis J J, van den Eijnden-Schrauwen Y, van den Hoogen C M, de Priester W, Westmuckett A and Lupu F 1997 An endothelial storage granule for tissue-type plasminogen activator *J. Cell Biol.* **139** 245–56

[263] Zhang Z, Wu Y, Wang Z, Dunning F M, Rehfuss J and Ramanan D *et al* 2011 Release mode of large and small dense-core vesicles specified by different synaptotagmin isoforms in PC12 cells *Mol. Biol. Cell* **22** 2324–36

[264] Bendahmane M, Morales A, Kreutzberger A J B, Schenk N A, Mohan R and Bakshi S *et al* 2020 Synaptotagmin-7 enhances calcium-sensing of chromaffin cell granules and slows discharge of granule cargos *J. Neurochem.* **154** 598–617

[265] Tawfik B, Martins J S, Houy S, Imig C, Pinheiro P S and Wojcik S M *et al* 2021 Synaptotagmin-7 places dense-core vesicles at the cell membrane to promote Munc13-2- and Ca^{2+}-dependent priming *eLife* **10** e64527

[266] Marsh B J, Mastronarde D N, McIntosh J R and Howell K E 2001 Structural evidence for multiple transport mechanisms through the Golgi in the pancreatic beta-cell line, HIT-T15 *Biochem. Soc. Trans.* **29** 461–7

[267] Pagliuso A, Valente C, Giordano L L, Filograna A, Li G and Circolo D *et al* 2016 Golgi membrane fission requires the CtBP1-S/BARS-induced activation of lysophosphatidic acid acyltransferase δ *Nat. Commun.* **7** 12148

[268] Obermüller S, Calegari F, King A, Lindqvist A, Lundquist I and Salehi A *et al* 2010 Defective secretion of islet hormones in chromogranin-B deficient mice *PLoS One* **5** e8936

[269] Hosaka M, Watanabe T, Sakai Y, Uchiyama Y and Takeuchi T 2002 Identification of a chromogranin A domain that mediates binding to secretogranin III and targeting to secretory granules in pituitary cells and pancreatic beta-cells *Mol. Biol. Cell* **13** 3388–99

[270] Geumann U, Schäfer C, Riedel D, Jahn R and Rizzoli S O 2010 Synaptic membrane proteins form stable microdomains in early endosomes *Microsc. Res. Tech.* **73** 606–17

[271] Hoppa M B, Collins S, Ramracheya R, Hodson L, Amisten S and Zhang Q *et al* 2009 Chronic palmitate exposure inhibits insulin secretion by dissociation of Ca(2+) channels from secretory granules *Cell Metab.* **10** 455–65

[272] Ferdaoussi M, Dai X, Jensen M V, Wang R, Peterson B S and Huang C *et al* 2015 Isocitrate-to-SENP1 signaling amplifies insulin secretion and rescues dysfunctional β cells. *J Clin Invest.* **125** 3847–60

[273] Fowler K T, Andrews N W and Huleatt J W 2007 Expression and function of synaptotagmin VII in CTLs *J. Immunol.* **178** 1498–504

[274] Wakana Y, Kotake R, Oyama N, Murate M, Kobayashi T and Arasaki K *et al* 2015 CARTS biogenesis requires VAP-lipid transfer protein complexes functioning at the endoplasmic reticulum-Golgi interface *Mol. Biol. Cell* **26** 4686–99

[275] Stephens S B, Edwards R J, Sadahiro M, Lin W-J, Jiang C and Salton S R *et al* 2017 The prohormone VGF regulates β cell function via insulin secretory granule biogenesis *Cell Rep.* **20** 2480–9

[276] Wu B, Wei S, Petersen N, Ali Y, Wang X and Bacaj T *et al* 2015 Synaptotagmin-7 phosphorylation mediates GLP-1-dependent potentiation of insulin secretion from β-cells *Proc. Natl Acad. Sci. USA* **112** 9996–10001

[277] Dai X-Q, Plummer G, Casimir M, Kang Y, Hajmrle C and Gaisano H Y *et al* 2011 SUMOylation regulates insulin exocytosis downstream of secretory granule docking in rodents and humans *Diabetes* **60** 838–47

[278] Henquin J C 2009 Regulation of insulin secretion: a matter of phase control and amplitude modulation *Diabetologia* **52** 739–51

[279] Henneberry A L, Wright M M and McMaster C R 2002 The major sites of cellular phospholipid synthesis and molecular determinants of fatty acid and lipid head group specificity *Mol. Biol. Cell* **13** 3148–61

[280] Peretti D, Dahan N, Shimoni E, Hirschberg K and Lev S 2008 Coordinated lipid transfer between the endoplasmic reticulum and the Golgi complex requires the VAP proteins and is essential for Golgi-mediated transport *Mol. Biol. Cell* **19** 3871–84

[281] Litvak V, Dahan N, Ramachandran S, Sabanay H and Lev S 2005 Maintenance of the diacylglycerol level in the Golgi apparatus by the Nir2 protein is critical for Golgi secretory function *Nat. Cell Biol.* **7** 225–34

[282] Pemberton J G, Kim Y J, Humpolickova J, Eisenreichova A, Sengupta N and Toth D J *et al* 2020 Defining the subcellular distribution and metabolic channeling of phosphatidylinositol *J. Cell Biol.* **219**

[283] Ashlin T G, Blunsom N J and Cockcroft S 2021 Courier service for phosphatidylinositol: PITPs deliver on demand *Biochim. Biophys. Acta, Mol. Cell. Biol. Lipids* **1866** 158985

IOP Publishing

Chapter 3

Exocytotic SNARE complex assembly, disassembly, and regulation

Ira Milosevic and Jefferson D Knight

Cell-to-cell communication features precise temporal and spatial control that is achieved by regulated exocytosis. It builds on the fusion of secretory vesicles with the plasma membrane, which is made possible by the presence of soluble N-ethylmaleimide sensitive factor attachment protein (SNAP) receptor (SNARE) proteins embedded in opposing membranes. SNARE proteins can assemble on their own to drive the fusion of membranes in reconstituted systems, yet this process is slow and uncoordinated in the absence of chaperone proteins such as Munc18 and Munc13. These and other regulatory proteins control the activity of SNARE proteins and drive the formation of the SNARE complex, including 'zippering' of the SNARE domains to promote membrane fusion. Following fusion, the SNARE proteins need to be recovered for the next rounds of membrane fusion; this is achieved by the action of the N-ethylmaleimide sensitive factor (NSF) and SNAP proteins. Here, we review scientific discoveries over the past few decades that reveal how SNAREs and their regulatory proteins drive membrane fusion in neuronal and neurosecretory cells. While the structures of the core SNAREpin complex proteins were resolved by the early 2000s, an understanding of their mechanism of chaperoned assembly has rapidly advanced over the past few years. As roles for the 'non-canonical' SNARE proteins continue to emerge, our knowledge of the temporal and spatial regulation of membrane fusion continues to grow toward a full understanding of how cells achieve precise control of this fundamental biological process.

3.1 Introduction to exocytotic SNARE proteins and fusion events at the neuronal plasma membrane

Like other cell types, neurons and (neuro)endocrine cells consist of subcellular compartments. Hence, they share trafficking processes that shuffle molecules (lipids and proteins) in membranous carriers, also called vesicles, between various compartments. These vesicles bud from a donor membrane and fuse with a target membrane.

doi:10.1088/978-0-7503-3771-7ch3

Neurons and (neuro)endocrine cells are best known for their ability to quickly release signals to the extracellular space in a regulated manner in order to communicate with their surroundings and with other cells. In the former case, secretory vesicles fuse with the neuronal plasma membrane to release signaling molecules (e.g. neurotransmitters) that are detected by receptors on the neighboring cells. Contact points between neighboring neurons where the most intense exchange of signals happens are called synapses.

Membrane fusion events that take place at the synaptic plasma membrane are usually synchronized and cooperative processes that take part in three steps [1]. First, two membranes come into close contact (~1 nm). Second, a specialized structure is created that connects two membranes. Third, the structure thus formed matures into a funnel-like assembly, termed the fusion pore, to join the membranes and allow a flux of signaling molecules. Many essential cell functions are contingent on intracellular transport processes and the fusion of membrane bilayers, prompting extensive studies of these processes over the past four decades.

The most studied and best understood example of the fusion of two biological membranes is the mechanism of neurotransmitter release at the synapse. Here, small synaptic vesicles (~40 nm in diameter) filled with neurotransmitters fuse with the plasma membrane at the presynaptic terminus. The basic machinery mediating this process, a family of SNAREs, is found to be universally present in other examples of intracellular membrane fusion [2, 3]. All SNARE family members are characterized by the SNARE motif, a conserved domain 60 to 70 amino acids long [4–6]. Four motifs can spontaneously assemble into a stable coiled-coil bundle with 16 layers of interacting amino-acid residues, called the SNARE complex [7–10]. While most layers are made by hydrophobic amino acids, the central '0-layer' is hydrophilic and composed of one arginine (R) and three glutamines (Q) [7]. Hence, SNARE proteins are classified as Q_a-, Q_b-, Q_c-, and R-SNAREs [11, 12]. Many SNARE complexes are generated by four distinct SNARE proteins and follow a $Q_aQ_bQ_cR$ pattern. However, other systems including the well-studied synaptic SNARE complex comprise three proteins only: synaptobrevin-2/vesicle-associated membrane protein (VAMP) 2 [13, 14], synaptosome-associated protein of 25 kDa (SNAP-25) [15] and syntaxin-1 [16]. Such tripartite complexes still constitute a four-helix bundle because the SNAP-25 protein contains two SNARE motifs, Q_b and Q_c [7, 11].

Historically, neurotoxin-based studies helped to reveal the composition and importance of the SNARE complex [17, 18]. The botulinum neurotoxins BoNTA, BoNTB, and BoNTC were found to cut SNAP-25, synaptobrevin-2/VAMP2 (henceforth synaptobrevin-2) and syntaxin-1, respectively, while tetanus toxin cleaves synaptobrevin-2. Recombinantly expressed and purified SNAP-25, syntaxin-1, and synaptobrevin-2 inserted into liposomes were observed to be sufficient to fuse membrane bilayers, and are hence considered a 'minimal' machinery for membrane fusion [19]. When added to the cellular surface of cells, spontaneous cell-to-cell fusion was observed, demonstrating that these three SNARE proteins are sufficient to induce fusion of biological membranes [20].

Synaptobrevin-2 is a main SNARE protein on the secretory vesicles, while syntaxin-1 is inserted into the plasma membrane and SNAP-25 associates with the plasma membrane peripherally. Syntaxin-1, synaptobrevin-2, and SNAP-25 form a ternary 1:1:1 SNARE complex [8, 21]. A formation of this stable complex between the vesicle membrane and the plasma membrane, also called the *trans*-SNARE complex, fuses membranes. When membranes are merged, the SNARE proteins become a part of the same membrane, forming a so-called *cis*-SNARE complex. Post-fusion, the *cis*-SNARE complexes that result from the fusion of multiple secretory vesicles with the plasma membrane during exocytosis need to be disassembled, i.e. recovered for new rounds of exocytosis (see below).

3.2 Mechanisms of SNARE complex assembly

To comprehend how the neuronal SNARE complex is assembled, it is useful to get to know the main players first (figure 3.1a).

The Q_a-SNARE protein syntaxin-1 contributes the first helix in the neuronal SNARE complex [7, 11]. In addition to its SNARE motif, this 35 kDa protein contains a transmembrane domain at its C-terminus and a regulatory region at its N-terminus comprising the N-peptide and H_{abc} domain [16]. The H_{abc} domain has been found to control protein accessibility in the plasma membrane by interacting with the internal SNARE motif. Thus, syntaxin-1 maintains two distinct protein conformations: the 'closed' conformation (H_{abc}-domain interacts with the SNARE motif) does not allow a functional SNARE complex to form, while the 'open' conformation (H_{abc}-domain does not interact with the SNARE motif) does [22]. Neurons and (neuro)endocrine cells express two syntaxin paralogs, 1A and 1B [23]. The ablation of syntaxin-1A ablates neurotransmitter release in a fruit fly model [24, 25], while the roles of syntaxin-1A and syntaxin-1B complement each other in rodents [26, 27].

The Q_{bc}-SNARE protein SNAP-25 contributes two helices to the SNARE complex formed at the plasma membrane [7, 11]. Of note, the Q_b- and Q_c-SNARE motifs are more often provided by independent proteins in intracellular fusion reactions [28]. A linker domain connects these two SNARE motifs, which seems to be an evolutionary adaptation that supports a faster exocytotic process [29]. The linker domain contains four cysteines; palmitoylation of these cysteine residues efficiently targets SNAP-25 to the plasma membrane [30, 31]. Without SNAP-25, the fusion of secretory vesicles is almost abolished [32] and mice die at late embryonic stages [33]. SNAP-25 is a very conserved protein. Two isoforms, SNAP-25a and SNAP-25b, differ only by nine amino acids; both isoforms can rescue exocytosis in SNAP-25 *null* cells, yet SNAP-25b is more efficient [32, 34]. Two non-conservative substitutions in the N-terminal SNARE motif, rather than the different position of one palmitoylated cysteine, are found to cause the functional difference between these two isoforms [29]. They allow SNAP-25b to better bind the main calcium sensor in presynaptic exocytosis, synaptotagmin-1 [35]. Paralogs of SNAP-25 in mammals include SNAP-23, which functions in many non-neuronal regulated exocytosis pathways, as well as family members that provide the Q_{bc} structure for some intracellular membrane fusion reactions (recently reviewed in [36]).

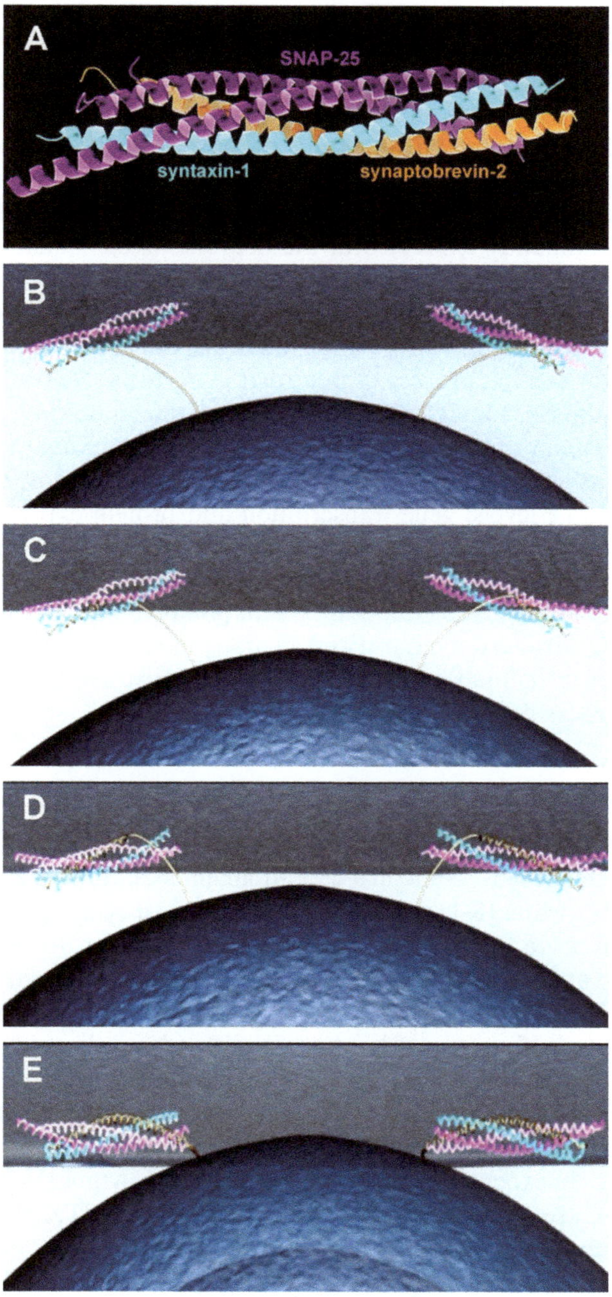

Figure 3.1. A synaptic vesicle (lower) is brought into close proximity with a plasma membrane (upper) through SNARE-mediated actions. (a) Schematic (based on the crystal structure) of the ternary SNARE complex that mediates synaptic vesicle fusion: SNAP-25 (violet), syntaxin-1 (cyan), and synaptobrevin-2 (orange). A similar color coding for these proteins is used in all figure panels. (b) Syntaxin-1 (Q_a) and SNAP-25 (Q_{bc}) are present in the plasma membrane, while synaptobrevin-2 (R) is in the synaptic vesicle. (c and d) Schematic of SNARE complex 'zippering'; a secondary structure is introduced into synaptobrevin-2's SNARE domain. (e) Formation of the SNARE complex composed of syntaxin-1, SNAP-25, and synaptobrevin-2 brings membranes into close proximity and drives fusion of the vesicular membrane with the plasma membrane.

Synaptobrevin-2 is an R-SNARE that provides the fourth helix of the SNARE complex at the neuronal plasma membrane [7, 11]. In addition to the SNARE motif, this abundant brain protein contains an N-terminal proline-rich region and a C-terminal transmembrane region. Ablation of synaptobrevin-2 is lethal: fusion in hippocampal neurons is decreased, yet not abolished [37]. This is probably due the presence of synaptobrevin-1/VAMP1 that can, in part, substitute for its function. VAMP1 is predominantly expressed in the neurons of the somatic motor system, while synaptobrevin-2 displays more uniform distribution in the brain [38].

Notably, for reasons discussed below, fusion in reconstituted systems driven only by the SNARE proteins operates at the timescale of several seconds to minutes, which is very slow for the process of neuronal exocytosis. Neurons require exceptionally high temporal activity on submillisecond timescales [39, 40] and had to develop sophisticated ways to regulate the location and functionality of the SNARE proteins as well as assembly of the SNARE complex. In neurons, these aspects are regulated at multiple levels: gene transcription and protein expression, post-translational modifications (e.g. palmitoylation, phosphorylation), distribution into the proper target membranes, protein activity, SNARE complex assembly and disassembly, and sensitivity to calcium ions are all points of regulation (for details, see [41]). In addition, advanced temporal and spatial control of membrane fusion at the synapse, which is a hallmark of fast neuronal activity, requires many accessory regulatory factors, as detailed below.

3.2.1 Act I: closed syntaxin-1 and Munc18

The protein that stabilizes syntaxin-1 in its 'closed' form is Munc18-1/nsec-1, a large soluble protein that binds syntaxin's H_{abc}-domain [42]. Like the aforementioned SNARE proteins, Munc18-1 is required for fusion in cells [43–45]. A detailed understanding of how the Munc18 and SNARE proteins act together to drive membrane fusion has begun to emerge over the last decade (reviewed in [46] and [47]).

Historically, the *Caenorhabditis elegans* protein UNC-18 was discovered in the early 1970s in a genetic screen for mutants with uncoordinated movement [48]. Several years later, the *Saccharomyces cerevisiae* ortholog Sec1 was described [49]. The mammalian ortholog Munc18-1 was reported in the early 1990s [50]. Sec1 and Munc18 together were subsequently named the SM proteins. There are three SM proteins with roles in exocytosis in mammals: Munc18-1, Munc18-2, and Munc18-3. (In humans, these are also called syntaxin-binding proteins STXBP1, STXBP2, and STXBP3, respectively.)

Munc18-1 has several actions along the exocytotic pathway. First, Munc18-1 may function as a key molecular chaperone for syntaxin-1 in the secretory pathway. Specifically, during protein trafficking to the plasma membrane, Munc18-1 binds syntaxin-1 in the closed configuration and protects ectopic SNARE complex formation [51, 52]. Rodents without Munc18-1 have reduced levels of syntaxin-1: only ~30% of syntaxin-1 localizes properly at neuronal synapses in this model [53]. Munc18-1 itself dissociates a misformed 2:1 syntaxin-1/SNAP-25 heterotrimer *in vitro* [54], and it was recently shown that other off-pathway complexes and 1:1

syntaxin-1/SNAP-25 heterodimers are susceptible to dissociation by the NSF/α-SNAP machinery that also disassembles *cis*-SNARE complexes after fusion [55, 56]. Munc18-1, along with additional key exocytosis proteins Munc13-1 and complexin, protects against this NSF/α-SNAP-mediated disassembly *in vitro* [57]. Thus, NSF/α-SNAP is thought to act as a quality control mechanism against ectopic SNARE complex assembly and misassembly, and Munc18 binding stabilizes syntaxin-1 in a closed state during its own trafficking to the plasma membrane. Without UNC-18 or Munc18-1, docking of secretory vesicles to the plasma membrane is nearly absent [58, 59]. Because docking also depends on syntaxin-1 levels [60, 61], the Munc18-1 chaperone properties may explain this phenotype.

Second, recruitment of syntaxin-1 from its dense protein clusters on the plasma membrane seems to be regulated by Munc18-1. While this may happen spontaneously through lateral diffusion of syntaxin-1 from the rim of its clusters, leading to the creation of reactive syntaxin-1 molecules in the membrane [62–64], this process is likely controlled. Once outside its clusters, syntaxin-1 can associate with SNAP-25 [65–71]. Of note, it has also been suggested that the secretory vesicles themselves may induce recruitment of plasmalemmal SNARE proteins to the underlying membrane [72, 73].

Third, structural and mechanistic evidence has recently emerged that supports a model in which Munc18 acts as a chaperone for SNARE complex assembly, supported by the action of Munc13-1 (see below). This concept has been built on a demonstration by the Rizo group that reconstituted proteoliposome fusion is more efficient when it starts with a Munc18-1–syntaxin-1 complex, rather than starting with a syntaxin-1—SNAP-25 complex, as had been typical in earlier liposome fusion assays [55]. A possible structural explanation was suggested by Baker, Hughson, and colleagues, who crystallized the yeast Munc18-1 homolog Vps33 bound to either a Q_a-SNARE (i.e. syntaxin homolog) or R-SNARE (i.e. synaptobrevin homolog) in neighboring sites [74]. Although the ternary complex was not observed, the authors speculated that Munc18-1 could act as a template for initial association between these two SNARE motifs. Subsequently, Jiao *et al* synthesized a complex of syntaxin-1 and synaptobrevin-2 covalently linked through a disulfide bond near the N-terminal ends of their respective SNARE motifs, which formed a stable SNARE complex with SNAP-25 [75]. By connecting the C-terminal ends of syntaxin-1 and synaptobrevin-2 to a single-molecule force apparatus in the presence or absence of Munc18-1, the team observed that Munc18-1 could catalyze the refolding of the syntaxin-1 and synaptobrevin-2 SNARE motifs after the full SNARE complex was pulled apart, supporting the idea of Munc18 as a possible chaperone for complex assembly. The syntaxin-1/synaptobrevin/Munc18-1 complex was further stabilized by Munc13-1, another important chaperone discussed further below [76]. Very recently, the same crosslinking strategy was used to generate stable ternary complexes of an 'open' mutant of syntaxin-1 with Munc18-1 and synapobrevin-2 and solve their structure in two poses using cryo-electron microscopy [77]. Both poses show the C-termini of the two SNARE motifs associated with different regions of Munc18-1 while the N-terminal ends associate with each other, and together these results suggest possible conformational changes that are proposed to template this initial SNARE assembly. The model that emerges here is that after the Munc18-1:syntaxin-1:synaptobrevin-2 complex is formed, SNAP-25 binds it rapidly to form the full SNARE complex. When the assembly of the SNARE complex is complete, Munc18-1 is presumed to dissociate from the four-helix bundle.

Fourth, in neurosecretory cells, Munc18-1 seems to support events downstream of docking and distinct from binding to closed syntaxin-1. For example, the expression of a Munc18 mutant defective in syntaxin-1 binding stimulated exocytosis in neurosecretory PC12 cells [78]. Further, the calcium sensor synaptotagmin-1, rather than synaptobrevin-2, seems to be involved in docking in chromaffin cells [79]. Hence, synaptotagmin-1 may act upstream of Munc18-1 to promote docking in this system, possibly by binding to SNAP-25 [35]. Curiously, overexpression of SNAP-25 fully rescued docking, although not exocytosis, in Munc18-1 knockout chromaffin cells [80].

The multiple roles of Munc18-1 in exocytosis likely correlate with its various binding arrangements with the SNARE proteins. One arrangement is the aforementioned binding to closed syntaxin-1, a second is the templating of the SNARE assembly discussed above, and a third is binding to the fully assembled SNARE complex [81–84]. All three arrangements involve interaction with the N-terminus of syntaxin-1 [81, 82, 85]; hence, it is likely that the syntaxin-1 N-terminus recruits Munc18-1 to the assembling SNARE complex to fulfill its fusion-promoting role by promoting SNARE complex assembly [81, 86]. Notably, the ternary SNARE complex binds to the same part of Munc18 [87], although such binding has a lower affinity than binding to the closed form of syntaxin-1 [85, 87]. When syntaxin-1 or mutants were reintroduced into syntaxin-1 knockout mice, the H_{abc} domain near the N-terminus of syntaxin-1 was found to be essential for fusion in murine knock-in neurons, while the N-peptide at the immediate N-terminus plays a modulatory role [88].

It should be noted that Munc18-1 is not required for fusion *in vitro*, where the three canonical neuronal SNARE proteins assemble and are sufficient to drive membrane fusion even without Munc18, at least in the absence of NSF/αSNAP [19, 89]. It has also been proposed that Munc18 may catalyze lipid mixing once it is brought into contact with the fusion site [90], although the extent to which such an activity contributes to fusion remains unclear. Last but not least, the role of Munc18 in activating SNARE-mediated fusion is not observed in a complexin/synaptotagmin-1-regulated fusion process [91], raising the question of the extent to which the Munc18 activation function is needed under physiological conditions. Otherwise, it is well established that Munc18 proteins, and SM proteins generally, are needed for all SNARE-mediated fusion events.

3.2.2 Act II: opening syntaxin-1 with Munc13 and/or CAPS

The opening of syntaxin-1 is an essential step towards the fusion of secretory vesicles with the plasma membrane, given that closed syntaxin-1 blocks the SNARE complex assembly. The substitution of syntaxin-1 with its mutated version that readily adopts the open conformation increases release probability and promotes spontaneous release [26]. The Munc13 protein has been found to have a key role in opening syntaxin-1 in fruit flies, where the presence of the mutated open syntaxin-1 form partly overcomes the secretion defect in a Munc13-deficient model [92, 93]. Notably, synaptic transmission is completely abolished without Munc13 proteins [94–96], while Munc13 overexpression promotes secretory vesicle priming [97]. Mammals have four Munc13 proteins, out of which Munc13-1 is the most common in the brain. All Munc13 family members are known to promote or mediate the opening of syntaxin-1, likely in concert with Munc18.

In addition to syntaxin-1, Munc13 also interacts with Munc18 and the SNARE complex [98]. Originally, biochemical studies suggested that Munc18-1 catalyzes opening by keeping syntaxin-1 in a half-open configuration [85]. Subsequent structural studies led to the idea that a conformational change in Munc18 drives the formation of a SNAP-25:syntaxin-1 acceptor complex while syntaxin-1 remains bound to Munc18-1 [99–101]. However, this model has largely fallen out of favor in light of the observation that SNAP-25:syntaxin-1 complexes are readily disassembled by α-SNAP [55]. Rather, it appears that Munc13-1 and Munc18-1 cooperate to inhibit inadequate SNARE complex assembly and act as chaperones that enhance the speed and the accuracy of proper SNARE assembly. This activity leads to more efficient fusion of proteoliposomes in the presence of Munc13-1, particularly when starting from preformed syntaxin-1—Munc18-1 complexes [102]. Notably, the requirement for Munc13-1 in cellular exocytosis is eliminated when a mutant form of Munc18-1 is used that promotes syntaxin-1 opening, which also supports the hypothesis that these two proteins work together. [103, 104]. An overall mechanism for Munc18-1 and Munc13-1 cooperativity has recently been put forward by Zhang and Hughson, although the precise order of assembly is still unclear [46]. Some of the major findings that led to this model are summarized below.

Several of the key residues for the priming function of Munc13-1 are found in a hydrophobic pocket of its central MUN domain, suggesting that it may function as a molecular chaperone that promotes assembly of a primed complex [105]. Indeed, the interaction between Munc13-1 and the syntaxin-1—Munc18-1 complex that promotes synaptic vesicle priming and proteoliposome fusion involves two conserved residues in the linker region of syntaxin-1 (between its H_{abc} domain and its H3 SNARE motif): hydrophobic I155 and polar R151 [106]. The interface also includes key residues on domain 3a of Munc18-1 that interact with the MUN domain of Munc13-1 [107]. Together, these interactions stabilize the open state of the syntaxin-1—Munc18 complex. In addition, an interaction has been found between the MUN domain of Munc13-1 and a linker region in synaptobrevin-2 between its SNARE motif and transmembrane helix [108]. Thus, it has been proposed that Munc13-1 bridges the secretory vesicle and the plasma membrane and stabilizes the syntaxin-1/Munc18-1/synaptobrevin-2 ternary complex [46, 47, 76]. Munc13 also aids in the recruitment of SNAP-25 to this ternary complex *in vitro*, a reaction which presumably occurs rapidly *in vivo* following the initiation of SNARE complex formation [109]. The precise structures of these interactions are not yet known.

In addition to the protein–protein interactions of its MUN domain, Munc13 is activated by diacylglycerol, calcium, and phosphatidylinositol(4,5)bisphosphate [PI(4,5)P_2] [110–112]. These interactions are mediated by the C1 and C2B domains on the N-terminal portion of the protein and stabilize a calcium-dependent interaction with the plasma membrane [113]. In addition, the C2C domain near the C-terminus of Munc13-1 binds secretory vesicle membranes [114]. Thus, in addition to its protein–protein interactions that activate SNARE assembly, Munc13-1 can also facilitate priming by direct lipid bridging between secretory vesicles and the plasma membrane [115]. The calcium-dependent binding of the C2B domain to plasma membrane lipids is required for Munc13's roles in priming and short-term plasticity, as multiple different mutations that impair lipid or calcium binding severely diminish these functions [116, 117].

Interestingly, Munc13-1 forms nanoclusters on the plasma membrane that enhance fusion probability in neurons at the locations of the clusters [118]. Such nanoclusters have been observed *in vitro* on supported lipid bilayers as well, ranging in size from two to twenty copies per oligomer [119]. In this system, clusters of hexamer size and larger were capable of capturing small vesicles introduced into the system, an activity which required a functional C2C domain [119]. Interestingly, Munc13-1 sandwiched between two lipid bilayers readily forms two-dimensional crystals of trimers and hexamers, whose structures have recently been determined via cryo-electron microscopy [120]. While a role for these particular structures *in vivo* remains to be demonstrated, it is intriguing to note their similarity to the ring-like structures of synaptotagmin-1 and related synaptic proteins that have been observed in recent years [121].

In addition to the Munc13 family, calcium-dependent activator proteins for secretion (CAPS) family proteins have a similar role in opening syntaxin-1. Both families contain a MUN domain, which mediates vesicle priming [122, 123]. While Munc13-1 binds plasma membranes via its C1 and C2B domains, CAPS-1 and CAPS-2 instead contain a PI(4,5)P_2-binding pleckstrin homology (PH) domain, which is essential for their function [124, 125]. Ablation of both CAPS proteins produces a mild phenotype in neurons: vesicles can be primed by increasing the basal calcium in the cytosol, but then soon revert back to the non-primed form [126]. Thus, cells may have two ways to prime secretory vesicles, one driven by CAPS and the other by Munc13 proteins.

In neurosecretory cells, CAPS binds SNARE proteins, mediates SNARE complex formation, and works in vesicle priming [127–129]. In the absence of CAPS, fusion of secretory vesicles can be promoted by the expression of open syntaxin-1, indicating the similarities between CAPS and Munc13 in syntaxin-1 opening properties [130, 131]. While these large peripheral proteins likely have overlapping functions, the presence of C1 and C2 domains in Munc13 and a PH domain in CAPS suggests that these two protein families may be regulated differently. It is notable that CAPS expression does not compensate for Munc13 deficiency, and vice versa [126, 130]; thus, Munc13 and CAPS proteins perform some distinct functions in exocytosis.

Interestingly, syntaxin-1 without its N-terminal H_{abc} domain can support synchronous (evoked) release in neurons without syntaxin-1, but this deletion reduces spontaneous release [132]. This observation, together with the observed partial rescue of CAPS or Munc13-deficiencies by mutant open syntaxin-1, suggests that opening of syntaxin-1 has a positive role in vesicle priming, while the closed syntaxin-1 conformation is inhibitory for the SNARE complex formation *per se.* Thus, the role of Munc13 in the opening of syntaxin-1 may be an essential regulatory point in exocytosis. Furthermore, Munc13 regulatory domains may have developed to link the calcium levels to the synaptic priming speed and the activity of G-protein-regulated receptors [41]. Munc13 and its homologs perform a variety of other functions as well, including linking the fusion apparatus to calcium channels; we refer the reader to a recent review summarizing the many functions of this protein [133].

Of note, the essential functions of the SNARE accessory proteins Munc13 and Munc18 have been reconstituted in a fusion assay *in vitro* [55]. Here, it was observed that Munc18 could displace SNAP-25 from syntaxin-1, making the fusion reaction dependent on Munc13. Importantly, Munc18:syntaxin-1 dimers were able to bind synaptobrevin-2 and SNAP-25, and the complex was resistant to NSF and α-SNAP addition. Hence, as mentioned before, Munc13 and Munc18 likely act as chaperones which enhance the speed and the accuracy of SNARE assembly and inhibit inadequate SNARE complex formation [75, 102, 108]. In neurons, the function of Munc13 in synaptic vesicle docking is likely augmented by an additional complex that consists of Rab3, Rab3-interacting molecule (RIM), and RIM-binding protein (RIM-BP) [3, 134].

3.2.3 Act III: zippering up the SNARE complex

Following the understanding that the SNARE domains align in parallel to produce the ternary SNARE complex, the SNARE 'zipper' hypothesis was created [7, 135]. This hypothesis places stiff transmembrane domains of syntaxin-1 and synaptobrevin-2 at the same end and builds on 'zippering' the complex from an N- to C-terminus in order to bring the membrane bilayers closer and closer together. Numerous experiments based on peptide interference, fusion assays, and deletion studies support this idea [136–141]. When assembled *in vitro* or *in silico* starting from a syntaxin-1—SNAP-25 heterodimer, the ternary SNARE complex formation results in the introduction of a secondary structure into the previously unstructured synaptobrevin-2 SNARE domain [21] (figures 3.1(b)–(e)). *In vivo*, it is not yet known whether all four SNARE domains zipper in parallel or sequentially following initiation and recruitment of all three proteins. We also note that the helical SNARE domains are readily interactive, which may lead to the formation of other complexes, particularly when the work is done *in vitro* (some of these complexes may potentially be *in vitro* artefacts rather than proper prefusion assemblies). In any case, the cell seems to tightly control the process of ternary SNARE complex formation through Munc13 and Munc18 actions [75, 90, 102].

The SNARE complex 'zippering' idea has also been tested in living cells by antibody infusion into neurosecretory cells, which also raised the suggestion that the complex may exist in two states: a 'tight' state and a 'loose' state [141]. Experiments with neurotoxins published the same year revealed that the N-terminal portion of the forming SNARE complex is shielded [142]. More definitive data came from mutagenesis-based studies [143–146]. Here, mutations were introduced into the hydrophobic layers of the ternary SNARE complex. The mutations caused local destabilization and had various consequences, depending on where along the complex the mutations were placed. Mutations towards the N-terminal end or in the center reduced priming rates, while mutations near the C-terminal end reduced fusion speed [143]. Basically, the fusion rates correlated with the stability of the C-terminal end of the ternary SNARE complex. In addition, unfolding experiments revealed that the complex with the C-terminal mutations disassembles at lower temperatures than the complex with the N-terminal mutations [143]. Further experiments capitalizing on the use of magnetic and optical tweezers that forced the 'zippering' and 'unzippering' of the single complex to take place (without other proteins besides SNAREs) revealed that the complex folds and unfolds in discrete steps [147, 148]. SNARE protein transmembrane and linker domains contribute to this process [149]. Further, the N-terminal complex assembly yields a stable intermediate and precedes subsequent C-terminal assembly [150]. It is possible that the complex is kept partially assembled by arresting the complete ternary SNARE complex due to the repulsion energy between membrane bilayers.

Given that C-terminal SNARE complex assembly controls the fusion of a secretory vesicle, while the N-terminal complex assembly drives vesicle priming, it became clear that cells employ mechanism(s) to keep the SNARE complex partially assembled until the exocytotic trigger, i.e. a calcium signal, arrives. Complexin-1, a soluble protein with a SNARE domain, is thought to bind to the forming SNARE complex by inserting its 'accessory' SNARE domain into a partially assembled complex and blocking assembly until calcium binding to synaptotagmin-1 relieves the block [151–155] (see also section 3.3). In addition, the complexin-1 binding site seems to be created only during SNARE complex formation [156, 157]. The role of synaptotagmin-1 in this process has been controversial, as this small protein appears to have several different functions in neurotransmitter release, some of which are mediated by the same polybasic region. Nevertheless, several lines of evidence suggest that synaptotagmin-1 acts as an inhibitor of membrane fusion prior to calcium entry [158]. Over the past few years, several reported structures show that synaptotagmin-1 binds SNARE complexes, including multiple binding interfaces mainly centered on the C2B domain [159–161]. Although these interactions are highly sensitive to ionic strength [162], interactions involving key lysine and arginine clusters of the C2B domain have been observed *in vitro* under near-physiological conditions [163]. These structural studies with full SNARE complexes clarify the findings of earlier reports that the synaptotagmin-1 C2B domain interacts with syntaxin-1—SNAP-25 dimers [164, 165]. However, structure–function studies are difficult to interpret, because the same polybasic regions also interact with PI(4,5)P2 lipids in a partially calcium-independent manner [166, 167]. The balance of the SNARE- and lipid-binding functions of synaptotagmin-1, as well as possible homo-oligomerization (as discussed below), remain unclear at this point, but it is clear that calcium influx promotes the release of synaptotagmin-1 from SNARE complexes and the insertion of its two C2 domains into membranes.

Whether the partial assembly of the SNARE complex indeed happens in the cell is unclear. Not all published data back the interpretation that a partial SNARE complex assembly is needed to fuse the membranes. SNARE complexes with only slightly distressed C-termini or fully assembled ternary SNARE complexes have been identified between membranes undergoing fusion [167, 168]. Therefore, the full assembly of the SNARE complex may not be enough to fuse the membrane bilayers. Contributions may be needed from the transmembrane anchors and/or linker domains, aided by the binding of calcium ions to synaptotagmin-1 and its subsequent membrane insertion.

3.2.4 Finale: membrane fusion

While much knowledge has been accumulated about the steps preceding membrane bilayer fusion, there are many open questions regarding the steps that lead to a merger of two membranes. During the last two decades, significant progress has been made toward understanding how the fusion of secretory vesicles with the plasma membrane occurs. Because this process is difficult to study directly in living neurons, much knowledge was obtained from *in vitro* models, which are often just an approximation for the complexity of this process. While the original work was predominantly based on indirect methods and molecular dynamics simulations, recent efforts have been focused on advancing high-temporal-resolution measurements of lipid mixing in cellular models. In general, SNARE-mediated membrane bilayer fusion seems to follow the same rules as the fusion of membrane bilayers in the absence of proteins [169]. Upon close proximity between membrane bilayers, their lipids start to spread out and build a lipid stalk, which prompts the fusion of proximal leaflets. Subsequently, the stalk either rapidly creates a fusion pore, or expands further and forms a hemi-fused arrangement which then develops into a fusion pore [170–174]. The formation of such a fusion pore may already be sufficient to release neurotransmitters from synaptic vesicles [175]. Nonetheless, the fusion pore usually expands, leading to fully fused membrane bilayers.

How does the assembly of SNARE complexes drive fusion of the membrane bilayers? In the case of large secretory vesicles, SNARE-mediated membrane fusion is a well-organized process, often composed of a several SNARE complexes (5–15) distributed around the nascent fusion pore in a radially symmetric arrangement [176, 177]. In the case of small synaptic vesicles, the number of SNARE complexes needed for fusion is much smaller (three to five; see [178]). While many scientists in this field presume that lipids line the fusion pore, some models posit that a fusion pore has proteinaceous nature and is lined by the transmembrane domains of synaptobrevin-2 and syntaxin-1 [179]. Several studies have demonstrated that membrane fusion can be mediated by a single or few SNARE complexes, while three complexes are required to keep the fusion pore open [180]. These observations are consistent with data obtained from neurosecretory PC12 cells [181] and findings that two to three SNARE complexes are sufficient to trigger fusion in living neurons [178, 182, 183].

These results are compatible with a model in which SNARE complexes act as independent mini-machines to fuse membranes and catalyze multiple kinetic steps that drive exocytosis. More experiments are needed to establish whether the number of SNARE complexes is variable or fixed, and whether their zippering occurs in a coordinated or stochastic manner. A mechanism that organizes ring-shaped SNARE structures is starting to be addressed by both modeling and experimental approaches. For example, Rothman *et al* presented a 'buttressed-ring hypothesis' in which an outer ring of up to six curved Munc13 MUN domains (transiently anchored to the plasma membrane via its flanking domains) surrounds an inner ring composed of synaptotagmin C2 domains, which together act as a workbench on which SNARE complexes are templated [184]. Some experimental evidence has been provided for the organization of SNARE complexes via binding to synaptotagmin-1 and Munc13, both of which have been shown to oligomerize into ring-like structures under resting conditions [120, 121, 185]. An alternative model that builds on a synaptophysin/synaptobrevin complex has also been proposed [186]. An earlier mechanism, suggested almost two decades ago, involves SNARE domain sharing through SNAP-25 proteins, i.e. the two SNARE domains of SNAP-25 might join neighboring yet different complexes, which might promote the organization of multiple SNARE complexes in a ring-formed structure [187].

It is pivotal to understand whether assembly of the N-terminal end of the SNARE helices yields a tense structure capable of storing energy that can be used for the assembly of the C-termini. However, fusion speeds in neurosecretory cells were not compromised by mutations that destabilized the N-terminal end [143, 144, 188], while such mutations in neurons caused a slight increase in release probability [146]. These observations are better reconciled by a model in which the assembled N-terminal end of the SNARE complex is a rather stable structure (i.e. a low point in the energy landscape), resulting in an effective magnitude increase of the energy barrier for fusion. Therefore, increasing the fusion energy threshold by partial assembly of the SNARE complex would allow fast Ca^{2+}-triggered fusion. The model 'complexin-1 as a clamp' (see above) proposes exactly this: the tight assembly of the SNARE complex N-terminus triggers binding to complexin-1 and clamping of the SNARE complex C-terminal assembly. Alternatively, a similar feature could be encoded in the SNARE complex itself, which is then magnified through the interaction with complexin-1/synaptotagmin-1.

Of note, the theoretical estimations for the fusion energy have been recently verified by experimental measurements, showing the activation energy for fusing vesicles to be ~30 $k_B T$ [189]. Considering that a single SNARE complex releases about ~35 $k_B T$ energy [190] and that zippering of the complex's C-terminus is the power stroke of the fusion process [148], it is feasible that, given sufficient time, a single SNARE complex can drive lipid mixing, and two to three SNARE complexes can create a fusion pore. Further, in living cells, the SNARE proteins likely interact with other factors which might act to lower the energy barrier(s) for fusion. Assembly of SNARE complexes would likely surpass the first energy barrier, allowing apposition of the membranes. In addition, Munc18 has been proposed to induce mixing of lipids directly [90], thus helping to surpass the second energy threshold. Likewise, synaptotagmin family proteins help to enlarge the fusion pore [165, 191, 192]. Thus, SNAREs can be considered as components of a wider fusion machinery, able to surpass several different energy thresholds.

In the scenario in which the energy released during SNARE complex formation is channeled towards membrane fusion, the linkers connecting the SNARE domains and the transmembrane domains of syntaxin-1 and synaptobrevin-2 must transduce force to the membrane. This might occur through the continuous formation of helices across the linker and transmembrane domains, as observed in a crystal structure comprising those regions [149]. Due of the stiffness of α-helices, the formation of a progressive, continuous α-helix from the N- to the C-terminus would tie together membrane deformation with the conformation of synaptobrevin-2. Mutagenesis experiments showed, however, that α-helical continuity is not, *per se,* necessary for membrane fusion. In assays of membrane fusion *in vitro*, the insertion of helix breakers (composed of five amino acids, of which two were prolines) into synaptobrevin-2 did not cause lipid mixing inhibition, but their insertion into syntaxin-1 decreased the rate of fusion by a factor of two [193]. Lipid mixing was also progressively reduced by inserting flexible domains into the linkers, and syntaxin-1 showed more susceptibility again than synaptobrevin-2, confirming the asymmetrical action of the SNARE complex on the opposing membranes. Simulation models showed considerable stiffness of the syntaxin-1 linker, which makes the protein assume an upright posture in the membrane [194], while the more flexible synaptobrevin-2 linker might be only partially inserted [195, 196]. Recent experiments with fused lipid nanodiscs have revealed that the transmembrane regions of syntaxin and synaptobrevin strongly influence the dynamics of fusion pore opening and closing; surprisingly, mutations that disrupt zippering prolonged the lifetime of the open state of the fusion pore [197]. This effect is believed to arise from a combination of protein flexibility and protein–lipid interactions and is an exciting area of ongoing study [198].

Experiments in cultured neurons and chromaffin cells have shown that the insertion of additional flexible sequences (two to eight amino acids) into the synaptobrevin-2 linker domain impairs vesicle priming, extends the delay preceding exocytosis, and retards release from individual vesicles [199–201]. Curiously, when a linker containing 22 additional amino acids was expressed, secretion was similar to synaptobrevin-2 knockout cells [199]. Inserting three or seven amino acids into the linker domain of syntaxin-1 in neurons ablated their capacity to support evoked release [202]. Two tryptophan residues in the linker domain of synaptobrevin-2 are essential to impose a fusion clamp, which promotes evoked rather than spontaneous fusion [203, 204]. It is also interesting to compare mutations in the linker domain with those in the C-terminal layers of the SNARE complex, where severe phenotypes result from various point mutations [143, 144, 205, 206]. In conclusion, the linker domains of syntaxin-1 and synaptobrevin-2 have important roles in transducing force to the fusing membranes in the synchronized release mechanism (spontaneous release is discussed below).

In mammalian neurons, it has been elegantly demonstrated that the replacement of the transmembrane domains of both syntaxin-1 and synaptobrevin-2 with lipid anchors (which insert only into one of the membrane leaflets) is compatible with fast evoked release [202]. This agrees with the concept that the main function of the SNARE transmembrane domains is the transfer of force to the membranes (the details of how this is achieved remain unclear). However, this outcome was somewhat unexpected because a previous *in vitro* study had found that fusion could not be reconstituted when SNARE domains were anchored to lipids [207]. Analysis of single vesicle fusion events also showed that SNAREs add force in two stages to drive membrane fusion: first during fusion pore formation, and then during its subsequent expansion [199–201]. Therefore, the role played by SNARE transmembrane domains in the formation and expansion of the fusion pore is likely important. Deleting of the C-terminal half of the synaptobrevin-2 transmembrane domains in PC12 cells suppresses secretion [208], possibly caused by arrest in the hemi-fused state. The transmembrane domains of synaptobrevin-2 dimerize [71, 209], but a mutation that eliminates dimerization does not inhibit fusion [208]. Adding one or two amino-acid residues to the C-terminal intravesicular end of synaptobrevin-2 inhibited fusion in chromaffin cells [210]. This result suggests that the C-terminal end of the transmembrane domain is pulled into the membrane, driven by formation of the SNARE complex, leading to disruption of the hemi-fused stalk and formation of the fusion pore.

Molecular dynamics simulations have yielded important insights into the final events leading up to membrane fusion (reviewed in [1, 211]). This approach has confirmed experimental findings that fusion can be driven by one, or a few, SNARE complexes. Those simulations also revealed that SNARE linkers and transmembrane domains have important roles in inducing lipid disorder, which subsequently leads to stalk formation [212]. Subsequent to the formation of the hemi-fused state, the C-termini of the transmembrane domains aid in the formation of the fusion pore. Consistent with the crystal structure [149], the pore releases SNARE bending stress by placing the charged C-terminal ends in the hydrophilic pore [212, 213]. Important roles for dimerization were also identified in one simulation [212]. SNARE mutants known to inhibit fusion did not form pores in these simulations [214]. Other simulations have suggested a complementary mechanism: that the entropy of SNAREpins around a fusing pore also significantly contributes to fusion and pore expansion [215].

In summary, there is compelling evidence to conclude that formation of the SNARE complex is exergonic, and the energy released is transferred to the membrane through the linkers and transmembrane domains as a force driving membrane fusion. This force promotes close alignment of the membranes, leading to dehydration of the lipid head groups, at which point the fusion process becomes governed by still poorly understood lipid–protein and lipid–lipid interactions. It seems most likely that the subsequent events are lipid splaying, formation of a stalk and a hemi-fusion intermediate, giving way to the formation of a fusion pore and release of the vesicular content (for a model incorporating a transient proteinaceous pore, see [216]). The SNARE proteins continue to drive fusion throughout these stages, but it seems likely that the last part of the process can proceed in multiple manners.

3.3 Negative regulation of the SNARE complex assembly by tomosyn-1 and amisyn

In addition to the previously discussed Munc13, Munc18, CAPS, and complexin proteins, as well as various phosphatidylinositols (for example, PI(4,5)P_2 [217]), the proteins tomosyn (STXBP5) and amisyn (STXBP6) have been suggested to have a regulatory role in SNARE complex formation and membrane fusion [218, 219].

Tomosyn is a large (130 kDa) brain-enriched protein that contains an N-terminal WD40 domain and a C-terminal SNARE domain. It also carries a SNARE motif and is therefore classified as an R-SNARE protein that can bind to the SNAP-25:syntaxin-1 complex to block or limit vesicle priming (reviewed in [218]). It is known as a presynaptic negative regulator of synaptic transmission [220, 221]. The WD40 domain at the N-terminus of tomosyn-1 is important for its inhibitory activity in exocytosis and is the region where several of its autism-spectrum-disorder-linked mutations are located [220, 222], supporting the notion that these mutations may cause alterations to the regulation of synaptic processes. This brain-enriched protein is present in hippocampus mossy fiber synapses and its loss significantly reduces long-term potentiation, which is proposed to have an impact on learning and memory processes [221]. Importantly, tomosyn-1 has been found to have the ability to dissociate Munc18 from syntaxin-1 and form a SNARE complex with it and with SNAP-25 [218]. The next step then requires synaptobrevin-2 to replace tomosyn, yet the mechanism of that reaction is not clear. In addition, tomosyn and Munc18 are suggested to antagonistically control synaptic vesicle priming downstream of Munc13 in *Caenorhabditis elegans* [103].

Amisyn was originally described as a small (24 kDa) brain-enriched cytosolic SNARE protein that interacts with syntaxin-1 [223]. The function of its N-terminal domain was unclear until recently, when it was found to be a PH domain [219]. The SNARE motif of amisyn is a synaptobrevin-2-like C-terminal coiled-coil domain, able to form a ternary complex with the neuronal SNAREs syntaxin-1 and SNAP-25 [219, 223]. However, the formed complex is referred to as a 'fusion-inactive' SNARE complex, since amisyn does not have a transmembrane domain [219]. Elevated levels of amisyn or the addition of recombinant amisyn inhibit exocytosis in neurosecretory cells [219, 223, 224].

The mechanisms by which amisyn inhibits exocytosis still remain to be elucidated. One hypothesis for tomosyn-1 and amisyn's inhibitory roles in exocytosis is that they may compete with secretory vesicles for their fusion site [218, 219]. This is an intriguing possibility that needs further studies.

3.4 Disassembly of the SNARE complex and reuse of the SNARE proteins

The SNARE proteins are present in vast amounts in the cell, yet they can quickly become the limiting factor in exocytosis if they are not recycled and reused. Hence, after each round of fusion, the ternary complexes that are now in the plasma membrane, so-called *cis*-SNARE complexes, are taken apart in order to recycle the SNARE proteins for new rounds of exocytosis. The process of disassembly is executed by the highly conserved ATPase NSF and its cofactor α-/β-/γ-SNAP, using the hydrolysis of ATP as an energy source [225]. Curiously, NSF and α-/β-/γ-SNAP are essential for all eukaryotic cells and have been found to operate in all intracellular processes that involve membrane fusion. α-SNAPs recognize the *cis*-SNAREs and then recruit and activate NSF, which frees individual SNARE proteins from the complex. Two homologs of α-SNAP, β-SNAP and γ-SNAP, can also recruit NSF to membranes [226, 227]. The activity of NSF is likely high: most SNARE proteins detected in the plasma membrane and living cells are considered not to be in a complex [62].

In the past, significant progress was made in elucidating the structure of NSF: SNARE complexes, and also in comprehending the mechanism of NSF action. Studies using single-molecule measurements and cryo-electron microscopy found that NSF and α-SNAP keep a tight grip on the SNARE complex. After ATP hydrolysis and phosphate release, NSF and α-SNAP build up mechanical tension that is resolved by ripping apart the SNARE complex in a single burst [228].

What happens if the SNARE complexes are not efficiently disassembled after each membrane fusion event? Synaptic depression during repetitive stimulation is observed when a fruit fly NSF homolog, dNSF-1 or *comatose*, is inactive [229, 230]. Mammalian neurons without α-SNAP showed rundown during high neuronal activity [231]. Notably, NSF is found not to be needed for the fusion process itself, but to recover the SNARE proteins [232, 233]. However, it is possible that α-/β-SNAP and NSF function prior to the fusion of secretory vesicles [231, 234].

Importantly, *trans*-SNARE complexes that mediate fusion of secretory vesicles with the plasma membrane must not be recognized by α-/β-SNAP and NSF, otherwise the *trans*-SNARE complex would be disassembled and fusion would be blocked. Various mechanisms are thought to guard the *trans*-SNARE complex from disassembly, such as membrane steric hindrance and conformations of partially assembled SNARE complexes that are not recognized by SNAPs [235]. Interaction between the *trans*-SNARE complex and the Munc13 and Munc18 proteins has also been found to guard *trans*-SNARE complexes from α-SNAP/NSF-driven disassembly [55].

3.5 SNARE-mediated membrane fusion beyond 2022: functional diversity of non-canonical SNARE proteins

In addition to the synchronous release of neurotransmitters triggered by stimulation (action potential), neurotransmitters can also be released asynchronously with a variable delay following action potentials, or as spontaneous events that occur randomly in the absence of action potentials (i.e. spontaneous release). The classical exocytotic SNARE proteins synaptobrevin-2, syntaxin-1, and SNAP-25 participate in asynchronous and spontaneous release minimally or only modestly. Specifically, ablation of synaptobrevin-2 or SNAP-25 eliminates synchronous release almost completely, yet asynchronous release and spontaneous release are not prominently affected by deletion of these key SNARE proteins [24, 33, 37, 236–238]. This likely happens due to a substitution of classical SNARE proteins by non-canonical SNARE family members. For example, cellubrevin can substitute for synaptobrevin-2 [79, 239], while SNAP-23 restores spontaneous secretion in SNAP-25 knockout cells [237]. In addition, the SNARE proteins vti1a, VAMP4, and VAMP7 are found to participate in synaptic vesicle fusion, in both asynchronous and spontaneous release [240–243]. Some SNARE proteins seem to be partly interchangeable [11], yet not syntaxin-1, whose absence reduces both synchronous and spontaneous release, probably due to its interactions with Munc13 and Munc18 [183, 202, 244].

We are only starting to understand the relevance that non-canonical SNARE proteins play in neuronal transmission: more research work is needed to comprehend the role that the non-canonical SNARE proteins play in spontaneous and asynchronous release under physiological conditions. Different vesicle pools exist at the synapse, but it is not known whether they can be distinguished based on the exocytotic machinery that they use. Likewise, it is not known what determines the relative participation of different SNARE proteins in exocytotic events. In addition, the expression levels of various SNARE proteins, both classical and non-canonical, may vary among neuronal populations. For example, the non-canonical SNARE proteins may have the primary role in neuron populations where they are expressed most strongly, like the CCK-interneurons that have predominant asynchronous release and abundantly express VAMP4 [245].

Research into SNARE proteins has reached an advanced phase in which we have grasped the composition of the fusion machinery and the basic principles of how it operates. Many discoveries have emerged from biochemical, reconstitution, and electrophysiological studies. More recently, advanced imaging (e.g. super-resolution microscopy, light sheet microscopy) combined with the expression of fluorescently labeled SNARE proteins and biophysical molecular modeling studies are starting to reveal new properties of SNARE complexes and membrane fusion. Some remaining questions include: what is the function of the SNARE complex biophysical features in the light of neuronal activity and the activity of neuronal network? How are the various levels of SNARE complex regulation, by both positive and negative factors, integrated? Are there cell-specific and tissue-specific aspects to the regulation of SNARE-mediated exocytosis, and if so, what are they? Importantly, how do alterations in SNARE complex formation and/or regulation lead to diseases? Much has been learnt so far; nonetheless, many exciting years of discoveries await.

Acknowledgements

I.M. thanks Ante Milosevic for help with the figure and the John Black Foundation, John Fell Fund, FCT (PTDC/MED-NEU/8030), and the Wellcome Trust for supporting her work. Apologies to the authors of relevant papers that were not cited here due to space constraints. The authors declare no conflict of interest.

References and further reading

[1] Risselada H J and Grubmüller H 2012 How SNARE molecules mediate membrane fusion: recent insights from molecular simulations *Curr. Opin. Struct. Biol.* **22** 187–96

[2] Jahn R and Fasshauer D 2012 Molecular machines governing exocytosis of synaptic vesicles *Nature* **490** 201–7

[3] Sudhof T C 2013 Neurotransmitter release: the last millisecond in the life of a synaptic vesicle *Neuron* **80** 675–90

[4] Terrian D M and White M K 1997 Phylogenetic analysis of membrane trafficking proteins: a family reunion and secondary structure predictions *Eur. J. Cell Biol.* **73** 198–204

[5] Weimbs T, Mostov K, Low S H and Hofmann K 1998 A model for structural similarity between different SNARE complexes based on sequence relationships *Trends Cell Biol.* **8** 260–2

[6] Kloepper T H, Kienle C N and Fasshauer D 2007 An elaborate classification of snare proteins sheds light on the conservation of the eukaryotic endomembrane system *Mol. Biol. Cell* **18** 3463–71

[7] Sutton R B, Fasshauer D, Jahn R and Brunger A T 1998 Crystal structure of a SNARE complex involved in synaptic exocytosis at 2.4 A resolution *Nature* **395** 347–53

[8] Fasshauer D, Eliason W K, Brunger A T and Jahn R 1998 Identification of a minimal core of the synaptic SNARE complex sufficient for reversible assembly and disassembly *Biochemistry* **37** 10354–62

[9] Antonin W, Holroyd C, Fasshauer D, Pabst S, Von Mollard G F and Jahn R 2000 A SNARE complex mediating fusion of late endosomes defines conserved properties of SNARE structure and function *EMBO J.* **19** 6453–64

[10] Sollner T, Whiteheart S W, Brunner M, Erdjument-Bromage H, Geromanos S, Tempst P and Rothman J E 1993 SNAP receptors implicated in vesicle targeting and fusion *Nature* **362** 318–24

[11] Fasshauer D, Sutton R B, Brunger A T and Jahn R 1998 Conserved structural features of the synaptic fusion complex: SNARE proteins reclassified as Q- and R-SNAREs *Proc. Natl. Acad. Sci. USA* **95** 15781–6

[12] Bock J B, Matern H T, Peden A A and Scheller R H 2001 A genomic perspective on membrane compartment organization *Nature* **409** 839–41

[13] Trimble W S, Cowan D M and Scheller R H 1988 VAMP-1: a synaptic vesicle-associated integral membrane protein *Proc. Natl. Acad. Sci. USA* **85** 4538–42

[14] Baumert M, Maycox P R, Navone F, De Camilli P and Jahn R 1989 Synaptobrevin: an integral membrane protein of 18,000 daltons present in small synaptic vesicles of rat brain *EMBO J.* **8** 379–84

[15] Oyler G A, Higgins G A, Hart R A, Battenberg E, Billingsley M, Bloom F E and Wilson M C 1989 The identification of a novel synaptosomal-associated protein, SNAP-25, differentially expressed by neuronal subpopulations *J. Cell Biol.* **109** 3039–52

[16] Bennett M K, Calakos N and Scheller R H 1992 Syntaxin: a synaptic protein implicated in docking of synaptic vesicles at presynaptic active zones *Science* **257** 255–9

[17] Niemann H, Blasi J and Jahn R 1994 Clostridial neurotoxins: new tools for dissecting exocytosis *Trends Cell Biol.* **4** 179–85

[18] Montecucco C and Schiavo G 1995 Structure and function of tetanus and botulinum neurotoxins *Q. Rev. Biophys.* **28** 423–72

[19] Weber T, Zemelman B V, McNew J A, Westermann B, Gmachl M, Parlati F, Sollner T H and Rothman J E 1998 SNAREpins: minimal machinery for membrane fusion *Cell* **92** 759–72

[20] Hu C, Ahmed M, Melia T J, Sollner T H, Mayer T and Rothman J E 2003 Fusion of cells by flipped SNAREs *Science* **300** 1745–9

[21] Fasshauer D, Otto H, Eliason W K, Jahn R and Brunger A T 1997 Structural changes are associated with soluble N-ethylmaleimide-sensitive fusion protein attachment protein receptor complex formation *J. Biol. Chem.* **272** 28036–41

[22] Dulubova I, Sugita S, Hill S, Hosaka M, Fernandez I, Sudhof T C and Rizo J 1999 A conformational switch in syntaxin during exocytosis: role of munc18 *EMBO J.* **18** 4372–82

[23] Ruiz-Montasell B, Aguado F, Majo G, Chapman E R, Canals J M, Marsal J and Blasi J 1996 Differential distribution of syntaxin isoforms 1A and 1B in the rat central nervous system *Eur. J. Neurosci.* **8** 2544–52
[24] Broadie K, Prokop A, Bellen H J, O'Kane C J, Schulze K L and Sweeney S T 1995 Syntaxin and synaptobrevin function downstream of vesicle docking in *Drosophila Neuron* **15** 663–73
[25] Schulze K L, Broadie K, Perin M S and Bellen H J 1995 Genetic and electrophysiological studies of *Drosophila* syntaxin-1A demonstrate its role in nonneuronal secretion and neurotransmission *Cell* **80** 311–20
[26] Gerber S H *et al* 2008 Conformational switch of syntaxin-1 controls synaptic vesicle fusion *Science* **321** 1507–10
[27] Fujiwara T, Mishima T, Kofuji T, Chiba T, Tanaka K, Yamamoto A and Akagawa K 2006 Analysis of knock-out mice to determine the role of HPC-1/syntaxin 1A in expressing synaptic plasticity *J. Neurosci.* **26** 5767–76
[28] Fukuda R, McNew J A, Weber T, Parlati F, Engel T, Nickel W, Rothman J E and Sollner T H 2000 Functional architecture of an intracellular membrane t-SNARE *Nature* **407** 198–202
[29] Nagy G, Milosevic I, Mohrmann R, Wiederhold K, Walter A M and Sorensen J B 2008 The SNAP-25 linker as an adaptation toward fast exocytosis *Mol. Biol. Cell* **19** 3769–81
[30] Hess D T, Slater T M, Wilson M C and Skene J H 1992 The 25 kDa synaptosomal-associated protein SNAP-25 is the major methionine-rich polypeptide in rapid axonal transport and a major substrate for palmitoylation in adult CNS *J. Neurosci.* **12** 4634–41
[31] Gonzalo S, Greentree W K and Linder M E 1999 SNAP-25 is targeted to the plasma membrane through a novel membrane-binding domain *J. Biol. Chem.* **274** 21313–8
[32] Sorensen J B, Nagy G, Varoqueaux F, Nehring R B, Brose N, Wilson M C and Neher E 2003 Differential control of the releasable vesicle pools by SNAP-25 splice variants and SNAP-23 *Cell* **114** 75–86
[33] Washbourne P *et al* 2002 Genetic ablation of the t-SNARE SNAP-25 distinguishes mechanisms of neuroexocytosis *Nat. Neurosci.* **5** 19–26
[34] Bark I C and Wilson M C 1994 Human cDNA clones encoding two different isoforms of the nerve terminal protein SNAP-25 *Gene* **139** 291–2
[35] Mohrmann R, de Wit H, Connell E, Pinheiro P S, Leese C, Bruns D, Davletov B, Verhage M and Sorensen J B 2013 Synaptotagmin interaction with SNAP-25 governs vesicle docking, priming, and fusion triggering *J. Neurosci.* **33** 14417–30
[36] Kadkova A, Radecke J and Sorensen J B 2019 The SNAP-25 protein family *Neuroscience* **420** 50–71
[37] Schoch S, Deak F, Konigstorfer A, Mozhayeva M, Sara Y, Sudhof T C and Kavalali E T 2001 SNARE function analyzed in synaptobrevin/VAMP knockout mice *Science* **294** 1117–22
[38] Trimble W S, Gray T S, Elferink L A, Wilson M C and Scheller R H 1990 Distinct patterns of expression of two VAMP genes within the rat brain *J. Neurosci.* **10** 1380–7
[39] Schneggenburger R and Neher E 2000 Intracellular calcium dependence of transmitter release rates at a fast central synapse *Nature* **406** 889–93
[40] Sabatini B L and Regehr W G 1996 Timing of neurotransmission at fast synapses in the mammalian brain *Nature* **384** 170–2
[41] Milosevic I and Sørensen J B 2015 *Presynaptic Terminals* ed S Mochida (Tokyo: Springer Japan) pp 87–127

[42] Misura K M, Scheller R H and Weis W I 2000 Three-dimensional structure of the neuronal-Sec1-syntaxin 1a complex *Nature* **404** 355–62

[43] Verhage M *et al* 2000 Synaptic assembly of the brain in the absence of neurotransmitter secretion *Science* **287** 864–9

[44] Harrison S D, Broadie K, van de Goor J and Rubin G M 1994 Mutations in the *Drosophila* Rop gene suggest a function in general secretion and synaptic transmission *Neuron* **13** 555–66

[45] Hosono R, Hekimi S, Kamiya Y, Sassa T, Murakami S, Nishiwaki K, Miwa J, Taketo A and Kodaira K I 1992 The unc-18 gene encodes a novel protein affecting the kinetics of acetylcholine metabolism in the nematode *Caenorhabditis elegans J. Neurochem.* **58** 1517–25

[46] Zhang Y and Hughson F M 2021 Chaperoning SNARE folding and assembly *Annu. Rev. Biochem.* **90** 581–603

[47] Wang S and Ma C 2022 Neuronal SNARE complex assembly guided by Munc18-1 and Munc13-1 *FEBS Open Bio.*

[48] Brenner S 1974 The genetics of *Caenorhabditis elegans Genetics* **77** 71–94

[49] Novick P and Schekman R 1979 Secretion and cell-surface growth are blocked in a temperature-sensitive mutant of *Saccharomyces cerevisiae Proc. Natl Acad. Sci. USA* **76** 1858–62

[50] Hata Y, Slaughter C A and Sudhof T C 1993 Synaptic vesicle fusion complex contains unc-18 homologue bound to syntaxin *Nature* **366** 347–51

[51] Medine C N, Rickman C, Chamberlain L H and Duncan R R 2007 Munc18-1 prevents the formation of ectopic SNARE complexes in living cells *J. Cell Sci.* **120** 4407–15

[52] Rowe J, Corradi N, Malosio M L, Taverna E, Halban P, Meldolesi J and Rosa P 1999 Blockade of membrane transport and disassembly of the golgi complex by expression of syntaxin 1A in neurosecretion-incompetent cells: prevention by rbSEC1 *J. Cell Sci.* **112** 1865–77

[53] Toonen R F, de Vries K J, Zalm R, Sudhof T C and Verhage M 2005 Munc18-1 stabilizes syntaxin 1, but is not essential for syntaxin 1 targeting and SNARE complex formation *J. Neurochem.* **93** 1393–400

[54] Dawidowski D and Cafiso D S 2016 Munc18-1 and the syntaxin-1 N terminus regulate open-closed states in a t-SNARE complex *Structure* **24** 392–400

[55] Ma C, Su L, Seven A B, Xu Y and Rizo J 2013 Reconstitution of the vital functions of Munc18 and Munc13 in neurotransmitter release *Science* **339** 421–5

[56] Choi U B, Zhao M, White K I, Pfuetzner R A, Esquivies L, Zhou Q and Brunger A T 2018 NSF-mediated disassembly of on- and off-pathway SNARE complexes and inhibition by complexin *Elife* **7** e36497

[57] Prinslow E A, Stepien K P, Pan Y Z, Xu J and Rizo J 2019 Multiple factors maintain assembled trans-SNARE complexes in the presence of NSF and alphaSNAP *Elife* **8** e38880

[58] Voets T, Toonen R F, Brian E C, de Wit H, Moser T, Rettig J, Sudhof T C, Neher E and Verhage M 2001 Munc18-1 promotes large dense-core vesicle docking *Neuron* **31** 581–91

[59] Weimer R M, Richmond J E, Davis W S, Hadwiger G, Nonet M L and Jorgensen E M 2003 Defects in synaptic vesicle docking in unc-18 mutants *Nat. Neurosci.* **6** 1023–30

[60] de Wit H, Cornelisse L N, Toonen R F and Verhage M 2006 Docking of secretory vesicles is syntaxin dependent *PLoS One* **1** e126

[61] Hammarlund M, Palfreyman M T, Watanabe S, Olsen S and Jorgensen E M 2007 Open syntaxin docks synaptic vesicles *PLoS Biol.* **5** e198

[62] Lang T, Margittai M, Holzler H and Jahn R 2002 SNAREs in native plasma membranes are active and readily form core complexes with endogenous and exogenous SNAREs *J. Cell Biol.* **158** 751–60
[63] Bar-On D, Gutman M, Mezer A, Ashery U, Lang T and Nachliel E 2009 Evaluation of the heterogeneous reactivity of the syntaxin molecules on the inner leaflet of the plasma membrane *J. Neurosci.* **29** 12292–301
[64] Bar-On D, Wolter S, van de Linde S, Heilemann M, Nudelman G, Nachliel E, Gutman M, Sauer M and Ashery U 2012 Super-resolution imaging reveals the internal architecture of nano-sized syntaxin clusters *J. Biol. Chem.* **287** 27158–67
[65] An S J and Almers W 2004 Tracking SNARE complex formation in live endocrine cells *Science* **306** 1042–6
[66] Rickman C, Medine C N, Dun A R, Moulton D J, Mandula O, Halemani N D, Rizzoli S O, Chamberlain L H and Duncan R R 2010 t-SNARE protein conformations patterned by the lipid microenvironment *J. Biol. Chem.* **285** 13535–41
[67] Misura K M, Gonzalez L C Jr., May A P, Scheller R H and Weis W I 2001 Crystal structure and biophysical properties of a complex between the N-terminal SNARE region of SNAP25 and syntaxin 1a *J. Biol. chem.* **276** 41301–9
[68] Margittai M, Fasshauer D, Pabst S, Jahn R and Langen R 2001 Homo- and heterooligomeric SNARE complexes studied by site-directed spin labeling *J. Biol. Chem.* **276** 13169–77
[69] Freedman S J, Song H K, Xu Y, Sun Z Y and Eck M J 2003 Homotetrameric structure of the SNAP-23 N-terminal coiled-coil domain *J. Biol. Chem.* **278** 13462–7
[70] Halemani N D, Bethani I, Rizzoli S O and Lang T 2010 Structure and dynamics of a two-helix SNARE complex in live cells *Traffic* **11** 394–404
[71] Laage R, Rohde J, Brosig B and Langosch D 2000 A conserved membrane-spanning amino acid motif drives homomeric and supports heteromeric assembly of presynaptic SNARE proteins *J. Biol. Chem.* **275** 17481–7
[72] Barg S, Knowles M K, Chen X, Midorikawa M and Almers W 2010 Syntaxin clusters assemble reversibly at sites of secretory granules in live cells *Proc. Natl. Acad. Sci. USA* **107** 20804–9
[73] Knowles M K, Barg S, Wan L, Midorikawa M, Chen X and Almers W 2010 Single secretory granules of live cells recruit syntaxin-1 and synaptosomal associated protein 25 (SNAP-25) in large copy numbers *Proc. Natl. Acad. Sci. USA* **107** 20810–5
[74] Baker R W, Jeffrey P D, Zick M, Phillips B P, Wickner W T and Hughson F M 2015 A direct role for the Sec1/Munc18-family protein Vps33 as a template for SNARE assembly *Science* **349** 1111–4
[75] Jiao J *et al* 2018 Munc18-1 catalyzes neuronal SNARE assembly by templating SNARE association *Elife* **7** e41771
[76] Shu T, Jin H, Rothman J E and Zhang Y 2020 Munc13-1 MUN domain and Munc18-1 cooperatively chaperone SNARE assembly through a tetrameric complex *Proc. Natl. Acad. Sci. USA* **117** 1036–41
[77] Stepien K P, Xu J, Zhang X, Bai X C and Rizo J 2022 SNARE assembly enlightened by cryo-EM structures of a synaptobrevin-Munc18-1-syntaxin-1 complex *Sci Adv* **8** eabo5272
[78] Schutz D, Zilly F, Lang T, Jahn R and Bruns D 2005 A dual function for Munc-18 in exocytosis of PC12 cells *Eur. J. Neurosci.* **21** 2419–32
[79] Borisovska M, Zhao Y, Tsytsyura Y, Glyvuk N, Takamori S, Matti U, Rettig J, Sudhof T and Bruns D 2005 v-SNAREs control exocytosis of vesicles from priming to fusion *EMBO J.* **24** 2114–26

[80] de Wit H, Walter A M, Milosevic I, Gulyas-Kovacs A, Riedel D, Sorensen J B and Verhage M 2009 Synaptotagmin-1 docks secretory vesicles to syntaxin-1/SNAP-25 acceptor complexes *Cell* **138** 935–46
[81] Shen J, Tareste D C, Paumet F, Rothman J E and Melia T J 2007 Selective activation of cognate SNAREpins by Sec1/Munc18 proteins *Cell* **128** 183–95
[82] Dulubova I, Khvotchev M, Liu S, Huryeva I, Sudhof T C and Rizo J 2007 Munc18-1 binds directly to the neuronal SNARE complex *Proc. Natl. Acad. Sci. USA* **104** 2697–702
[83] Khvotchev M, Dulubova I, Sun J, Dai H, Rizo J and Sudhof T C 2007 Dual modes of Munc18-1/SNARE interactions are coupled by functionally critical binding to syntaxin-1 N terminus *J. Neurosci.* **27** 12147–55
[84] Carr C M, Grote E, Munson M, Hughson F M and Novick P J 1999 Sec1p binds to SNARE complexes and concentrates at sites of secretion *J. Cell Biol.* **146** 333–44
[85] Burkhardt P, Hattendorf D A, Weis W I and Fasshauer D 2008 Munc18a controls SNARE assembly through its interaction with the syntaxin N-peptide *EMBO J.* **27** 923–33
[86] Schollmeier Y, Krause J M, Kreye S, Malsam J and Sollner T H 2011 Resolving the function of distinct Munc18-1/SNARE protein interaction modes in a reconstituted membrane fusion assay *J. Biol. Chem.* **286** 30582–90
[87] Xu Y, Su L and Rizo J 2010 Binding of Munc18-1 to synaptobrevin and to the SNARE four-helix bundle *Biochemistry* **49** 1568–76
[88] Vardar G, Salazar-Lazaro A, Brockmann M, Weber-Boyvat M, Zobel S, Kumbol V W, Trimbuch T and Rosenmund C 2021 Reexamination of N-terminal domains of syntaxin-1 in vesicle fusion from central murine synapses *eLife* **10** e69498
[89] Schuette C G, Hatsuzawa K, Margittai M, Stein A, Riedel D, Kuster P, Konig M, Seidel C and Jahn R 2004 Determinants of liposome fusion mediated by synaptic SNARE proteins *Proc. Natl. Acad. Sci. USA* **101** 2858–63
[90] Rizo J and Sudhof T C 2012 The membrane fusion enigma: SNAREs, Sec1/Munc18 proteins, and their accomplices--guilty as charged? *Annu. Rev. Cell Dev. Biol.* **28** 279–308
[91] Zhang Y *et al* 2015 Munc18a does not alter fusion rates mediated by neuronal SNAREs, synaptotagmin, and complexin *J. Biol. Chem.* **290** 10518–34
[92] Richmond J E, Weimer R M and Jorgensen E M 2001 An open form of syntaxin bypasses the requirement for UNC-13 in vesicle priming *Nature* **412** 338–41
[93] McEwen J M, Madison J M, Dybbs M and Kaplan J M 2006 Antagonistic regulation of synaptic vesicle priming by Tomosyn and UNC-13 *Neuron* **51** 303–15
[94] Varoqueaux F, Sigler A, Rhee J S, Brose N, Enk C, Reim K and Rosenmund C 2002 Total arrest of spontaneous and evoked synaptic transmission but normal synaptogenesis in the absence of Munc13-mediated vesicle priming *Proc. Natl. Acad. Sci. USA* **99** 9037–42
[95] Augustin I, Rosenmund C, Sudhof T C and Brose N 1999 Munc13-1 is essential for fusion competence of glutamatergic synaptic vesicles *Nature* **400** 457–61
[96] Richmond J E, Davis W S and Jorgensen E M 1999 UNC-13 is required for synaptic vesicle fusion in *C. elegans Nat. Neurosci.* **2** 959–64
[97] Ashery U, Varoqueaux F, Voets T, Betz A, Thakur P, Koch H, Neher E, Brose N and Rettig J 2000 Munc13-1 acts as a priming factor for large dense-core vesicles in bovine chromaffin cells *EMBO J.* **19** 3586–96
[98] Ma C, Li W, Xu Y and Rizo J 2011 Munc13 mediates the transition from the closed syntaxin-Munc18 complex to the SNARE complex *Nat. Struct. Mol. Biol.* **18** 542–9

[99] Christie M P *et al* 2012 Low-resolution solution structures of Munc18:Syntaxin protein complexes indicate an open binding mode driven by the Syntaxin N-peptide *Proc. Natl. Acad. Sci. USA* **109** 9816–21

[100] Colbert K N, Hattendorf D A, Weiss T M, Burkhardt P, Fasshauer D and Weis W I 2013 Syntaxin1a variants lacking an N-peptide or bearing the LE mutation bind to Munc18a in a closed conformation *Proc. Natl. Acad. Sci. USA* **110** 12637–42

[101] Hu S H, Christie M P, Saez N J, Latham C F, Jarrott R, Lua L H, Collins B M and Martin J L 2011 Possible roles for Munc18-1 domain 3a and Syntaxin1 N-peptide and C-terminal anchor in SNARE complex formation *Proc. Natl. Acad. Sci. USA* **108** 1040–5

[102] Lai Y *et al* 2017 Molecular mechanisms of synaptic vesicle priming by Munc13 and Munc18 *Neuron* **95** 591–607 e10

[103] Park S *et al* 2017 UNC-18 and Tomosyn antagonistically control synaptic vesicle priming downstream of UNC-13 in *Caenorhabditis elegans J. Neurosci.* **37** 8797–815

[104] Sitarska E, Xu J, Park S, Liu X, Quade B, Stepien K, Sugita K, Brautigam C A, Sugita S and Rizo J 2017 Autoinhibition of Munc18-1 modulates synaptobrevin binding and helps to enable Munc13-dependent regulation of membrane fusion *eLife* **6** e24278

[105] Yang X *et al* 2015 Syntaxin opening by the MUN domain underlies the function of Munc13 in synaptic-vesicle priming *Nat. Struct. Mol. Biol.* **22** 547–54

[106] Wang S, Choi U B, Gong J, Yang X, Li Y, Wang A L, Yang X, Brunger A T and Ma C 2017 Conformational change of syntaxin linker region induced by Munc13s initiates SNARE complex formation in synaptic exocytosis *EMBO J.* **36** 816–29

[107] Wang X, Gong J, Zhu L, Wang S, Yang X, Xu Y, Yang X and Ma C 2020 Munc13 activates the Munc18-1/syntaxin-1 complex and enables Munc18-1 to prime SNARE assembly *EMBO J.* **39** e103631

[108] Wang S, Li Y, Gong J, Ye S, Yang X, Zhang R and Ma C 2019 Munc18 and Munc13 serve as a functional template to orchestrate neuronal SNARE complex assembly *Nat. Commun.* **10** 69

[109] Kalyana Sundaram R V, Jin H, Li F, Shu T, Coleman J, Yang J, Pincet F, Zhang Y, Rothman J E and Krishnakumar S S 2021 Munc13 binds and recruits SNAP25 to chaperone SNARE complex assembly *FEBS Lett.* **595** 297–309

[110] Shin O H *et al* 2010 Munc13 C2B domain is an activity-dependent Ca^{2+} regulator of synaptic exocytosis *Nat. Struct. Mol. Biol.* **17** 280–8

[111] Lipstein N, Sakaba T, Cooper B H, Lin K H, Strenzke N, Ashery U, Rhee J S, Taschenberger H, Neher E and Brose N 2013 Dynamic control of synaptic vesicle replenishment and short-term plasticity by Ca(2+)-calmodulin-Munc13-1 signaling *Neuron* **79** 82–96

[112] Rhee J S *et al* 2002 Beta phorbol ester- and diacylglycerol-induced augmentation of transmitter release is mediated by Munc13s and not by PKCs *Cell* **108** 121–33

[113] Xu J *et al* 2017 Mechanistic insights into neurotransmitter release and presynaptic plasticity from the crystal structure of Munc13-1 C1C2BMUN *eLife* **6** e22567

[114] Quade B, Camacho M, Zhao X, Orlando M, Trimbuch T, Xu J, Li W, Nicastro D, Rosenmund C and Rizo J 2019 Membrane bridging by Munc13-1 is crucial for neurotransmitter release *eLife* **8** e42806

[115] Magdziarek M, Bolembach A A, Stepien K P, Quade B, Liu X and Rizo J 2020 Re-examining how Munc13-1 facilitates opening of syntaxin-1 *Protein Sci.* **29** 1440–58

[116] Lipstein N, Chang S, Lin K H, Lopez-Murcia F J, Neher E, Taschenberger H and Brose N 2021 Munc13-1 is a Ca(2+)-phospholipid-dependent vesicle priming hub that shapes synaptic short-term plasticity and enables sustained neurotransmission *Neuron* **109** 3980–4000 e7

[117] Camacho M, Quade B, Trimbuch T, Xu J, Sari L, Rizo J and Rosenmund C 2021 Control of neurotransmitter release by two distinct membrane-binding faces of the Munc13-1 C1C2B region *eLife* **10** e72030

[118] Sakamoto H, Ariyoshi T, Kimpara N, Sugao K, Taiko I, Takikawa K, Asanuma D, Namiki S and Hirose K 2018 Synaptic weight set by Munc13-1 supramolecular assemblies *Nat. Neurosci.* **21** 41–9

[119] Li F, Kalyana Sundaram R V, Gatta A T, Coleman J, Ramakrishnan S, Krishnakumar S S, Pincet F and Rothman J E 2021 Vesicle capture by membrane-bound Munc13-1 requires self-assembly into discrete clusters *FEBS Lett.* **595** 2185–96

[120] Grushin K, Kalyana Sundaram R V, Sindelar C V and Rothman J E 2022 Munc13 structural transitions and oligomers that may choreograph successive stages in vesicle priming for neurotransmitter release *Proc. Natl. Acad. Sci. USA* **119** e2121259119

[121] Wang J, Bello O, Auclair S M, Wang J, Coleman J, Pincet F, Krishnakumar S S, Sindelar C V and Rothman J E 2014 Calcium sensitive ring-like oligomers formed by synaptotagmin *Proc. Natl. Acad. Sci. USA* **111** 13966–71

[122] Stevens D R, Wu Z X, Matti U, Junge H J, Schirra C, Becherer U, Wojcik S M, Brose N and Rettig J 2005 Identification of the minimal protein domain required for priming activity of Munc13-1 *Curr. Biol.* **15** 2243–8

[123] Basu J, Shen N, Dulubova I, Lu J, Guan R, Guryev O, Grishin N V, Rosenmund C and Rizo J 2005 A minimal domain responsible for Munc13 activity *Nat. Struct. Mol. Biol.* **12** 1017–8

[124] Grishanin R N, Klenchin V A, Loyet K M, Kowalchyk J A, Ann K and Martin T F 2002 Membrane association domains in Ca^{2+}-dependent activator protein for secretion mediate plasma membrane and dense-core vesicle binding required for Ca^{2+}-dependent exocytosis *J. Biol. Chem.* **277** 22025–34

[125] Khodthong C, Kabachinski G, James D J and Martin T F 2011 Munc13 homology domain-1 in CAPS/UNC31 mediates SNARE binding required for priming vesicle exocytosis *Cell Metab.* **14** 254–63

[126] Jockusch W J, Speidel D, Sigler A, Sorensen J B, Varoqueaux F, Rhee J S and Brose N 2007 CAPS-1 and CAPS-2 are essential synaptic vesicle priming proteins *Cell* **131** 796–808

[127] Liu Y, Schirra C, Stevens D R, Matti U, Speidel D, Hof D, Bruns D, Brose N and Rettig J 2008 CAPS facilitates filling of the rapidly releasable pool of large dense-core vesicles *J. Neurosci.* **28** 5594–601

[128] Elhamdani A, Martin T F, Kowalchyk J A and Artalejo C R 1999 Ca(2+)-dependent activator protein for secretion is critical for the fusion of dense-core vesicles with the membrane in calf adrenal chromaffin cells *J. Neurosci.* **19** 7375–83

[129] Speidel D, Salehi A, Obermueller S, Lundquist I, Brose N, Renstrom E and Rorsman P 2008 CAPS1 and CAPS2 regulate stability and recruitment of insulin granules in mouse pancreatic beta cells *Cell Metab.* **7** 57–67

[130] Liu Y *et al* 2010 Two distinct secretory vesicle-priming steps in adrenal chromaffin cells *J. Cell Biol.* **190** 1067–77

[131] Hammarlund M, Watanabe S, Schuske K and Jorgensen E M 2008 CAPS and syntaxin dock dense core vesicles to the plasma membrane in neurons *J. Cell Biol.* **180** 483–91

[132] Zhou P, Pang Z P, Yang X, Zhang Y, Rosenmund C, Bacaj T and Sudhof T C 2013 Syntaxin-1 N-peptide and Habc-domain perform distinct essential functions in synaptic vesicle fusion *Embo J.* **32** 159–71

[133] Dittman J S 2019 Unc13: a multifunctional synaptic marvel *Curr. Opin. Neurobiol.* **57** 17–25

[134] Brockmann M M, Zarebidaki F, Camacho M, Grauel M K, Trimbuch T, Sudhof T C and Rosenmund C 2020 A Trio of active zone proteins comprised of RIM-BPs, RIMs, and Munc13s governs neurotransmitter release *Cell Rep.* **32** 107960

[135] Hanson P I, Roth R, Morisaki H, Jahn R and Heuser J E 1997 Structure and conformational changes in NSF and its membrane receptor complexes visualized by quick-freeze/deep-etch electron microscopy *Cell* **90** 523–35

[136] Pobbati A V, Stein A and Fasshauer D 2006 N- to C-terminal SNARE complex assembly promotes rapid membrane fusion *Science* **313** 673–6

[137] Fasshauer D and Margittai M 2004 A transient N-terminal interaction of SNAP-25 and syntaxin nucleates SNARE assembly *J. Cell Biol. Chem.* **279** 7613–21

[138] Melia T J, Weber T, McNew J A, Fisher L E, Johnston R J, Parlati F, Mahal L K, Sollner T H and Rothman J E 2002 Regulation of membrane fusion by the membrane-proximal coil of the t-SNARE during zippering of SNAREpins *J. Cell Biol.* **158** 929–40

[139] Matos M F, Mukherjee K, Chen X, Rizo J and Sudhof T C 2003 Evidence for SNARE zippering during Ca^{2+}-triggered exocytosis in PC12 cells *Neuropharmacology* **45**

[140] Chen Y A, Scales S J and Scheller R H 2001 Sequential SNARE assembly underlies priming and triggering of exocytosis *Neuron* **30** 161–70

[141] Xu T, Rammner B, Margittai M, Artalejo A R, Neher E and Jahn R 1999 Inhibition of SNARE complex assembly differentially affects kinetic components of exocytosis *Cell* **99** 713–22

[142] Hua S Y and Charlton M P 1999 Activity-dependent changes in partial VAMP complexes during neurotransmitter release *Nat. Neurosci.* **2** 1078–83

[143] Sorensen J B, Wiederhold K, Muller E M, Milosevic I, Nagy G, de Groot B L, Grubmuller H and Fasshauer D 2006 Sequential N- to C-terminal SNARE complex assembly drives priming and fusion of secretory vesicles *EMBO J.* **25** 955–66

[144] Walter A M, Wiederhold K, Bruns D, Fasshauer D and Sorensen J B 2010 Synaptobrevin N-terminally bound to syntaxin-SNAP-25 defines the primed vesicle state in regulated exocytosis *J. Cell Biol.* **188** 401–13

[145] Wei S, Xu T, Ashery U, Kollewe A, Matti U, Antonin W, Rettig J and Neher E 2000 Exocytotic mechanism studied by truncated and zero layer mutants of the C-terminus of SNAP-25 *EMBO J.* **19** 1279–89

[146] Weber J P, Reim K and Sorensen J B 2010 Opposing functions of two sub-domains of the SNARE-complex in neurotransmission *EMBO J.* **29** 2477–90

[147] Min D, Kim K, Hyeon C, Cho Y H, Shin Y K and Yoon T Y 2013 Mechanical unzipping and rezipping of a single SNARE complex reveals hysteresis as a force-generating mechanism *Nat. Commun.* **4** 1705

[148] Gao Y, Zorman S, Gundersen G, Xi Z, Ma L, Sirinakis G, Rothman J E and Zhang Y 2012 Single reconstituted neuronal SNARE complexes zipper in three distinct stages *Science* **337** 1340–3

[149] Stein A, Weber G, Wahl M C and Jahn R 2009 Helical extension of the neuronal SNARE complex into the membrane *Nature* **460** 525–8

[150] Li F, Kummel D, Coleman J, Reinisch K M, Rothman J E and Pincet F 2014 A half-zippered SNARE complex represents a functional intermediate in membrane fusion *J. Am. Chem. Soc.* **136** 3456–64

[151] Tang J, Maximov A, Shin O H, Dai H, Rizo J and Sudhof T C 2006 A complexin/synaptotagmin 1 switch controls fast synaptic vesicle exocytosis *Cell* **126** 1175–87

[152] Kummel D, Krishnakumar S S, Radoff D T, Li F, Giraudo C G, Pincet F, Rothman J E and Reinisch K M 2011 Complexin cross-links prefusion SNAREs into a zigzag array *Nat. Struct. Mol. Biol.* **18** 927–33

[153] Giraudo C G, Eng W S, Melia T J and Rothman J E 2006 A clamping mechanism involved in SNARE-dependent exocytosis *Science* **313** 676–80

[154] Xue M, Reim K, Chen X, Chao H T, Deng H, Rizo J, Brose N and Rosenmund C 2007 Distinct domains of complexin I differentially regulate neurotransmitter release *Nat. Struct. Mol. Biol.* **14** 949–58

[155] Malsam J, Parisotto D, Bharat T A, Scheutzow A, Krause J M, Briggs J A and Sollner T H 2012 Complexin arrests a pool of docked vesicles for fast Ca^{2+}-dependent release *EMBO J.* **31** 3270–81

[156] Pabst S, Hazzard J W, Antonin W, Sudhof T C, Jahn R, Rizo J and Fasshauer D 2000 Selective interaction of complexin with the neuronal SNARE complex. Determination of the binding regions *J. Biol. Chem.* **275** 19808–18

[157] Chen X, Tomchick D R, Kovrigin E, Arac D, Machius M, Sudhof T C and Rizo J 2002 Three-dimensional structure of the complexin/SNARE complex *Neuron* **33** 397–409

[158] Tawfik B, Martins J S, Houy S, Imig C, Pinheiro P S, Wojcik S M, Brose N, Cooper B H and Sorensen J B 2021 Synaptotagmin-7 places dense-core vesicles at the cell membrane to promote Munc13-2- and Ca(2+)-dependent priming *eLife* **10** e64527

[159] Zhou Q *et al* 2015 Architecture of the synaptotagmin-SNARE machinery for neuronal exocytosis *Nature* **525** 62–7

[160] Zhou Q, Zhou P, Wang A L, Wu D, Zhao M, Sudhof T C and Brunger A T 2017 The primed SNARE-complexin-synaptotagmin complex for neuronal exocytosis *Nature* **548** 420–5

[161] Brewer K D *et al* 2015 Dynamic binding mode of a synaptotagmin-1-SNARE complex in solution *Nat. Struct. Mol. Biol.* **22** 555–64

[162] Park Y *et al* 2015 Synaptotagmin-1 binds to PIP(2)-containing membrane but not to SNAREs at physiological ionic strength *Nat. Struct. Mol. Biol.* **22** 815–23

[163] Voleti R, Jaczynska K and Rizo J 2020 Ca(2+)-dependent release of synaptotagmin-1 from the SNARE complex on phosphatidylinositol 4,5-bisphosphate-containing membranes *eLife* **9** e57154

[164] Rickman C, Archer D A, Meunier F A, Craxton M, Fukuda M, Burgoyne R D and Davletov B 2004 Synaptotagmin interaction with the syntaxin/SNAP-25 dimer is mediated by an evolutionarily conserved motif and is sensitive to inositol hexakisphosphate *J. Biol. Chem.* **279** 12574–9

[165] Bai J, Wang C T, Richards D A, Jackson M B and Chapman E R 2004 Fusion pore dynamics are regulated by synaptotagmin*t-SNARE interactions *Neuron* **41** 929–42

[166] Ubach J, Lao Y, Fernandez I, Arac D, Sudhof T C and Rizo J 2001 The C2B domain of synaptotagmin I is a Ca^{2+}-binding module *Biochemistry* **40** 5854–60

[167] Hernandez J M, Stein A, Behrmann E, Riedel D, Cypionka A, Farsi Z, Walla P J, Raunser S and Jahn R 2012 Membrane fusion intermediates via directional and full assembly of the SNARE complex *Science* **336** 1581–4

[168] Shin J, Lou X, Kweon D H and Shin Y K 2014 Multiple conformations of a single SNAREpin between two nanodisc membranes reveal diverse pre-fusion states *Biochem. J.* **459** 95–102

[169] Chernomordik L V and Kozlov M M 2003 Protein–lipid interplay in fusion and fission of biological membranes *Annu. Rev. Biochem.* **72** 175–207

[170] Giraudo C G, Hu C, You D, Slovic A M, Mosharov E V, Sulzer D, Melia T J and Rothman J E 2005 SNAREs can promote complete fusion and hemifusion as alternative outcomes *J. Biol. Chem.* **170** 249–60

[171] Abdulreda M H, Bhalla A, Chapman E R and Moy V T 2008 Atomic force microscope spectroscopy reveals a hemifusion intermediate during soluble N-ethylmaleimide-sensitive factor-attachment protein receptors-mediated membrane fusion *Biophys. J.* **94** 648–55

[172] Wong J L, Koppel D E, Cowan A E and Wessel G M 2007 Membrane hemifusion is a stable intermediate of exocytosis *Dev. Cell* **12** 653–9

[173] Lu X, Zhang F, McNew J A and Shin Y K 2005 Membrane fusion induced by neuronal SNAREs transits through hemifusion *J. Biol. Chem.* **280** 30538–41

[174] Reese C, Heise F and Mayer A 2005 Trans-SNARE pairing can precede a hemifusion intermediate in intracellular membrane fusion *Nature* **436** 410–4

[175] Bruns D and Jahn R 1995 Real-time measurement of transmitter release from single synaptic vesicles *Nature* **377** 62–5

[176] Montecucco C, Schiavo G and Pantano S 2005 SNARE complexes and neuroexocytosis: how many, how close? *Trends Biochem. Sci.* **30** 367–72

[177] Megighian A, Zordan M, Pantano S, Scorzeto M, Rigoni M, Zanini D, Rossetto O and Montecucco C 2013 Evidence for a radial SNARE super-complex mediating neurotransmitter release at the *Drosophila* neuromuscular junction *J. Cell Sci.* **126** 3134–40

[178] Mohrmann R, de Wit H, Verhage M, Neher E and Sorensen J B 2010 Fast vesicle fusion in living cells requires at least three SNARE complexes *Science* **330** 502–5

[179] Han X, Wang C T, Bai J, Chapman E R and Jackson M B 2004 Transmembrane segments of syntaxin line the fusion pore of Ca^{2+}-triggered exocytosis *Science* **304** 289–92

[180] Shi L, Shen Q T, Kiel A, Wang J, Wang H W, Melia T J, Rothman J E and Pincet F 2012 SNARE proteins: one to fuse and three to keep the nascent fusion pore open *Science* **335** 1355–9

[181] Hua Y and Scheller R H 2001 Three SNARE complexes cooperate to mediate membrane fusion *Proc. Natl Acad. Sci. USA* **98** 8065–70

[182] Sinha R, Ahmed S, Jahn R and Klingauf J 2011 Two synaptobrevin molecules are sufficient for vesicle fusion in central nervous system synapses *Proc. Natl Acad. Sci. USA* **108** 14318–23

[183] Arancillo M *et al* 2013 Titration of Syntaxin1 in mammalian synapses reveals multiple roles in vesicle docking, priming, and release probability *J. Neurosci.* **33** 16698–714

[184] Rothman J E, Krishnakumar S S, Grushin K and Pincet F 2017 Hypothesis – buttressed rings assemble, clamp, and release SNAREpins for synaptic transmission *FEBS Lett.* **591** 3459–80

[185] Tagliatti E, Bello O D, Mendonca P R F, Kotzadimitriou D, Nicholson E, Coleman J, Timofeeva Y, Rothman J E, Krishnakumar S S and Volynski K E 2020 Synaptotagmin 1 oligomers clamp and regulate different modes of neurotransmitter release *Proc. Natl. Acad. Sci. USA* **117** 3819–27

[186] Adams D J, Arthur C P and Stowell M H 2015 Architecture of the synaptophysin/synaptobrevin complex: structural evidence for an entropic clustering function at the synapse *Sci. Rep.* **5** 13659
[187] Tokumaru H, Umayahara K, Pellegrini L L, Ishizuka T, Saisu H, Betz H, Augustine G J and Abe T 2001 SNARE complex oligomerization by synaphin/complexin is essential for synaptic vesicle exocytosis *Cell* **104** 421–32
[188] Wiederhold K, Kloepper T H, Walter A M, Stein A, Kienle N, Sorensen J B and Fasshauer D 2010 A coiled coil trigger site is essential for rapid binding of synaptobrevin to the SNARE acceptor complex *J. Biol. Chem.* **285** 21549–59
[189] Francois-Martin C, Rothman J E and Pincet F 2017 Low energy cost for optimal speed and control of membrane fusion *Proc. Natl Acad. Sci. USA* **114** 1238–41
[190] Li F, Pincet F, Perez E, Eng W S, Melia T J, Rothman J E and Tareste D 2007 Energetics and dynamics of SNAREpin folding across lipid bilayers *Nat. Struct. Mol. Biol.* **14** 890–6
[191] Wang C T, Grishanin R, Earles C A, Chang P Y, Martin T F, Chapman E R and Jackson M B 2001 Synaptotagmin modulation of fusion pore kinetics in regulated exocytosis of dense-core vesicles *Science* **294** 1111–5
[192] Wang C T, Bai J, Chang P Y, Chapman E R and Jackson M B 2006 Synaptotagmin-Ca^{2+} triggers two sequential steps in regulated exocytosis in rat PC12 cells: fusion pore opening and fusion pore dilation *J. Physiol.* **570** 295–307
[193] McNew J A, Weber T, Engelman D M, Sollner T H and Rothman J E 1999 The length of the flexible SNAREpin juxtamembrane region is a critical determinant of SNARE-dependent fusion *Mol. Cell* **4** 415–21
[194] Knecht V and Grubmuller H 2003 Mechanical coupling via the membrane fusion SNARE protein syntaxin 1A: a molecular dynamics study *Biophys. J.* **84** 1527–47
[195] Kweon D H, Kim C S and Shin Y K 2003 Insertion of the membrane-proximal region of the neuronal SNARE coiled coil into the membrane *J. Biol. Chem.* **278** 12367–73
[196] Ellena J F, Liang B, Wiktor M, Stein A, Cafiso D S, Jahn R and Tamm L K 2009 Dynamic structure of lipid-bound synaptobrevin suggests a nucleation-propagation mechanism for trans-SNARE complex formation *Proc. Natl Acad. Sci. USA* **106** 20306–11
[197] Wu Z, Auclair S M, Bello O, Vennekate W, Dudzinski N R, Krishnakumar S S and Karatekin E 2016 Nanodisc-cell fusion: control of fusion pore nucleation and lifetimes by SNARE protein transmembrane domains *Sci. Rep.* **6** 27287
[198] Wu Z, Thiyagarajan S, O'Shaughnessy B and Karatekin E 2017 Regulation of exocytotic fusion pores by SNARE protein transmembrane domains *Front. Mol. Neurosci.* **10** 315
[199] Kesavan J, Borisovska M and Bruns D 2007 v-SNARE actions during Ca(2+)-triggered exocytosis *Cell* **131** 351–63
[200] Guzman R E, Schwarz Y N, Rettig J and Bruns D 2010 SNARE force synchronizes synaptic vesicle fusion and controls the kinetics of quantal synaptic transmission *J. Neurosci.* **30** 10272–81
[201] Bretou M, Anne C and Darchen F 2008 A fast mode of membrane fusion dependent on tight SNARE zippering *J. Neurosci.* **28** 8470–6
[202] Zhou P, Bacaj T, Yang X, Pang Z P and Sudhof T C 2013 Lipid-anchored SNAREs lacking transmembrane regions fully support membrane fusion during neurotransmitter release *Neuron* **80** 470–83
[203] Fang Q, Zhao Y and Lindau M 2013 Juxtamembrane tryptophans of synaptobrevin 2 control the process of membrane fusion *FEBS Lett.* **587** 67–72

[204] Maximov A, Tang J, Yang X, Pang Z P and Sudhof T C 2009 Complexin controls the force transfer from SNARE complexes to membranes in fusion *Science* **323** 516–21

[205] Gil A, Gutierrez L M, Carrasco-Serrano C, Alonso M T, Viniegra S and Criado M 2002 Modifications in the C terminus of the synaptosome-associated protein of 25 kDa (SNAP-25) and in the complementary region of synaptobrevin affect the final steps of exocytosis *J. Biol. Chem.* **277** 9904–10

[206] Criado M, Gil A, Viniegra S and Gutierrez L M 1999 A single amino acid near the C terminus of the synaptosomeassociated protein of 25 kDa (SNAP-25) is essential for exocytosis in chromaffin cells *Proc. Natl Acad. Sci. USA* **96** 7256–61

[207] McNew J A, Weber T, Parlati F, Johnston R J, Melia T J, Sollner T H and Rothman J E 2000 Close is not enough: SNARE-dependent membrane fusion requires an active mechanism that transduces force to membrane anchors *J. Cell Biol.* **150** 105–17

[208] Fdez E, Martinez-Salvador M, Beard M, Woodman P and Hilfiker S 2010 Transmembrane-domain determinants for SNARE-mediated membrane fusion *J. Cell Sci.* **123** 2473–80

[209] Laage R and Langosch D 1997 Dimerization of the synaptic vesicle protein synaptobrevin (vesicle-associated membrane protein) II depends on specific residues within the trans-membrane segment *Eur. J. Biochem.* **249** 540–6

[210] Ngatchou A N, Kisler K, Fang Q, Walter A M, Zhao Y, Bruns D, Sorensen J B and Lindau M 2010 Role of the synaptobrevin C terminus in fusion pore formation *Proc. Natl Acad. Sci. USA* **107** 18463–8

[211] Markvoort A J and Marrink S J 2011 Lipid acrobatics in the membrane fusion arena *Curr. Top. Membr.* **68** 259–94

[212] Risselada H J, Kutzner C and Grubmuller H 2011 Caught in the act: visualization of SNARE-mediated fusion events in molecular detail *ChemBioChem* **12** 1049–55

[213] Lindau M, Hall B A, Chetwynd A, Beckstein O and Sansom M S 2012 Coarse-grain simulations reveal movement of the synaptobrevin C-terminus in response to piconewton forces *Biophys. J.* **103** 959–69

[214] Sharma S and Lindau M 2018 Molecular mechanism of fusion pore formation driven by the neuronal SNARE complex *Proc. Natl Acad. Sci. USA* **115** 12751–6

[215] McDargh Z A, Polley A and O'Shaughnessy B 2018 SNARE-mediated membrane fusion is a two-stage process driven by entropic forces *FEBS Lett.* **592** 3504–15

[216] Jackson M B 2010 SNARE complex zipping as a driving force in the dilation of proteinaceous fusion pores *J. Membrane Biol.* **235** 89–100

[217] Milosevic I, Sorensen J B, Lang T, Krauss M, Nagy G, Haucke V, Jahn R and Neher E 2005 Plasmalemmal phosphatidylinositol-4,5-bisphosphate level regulates the releasable vesicle pool size in chromaffin cells *J. Neurosci.* **25** 2557–65

[218] Ashery U, Bielopolski N, Barak B and Yizhar O 2009 Friends and foes in synaptic transmission: the role of tomosyn in vesicle priming *Neurosci.* **32** 275–82

[219] Kondratiuk I *et al* 2020 PI(4,5)P2-dependent regulation of exocytosis by amisyn, the vertebrate-specific competitor of synaptobrevin 2 *Proc. Natl Acad. Sci. USA* **117** 13468–79

[220] Yizhar O, Lipstein N, Gladycheva S E, Matti U, Ernst S A, Rettig J, Stuenkel E L and Ashery U 2007 Multiple functional domains are involved in tomosyn regulation of exocytosis *J. Neurochem.* **103** 604–16

[221] Ben-Simon Y, Rodenas-Ruano A, Alvina K, Lam A D, Stuenkel E L, Castillo P E and Ashery U 2015 A combined optogenetic-knockdown strategy reveals a major role of tomosyn in mossy fiber synaptic plasticity *Cell Rep.* **12** 396–404

[222] Geerts C J, Mancini R, Chen N, Koopmans F T W, Li K W, Smit A B, van Weering J R T, Verhage M and Groffen A J A 2017 Tomosyn associates with secretory vesicles in neurons through its N- and C-terminal domains *PLoS One* **12** e0180912

[223] Scales S J, Hesser B A, Masuda E S and Scheller R H 2002 Amisyn, a novel syntaxin-binding protein that may regulate SNARE complex assembly *J. Biol. Chem.* **277** 28271–9

[224] Constable J R, Graham M E, Morgan A and Burgoyne R D 2005 Amisyn regulates exocytosis and fusion pore stability by both syntaxin-dependent and syntaxin-independent mechanisms *J. Biol. Chem.* **280** 31615–23

[225] Sollner T, Bennett M K, Whiteheart S W, Scheller R H and Rothman J E 1993 A protein assembly-disassembly pathway *in vitro* that may correspond to sequential steps of synaptic vesicle docking, activation, and fusion *Cell* **75** 409–18

[226] Clary D O and Rothman J E 1990 Purification of three related peripheral membrane proteins needed for vesicular transport *J. Biol. Chem.* **265** 10109–17

[227] Whiteheart S W, Schraw T and Matveeva E A 2001 N-ethylmaleimide sensitive factor (NSF) structure and function *Int. Rev. Cytol.* **207** 71–112

[228] Ryu J K, Jahn R and Yoon T Y 2016 Review: progresses in understanding N-ethylmaleimide sensitive factor (NSF) mediated disassembly of SNARE complexes *Biopolymers* **105** 518–31

[229] Littleton J T, Barnard R J, Titus S A, Slind J, Chapman E R and Ganetzky B 2001 SNARE-complex disassembly by NSF follows synaptic-vesicle fusion *Proc. Natl Acad. Sci. USA* **98** 12233–8

[230] Littleton J T, Chapman E R, Kreber R, Garment M B, Carlson S D and Ganetzky B 1998 Temperature-sensitive paralytic mutations demonstrate that synaptic exocytosis requires SNARE complex assembly and disassembly *Neuron* **21** 401–13

[231] Burgalossi A *et al* 2010 SNARE protein recycling by alphaSNAP and betaSNAP supports synaptic vesicle priming *Neuron* **68** 473–87

[232] Xu T, Ashery U, Burgoyne R D and Neher E 1999 Early requirement for alpha-SNAP and NSF in the secretory cascade in chromaffin cells *EMBO J.* **18** 3293–304

[233] Banerjee A, Barry V A, DasGupta B R and Martin T F 1996 N-Ethylmaleimide-sensitive factor acts at a prefusion ATP-dependent step in Ca^{2+}-activated exocytosis *J. Biol. Chem.* **271** 20223–6

[234] Kuner T, Li Y, Gee K R, Bonewald L F and Augustine G J 2008 Photolysis of a caged peptide reveals rapid action of N-ethylmaleimide sensitive factor before neurotransmitter release *Proc. Natl Acad. Sci. USA* **105** 347–52

[235] Weber T, Parlati F, McNew J A, Johnston R J, Westermann B, Sollner T H and Rothman J E 2000 SNAREpins are functionally resistant to disruption by NSF and alphaSNAP *J. Cell Biol.* **149** 1063–72

[236] Bronk P, Deak F, Wilson M C, Liu X, Sudhof T C and Kavalali E T 2007 Differential effects of SNAP-25 deletion on Ca^{2+}-dependent and Ca^{2+}-independent neurotransmission *J. Neurophysiol.* **98** 794–806

[237] Delgado-Martinez I, Nehring R B and Sorensen J B 2007 Differential abilities of SNAP-25 homologs to support neuronal function *J. Neurosci.* **27** 9380–91

[238] Deitcher D L, Ueda A, Stewart B A, Burgess R W, Kidokoro Y and Schwarz T L 1998 Distinct requirements for evoked and spontaneous release of neurotransmitter are revealed by mutations in the *Drosophila* gene neuronal-synaptobrevin *J. Neurosci.* **18** 2028–39

[239] Deak F, Shin O H, Kavalali E T and Sudhof T C 2006 Structural determinants of synaptobrevin 2 function in synaptic vesicle fusion *J. Neurosci.* **26** 6668–76

[240] Takamori S *et al* 2006 Molecular anatomy of a trafficking organelle *Cell* **127** 831–46

[241] Antonin W, Riedel D and von Mollard G F 2000 The SNARE Vti1a-beta is localized to small synaptic vesicles and participates in a novel SNARE complex *J. Neurosci.* **20** 5724–32

[242] Ramirez D M, Khvotchev M, Trauterman B and Kavalali E T 2012 Vti1a identifies a vesicle pool that preferentially recycles at rest and maintains spontaneous neurotransmission *Neuron* **73** 121–34

[243] Raingo J *et al* 2012 VAMP4 directs synaptic vesicles to a pool that selectively maintains asynchronous neurotransmission *Nat. Neurosci.* **15** 738–45

[244] Stewart B A, Mohtashami M, Trimble W S and Boulianne G L 2000 SNARE proteins contribute to calcium cooperativity of synaptic transmission *Proc. Natl Acad. Sci. USA* **97** 13955–60

[245] Hefft S and Jonas P 2005 Asynchronous GABA release generates long-lasting inhibition at a hippocampal interneuron-principal neuron synapse *Nat. Neurosci.* **8** 1319–28

[246] Nyenhuis S B, Karandikar N, Kiessling V, Kreutzberger A J B, Thapa A, Liang B, Tamm L K and Cafiso D S 2021 Conserved arginine residues in synaptotagmin 1 regulate fusion pore expansion through membrane contact *Nat. Commun.* **12** 761

Chapter 4

Secretory granule motions adjacent to the plasma membrane and granule membrane protein mobility: implications for exocytosis

Ronald W Holz and Daniel Axelrod

This chapter focuses on studies in which we applied quantitative techniques to analyze total internal fluorescence microscopy (TIRFM) movies of granules and the plasma membrane of secreting cells. TIRFM is ideally suited to investigate the behavior of granules in the critical environment adjacent to the plasma membrane before fusion and to determine the steps that enable granules to undergo Ca^{2+}-dependent exocytosis. At an almost molecular scale, TIRFM revealed a highly unusual environment immediately adjacent to the plasma membrane from which granules undergo exocytosis. Translational and rotational motions are highly restricted, suggesting that granules are tethered and/or caged. Granules rarely move further than their diameter over many minutes, but rather continuously undergo jittering motions across tens of nanometers, much less than their diameter (300 nm). ATP and Ca^{2+} both tend to increase the jittering motion. Remarkably, granules that are destined to undergo fusion specifically *increase* their motion by two to three times and tend to travel tens of nanometers to new areas that they did not visit, or rarely visited, 100 ms before fusion. In additional, approximately 20% of granules that fused with the plasma membrane were not evident in the evanescent field within 300 ms and sometimes were not even observed before fusion, suggesting that they moved several hundreds of nanometers closer to the plasma membrane immediately before fusion. These granule motions before fusion indicate that *trans*-SNARE and other interactions that catalyze fusion can be initiated within a hundred milliseconds of fusion. TIRFM also permitted the submicroscopic adaptation of a fluorescence recovery after photobleaching (FRAP) technique to measure the mobility of granule membrane proteins of individual secretory granules. These analyses raise the possibility that the diffusion characteristics of VAMP and synaptotagmin could be factors that influence the rate of exocytosis and fusion pore expansion.

doi:10.1088/978-0-7503-3771-7ch4

After emerging from the *trans*-Golgi network, protein-containing secretory granules mature over many hours, during which time proteins are concentrated and processed, and low-molecular-weight factors and neurotransmitters are accumulated. Mature granules are typically maintained for days in chromaffin cells, an intensively studied neuroendocrine cell that is the subject of this chapter. There are 20,000–30,000 secretory granules (chromaffin granules) per cell, only ~500 of which are within a very close approximation of the plasma membrane as determined by quantitative electron microscopy of cryofixed, cultured chromaffin cells (Plattner *et al* 1997). It has been difficult to study the local environment and the movement of granules immediately before fusion. Electron microscopy gives a static view of this critical interface. Secretion measurements probe the final events of fusion and the release of granule contents. Only inferences can be made about the preceding steps, including the freedom or restriction of granule motion near the membrane before secretion. TIRFM (Axelrod 1981) is ideally suited to investigating the behavior of granules in the critical environment in which they reside before fusion and to determining the steps that enable granules to undergo Ca^{2+}-dependent exocytosis. In TIRFM, excitation light incident to a glass coverslip at greater than the critical angle (defined by Snell's law, figure 4.1(A)) produces exponentially decaying excitation (the evanescent field) with a decay constant that, under most experimental conditions, is 60–120 nm. Granules appear brighter when closer to the glass interface. Because of the short decay distance of the evanescent field, granules imaged by the evanescent field are within tens to hundreds of nanometers (significantly less than the 300 nm granule diameter) away from the plasma membrane (figure 4.1(B)).

This chapter will mainly focus on our work in which we applied quantitative techniques to analyze TIRFM movies of granules and the plasma membranes of secreting cells. We chose to work with highly differentiated, primary, regulated

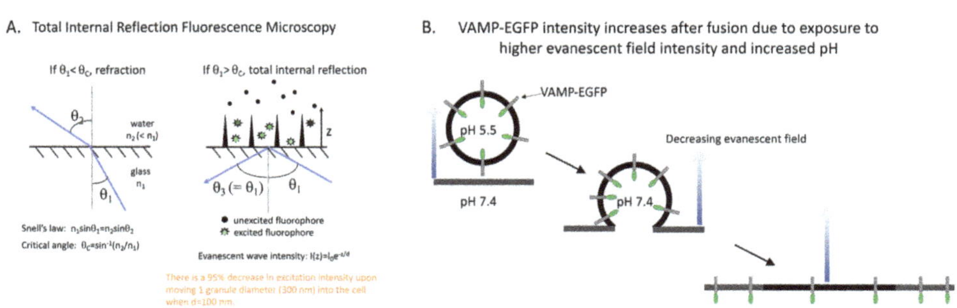

Figure 4.1. TIRFM and detection of exocytosis. (A) Incident light is refracted at a glass/water interface according to Snell's law. When the incident angle Θ_1 is greater than the critical angle, the incident light is totally reflected and an evanescent field results that decays exponentially into the aqueous phase. (B) A secretory granule membrane is labeled with VAMP-EGFP with EGFP in the lumen. The low pH of the granule lumen partially quenches the fluorescence. Upon fusion, the fluorophore dequenches because of the increase in pH. An even larger increase in fluorescence occurs because some VAMP-EGFP molecules are exposed to a greater evancescent field intensity when the granule membrane partially or fully flattens into the plasma membrane and VAMP-EGFP diffuses into the plasma membrane. The total intensity increases by an average of about nine times upon fusion (Allersma *et al* 2004).

secretory cells, bovine adrenal medullary chromaffin cells, rather than with a continuous cell line or embryonic cells. We thereby avoided imaging granule motions that reflect the burden of continuously providing new secretory granules to continually multiplying (Lang *et al* 2000, Ivarsson *et al* 2004, Huet *et al* 2006) or maturing cells (Toonen *et al* 2006). The characteristics of small motions and the rare appearance of new granules in the evanescent field of adult chromaffin cells (described below) are similar to those of insulin granules in healthy rat and human pancreatic islet beta cells (Michael *et al* 2007). Experiments were performed at least four days after transfection with plasmids encoding fluorescent granule proteins, thereby allowing granule visualization that reflected the population of endogenous granules. Indeed, the density of granules visualized by TIRFM was $0.76/\mu m^2$ (Johns *et al* 2001), similar to the density of granules closely opposed to the plasma membrane as detected by electron microscopy ($0.55–0.78\ \mu m^2$) (Plattner *et al* 1997).

One of the motivations for the studies was the possibility that the motion of granules adjacent to the plasma membrane before fusion might provide insights into the fusion reaction itself. Granules must move into intimate contact with the plasma membrane to enable granule membrane (VAMP2 and synaptotagmin) and plasma membrane proteins (syntaxin and SNAP25) to engage with cytosolic proteins (complexin, Munc18, and Munc13) and initiate the fusion reaction (Sudhof 2013, Zhou *et al* 2017). It is often assumed that the engagement ('docking') is relatively long-lived, reflecting a metastable state awaiting the Ca^{+2} trigger to initiate fusion. The high-spatial-resolution and high-temporal-resolution measurements of granule motions immediately before fusion that are discussed below indicate that the interactions can be short (less than 100 ms), suggesting that there can be a rapid progression of granule and plasma membrane protein interactions leading to fusion.

Granules visualized by TIRFM are in an unusual environment in which translational and rotational motions are highly restricted. We initially imaged ANP-GFP-labeled granules (Johns *et al* 2001). In subsequent granule motion studies, granules were also labeled with VAMP2-EGFP (Allersma *et al* 2004, 2006, Degtyar *et al* 2007). Granule tracking software was developed that enabled analysis of many thousands of motions. Granules imaged with TIRFM have limited mobility; less than 1% move a granule diameter or more in the *xy*-plane in 4 min. However, granules undergo small jittering-type motions in both the *xy*-plane (detected by centroid analysis) and in the *z*-direction perpendicular to the glass interface/plasma membrane (detected by changing fluorescence intensity in the evanescent field). The precision of the measurements was 10 nm or less. These motions are much smaller than the diameter of a chromaffin granule, on the order of tens of nanometers or less in 0.1–0.5s (Johns *et al* 2001, Allersma *et al* 2006). The small translational motions are not compatible with free diffusion because the average travel distance increases more slowly than proportionally to the square root of the time, which would be expected for free diffusion. In other words, travel over longer distances and times is less than one would expect from diffusion theory based on the travel distances observed through shorter times; longer distances appear to be confined.

The motions are dependent upon distance from the plasma membrane. The jittering motions (both in the *xy*-plane and the *z*-direction) of the brighter-appearing granules (on average closer to the plasma membrane, smaller apparent z-position) are many times smaller than those of dimmer-appearing granules (which, on average, are further away from the glass interface and have larger apparent z-positions) (Johns *et al* 2001, Allersma *et al* 2006) (figure 4.2). The actual distance of a granule to the plasma membrane is unknown.

It is likely that granules do not freely rotate based upon experiments that combined FRAP and TIRFM to measure the diffusion of lumenal proteins (Weiss *et al* 2014) and granule membrane proteins (including VAMP-EGFP, see below) (Abbineni *et al* 2021). A bright flash of strongly decaying evanescent field produced by total internal reflection was used to photobleach EGFP-labeled proteins. The measurements were successful in estimating diffusion coefficients because of the ability of this approach to imprint and detect a gradient of bleached, EGFP-labeled proteins over a timescale of approximately many hundreds of milliseconds on *individual* granules. The success of the method indicates that the rotational mobility of the entire granule is restricted over the timescale of the experiment (hundreds of milliseconds).

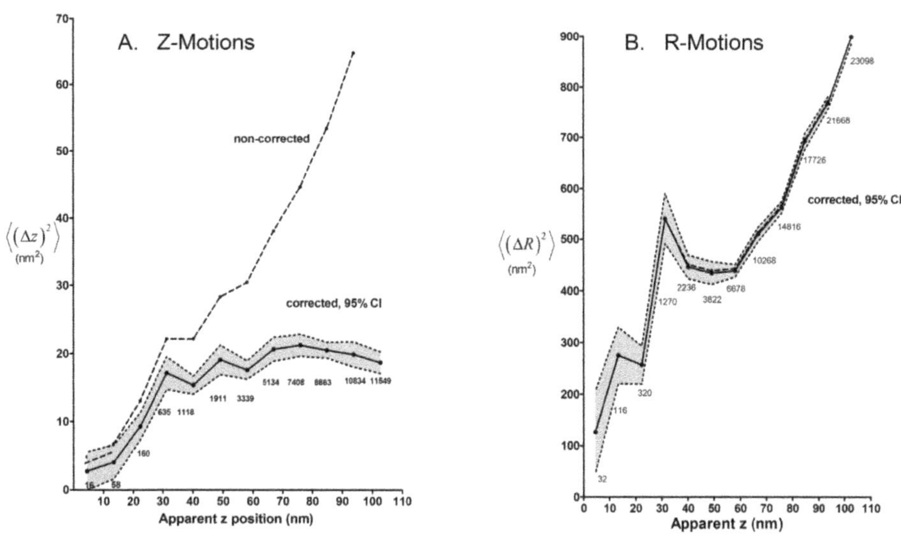

Figure 4.2. Z-motions (perpendicular to the glass interface) and R-motions (in the XY-plane) are more restricted close to the plasma membrane. The apparent z-position is based on the observed average intensity between every successive pair of z or R measurements, with the corresponding Δz or ΔR placed in the appropriate uniformly spaced z bin. The brightest granule in the cell is used to define $z = 0$. The apparent z-position is proportional to the negative of the log of the fluorescence intensity in the evanescent field. The apparent z-position and, in particular, the corresponding $\langle(\Delta z)^2\rangle$ values are underestimates because of the contribution of scattered light to the excitation. The 'corrected value' z- and R-motions take into account the part of the uncorrected values that arises from instrumental noise. The gray band around the corrected values represents the ±2×standard error (SE) (i.e. 95% confidence) values. The numbers in the graph correspond to the number of motions in each bin. The real motion for both Δz or ΔR is significant at all z; they increase with z. There is a strong tendency for the z- and R-motions to be more restricted close to the plasma membrane. Adapted from Allersma *et al* (2006). Copyright 2006 ASCB.

The low rotational mobility could reflect tethering and/or viscous drag. If the (assumed spherical) granules are not tethered and are free to rotate, i.e. they are limited only by the surrounding cytoplasmic viscosity, the characteristic time for rotation can be calculated from the relevant Einstein hydrodynamic equation and the surrounding liquid viscosity (assumed to be infinite in extent). Based on the viscosity deduced from the experimentally measured tumbling time of a small molecule (Kao *et al* 1993) within the liquid interstices of a cell's cytoplasm, the tumbling time for a 300 nm diameter spherical granule can be predicted to be 30 ms. Such rapid tumbling (relative to the bleaching duration) would not permit a photobleached gradient to be imprinted. Since we do indeed detect evidence of such a gradient, the granules must be tumbling much more slowly, if at all. It is possible that the effective local viscosity is greatly increased for a granule trapped in a cage barely larger than the granule itself.

It is tempting to speculate that the restricted motions of granules observed using TIRFM reflect caging or tethering by the actin cytoskeleton. It is well established that chromaffin cells as well as other secretory cells have a layer of f-actin adjacent to the plasma membrane that can be as large as several hundred nanometers and has occasional embedded secretory granules (Nakata and Hirokawa 1992). PC12 cells also have a cortical actin layer in which most granules are embedded (Lang *et al* 2000, Rudolf *et al* 2001b). Remarkably, disruption of the actin cytoskeleton using a variety of robust techniques has little effect on the highly restricted motions of chromaffin granules visualized in TIRFM. For example, granule motion remains highly restricted even after the almost complete disruption of chromaffin cell cortical actin by a protein kinase expressed by *Yersinia*, YpkA (Allersma *et al* 2006). YpkA is activated by and phosphorylates actin (Juris *et al* 2000). Similarly, latrunculin B, which binds g-actin, does not alter the motion of tracked granules. In fact, interference with myosin II function reduces granule motion, suggesting a role for myosin and actin interactions in causing granule motion (Berberian *et al* 2009, Neco *et al* 2004). The findings for PC12 cells are consistent with these findings. Disruption of cortical actin in PC12 cells does not increase granule motion (Lang *et al*, 2000). Instead, actin disruption decreases long-range motion, suggesting that granules close to the plasma membrane in PC12 cells travel on actin filaments. (Such long-range motion is rarely observed in chromaffin cells.) Similarly, the myosin Va tail region, an inhibitor of myosin Va function, disrupts the cortical localization of recently synthesized granules in PC12 cells (Rudolf *et al* 2001a, Rudolf *et al* 2001b). Thus, despite the logic of sub-plasma-lemma cortical actin being responsible for the highly restricted motion of chromaffin granules, there is little experimental evidence for the notion. The cause of the restricted motion is unknown. It should be noted that that in growth cones of nerve-growth-factor-differentiated PC12 cells, actin disruption does increase dense core granule mobility (Ng *et al* 2002).

Translational motions are regulated by ATP and Ca^{2+}, decrease significantly within tens of nanometers of the plasma membrane and partially reflect granule/plasma membrane interactions through SNARE proteins. The small jittering motions of granules are affected by their biochemical environment. We were able to track individual granules and determine changes in motion during an experiment. Increased cytosolic Ca^{2+} during nicotinic stimulation of intact chromaffin cells (Allersma *et al* 2006) increased jittering of non-fusing granules and granules about to undergo fusion (discussed below). A similar effect in which cytosolic Ca^{2+} increased dense core granule mobility was observed in the *Drosophila* neuromuscular junction (Shakiryanova *et al* 2007). (The latter effect occurred through activation of calmodulin kinase II.) Removal of cytosolic ATP in digitonin-permeabilized chromaffin cells reduced motion by 20%–30%. This latter effect is illustrated by analyzing the differences in successive TIRF images in the presence or absence of ATP (figure 4.3). The ATP-dependent motions may reflect the function of an ATP-dependent myosin motor when bound to actin (Berberian *et al* 2009). A similar effect in which ATP removal reduced granule motion was found during TIRFM imaging of PC12 cells (Lang *et al* 2000).

Motions of granules close to the membrane may, in part, reflect reversible interactions between the granule and plasma membrane through SNARE proteins. Expression of either tetanus toxin light chain (cleaves VAMP2 on secretory granule membrane) or Botulinum toxin type A (cleaves SNAP25 on the plasma membrane) reduces the time that individual tracked granules are closest to the plasma membrane (Johns *et al* 2001) (figure 4.4).

Granules tend to increase their *z*- and *xy*-motions within hundreds of milliseconds of fusion and have a tendency to move closer to the plasma membrane. As discussed above, granules in chromaffin cells tend to jitter in place for many seconds to minutes, rarely moving by more than a granule diameter. We expected that granule motion would decrease if granules engaged the plasma membrane well before fusion. Just the opposite was observed (Allersma *et al* 2006, Degtyar *et al* 2007). There was a significant *increase* in both *z* and *xy* granule motions in granules about to fuse (Allersma *et al* 2004). Some granules moved closer to the plasma membrane within a few hundred milliseconds of fusion. Approximately 20% of the fusion events of VAMP-GFP-labeled granules resulted from granules that appeared in the evanescent field within 300 ms of fusion; many were not evident in the evanescent field even 100 ms before fusion (Allersma *et al* 2004). Their post-fusion intensities (proportional to the total amount of VAMP-EGFP in the granule before fusion) (figure 4.1(B)) overlapped with those of granules that were present in the evanescent field before fusion. Thus, most of the granules that suddenly appeared would have been detected if they were present within the evanescent field for longer periods. They were likely distant from the plasma membrane where the evanescent field was weak and moved towards the plasma membrane within 300 ms of fusion. Granules not undergoing exocytosis rarely move into or out of the evanescent field (less than 1% in 300 ms). Thus, sudden motion through the entire evanescent field depth towards the plasma membrane is associated with a greater tendency of granules to fuse. Similarly, approximately 20% of VAMP-EGFP-labeled granules present in the evanescent field before fusion had much larger than expected intensity increases upon fusion, suggesting that these granules also move closer to the plasma membrane within 100 ms of fusion (Allersma *et al* 2004).

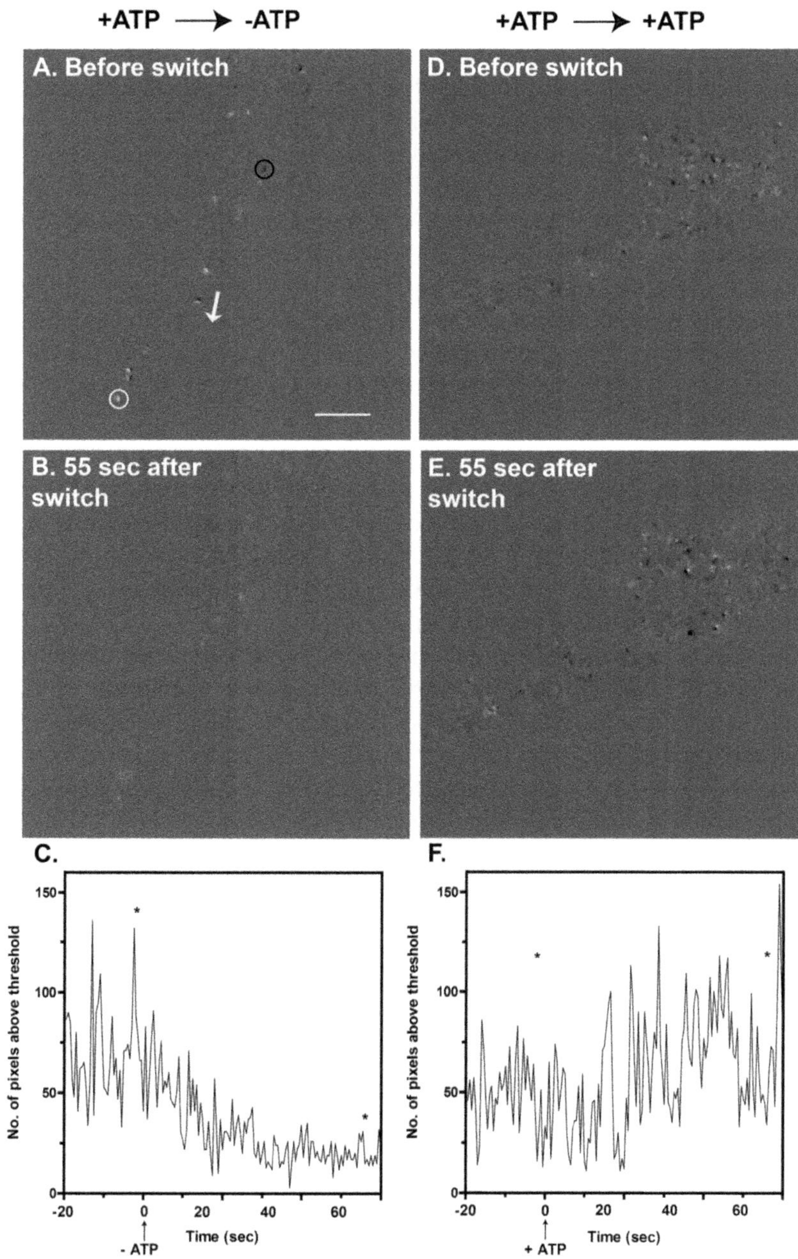

Figure 4.3. ATP-dependent granule motion: difference-image analysis of granule movement before and after the removal of ATP reveals. Chromaffin cells were labeled with transiently expressed ANP-GFP, which is targeted to the secretory granule. Cells were visualized by TIRFM and followed for 20 s before being permeabilized for 30 s with 20 μM digitonin in an NaGEP buffer containing 2 mM MgATP, incubated for an additional 60 s in the continuing presence of MgATP, and finally incubated for 60 s in the absence (A–C) or presence (D–F) of MgATP. These movies were subjected to difference-image analysis. Each image was subtracted from the subsequent image on a pixel-by-pixel basis. A dark spot indicates a negative change

(i.e. the disappearance of a granule), an example of which is circled in black in panel (a). A bright white spot (white circle in panel (a)) indicates a positive change. A black/white pair (adjacent to white arrow, panel (A)) indicates a granule moving laterally in the direction of black to white (direction indicated by white arrow). Panels (C) and (F): for each difference image, the number of pixels above a threshold (or below a negative threshold) was counted. The threshold values were large enough to exclude pixel differences which were in the background, but not so large as to exclude all pixels in the light or dark spots. The asterisks in panels (C) and (F) mark the locations of the individual difference images shown in panels (A) and (B), and (D) and (E), respectively. Scale bar, 5 μm. Adapted from Allersma *et al* (2006). Copyright 2006 ASCB.

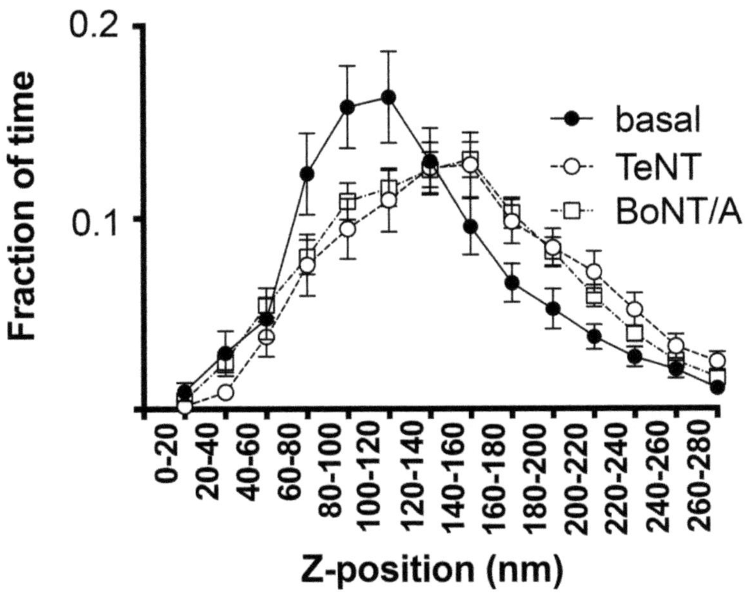

Figure 4.4. Botulinum toxins reduce the residence time of tracked granules close to the plasma membrane. For granules that can be continually tracked over at least 40 frames and which covered a total distance of at least 100 nm during this time, the fraction of the number of frames that the granule resides within a particular *z*-position window is plotted vs. *z*-position (divided into 20 nm wide bins). The brightest granule in a cell is assumed to be closest to the glass interface and the plasma membrane and is assigned a *z*-position of 0. Values are the mean ± SEM for: basal cells (57 granules from 12 cells); TeNT-expressing cells (54 granules from six cells); and BoNT/A expressing cells (149 granules from eight cells). Copyright 2001 Johns *et al.* Originally published by The Rockefeller University Press (Johns *et al* 2001).

Not only does the jittering increase, but there is also an increase in the *travel* of secretory granules parallel to the plasma membrane shortly before fusion, often to regions not previously visited (Degtyar *et al* 2007). Because neighboring, non-fusing granules analyzed at the same time do not have this tendency, the movement likely reflects important steps that make fusion imminent. The results are contrary to the notion of a necessary, long-lived docked state as a prelude to fusion. The actual data were quite dramatic. Individual *xy* motions were measured as a function of time before fusion. Since there is considerable variability in the motions of individual granules, frame-to-frame motions (10 Hz imaging) were expressed for each granule relative to the average before stimulation (figures 4.5(C) and (D)). On average, there

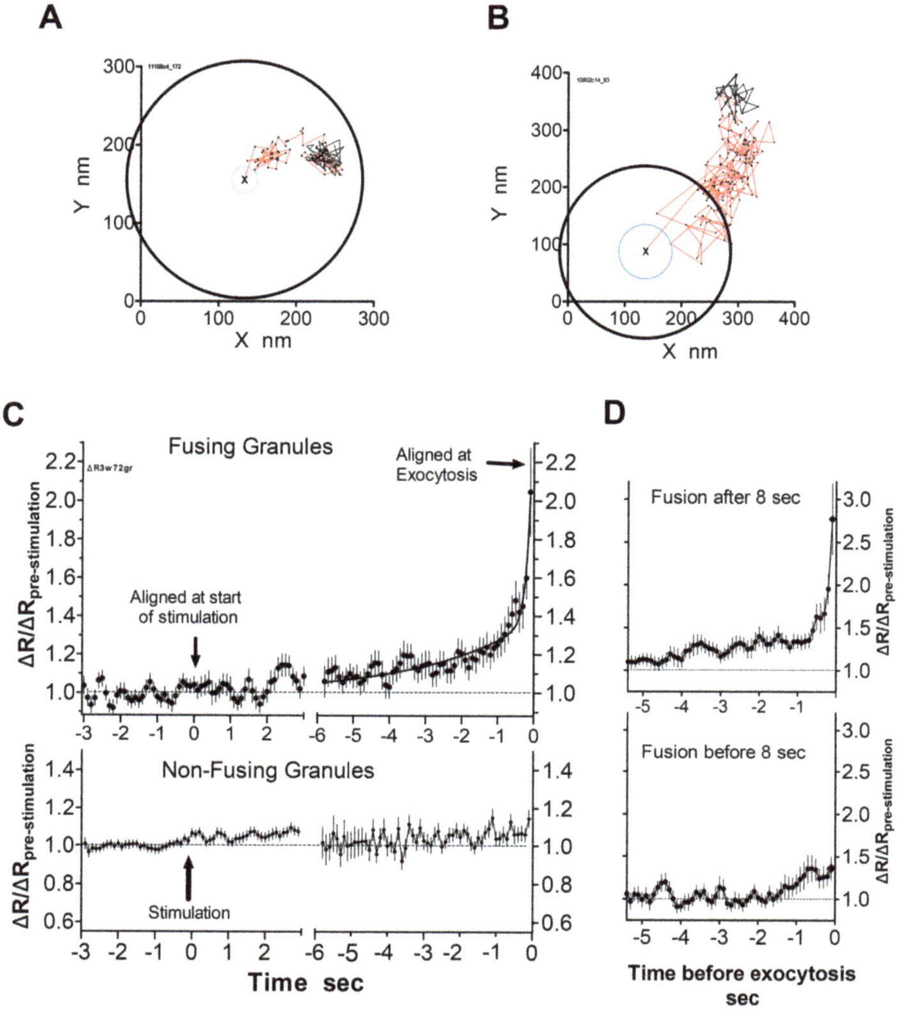

Figure 4.5. Granules increase their motions just before exocytosis. (A) and (B) Tracks of two granules before stimulation (black path) and during stimulation with DMPP (red path) until fusion (marked with X). The large black circle in each panel represents a 300 nm granule centered on the last position before fusion. The blue circles in A and B are used in the visit density analysis and have radii of 17 nm and 47 nm, respectively. (C) Top: the motion of granules destined to fuse during stimulation was expressed relative to the average motion before stimulation, $\Delta R/\Delta R_{\text{pre-stimulation}}$. The paths of 72 granules immediately before fusion were aligned at the time of fusion (right side of panel) or at the time of stimulation (left side of panel). The average ± SEM of $\Delta R_{\text{pre-stimulation}}$ was 25.3 ± 1.1 nm. The time course was subjected to a sliding three-position smoothing algorithm. The last motion before fusion was not smoothed. There was an increase in motion in the last few hundred milliseconds of fusion, and the motion during the last 100 ms of fusion was twice as large as the pre-stimulation motions. Bottom: the same analysis was performed on non-fusing granules. Non-fusing granules were selected from the same cells as those in C. Their motions were aligned during stimulation to the average time at which granules were fusing in the same cell. Non-fusing granules did not substantially increase their motion when neighboring granules were fusing. (D) Time courses of increases in $\Delta R/\Delta R_{\text{pre-stimulation}}$ for 35 granules that fused after 8 sec of stimulation (top) and for 37 granules that fused during the first 8 sec of stimulation. From Degtyar *et al* (2007),

was a two-fold increase in motion, which was most evident within 1 sec of fusion. Such a large increase in motion was specific for granules undergoing secretion. Non-fusing granules increased their motion, but to a much smaller degree (10%–20%, figure 4.5(C)). The increase in motion was greatest for the 50% of the granules that fused later during stimulation (after 8 s) (figure 4.5(D)), and the largest increases in motion occurred within 200 ms of fusion.

If the probability of fusion is independent of position and purely stochastic, then granules would tend to fuse in areas where they spend the most time. However, inspection of a large number of granule tracks, including those in figures 4.5(A) and (B), suggested that this was not the case. Not only was there an increase in motion, but there was also a tendency for the last position before fusion to be separated by tens of nanometers from positions that the granule had visited previously. This impression was verified by counting for each position the number of times in the track that the granule was found within a specified radius of that position ('visit density,' figure 4.6). The smaller the number on the abscissa, the less often the granule visited a location on the path before fusion. For granules that fused after 8 s, ~60% of fusion events occurred for granules that had suddenly moved to an area that they had rarely or never visited before (zero to two visits, bin size three) (figure 4.6(A), filled circles). In contrast, prior to the last position (29 positions), granules tended to visit areas that they had occupied before (figure 4.6(A), open circles). The decreased visit density immediately before fusion is not an artifact of the analysis, since it was not observed when the visit density analysis for the same granules was performed for 30 consecutive positions before stimulation (figure 4.6(B)). The analysis of non-fusing granules during stimulation hardly showed this effect (figure 4.6(C)). The behaviors of actual granules were compared to that of a hypothetical granule undergoing simple Brownian motion. A granule undergoing Brownian motion would routinely visit new areas (figure 4.6(D)) both in the last step and in the preceding steps. The tendency of actual granules to visit new areas in the previous 29 positions was less than that of a granule moving by Brownian motion (compare the open circles in figures 4.6(A)–(C) with those in figure 4.6(D)). These analyses suggest that granules about to fuse are highly restricted in their motion, consistent with being tethered or caged, and that immediately before fusion they tend to travel as if they have become less restricted.

The finding that granules tend to move immediately before fusion is supported by independent experiments in which membrane curvature changes upon fusion were imaged by a combination of polarized optics and TIRFM (pTIRFM) (Anantharam *et al* 2010). The lateral position of the curvature (*P*/*S*) changes immediately *after* fusion was, on average, 110 nm from the position of NYP-GFP labeled granules immediately *before* fusion (figure 4.7). These curvature changes often persist for many seconds and are modulated by dynamin (Anantharam *et al* 2010, 2011) and lumenal granule contents (Bohannon *et al* 2017, Abbineni *et al* 2018).

Although the lateral distances traveled (many tens to 100 nm) in the last second before fusion and even within 100 ms of fusion are less than a diameter of the secretory granule (300 nm), they are much greater than the interaction distance between proteins. For example, the distances are much greater than the ~10 nm

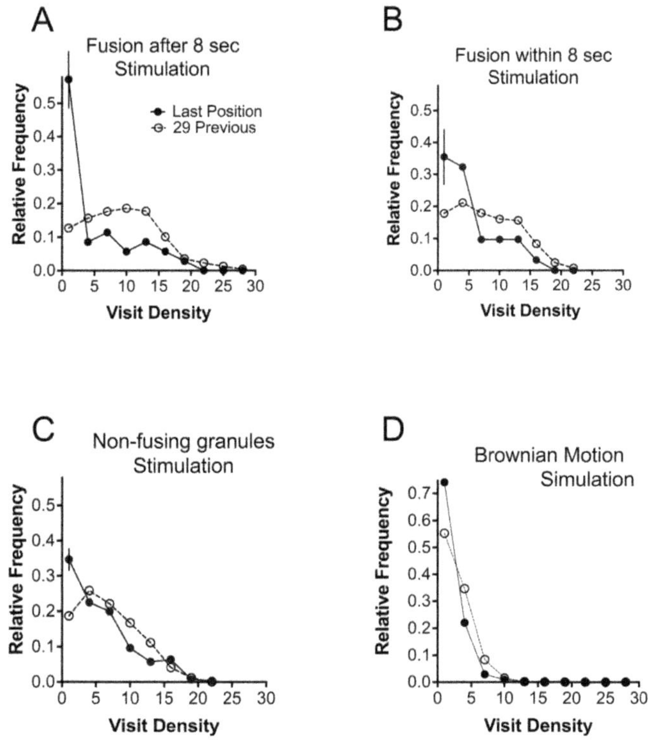

Figure 4.6. Granules tend to move to new positions immediately before exocytosis. The number of times a granule was within a small radius of each position was determined (visit density). For every granule position within 3 sec of fusion or during a 3 s interval before stimulation, the distances from all other 29 positions (panels (A)–(C)) were calculated (at a 10 Hz frame rate). The frame-to-frame standard deviation of ΔX or ΔY of each granule during the 3 s interval was used as the visit radius. In practice, the average visit radii for the different populations of granules were 18–30 nm. The visit density histogram of the last position (filled circles) and the combined histogram of the visit densities of the previous positions are presented (open circles). (A) Granules that fused after the 8 s of stimulation. (B) Analysis before stimulation for the same granules as those shown in panel (A). The histograms of the visit densities of the last positions in A and B were significantly different (35 granule tracks with medians of two and five respectively, $p<0.005$, Mann–Whittney). (C) Non-fusing granules during stimulation. Paths (280 granules) were aligned during stimulation to the average time at which neighboring granules fused in the same cell. (D) Visit density simulation for a granule undergoing Brownian motion. The probabilities were determined from 1000 simulations of a 30 position track. The visit radius, as in the analysis of actual granule travel, was set to be equal to the standard deviation of a random set of ΔX (or ΔY) picked from a Gaussian distribution. The square of the standard deviation of ΔX (or ΔY) is proportional to the diffusion coefficient of the simulated path. Adapted from Degtyar *et al* (2007), Copyright 2007 National Academy of Sciences. All rights reserved.

length of the tetra-helical SNARE complex (Hanson *et al* 1997). There are at least two possible explanations for the observation of greater travel preceding fusion. One explanation is statistical. Some granules are favorably positioned at the time of stimulation to interact with key proteins or lipids, thus allowing fusion to occur relatively quickly with little travel. These granules may be part of the rapidly releasable pool detected after flash photolysis of caged Ca^{2+} (Rettig and Neher

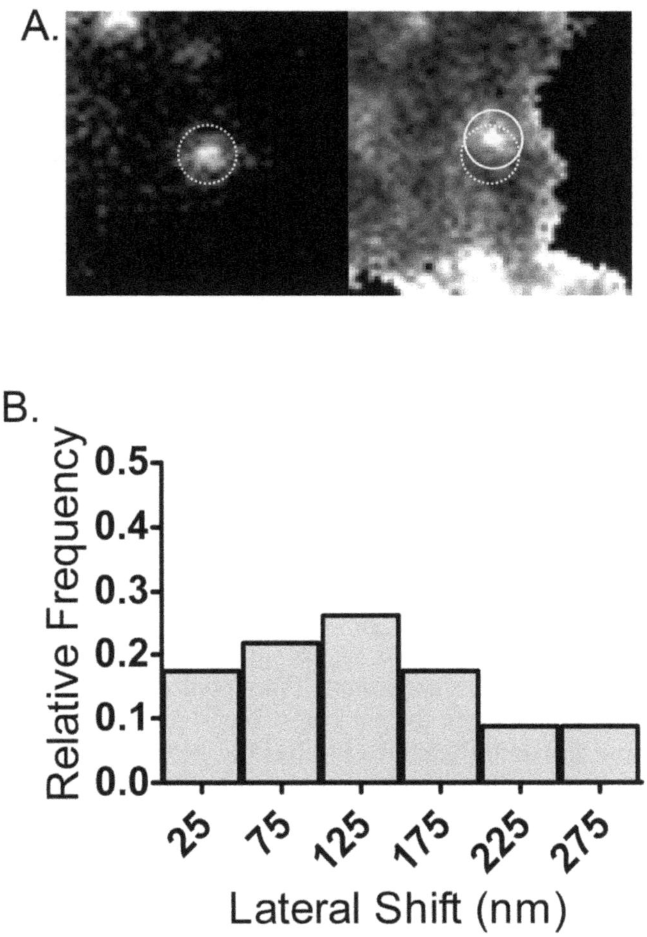

Figure 4.7. The position of the plasma membrane curvature change upon fusion is displaced from the last granule position. Imaging of the orientation of plasma membrane DiI (1,1′-dioctadecyl-3,3,3′,3′-tetramethylindocarbocyanine perchlorate) by polarized excitation using TIRF microscopy detects membrane curvature changes upon fusion. Curvature is detected by the ratio of emissions upon excitation by two different polarizations, P and S. (A) A dashed circle is centered on the last observed position of a fusing granule (left panel). A solid circle is centered on the first observed location of a *P*/*S* spot (right panel). Scale bar, 800 nm. (B) The lateral displacement from the center of mass of the fusing granule to the center of mass of the *P*/*S* puncta for 23 fusion events is plotted in a frequency histogram. The average displacement observed was 122 ± 15 nm. Copyright 2010 Anantharam *et al.* Originally published by The Rockefeller University Press (Anantharam *et al* (2010)).

2002). However, many granules are not in a favorable location for fusion. For such granules, sampling the plasma membrane in virgin territory rather than in domains that the granule had already unproductively visited increases the probability of subsequent fusion. Indeed, the greater tendency of late-fusing granules to travel to new areas in the last few hundred milliseconds before fusion supports this notion.

The increased travel could reflect the action of molecular motors (e.g. myosin Va or myosin II) or sudden granule untethering (consistent with the Brownian motion simulation, figure 4.6(D)). A second explanation is that the increased motion reflects part of the mechanism of the fusion process itself. The increased travel may reflect, in part, a well-documented Ca^{2+}-dependent priming step that precedes fusion (Bittner and Holz 1992, von Ruden and Neher 1993).

Regardless of its function and underlying mechanisms, the increased travel is strong evidence that that the formation of stable, *trans*-SNARE complexes is not a necessary prelude to fusion. Instead, these complexes can form within 100 ms before fusion. The key biochemical trigger of the fusion event may occur upstream of *trans*-SNARE complex formation (Zhao *et al* 2013), as has been postulated (Jahn and Fasshauer 2012). Complete N- to C-terminal zippering (Gao *et al* 2012) would have to start and finish within 100 ms of fusion (see Holz and Bittner 2020 for a more complete discussion).

How do these considerations apply to synaptic transmission? For fast synaptic vesicle fusion, which can begin within 100 μs of Ca^{2+} entry (Sabatini and Regehr 1996), events can only occur for those vesicles in the nerve terminal that are already engaged (docked) in *trans*-SNARE complexes with the plasma membrane. However, the fact that there is a rapid response to Ca^{2+} is not inconsistent with a synaptic vesicle having just arrived at the plasma membrane. Indeed, there is evidence that synaptic vesicles are mobile (Westphal *et al* 2008). In contrast with chromaffin granules (which are the subject of our studies), *in situ* synaptic vesicle/plasma membrane dynamics before fusion has not been investigated.

Mobility of granule membrane proteins and the kinetics of secretion. Granules must move into intimate contact with the plasma membrane in order for granule membrane and plasma membrane proteins to interact and initiate the fusion reaction. These proteins include the granule SNARE protein VAMP2 (synaptobrevin-2), the Ca^{2+} sensor synaptotagmin, and the plasma membrane acceptor complex composed of the t-SNARES, SNAP25, and syntaxin (Rizo and Südhof 2012, Jahn and Fasshauer 2012). It has not been established when these interactions occur. Close approximation of the granule and plasma membrane does not guarantee productive protein interactions, as discussed above. If granule membrane proteins are immobile, nanometer distance apposition of the two membranes would result in fusion only if, by chance, the cognate proteins are already present and properly oriented to enable interaction. On the other hand, if the secretory granule proteins are mobile, diffusion could facilitate interaction and the precise orientation of the granule and the plasma membrane would not be as critical. The short decay distance of the evanescent field (64 nm or 80 nm in the actual experiments) compared to the diameter of a chromaffin granule (~300 nm) enabled measurement of the lumenal (Weiss *et al* 2014) and membrane protein mobility (Abbineni *et al* 2021) of individual chromaffin granules by FRAP. Granule lumen and membrane components experience an excitation intensity that is many times greater when they are proximal rather than distal from the glass interface. A brief bleaching pulse imprints a gradient of fluorescent protein in individual granules, permitting the analysis of recovery. This method is not limited by standard optical resolution. Despite complication of the analysis by scattered light and reversible bleaching of the EGFP fluorophore, the method was able to estimate the mobility of VAMP2, Syt1, and Syt7 in the membranes of chromaffin granules in living cells (Abbineni *et al* 2021). We found that VAMP2-EGFP and Syt7-EGFP have a finite mobility of $D = 1.5\text{–}3 \times 10^{-10}\ \mathrm{cm^2\ s^{-1}}$. Syt1-EGFP has a mobility that is lower and not readily measurable. The bleaching characteristics of all three proteins are consistent with their being distributed over the granule membrane surface.

There is general agreement that fusion requires the formation of multiple *trans*-SNARE complexes at the fusion site, and there is evidence of at least three complexes in chromaffin cells (Mohrmann *et al* 2010). Synaptotagmin is also required for Ca^{2+}-triggered fusion. It is plausible that there are an equal number of synaptotagmin molecules and SNARE complexes at the fusion site. Assuming an initial homogeneous distribution of protein, we calculated the minimum time required for four copies of VAMP2 and Syt7 to reach an area encompassed by an early fusion pore that has a radius of 1 nm. (We speculated that four copies would be sufficient to enable both fusion pore opening and subsequent expansion.) In order to estimate the time required, we modeled the granule membrane as an infinite plane containing VAMP2 and Syt7 and modeled the fusion pore as a perfect sink that instantly and permanently traps molecules that it encounters. Because this method does not consider the depletion of the finite pool of molecules in the granule membrane, it estimates the *minimum* time required. Because the copy numbers of VAMP2 and synaptotagmin on large dense core secretory granules are unknown, a range of protein densities was investigated based upon the protein densities in synaptic vesicles (Takamori *et al* 2006) of 820–26,000 copies/μm^2 (which corresponds to 5–160 copies/synaptic vesicle). As expected, diffusion times decreased as protein abundance increased (figure 4.8(A)) and times ranged from 200 ms for the lowest protein density to 2 ms for the highest. If the densities of VAMP2 (black arrow in insert) and synaptotagmin (red arrow in main graph) in chromaffin granules were the same as in synaptic vesicles (70 and 15 copies of VAMP2 and synaptotagmin, respectively, in synaptic vesicles), the time required for four copies to be captured by a 1 nm radius pore complex would be greater than 7 ms and 50 ms, respectively (figure 4.8).

These diffusion times are comparable to measured fusion kinetics. Following a sudden elevation of intracellular Ca^{2+} in rat chromaffin cells, there is a burst of fusion events whose kinetics can be fitted to double exponentials (Heinemann *et al* 1994, Voets 2000, Chow *et al* 1992) with time constants of approximately 1 s and 100 ms at 10 μM Ca^{2+} (Voets 2000, Voets *et al* 2001). Fusion does not begin immediately, but with delays of ~10 ms and ~200 ms for the fast and slow components, respectively. It is, therefore, possible that if the formation of *trans*-SNARE complexes drives fusion, diffusion to the fusion site of VAMP2 and/or Syt could influence fusion kinetics. Furthermore, because fusion pore stability and the subsequent rate of fusion pore expansion in reconstituted fusion reactions increase with increasing numbers of SNARE complexes (Bao *et al* 2018, Wu *et al* 2017), diffusion may influence the fate of the fusion pore.

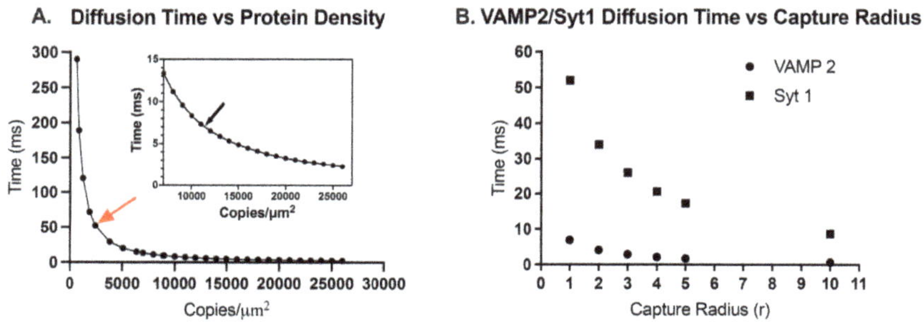

Figure 4.8. Influences of diffusion and protein abundance on the time required for proteins to reach a fusion pore of a defined radius. (A) The influence of vesicle membrane protein abundance on the time required for proteins to reach a fusion pore. Given a D of 3×10^{-10} cm^2 s^{-1}, we calculated the amount of time needed for four copies of a protein to reach a fusion pore 1 nm in radius, with protein density ranging from 820 to 26,000 copies/μm^2. The inset shows a magnified view of the 12,000–26,000 copies/μm^2 range. The red and black arrows show Syt1 and VAMP2 densities on synaptic vesicles, respectively, assuming a synaptic vesicle radius of 22 nm (Takamori *et al* 2006). (B) Time required for four copies of Syt1 or VAMP2 to arrive at a fusion pore of a defined radius. A diffusion coefficient of 3×10^{-10} cm^2 s^{-1} was used to describe the mobility of both VAMP2 and Syt7, and the concentrations of VAMP2 and Syt7 were set to the same densities as those on synaptic vesicles (equivalent to 70 copies of VAMP2 and 15 copies of Syt per synaptic vesicle). The analyses indicate that VAMP2 and synaptotagmin mobilities and abundance on the granule membrane may influence the rates of granule fusion and fusion pore expansion. Adapted from Abbineni *et al* (2021). Copyright 2021 ASCB.

4.1 Summary

1. At an almost molecular scale, TIRFM revealed a highly unusual environment immediately adjacent to the plasma membrane from which granules undergo exocytosis. Translational and rotational motions are highly restricted, suggesting that granules are tethered and/or caged. Granules rarely move by more than their diameter over many minutes. Instead, they continuously undergo jittering motions across tens of nanometers, i.e. much less than their diameter (300 nm). ATP and Ca^{2+} both tend to increase the jittering motion. These small motions decrease as granules move approximately 100 nm closer to the plasma membrane. The time spent close to the plasma membrane is reduced by cleavage of either VAMP2 on the granule membrane or SNAP25 on the plasma membrane, suggesting the influence of reversible *trans*-SNARE interactions on granule dynamics in close proximity to the plasma membrane under basal conditions.
2. Granules that are destined to undergo fusion specifically *increase* their motion by two to three times and tend to travel tens of nanometers to new areas that they have not visited, or rarely visited, in the 100 ms preceding fusion. The calculated motions and distances traveled are well above the noise level of the measurements and much greater than the interaction distances between proteins. Additionally, approximately 20% of the granules that fuse with the plasma membrane are not evident in the evanescent field 300 ms before fusion and are sometimes not even observed before fusion,

suggesting that they move several hundreds of nanometers closer to the plasma membrane immediately before fusion. Granule motions before fusion indicate that *trans*-SNARE and other interactions that catalyze fusion can be initiated within 100 ms of fusion.

3. TIRFM permitted the submicroscopic adaptation of a FRAP technique to measure the mobility of granule membrane proteins of individual secretory granules. Calculations based upon the mobilities of VAMP2 and synaptotagmin in the granule membrane suggest that the time required for a sufficient number of molecules required for fusion to arrive at a nascent fusion site could be milliseconds to tens of milliseconds. The analyses raise the possibility that the diffusion characteristics of VAMP and synaptotagmin could be factors that influence the rate of exocytosis and fusion pore expansion.

Acknowledgements

The work discussed in this chapter is the result of the efforts of many talented investigators with whom the authors have been privileged to work including Drs Mary Bittner, Arun Anantharam, Kevin Bohannon, Prabhodh Abbineni, Vadim Degtyar, Annita Ngatchou Weiss, and Laura Johns.

References

Abbineni P S, Bittner M A, Axelrod D and Holz R W 2018 Chromogranin A, the major lumenal protein in chromaffin granules, controls fusion pore expansion *J. Gen. Physiol.* **151** 118–30

Abbineni P S, Briguglio J S, Chapman E R, Holz R W and Axelrod D 2021 VAMP2 and synaptotagmin mobility in chromaffin granule membranes: implications for regulated exocytosis *Mol. Biol. Cell* **33** ar53

Allersma M W, Bittner M A, Axelrod D and Holz R W 2006 Motion matters: secretory granule motion adjacent to the plasma membrane and exocytosis *Mol. Biol. Cell* **17** 2424–38

Allersma M W, Wang L, Axelrod D and Holz R W 2004 Visualization of regulated exocytosis with a granule-membrane probe using total internal reflection microscopy *Mol. Biol. Cell* **15** 4658–68

Anantharam A, Bittner M A, Aikman R L, Stuenkel E L, Schmid S L, Axelrod D and Holz R W 2011 A new role for the dynamin GTPase in the regulation of fusion pore expansion *Mol. Biol. Cell* **22** 1907–18

Anantharam A, Onoa B, Edwards R H, Holz R W and Axelrod D 2010 Localized topological changes of the plasma membrane upon exocytosis visualized by polarized TIRFM *J. Cell Biol.* **188** 415–28

Axelrod D 1981 Cell-substrate contacts illuminated by total internal reflection fluorescence *J. Cell Biol.* **89** 141–45

Bao H, Das D, Courtney N A, Jiang Y, Briguglio J S, Lou X, Roston D, Cui Q, Chanda B and Chapman E R 2018 Dynamics and number of trans-SNARE complexes determine nascent fusion pore properties *Nature* **554** 260–63

Berberian K, Torres A J, Fang Q, Kisler K and Lindau M 2009 F-actin and myosin II accelerate catecholamine release from chromaffin granules *J. Neurosci.* **29** 863–70

Bittner M A and Holz R W 1992 Kinetic analysis of secretion from permeabilized adrenal chromaffin cells reveals distinct components *J. Biol. Chem.* **267** 16219–25

Bohannon K P, Bittner M A, Lawrence D A, Axelrod D and Holz R W 2017 Slow fusion pore expansion creates a unique reaction chamber for co-packaged cargo *J. Gen. Physiol.* **149** 921–34

Chow R H, Von Ruden L and Neher E 1992 Delay in vesicle fusion revealed by electrochemical monitoring of single secretory events in adrenal chromaffin cells *Nature* **356** 60–3

Degtyar V E, Allersma M W, Axelrod D and Holz R W 2007 Increased motion and travel, rather than stable docking, characterize the last moments before secretory granule fusion *Proc. Natl Acad. Sci. USA* **104** 15929–34

Gao Y, Zorman S, Gundersen G, Xi Z, Ma L, Sirinakis G, Rothman J E and Zhang Y 2012 Single reconstituted neuronal SNARE complexes zipper in three distinct stages *Science* **337** 1340–3

Hanson P I, Roth R, Morisaki H, Jahn R and Heuser J E 1997 Structure and conformational changes in NSF and its membrane receptor complexes visualized by quick-freeze/deep-etch electron microscopy *Cell* **90** 523–35

Heinemann C, Chow R H, Neher E and Zucker R S 1994 Kinetics of the secretory response in bovine chromaffin cells following flash photolysis of caged Ca^{2+} *Biophys. J.* **67** 2546–57

Holz R W and Bittner M A 2020 Roles for the SNAP25 linker domain in the fusion pore and a dynamic plasma membrane SNARE "acceptor" complex *J. Gen. Physiol.* **152** e202012619

Huet S, Karatekin E, Tran V S, Fanget I, Cribier S and Henry J P 2006 Analysis of transient behavior in complex trajectories: application to secretory vesicle dynamics *Biophys. J.* **91** 3542–59

Ivarsson R, Obermuller S, Rutter G A, Galvanovskis J and Renstrom E 2004 Temperature-sensitive random insulin granule diffusion is a prerequisite for recruiting granules for release *Traffic* **5** 750–62

Jahn R and Fasshauer D 2012 Molecular machines governing exocytosis of synaptic vesicles *Nature* **490** 201–7

Johns L M, Levitan E S, Shelden E S, Holz R W and Axelrod D 2001 Restriction of secretory granule motion near the plasma membrane of chromaffin cells *J. Cell Biol.* **153** 177–90

Juris S J, Rudolph A E, Huddler D, Orth K and Dixon J E 2000 A distinctive role for the Yersinia protein kinase: actin binding, kinase activation, and cytoskeleton disruption *Proc. Natl Acad. Sci. USA* **97** 9431–6

Kao H P, Abney J R and Verkman A S 1993 Determinants of the translational mobility of a small solute in cell cytoplasm *J. Cell Biol.* **120** 175–84

Lang T, Wacker I, Wunderlich I, Rohrbach A, Giese G, Soldati T and Almers W 2000 Role of actin cortex in the subplasmalemmal transport of secretory granules in PC-12 cells *Biophys. J.* **78** 2863–77

Michael D J, Xiong W, Geng X, Drain P and Chow R H 2007 Human insulin vesicle dynamics during pulsatile secretion *Diabetes* **56** 1277–88

Mohrmann R, De Wit H, Verhage M, Neher E and Sorensen J B 2010 Fast vesicle fusion in living cells requires at least three SNARE complexes *Science* **330** 502–5

Nakata T and Hirokawa N 1992 Organization of cortical cytoskeleton of cultured chromaffin cells and involvement in secretion as revealed by quick-freeze, deep-etching, and double-label immunoelectron microscopy *J. Neurosci.* **12** 2186–97

Neco P, Giner D, Viniegra S, Borges R, Villarroel A and Gutierrez L M 2004 New roles of myosin II during the vesicle transport and fusion in chromaffin cells *J. Biol. Chem.* **279** 27450–57

Ng Y K, Lu X and Levitan E S 2002 Physical mobilization of secretory vesicles facilitates neuropeptide release by nerve growth factor-differentiated PC12 cells *J. Physiol.* **542** 395–402

Plattner H, Artalejo A R and Neher E 1997 Ultrastructural organization of bovine chromaffin cell cortex-analysis by cryofixation and morphometry of aspects pertinent to exocytosis *J. Cell Biol.* **139** 1709–17

Rettig J and Neher E 2002 Emerging roles of presynaptic proteins in Ca++-triggered exocytosis *Science* **298** 781

Rizo J and Südhof T C 2012 The membrane fusion enigma: SNAREs, SM proteins, and their accomplices-guilty as charged? *Annu. Rev. Cell Dev. Biol.* **28** 279–308

Rudolf R, Kogel T, Kuznetsov S A, Salm T, Schlicker O, Hellwig A, Hammer J A and Gerdes H H 2001a Myosin Va facilitates the distribution of secretory granules in the F-actin rich cortex of PC12 cells *J. Cell Sci.* **116** 1339–48

Rudolf R, Salm T, Rustom A and Gerdes H H 2001b Dynamics of immature secretory granules: role of cytoskeletal elements during transport, cortical restriction, and F-actin-dependent tethering *Mol. Biol. Cell* **12** 1353–65

Sabatini B L and Regehr W G 1996 Timing of neurotransmission at fast synapses in the mammalian brain *Nature* **384** 170–2

Shakiryanova D, Klose M K, Zhou Y, Gu T, Deitcher D L, Atwood H L, Hewes R S and Levitan E S 2007 Presynaptic ryanodine receptor-activated calmodulin kinase II increases vesicle mobility and potentiates neuropeptide Release *J. Neurosci.* **27** 7799–806

Sudhof T C 2013 Neurotransmitter release: the last millisecond in the life of a synaptic vesicle *Neuron* **80** 675–90

Takamori S *et al* 2006 Molecular anatomy of a trafficking organelle *Cell* **127** 831–46

Toonen R F, Kochubey O, De W, Gulyas-Kovacs A, Konijnenburg B, Sorensen J B, Klingauf J and Verhage M 2006 Dissecting docking and tethering of secretory vesicles at the target membrane *EMBO J.* **25** 3725–37

Voets T 2000 Dissection of three Ca2+-dependent steps leading to secretion in chromaffin cells from mouse adrenal slices *Neuron* **28** 537–45

Voets T, Moser T, Lund P E, Chow R H, Geppert M, Sudhof T C and Neher E 2001 Intracellular calcium dependence of large dense-core vesicle exocytosis in the absence of synaptotagmin I *Proc. Natl Acad. Sci. USA* **98** 11680–5

Von Ruden L and Neher E 1993 A Ca-dependent early step in the release of catecholamines from adrenal chromaffin cells *Science* **262** 1061–5

Weiss A N, Bittner M A, Holz R W and Axelrod D 2014 Protein mobility within secretory granules *Biophys. J.* **107** 16–25

Westphal V, Rizzoli S O, Lauterbach M A, Kamin D, Jahn R and Hell S W 2008 Video-rate far-field optical nanoscopy dissects synaptic vesicle movement *Science* **320** 246–9

Wu Z, Bello O D, Thiyagarajan S, Auclair S M, Vennekate W, Krishnakumar S S, O'shaughnessy B and Karatekin E 2017 Dilation of fusion pores by crowding of SNARE proteins *eLife* **6** e22964

Zhao Y, Fang Q, Herbst A D, Berberian K N, Almers W and Lindau M 2013 Rapid structural change in synaptosomal-associated protein 25 (SNAP25) precedes the fusion of single vesicles with the plasma membrane in live chromaffin cells *Proc. Natl Acad. Sci. USA* **110** 14249–54

Zhou Q, Zhou P, Wang A L, Wu D, Zhao M, Sudhof T C and Brunger A T 2017 The primed SNARE-complexin-synaptotagmin complex for neuronal exocytosis *Nature* **548** 420–25

IOP Publishing

Exocytosis: From Molecules to Cells

Arun Anantharam and Jefferson Knight

Chapter 5

Fusion pore stability and dynamics

Meyer B Jackson

During exocytosis, secretion is initiated by the formation of a fusion pore that serves as a fluid conduit between the vesicle lumen and extracellular space. Fusion pore properties can be measured in real time by a variety of sensitive biophysical techniques. These measurements reveal the evolution of a fusion pore as it expands, contracts, and closes. These dynamic processes have important functional implications for control of the speed, extent, and composition of secretion. This chapter focuses on the diversity of fusion pore states and the dynamics of their interconversion. The investigations into these processes have illuminated the molecular drives and controls of exocytosis, and revealed the mechanisms by which fusion pores adapt to meet diverse biological challenges.

The fusion pore serves as an essential exit pathway for secretion. Its critical function makes the fusion pore a natural focus of research in exocytosis. The fusion pore has provided a window into the mechanisms by which proteins drive membrane fusion, and has provided a framework for defining intermediate states with distinct functions. Secretion begins with the opening of a fusion pore, but because the initial pore is very narrow, only small molecules can escape at a very slow rate. The nascent fusion pore can undergo dynamic transitions, which are essential for most of the biological functions of exocytosis. Fusion pores can expand, constrict, and close to shape a chemical signal and determine the biological outcome. Exocytosis can reach an endpoint of full fusion, in which a fusion pore expands to expel a vesicle's entire content, perhaps accompanied by collapse of the vesicle membrane into the plasma membrane. Alternatively, exocytosis can terminate in kiss-and-run, in which a vesicle maintains its integrity and releases a fraction of its content before the fusion pore closes.

Fusion pore researchers initially looked for general features and focused on common properties. However, it quickly became clear that fusion pores are very diverse. Not only do estimates of size vary widely, but the size can fluctuate. Fusion pores interconvert between states with different properties that enable different

doi:10.1088/978-0-7503-3771-7ch5

forms of secretion. Fusion pore dynamics defines a spectrum of potential outcomes. Exocytosis is tuned and regulated by controlling which fusion pore structures form and how they interconvert. The fusion pore should be viewed as a dynamic entity moving along a kinetic reaction pathway. Studying the trajectory of a fusion pore along this pathway illuminates many of the biological controls of secretion. This chapter considers the diversity of fusion pore states, the kinetics of interconversions, and the functional consequences.

5.1 Techniques for studying fusion pores

Different experimental techniques have provided pictures of fusion pores in living cells. A number of these techniques can measure signals arising from single vesicles, so they can be viewed as probing fusion pores at the unitary, 'single-channel' level. Unitary measurements have the advantage of directly visualizing molecular heterogeneity, which is generally obscured in measurements of ensemble averages. Single fusion pore measurements have the additional advantage that they can be connected to a molecular dimension to estimate the pore size. Different measurements focus on characteristic timescales, capturing fusion pores at particular stages in their evolution. This makes it hard to compare the fusion pore dimensions inferred by different methods. The brief survey of techniques given here will emphasize the nature of the information they provide, and how they have revealed heterogeneity and dynamics. Discussions of function, energetics, and transitions will then synthesize the results obtained with different methods and consider the broader framework.

Electrical recording. Phase-locked impedance measurements revealed stepwise increases in capacitance as single vesicles fused. Some upward steps were immediately followed by downward steps, as though a pore was flickering open and closed [1, 2]. This was reminiscent of ion-channel gating transitions and anticipated the advent of fusion pores. The complex impedance of a membrane could be resolved into in-phase and out-of-phase components, and the fusion of giant vesicles in beige mouse mast cells changed both components. With the aid of an equivalent circuit, the fusion pore conductance was then calculated from these measurements [3]. At the instant when these large vesicles fused, they produced a distinct current transient as the vesicle membrane potential discharged, and analysis in terms of the equivalent circuit yielded a time course for fusion pore expansion [4].

Of all the fusion pore measurements, conductance comes the closest to measuring the physical size. However, these estimates generally depend on assuming the pore has a cylindrical geometry, with a length that spans two lipid bilayers. Applying the formula for a cylinder to ion channels with known structures gives reasonable values for pores with diameters >1 nm, but fails badly for smaller channels [5]. The cylindrical approximation also breaks down for very large pores when high fluxes create concentration gradients around the pore. In this case, the flux scales with the pore diameter rather than the area [5–7]. Proteinaceous pores are more likely to be cylindrical, but lipid bilayer elasticity opposes the formation of sharp corners. Lipidic pores are presumed to form a shape that minimizes the elastic bending energy of a lipid bilayer [8]. Modeling of lipid bilayer elasticity indicates that lengthening a pore and reducing the meridian curvature lowers the membrane bending energy. These 'bowed' or 'teardrop' pores are approximately cylindrical [9, 10]. The cylindrical model may be a good approximation for the flux through this form of lipidic pore. It would be useful to know more about how physical forces shape the fusion pore, but experimental studies are not currently able to address this question.

The cylindrical model has probably been overused, but it is the best we have at this stage. These limitations should be borne in mind when considering pore dimensions estimated from conductance. Fusion pore conductances range from tens of picosiemens to a few nanosiemens [11], and the inferred diameters range from a few Angstroms to a few nanometers. Fusion pores associated with small synaptic-like vesicles have conductances about tenfold smaller than fusion pores associated with large endocrine vesicles [12]. The wide range of conductance values reported in the literature [11] speaks to the range of sizes fusion pores can assume. In summary, complex impedance measurements have provided insight into the dimensions of pores and revealed rich dynamic behavior. Impedance measurement is, however, not the fastest technique and cannot resolve rapid dynamic processes beyond the millisecond timescale.

Amperometry. Some secreted molecules, such as catecholamines and serotonin, oxidize upon contact with a carbon fiber electrode polarized to a degree greater the molecule's redox potential. This oxidization provides a signal for the electrochemical detection of these molecules. Carbon fiber electrodes with ~5 μm tips positioned at the surface of a cell detect catecholamine release as an amperometric current, and the fusion of a single vesicle produces a large, rapid, spike-like event [13]. Close inspection of single-vesicle spikes has revealed that they often begin with a small foot. This prespike foot represents the opening of a fusion pore, and ends with a spike as the fusion pore abruptly expands [14, 15]. The prespike foot is probably the earliest measurable signal of a fusion pore, and thus provides the clearest view of the nascent fusion pore. Mutations in a variety of proteins alter their permeability and stability [16]. Mutations in the soluble *N*-ethylmaleimide-sensitive factor attachment receptor (SNARE) protein transmembrane domains (TMDs) alter fusion pore flux in a manner suggesting that these domains line the nascent fusion pore [17–19] (see figure 5.3).

After the onset of a spike, the impact of a fusion pore becomes more difficult to discern. The time course of the spike cannot be accounted for by diffusion of the escaping catecholamine to the electrode surface [20]. Dissociation from the protein matrix within a dense-core vesicle has been considered to be a potential factor in slowing catecholamine loss [21, 22]. However, the various contributions of diffusion, dissociation, and fusion pore dynamics have been difficult to resolve. A method of analysis of amperometry spikes has been developed that offers a window into the dynamics of pore permeability [23]. One begins by integrating the amperometric current to obtain the amount of catecholamine lost before a given time. The number of molecules remaining in the vesicle is then calculated as the initial number minus the number lost. The concentration gradient through the pore can be taken to be the intravesicular concentration, because the extracellular concentration is very small. Thus, the number of moles remaining in the vesicle divided by the vesicle volume represents the driving force for the flux. Dividing amperometric current by this driving force gives the fusion pore permeability over the vesicle volume. This permeability can be calculated over the course of a spike to reveal rapid fusion pore dynamics (permeability/volume = α/V in figure 5.1). α/V increases rapidly at the end of a prespike foot, reaches a peak, and declines, before settling into a plateau that lasts tens of milliseconds. In this way amperometry can be used to probe the dynamic behavior of a fusion pore over the course of the spike.

Fluorescence destaining. Fluorescent dye and fluorescent cargo can label vesicles, and when a vesicle fuses, the fluorescence from the label declines. In synaptic vesicles the destaining of the lipid probe FM1–43 can be resolved into quantal steps corresponding to the fusion of single vesicles [24]. In a study of goldfish retinal neurons using total internal reflection fluorescence (TIRF) microscopy, vesicles destained completely in a few milliseconds, suggesting that the fusion pore expanded so rapidly that dye could escape unimpeded [25]. However, in hippocampal nerve terminals, a significant fraction of events were subquantal, and dye loss was

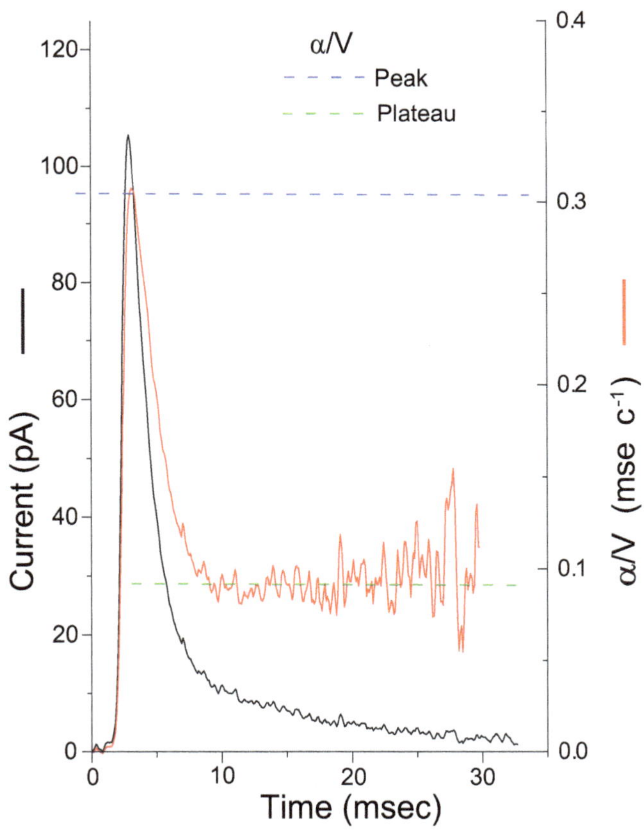

Figure 5.1. Amperometry current (black) and permeability (α/V, red). Permeability rises rapidly and then falls, settling to a stable plateau. Reproduced from [23]. Copyright 2000 Biophysical Society.

markedly slower [26, 27]. The slow events suggested dye loss by a process similar to that seen in liposomes containing ion channels. These events appeared to originate from exit through a stable fusion pore with a diameter of 1–2 nm [27]. Thus, single-vesicle fluorescence destaining measurements reveal two distinct modes of synaptic vesicle fusion, namely slow graded release through small fusion pores and rapid complete release through rapidly expanding fusion pores.

Dual fluorophore experiments permitted the simultaneous monitoring of cargo loss and membrane loss [28]. Vesicles lost the lipid label FM4–64 but retained the cargo enhanced green fluorescent protein (EGFP)-phogrin, indicating that the fusion pore closed after expanding to a lipidic state. Large hormones can remain within endocrine vesicles for seconds after the onset of fusion. This process, referred to as 'cavicapture,' indicates the presence of a narrow fusion pore that is stable for seconds [29].

Polarization TIRF. The perpendicular and in-plane components of polarization can be separated in TIRF, and the ratio of the two provides a signal closely related to the amount of lipid bilayer curvature at the site of fusion [30]. This technique revealed an increase in membrane curvature that occurred at the onset of loss of cerulean fluorescent protein-tagged neuropeptide Y (NPY). Increases prior to the onset of content loss report membrane deformations during, or in preparation of, nascent fusion pore opening [31]. Membranes became curved at the site of release within a few hundred milliseconds of the onset of cargo loss. Since high membrane curvature is a hallmark of a lipidic fusion pore, this technique links the lipidic pore to the efflux of peptide cargo.

Super-resolution microscopy. Stimulated emission depletion (STED) microscopy has been used to visualize the vesicle and plasma membrane in real time as fusion takes place [32, 33]. With a time resolution of ~30–100 ms and a spatial resolution of 65–150 nm, this technique follows the formation and expansion of omega-shaped figures and tracks the movement of plasma membrane label through the fusion pore and into the vesicle. STED microscopy reveals fusion pores at a very late stage after they have grown quite large (to a size of tens of nanometers). The technique lacks the temporal and spatial resolution to see the smaller, briefer pores observed with complex impedance and amperometry, which are estimated to have sizes of 1 nm or less. STED microscopy has revealed an especially rich dynamic repertoire of fusion pores with a wide range of trajectories. Pores expand and constrict; some vesicles preserve their integrity, while others collapse into the plasma membrane.

STED microscopy tracks aqueous and membrane labels and has the potential to test hypothetical mechanisms. During some fusion events, a label of the intracellular membrane face moved to vesicles before a label of the extracellular membrane face, and before a change in pH of the vesicle lumen [32]. The formation of a hemifusion diaphragm prior to fusion pore opening can account for this result. To label the inner surface of the plasma membrane, these results employed a green fluorescent protein (GFP)-tagged pleckstrin homology domain, which binds phosphatidylinositol 4,5-biphosphate (PIP_2). Abbineni *et al* [34] have pointed out that this reagent can inhibit exocytosis, further noting that Ca^{2+}-triggered PIP_2 synthesis on vesicles and the dynamics of this reagent's binding could lead to labeling of vesicles without diffusion through contiguous lipid monolayer.

Unitary synaptic release. The spontaneous fusion of a single synaptic vesicle elicits a miniature postsynaptic current (mPSC) that can be seen in whole-cell patch clamp recordings from a postsynaptic cell. In these single-vesicle fusion events, the postsynaptic receptor reports the time course of release, but because synaptic transmission is fast and dependent upon a sequence of rapid events, it is difficult to extract information about fusion pores from mPSCs. Molecular manipulations of the exocytotic apparatus of a nerve terminal alter the shape of an mPSC, suggesting that fusion pores control synaptic release [35–37]. Tryptophan substitutions in the TMD of synaptobrevin slow mPSCs in a manner consistent with a reduction in transmitter flux through the synaptic fusion pore (see figure 5.3) [38], and these effects are clearer in artificial synapses formed between neurons and human embryonic kidney (HEK) cells expressing appropriate postsynaptic proteins [39]. Modeling of synaptic release indicates that fusion pores must expand very rapidly to raise subsynaptic neurotransmitter concentrations to sufficient levels to activate ionotropic receptors [16, 40–43]. This aspect of fusion pore dynamics is functionally critical, but at present there are no direct observations of synaptic fusion pores. The mPSCs associated with kiss-and-run are clearly slower, indicating that when fusion pores fail to expand, the rate and extent of postsynaptic receptor activation are reduced [44, 45].

Nanodiscs. To study fusion pores *in vitro*, investigators have developed systems with proteins reconstituted into lipid bilayer membranes [46–48]. This enables the study of fusion in a system of defined composition using membranes containing recombinant proteins. Nanodiscs have emerged as an especially powerful reconstituted system for fusion pore studies. These small 6–50 nm lipid bilayer disks are surrounded by a membrane scaffold protein to protect the hydrophobic core at the edges from water [48, 49]. With nanodiscs one can control not only the protein composition but also the copy number. Nanodiscs containing SNARE proteins fuse with cells expressing reverse-oriented cognate SNAREs [50, 51], as well as with SNARE-containing black lipid bilayer membranes [37, 52]. In both cases, robust fusion pore signals were observed. The direct measurement of single-channel ionic current provides the highest resolution of fusion pore dynamics currently achievable, and reveals rapid opening and closing transitions reminiscent of ion-channel gating. However, the finite size of a nanodisc limits fusion pore expansion and leaves late-stage fusion pores out of reach.

5.2 Functional considerations

As fusion pores grow, constrict, and close, the speed and composition of secretion change. The distinct properties of fusion pore states have many important functional consequences.

Flux. Fusion pores control the speed with which molecules escape from a vesicle. Size has an obvious impact: larger pores support higher fluxes. The cylindrical pore discussed above in the context of conductance measurements is expected to produce a flux that is proportional to the area and inversely proportional to the length. A high flux produces a steep gradient outside the fusion pore, which creates high local concentrations at the pore mouth. If a secreting cell has receptors for the molecules it is releasing, then rapid flux can raise concentrations locally to activate these autoreceptors. Receptors activated cooperatively by multiple ligands are especially sensitive to a high local concentration created by high pore flux. If the cell surface has processing enzymes, high fluxes may saturate local catalytic sites and allow more substrate molecules to escape unchanged.

If surrounding extracellular structures impede escape, as in a synaptic cleft, then this transient confinement amplifies the impact of high flux. Local concentrations can then rise to a level sufficient to activate postsynaptic receptors. Rapid fusion pore flux is essential for the activation of ionotropic synaptic receptors, which generally require the binding of two or more ligands. This cooperative activation often depends on neurotransmitter concentrations reaching 10–100 μM. Modeling studies suggest that to achieve such concentrations, synaptic fusion pores must expand to diameters of several nanometers within ~50 μs of opening [16, 41]. If expansion is too slow then neurotransmitter diffuses away without reaching a high subsynaptic concentration, essentially losing the opportunity to activate postsynaptic receptors. At synapses, flux has a major impact on function, setting a requirement for fusion pore size, and for rapid expansion to reach that size.

Molecular size and filtering. Vesicles contain molecules varying in size from ~100 D for small neurotransmitters to ~100,000 D for large peptide hormones. Their molecular dimensions thus vary from a few Angstroms to nearly 10 nm. Molecules that are larger than a fusion pore cannot escape and are not secreted. Thus, fusion pores filter vesicle contents. This can serve as a very important form of biological control by determining the composition of a secretion. With vesicles containing mixed cargo, a stimulus can trigger the release of different molecules in different proportions by effecting the formation of small or large fusion pores [53, 54].

Charge selectivity. Molecular interactions between the secreted molecules and the surrounding pore could impact fusion pore flux, as is well documented for ion permeation through ion channels. Small fusion pores have been reported to show selectivity for cations over anions [55]. The selectivity is modest compared to that of ion channels, but nevertheless favors flux of cations such as catecholamines and acetylcholine over anions such as glutamate. As pores dilate they lose this selectivity.

Closure and kiss-and-run. Fusion pores can close to terminate an exocytotic event, in a process generally referred to as kiss-and-run. This results in subquantal release. If a fusion pore closes without reaching a size large enough for the passage of high-molecular-weight peptides, then only small molecules can escape. Fusion pores can close after reaching quite large sizes [56], in which case large cargoes can also escape. Kiss-and-run represents a distinct and well-defined mode of release, which at synapses produces synaptic potentials that are smaller and slower [44, 45].

Chemistry. The smallest nascent fusion pore can still allow virtually instantaneous equilibration of protons between the vesicle lumen and extracellular fluid. Thus, fusion pore opening produces an abrupt increase in the pH within a vesicle. This has been shown to have an impact on the activity of hormone-processing enzymes within a vesicle. In chromaffin cells the increase in pH activates the plasminogen activator inhibitor (PAI) contained within the same vesicle, which contains tissue plasminogen activator (tPA) [57]. When a fusion pore opens, the rise in pH activates PAI. If the fusion pore stays small then tPA is inactivated by the PAI as the two remain in close proximity. However, if the fusion pore expands rapidly then the tPA can escape in a fully active form. Thus, vesicles can serve as nanoscale chemical reaction chambers that alter the composition of secretion. Processing enzymes are often contained within the vesicles harboring their prohormone targets, and it is likely that fusion pores control the time and thus the extent of prohormone processing.

Membrane mixing. A fusion pore can serve as a conduit for the mixing of the vesicle and plasma membrane. A proteinaceous pore essentially blocks mixing, but a lipidic pore provides a fluid membrane pathway for the lateral diffusion of lipid and membrane protein. Single-vesicle capacitance steps have revealed the escape of vesicle membrane during transient fusion pore openings [58]. STED microscopy indicates that a plasma membrane label can flow into a vesicle [32]. For small lipidic pores, membrane mixing should be slow. Membrane mixing can be monitored by loss of membrane label, and efforts have been made to square the observed rate with that expected for fusion pores of different sizes [34]. The onset of lipid mixing pinpoints the time of appearance of lipidic pores and thus marks a critical structural transition in the fusion pore. Since vesicle membranes and plasma membranes differ in their lipid and protein composition, mixing of the membranes adds a challenge to the recycling of vesicle materials and the regeneration/replenishment of vesicles after secretion.

5.3 The energy landscape of membrane fusion

Fusion pore dynamics is shaped by an energy landscape that combines lipid bilayer mechanics and protein conformation. Lipid bilayer elasticity can be studied using continuum models that represent shape-related energies in terms of membrane curvature. Detailed molecular models of lipid bilayers complement this approach. The influence of protein conformation can be studied with the aid of experimental structures and computer models. The energy landscape determines the relative stability of different states and their rates of interconversion, thereby controlling the evolution of a fusion pore along the membrane fusion reaction pathway. Many of the energetic contributions are large and opposing, allowing small imbalances in one direction or the other to drive a fusion pore into different configurations.

Mean curvature of a membrane. Changing the shape of a lipid bilayer membrane changes its energy. Figure 5.2(a) illustrates the rearrangements of phospholipids within a lipid bilayer in the shape of a fusion pore. As the lipids rotate within each monolayer, their head-group and hydrocarbon tail distances change in opposite directions. These complex interactions between lipid molecules in a membrane can be simplified into a continuum description of membrane bending using the theory of elasticity [59, 60]. A point on a surface has two principle curvatures corresponding to two perpendicular arcs through that point (figure 5.2(a)). These arcs have radii R_1 and R_2; R_1 is for the meridian curvature, where the arc is in the plane of the illustration, and R_2 is for the parallel curvature, where the arc is perpendicular to the plane of the illustration (light blue contours in figure 5.2(a) are the arcs and the

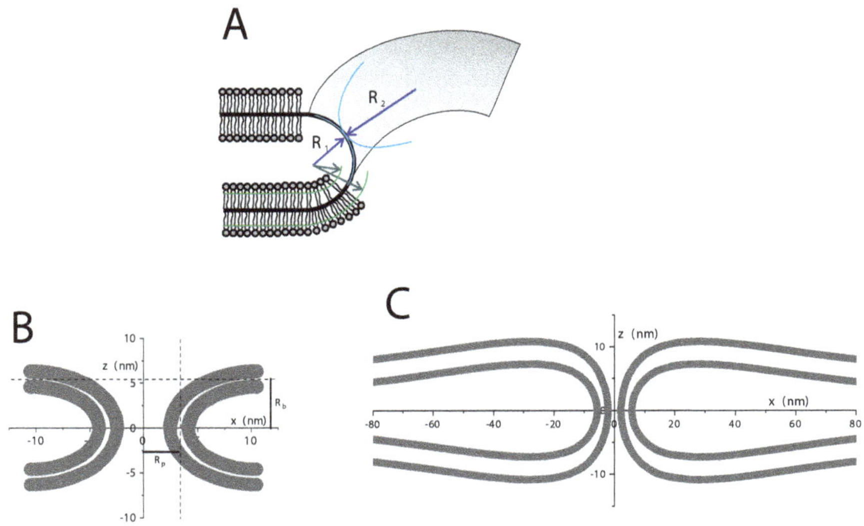

Figure 5.2. Membrane bending in fusion pores. (a) A toroidal lipidic fusion pore formed by a lipid bilayer. The circular in-plane arc has a radius of R_1 and the out-of-plane arc has a radius of R_2. The blue arcs are for the bilayer; the green arcs show the neutral surfaces of the meridian curvature in each monolayer. (b) A minimum bending energy shape of a fusion pore formed by a spatially restricted system. Reproduced with permission from [8]. Copyright © 2009, Springer Nature Switzerland AG. Part of Springer Nature. (c) A minimum bending energy shape of a fusion pore with a greater spatial extent allows bowing (lengthening) [10].

principal radii are dark blue). Figure 5.2(a) illustrates these features with a toroidal shape in which R_1 is constant within the pore. Other shapes are possible.

One of the forms of elastic bending energy at a point on a surface is proportional to the *mean curvature* squared, $\left(\frac{1}{R_1} + \frac{1}{R_1} - C_0\right)^2$, where C_0 is the spontaneous curvature. The total energy arising from mean curvature can be calculated by integrating this quantity over the surface and multiplying by the bending modulus. The lipid bilayer in a fusion pore has especially high curvature. In some regions, R_1 and R_2 can be smaller than the bilayer dimensions. Detailed calculations indicate that the magnitude of this energy is very sensitive to the shape of the fusion pore. C_0 also has an impact, and negative values reduce the energy of a fusion pore [7, 8, 61]. Lipids with negative C_0 promote the fusion of lipid vesicles [62, 63] and accelerate the transition from proteinaceous pores to lipidic pores during exocytosis [64].

For a toroidal fusion pore (figure 5.2(a)) the mean curvature bending energy can be calculated to be 75 kT [61, 65]. However, a toroid is not the shape that minimizes the bending energy. In fact, a well-known result from differential geometry states that a catenoid, which is the surface of revolution formed by a catenary (a hyperbolic cosine function), has zero mean curvature everywhere [66]. Indeed, soap films between two parallel coaxial rings spontaneously assume a catenoid shape [67]. A catenoid has a pore-like overall contour, suggesting that the mean curvature contribution to fusion pore energy can be dismissed as negligible. However, two factors lead to significant membrane bending energies in fusion pores. One issue is that plasma membranes are generally flat, but catenoids do not have a horizontal limiting surface, so the junction with the plasma membrane must depart from a catenoid shape. Another problem is that a lipid bilayer is formed from two monolayers, each of which has its own neutral surface (in figure 5.2(a), the monolayer neutral surfaces are light green; their meridian curvatures are dark green). The two surfaces cannot each be catenoids, because it is geometrically impossible for two catenoids to be parallel. For distances between the two monolayer neutral surfaces comparable to R_1 and R_2, the deviations from a catenoidal shape must be significant. Minimizing the bending energy of two coupled monolayers yielded values ranging from 11 to 120 kT, with lower values for negative monolayer C_0 [8]. Figure 5.2(b) illustrates a shape that minimizes bending energy with a limited lateral dimension and a basis function that limits shapes to monotonic gradation between vertical and horizontal. Biological membranes have an asymmetric lipid distribution between the two leaflets and this can also reduce the energy cost of forming a fusion pore. Further analysis of energy minimization using both continuum elasticity and coarse-grained molecular dynamics showed that that the energy can be reduced further by a modification of shape in which the pore elongates vertically. Figure 5.2(c) illustrates the minimal shape for minimization of a larger system in which the elongated shape is more striking. This change in shape can reduce the bending energy to low values [9, 10].

Theoretical studies of the membrane bending energy landscape have reached a consensus that a lipidic fusion pore has a minimum energy at a particular diameter, and that expansion of the fusion pore encounters a large energy barrier [7, 8, 61, 68]. This stable pore is supported by the observation of similar sizes when fusion is driven by different numbers of SNARE complexes [51] and by the observation of a metastable plateau in the permeability of late-stage fusion pores seen in amperometry recordings (figure 5.1; [23]). This minimum bending energy state is also likely responsible for the stable fusion pores often seen in capacitance recordings.

The variation in energy with shape has a major impact on fusion pore dynamics. To form a lipidic fusion pore, the two fusing membranes must initially directly contact one another. This means that it is not possible to directly proceed to the lowest-energy shape. An initial fusion pore has a shape that is closer to what is seen when the two membranes are touching (figure 5.2(b)). This initial lipidic pore cannot have a bowing/teardrop shape and must rearrange to reach a global minimum on the membrane bending energy surface (figure 5.2(c)). Thus, the energy landscape defined by mean curvature provides an appealing mechanism for the structural transition of lipidic fusion pores.

Gaussian curvature. In addition to the mean curvature just discussed, elastic membrane bending energy includes another term, Gaussian curvature, equal to $\frac{1}{R_1R_2}$. Like mean curvature, this quantity can be integrated over a surface and multiplied by a modulus to obtain a contribution to the energy of membrane bending. However, the Gaussian curvature has a remarkable mathematical property: for a given topology, its integral over a surface is invariant with respect to shape [66]. For example, a sphere can be contorted into any arbitrary shape; the mean curvature integral will vary, but as long as the surface completely separates the inside from the outside, the integral over the Gaussian curvature will not change. This makes Gaussian curvature irrelevant to the above discussion of fusion pore shapes (figure 5.2). However, Gaussian curvature is very relevant to the energetics of fusion pore formation. When a vesicle fuses with the plasma membrane, two spheres become one. The change in topology produces a precise change in the integral over the Gaussian curvature. The change in membrane bending energy is large and positive [69, 70]. This constitutes an important feature of the fusion energy landscape and is predicted to make the energy barrier to the formation of a lipidic fusion pore very high. Fusing the two monolayers sequentially offers an appealing hypothetical mechanism that breaks the process into two easier steps. Because of its topological origin, it has proven difficult to adapt the continuum concept of Gaussian curvature to the molecular scale and incorporate it into models for the rearrangements associated with membrane fusion. At present, there is no satisfactory explanation for how Gaussian curvature impacts the changes at the molecular level that accompany the membrane fusion transition.

Molecular models. Continuum elasticity suffers from the shortcoming of ignoring molecular details. One can overcome this weakness with computational molecular dynamics, but modeling membrane fusion requires a system with a large number of lipid molecules, and this presents a serious obstacle to computer simulations. Although all-atom models have been used to model membrane fusion in very small systems, coarse-grained models of lipid bilayers have been developed for work with larger systems. These studies succeeded in simulating membrane fusion and in capturing intermediates in various stages of exocytosis, including lipidic fusion pores [71–74]. A comparison of coarse-grained, all-atom, and continuum models [10] showed that as energy minimization progressed, each model approached a similar shape, recapitulating the axial lengthening described above for the continuum model (figure 5.2(c)). The convergence of these different theories validated the distinct approaches. A coarse-grained simulation has incorporated the SNARE complex to model the formation of a lipidic fusion pore containing the C-terminal residues of the SNARE TMDs [74]. Coarse-grained models of the fusion of small vesicles led to a fusion pore that expanded rapidly and was not metastable [71, 73, 74]. This contrasts with the theoretical and experimental studies summarized above, which indicated that a narrow pore is a metastable state at a local energy minimum. It is not clear whether the rapid expansion seen in stimulations is behavior unique to small vesicles or a shortcoming of the theory.

SNARE complex zipping. SNARE proteins have general membrane trafficking functions. Most forms of Ca^{2+}-triggered exocytosis are mediated by the vesicle SNARE synaptobrevin and the plasma membrane SNAREs syntaxin and SNAP-25. These three proteins form a complex of exceptionally high stability [75]. SNAREs harbor domains of ~65 amino acids, known as SNARE motifs, which are capable of assembling into a tight bundle of four α-helices (one each from synaptobrevin and syntaxin and two from SNAP-25). With SNARE proteins tethered to the vesicle and plasma membranes, zipping from the membrane's distal N-terminus to the membrane's proximal C-terminus draws the two membranes together [76–78]. The energy released during the formation of this complex is thought to be strong enough to overcome inter-membrane repulsion, and the stresses produced by assembly can bend and reshape lipid membranes. SNARE complex assembly proceeds in multiple stages [79, 80]. Assembly of the N-terminal, C-terminal, and linker domains of the SNARE motifs are estimated to release an energy of 65 kT [79]. Association of the two TMDs of syntaxin and synaptobrevin could contribute additional energy [81, 82], but this would have to be a late step, as it ends with a fully lipidic pore and places both TMDs in the same lipid membrane. In a proteinaceous pore the TMDs are still in separate membranes, so SNAREs cannot be fully zipped at this point. The distinct stages of SNARE complex assembly could provide driving forces for multiple steps, either leading to the formation of an initial fusion pore or to later pore states. Once assembly has completed, it can play no further role in subsequent transitions involving shape changes of the lipidic pore.

Helix completion. Syntaxin and synaptobrevin are anchored in membranes by a presumably helical TMD. Linkers made from ~11 amino acids connect the TMDs to the ~65 amino acid α-helical SNARE motifs. Assembly of the SNARE motifs up to the linkers leaves two nearly perpendicular α-helical segments in each protein. This creates a disordered break between two helical domains. The cooperativity of the helix–coil transition depends on positive free energy at the helix–coil boundary [5]. This gives rise to a driving force that joins the two helical segments into a single contiguous helix. The two lipid bilayers within which the two TMDs are embedded must merge to complete the helices, so helix completion can be harnessed to drive membrane fusion. This process has been incorporated into a theory for SNARE-mediated membrane fusion that produced rates in the millisecond range [83]. A SNARE complex anchored in a lipid bilayer by the TMD of syntaxin rotates from an orientation parallel to the lipid bilayer to a perpendicular one upon the addition of Ca^{2+}-synaptotagmin. This indicates regulation of the helix completion force [84]. It has been proposed that the rotation of synaptobrevin's TMD encounters an energy barrier due to its interaction with the lipid bilayer, in which the rotation pulls the C-terminus into the hydrocarbon interior and disrupts the bilayer to initiate fusion [85].

SNARE complex number. There have been many efforts to determine how many SNARE complexes participate in exocytosis, and numbers vary widely [16, 86]. Different numbers of SNARE complexes have been implicated in dictating whether a fusion pore is proteinaceous or lipidic [37]. With only one complex, no fusion pore forms and the outcome is hemifusion [87]. A proteinaceous initial fusion pore is anticipated to be larger when greater numbers of SNARE complexes provide more TMDs to line the pore, and the transition rates between different states are also strongly affected [51, 52]. Thus, the SNARE complex number plays an important role in the state of a fusion pore, and may well account for some of the diversity encountered in different studies. This raises interesting questions about whether additional SNARE complexes are recruited as fusion progresses and fusion pores grow, and whether cells control the number of SNARE complexes to choose different modes of secretion. Given the high energetic barrier to fusion pore dilation discussed above, one can speculate that recruiting more SNARE complexes may provide the essential drive for full fusion. Fewer SNARE complexes leave kiss-and-run as the favored option.

Entropy of SNARE pinwheels. The geometric arrangement of SNARE complexes extending outward from a central fusion axis has been termed a 'SNAREpin,' and likened to a pinwheel [46]. An outward, radial movement of the SNARE complexes in this arrangement forces the fusion pore to expand. This outward radial movement increases the rotational entropy, thus providing a driving force for SNARE complex movement away from the fusion pore center [88]. Incorporating this rotational entropy term into a model for fusion pore dilation recapitulated the increased rate of expansion with SNARE complex number [51]. This force is not relevant to the early steps of fusion pore dynamics, but can play an important role in later expansion steps.

Hydrophobic energy. The hydrophobic effect presents an energetic challenge to membrane trafficking. Avoiding hydrocarbon–water contact requires the interiors of two separate bilayers to merge with one another without being exposed to the water between the bilayers. This consideration motivated the introduction of the lipid-stalk model, in which fusion can occur with virtually no hydrocarbon–water contact [65, 89]. Any mechanism for fusion requires careful attention to how much hydrocarbon–water exposure occurs. The hydrophobic energy can be calculated by multiplying the area of water–oil contact by an empirical parameter. Surface tension measurements suggested a value for this parameter of 50 ergs cm^{-2} ~75 cal $Å^{-2}$. Using this value for a cylindrical water-filled pore lined entirely by hydrocarbon gave extremely high energies [89]. Solvent-partitioning measurements suggest a much lower value of 25 cal $Å^{-2}$ for this parameter [90, 91], and of course, using this value, the hydrophobic barrier to membrane fusion was much lower. Limited hydrocarbon–water contact was incorporated into a model of SNARE-mediated fusion driven by helix completion, and the resulting energy barriers were readily surmountable [83]. During the dynamic rearrangements of a lipidic fusion pore, the bilayer maintains its integrity, so the hydrophobic effect is irrelevant. However, fusion pore closure requires a process that reverses the formation of a lipidic pore. The walls of a constricting pore fuse [92], and the advantages of the lipid-stalk model in minimizing hydrophobic contacts are again likely to be relevant.

Hydration forces. Lipid bilayers repel one another due to the hydration energy of the phospholipid head groups [93]. Membranes must overcome this barrier in order to fuse. Puckering of the membrane can reduce this energy by limiting the contact area. Because this force originates from hydration, osmoticants have proven broadly effective in promoting many forms of membrane fusion [94]. The fact that osmoticants are so fusogenic indicates that the hydration force is a major barrier to membrane fusion. Modeling has suggested that membrane repulsion plays a role in opposing the fusion of membranes mediated by SNAREs [51].

Membrane tension. Lateral forces within membranes connect area changes to energy changes and drive lateral membrane movement. Membrane tension can drive fusion pore expansion by changing the area of the cell surface [95]. Tension promotes the fusion of lipid bilayers [96] and fusion pore expansion during exocytosis [97]. However, the initial fusion pores are small, and their very small area limits the impact of tension on their dynamics. Thus, membrane tension is likely to play a role in the later steps of fusion pore dynamics, rather than the earlier steps.

5.4 Fusion pore states

Fusion pores change their properties as they advance along the fusion reaction pathway. In many cases the boundaries between states seem blurred and methodological differences leave questions about how to make comparisons. Amperometry reveals the earliest, briefest states, while capacitance and imaging techniques reveal later, more stable states. Different techniques often reveal roughly similar sequences, albeit with different timescales. Pores assume an initial state, expand, and constrict; the final outcomes of closure, prolonged stasis, expansion, or collapse appear to vary widely. Some consistent trends can be discerned, some of the changes can be interpreted in terms of the energy landscape, and some classifications of different fusion pore states can be attempted.

Nascent pore. The critical opening transition generates a nascent fusion pore. This initial pore is necessarily small, as fusion starts out as a highly focused localized event. As noted above, the nascent pore produces a prespike foot in amperometry recordings [14, 15, 98, 99]. The prespike foot appears to be stable for milliseconds, but it is not entirely clear how stable it really is. Published recordings show prespike feet as approximately flat, with varying degrees of upward slope. The mean amplitudes of kiss-and-run events are uncorrelated with duration, arguing against growth [100, 101]. If expansion is indeed occurring, it is much slower than the expansion that marks the onset of the amperometric spike. The lifetimes of prespike feet vary stochastically, and distributions are exponential [99, 100, 102] or an exponential–Gaussian convolution [64]. Thus, the early transitions are Markov processes, recalling the kinetics of ion-channel gating [103]. The Markov picture supported by exponential lifetime distributions is not consistent with the slow expansion just discussed, as a Markov process implies stability prior to exit.

Fusion pore conductances fall within the range of ion-channel conductances, supporting the view of proteinaceous fusion pores [104, 105]. The ion-channel analogy is further supported by studies of prespike feet and SNARE TMD mutations. Tryptophan substitutions at key TMD residues reduce catecholamine flux through the nascent fusion pore, glycine substitution increases catecholamine flux, and charge substitutions alter flux in a manner consistent with electrostatic interactions [17–19]. These results indicate an intimate relation between the fusion pore and the SNARE TMDs. Figure 5.3 highlights the residues within the syntaxin and synaptobrevin TMDs implicated by amperometry experiments as pore liners. Flux perturbation studies in other experimental systems implicate these and other residues in fusion pores [37, 38], and these are also highlighted in figure 5.3. However, in these other studies, it is not as clear that the fusion pores are nascent, i.e. at the earliest time after opening.

Larger vesicles produce prespike feet with longer lifetimes. The greater stability of the nascent pore can be accounted for by variations in the amount of curved lipid bilayer created within the lipidic fusion pore (figure 5.2) when a nascent fusion pore expands [64]. This indicates that there is very little lipid bilayer curvature in the nascent pore, thus supporting a greater role for protein than lipid.

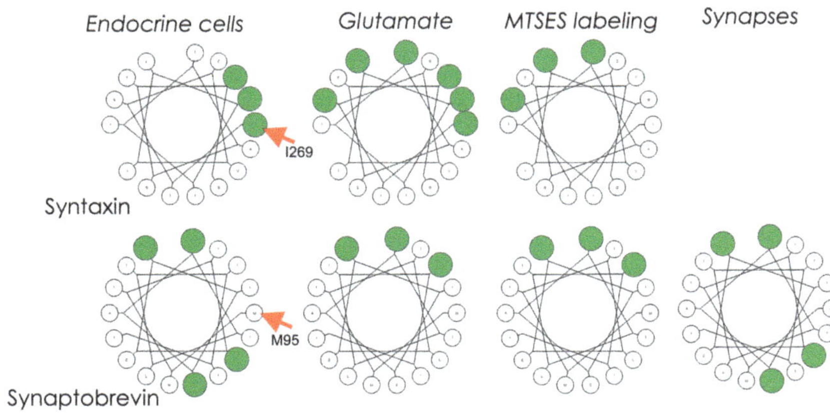

Figure 5.3. Helical wheels highlight SNARE TMD residues in initial fusion pores. The residues were implicated by amperometry and capacitance with mutants of syntaxin [19] and synaptobrevin [17], by glutamate flux and cysteine accessibility (MTSES labeling) in reconstituted fusion pores [37], and by glutamate flux through synaptic fusion pores [38]. Red arrowheads indicate the N-terminal residue of the TMDs. Copyright © 2017 Chang *et al.* Originally published by The Rockefeller University Press. Modified from [16].

Capacitance recording reveals opening transitions, but with lower time resolution than amperometry recording. Nevertheless, tryptophan mutations in SNARE TMDs alter the conductance of fusion pores [17, 19]. The initial fusion pores seen in capacitance recordings appear more stable than prespike feet, with lifetimes that are one or two orders of magnitude longer. Because of the difference in the time resolutions of capacitance and amperometry, it is not clear that they are viewing the same structure. Indeed, amperometry recording suggests that the capacitance measurements correspond to the later plateau in permeability (figure 5.1) [23].

Nascent fusion pores can close, indicating that this state is an important bifurcation point. Expansion is the other possible outcome. The initial transition from a closed pore to an open pore is a clear and functionally important transition that can be controlled by a host of molecules and conditions [39]. Likewise, the branch point between expansion and closure and the stability of the nascent pore are influenced by many factors. Thus, the initial state of the fusion pathway is an important focus of biological control.

Early expansion. A nascent fusion pore can expand to accelerate flux, and this expansion is especially clear at the onset of an amperometric spike. This expansion likely entails infiltration of a proteinaceious pore with lipid [83, 104] and delivery of the fusion pore into the lipid membrane bending energy surface. However, as discussed above, the first lipidic form taken by a fusion pore is not necessarily at a minimum in membrane bending energy. This depends on the nature of the transition and the shape that is most easily accessed [8]. Thus, early expansion generates an unstable transient structure, and the shape then moves to a minimum on the membrane bending energy surface (figure 5.1) [23]. The early permeability peak in this plot can be viewed as an unstable transition state between two metastable states. The expanded pore may have very high permeability, but its short duration limits the amount of vesicle content loss. Brief spike-like upticks have been observed during prespike feet, which are likely to be abortive starts to this initial expansion process [98].

Constriction. The rapid constriction following the permeability peak brings a fusion pore into a new stable state (figure 5.1). This dynamic process likely reflects the movement of the lipidic pore along the membrane bending landscape, and the pore becomes stable upon arriving at the energy minimum. The plateau in permeability can thus provide information about the nature of this lipidic pore and offers the hope of pointing to the shape a lipidic pore actually assumes (such as figures 5.2(b) and (c)). Amperometry indicates that this state remains stable for a minimum of several tens of milliseconds. During this time the entire content of catecholamine escapes. The retention of fluorescent content [29], preservation of vesicle shape [32], and sustained membrane deformation [106] for times of the order of seconds all support the existence of a stable constricted pore at the membrane bending energy minimum.

The constricted pore is an important focus of biological control [23]. Of the many candidate regulators, dynamin merits special mention. It forms collars around lipid tubules [107, 108] and influences the decay of amperometric spikes [109–111]. This action of dynamin suggests that it bends the membrane to regulate the shape of the narrow lipidic fusion pore. The action of dynamin is regulated by calcineurin dephosphorylation and syndapin/N-WASP signaling [112]. Other proteins that influence spike shape and presumably regulate late-stage fusion pores include myosin [113], Cdc42 [97], complexin [114], and Munc18 [115]. Focusing so many forms of regulation on this state speaks to its importance in controlling the speed, selectivity, and extent of content release.

Late-stage expansion and vesicle collapse. Electron micrographs reveal fusing vesicles in various stages of collapse [116–118]. Tracking the time course of conductance indicates that fusion pores can expand after opening [56]. In human neutrophils, once the conductance growth surpasses ~ 1 nS the pores pass a point of no return and commit to full fusion [119]. With large vesicles of beige mouse mast cells, membrane flows through a fluid lipidic fusion pore [58], indicating the onset of collapse of the vesicle membrane into the plasma membrane. In pancreatic islets the growth of fusion pores can be tracked as a vesicle collapses by following the passage of markers with increasing sizes [120]. Polarization TIRF microscopy [30] and STED microscopy [32] indicate that some fusion events progress to vesicle collapse. The transition to full fusion varies with biological conditions. Intense electrical activity increases the likelihood of full fusion in chromaffin cells [53], while cAMP controls this transition in pancreatic β cells [54]. Dynamin [121] and actin [122] also play a role in destabilizing the constricted pore so that it can expand to enable vesicle collapse. Full fusion and vesicle collapse play an important role in determining whether large peptides are retained or released, and the many forms of control over this process attest to the importance of vesicle collapse in controlling secretion.

Closure. Essentially all of the techniques for visualizing single fusion pores have detected some form of partial, subquantal release, thus implying fusion pore closure. Membrane capacitance increases in steps during vesicle fusion, and upward steps are often followed within a second or less by a downward step of the same size [1, 2, 12, 123]. The tight temporal correlation between the upward and downward steps suggests that the fusion pore can close soon after opening in a kiss-and-run process. Amperometry indicates that very small, nascent fusion pores can close [100, 101, 124]. Some spikes might be terminated by the closure of larger, late-stage pores [125]. Fluorescent content loss [126] and membrane label loss [26, 27] often appear limited. Pores can reach quite large sizes before closing without vesicle collapse [32]. Thus, closure is not limited to a single fusion pore state. Multiple states appear to be capable of closing and it is likely that the underlying mechanisms differ widely.

Small amperometric events with amplitudes roughly comparable to prespike feet often terminate without a spike. These 'stand-alone feet' are thought to report kiss-and-run events in which a fusion pore closes rather than expands [100, 101, 127]. Endocrine cells can also produce very small stand-alone feet that never terminate with a spike [124]. These events may represent a form of exocytosis controlled by a small fusion pore that cannot expand. The closure of these nascent proteinaceous pores is reminiscent of ion-channel gating, but late-stage closure is very different because the pore is lipidic. This form of closure can actually be visualized in mechanistic terms similar to those hypothesized for lipid bilayer fusion. As the lipid bilayer surrounding a fission pore narrows, the walls approach one another, push head groups away, and drive the formation of a stalk-like structure similar to that hypothesized for fusion [92]. In a sense, during fission, the pore walls fuse. This process may be driven by the compression of a surrounding protein collar such as that formed by dynamin.

5.5 Conclusions

Studies of fusion pores have provided insight into the mechanism of exocytosis. Although the divergent fusion pore behavior in the literature can appear confusing, the diversity actually offers the key to unlocking the mysteries of protein control. The different states of fusion pores highlight pathways taken by biological membranes during fusion. The membrane fusion reaction pathway proceeds through a sequence of fusion pore structures, each with distinct properties and functions. With the discoveries of multiple fusion pore states that interconvert dynamically, a framework is starting to emerge that can guide researchers in further studies of the mechanism of exocytosis.

References

[1] Neher E and Marty A 1982 Discrete changes of cell membrane capacitance observed under conditions of enhanced secretion in bovine adrenal chromaffin cells *Proc. Natl Acad. Sci. USA* **79** 6712–6

[2] Fernandez J M, Neher E and Gomperts B D 1984 Capacitance measurements reveal stepwise fusion events in degranulating mast cells *Nature* **312** 453–5

[3] Zimmerberg J, Curran M, Cohen F S and Brodwick M 1987 Simultaneous electrical and optical measurements show that membrane fusion precedes secretory granule swelling during exocytosis of beige mouse mast cells *Proc. Natl Acad. Sci. USA* **84** 1585–9

[4] Breckenridge L J and Almers W 1987 Currents through the fusion pore that forms during exocytosis of a secretory granule *Nature* **328** 814–7
[5] Jackson M B 2006 *Molecular and Cellular Biophysics* (Cambridge: Cambridge University Press)
[6] Hille B 1992 *Ion Channels of Excitable Membranes* (Sunderland: Sinauer)
[7] Nanavati C, Markin V S, Oberhauser A F and Fernandez J M 1992 The exocytotic fusion pore modeled as a lipidic pore *Biophys. J.* **63** 1118–32
[8] Jackson M B 2009 Minimum membrane bending energies of fusion pores *J. Membr. Biol.* **231** 101–15
[9] Ryham R J, Ward M A and Cohen F S 2013 Teardrop shapes minimize bending energy of fusion pores connecting planar bilayers *Phys. Rev. E Stat. Nonlin. Soft Matter Phys.* **88** 062701
[10] Yoo J, Jackson M B and Cui Q 2013 A comparison of coarse-grained and continuum models for membrane bending in lipid bilayer fusion pores *Biophys. J.* **104** 841–52
[11] Lindau M and Alvarez de Toledo G 2003 The fusion pore biochim *Biophys. Acta* **1641** 167–73
[12] Klyachko V A and Jackson M B 2002 Capacitance steps and fusion pores of small and large-dense-core vesicles in nerve terminals *Nature* **418** 89–92
[13] Wightman R M, Jankowski J A, Kennedy R T, Kawagoe K T, Schroeder T J, Leszczyszyn D J, Near J A, Diliberto D J Jr and Viveros O H 1991 Temporally resolved catecholamine spikes correspond to single vesicle release from individual chromaffin cells *Proc. Natl Acad. Sci. USA* **88** 10754–8
[14] Chow R H, von Rüden L and Neher E 1992 Delay in vesicle fusion revealed by electrochemical monitoring of single secretory events in adrenal chromaffin cells *Nature* **356** 60–3
[15] Jankowski J A, Schroeder T J, Ciolkowski E L and Wightman R M 1993 Temporal characteristics of quantal secretion of catecholamines from adrenal medullary cells *J. Biol. Chem.* **268** 14694–700
[16] Chang C W, Chiang C W and Jackson M B 2017 Fusion pores and their control of neurotransmitter and hormone release *J. Gen. Physiol.* **149** 301–22
[17] Chang C W, Hui E, Bai J, Bruns D, Chapman E R and Jackson M B 2015 A structural role for the synaptobrevin 2 transmembrane domain in dense-core vesicle fusion pores *J. Neurosci.* **35** 5772–80
[18] Han X and Jackson M B 2005 Electrostatic interactions between the syntaxin membrane anchor and neurotransmitter passing through the fusion pore *Biophys. J.* **88** L20–2
[19] Han X, Wang C T, Bai J, Chapman E R and Jackson M B 2004 Transmembrane segments of syntaxin line the fusion pore of Ca^{2+}-triggered exocytosis *Science* **304** 289–92
[20] Wightman R M, Schroeder T J, Finnegan J M, Ciolkowski E L and Pihel K 1995 Time course of release of catecholamines from individual vesicles during exocytosis at adrenal medullary cells *Biophys. J.* **68** 383–90
[21] Schroeder T J, Borges R, Finnegan J M, Pihel K, Amatore C and Wightman R M 1996 Temporally resolved, independent stages of individual exocytotic secretion events *Biophys. J.* **70** 1061–8
[22] Wightman R M, Dominguez N and Borges R 2018 How intravesicular composition affects exocytosis *Pflugers Arch.* **470** 135–41
[23] Jackson M B, Hsiao Y T and Chang C W 2020 Fusion pore expansion and contraction during catecholamine release from endocrine cells *Biophys. J.* **119** 219–31

[24] Ryan T A, Reuter H and Smith S J 1997 Optical detection of a quantal presynaptic membrane turnover *Nature* **388** 478–82
[25] Zenisek D, Steyer J A, Feldman M E and Almers W 2002 A membrane marker leaves synaptic vesicles in milliseconds after exocytosis in retinal bipolar cells *Neuron* **35** 1085–97
[26] Aravanis A M, Pyle J L and Tsien R W 2003 Single synaptic vesicles fusing transiently and successively without loss of identity *Nature* **423** 643–7
[27] Richards D A, Bai J and Chapman E R 2005 Two modes of exocytosis at hippocampal synapses revealed by rate of FM1–43 efflux from individual vesicles *J. Cell Biol.* **168** 929–39
[28] Taraska J W and Almers W 2004 Bilayers merge even when exocytosis is transient *Proc. Natl Acad. Sci. USA* **101** 8780–5
[29] Tsuboi T, McMahon H T and Rutter G A 2004 Mechanisms of dense core vesicle recapture following 'kiss and run' ('cavicapture') exocytosis in insulin-secreting cells *J. Biol. Chem.* **279** 47115–24
[30] Anantharam A, Onoa B, Edwards R H, Holz R W and Axelrod D 2010 Localized topological changes of the plasma membrane upon exocytosis visualized by polarized TIRFM *J. Cell Biol.* **188** 415–28
[31] Anantharam A, Axelrod D and Holz R W 2012 Real-time imaging of plasma membrane deformations reveals pre-fusion membrane curvature changes and a role for dynamin in the regulation of fusion pore expansion *J. Neurochem.* **122** 661–71
[32] Shin W, Ge L, Arpino G, Villarreal S A, Hamid E, Liu H, Zhao W D, Wen P J, Chiang H C and Wu L G 2018 Visualization of membrane pore in live cells reveals a dynamic-pore theory governing fusion and endocytosis *Cell* **173** 934–45 e12
[33] Zhao W D, Hamid E, Shin W, Wen P J, Krystofiak E S, Villarreal S A, Chiang H C, Kachar B and Wu L G 2016 Hemi-fused structure mediates and controls fusion and fission in live cells *Nature* **534** 548–52
[34] Abbineni P S, Axelrod D and Holz R W 2018 Visualization of expanding fusion pores in secretory cells *J. Gen. Physiol.* **150** 1640–6
[35] Pawlu C, DiAntonio A and Heckmann M 2004 Postfusional control of quantal current shape *Neuron* **42** 607–18
[36] Guzman R E, Schwarz Y N, Rettig J and Bruns D 2010 SNARE force synchronizes synaptic vesicle fusion and controls the kinetics of quantal synaptic transmission *J. Neurosci.* **30** 10272–81
[37] Bao H, Goldschen-Ohm M, Jeggle P, Chanda B, Edwardson J M and Chapman E R 2016 Exocytotic fusion pores are composed of both lipids and proteins *Nat. Struct. Mol. Biol.* **23** 67–73
[38] Chiang C W, Chang C W and Jackson M B 2018 The transmembrane domain of synaptobrevin influences neurotransmitter flux through synaptic fusion pores *J. Neurosci.* **38** 7179–91
[39] Chiang C W, Shu W C, Wan J, Weaver B A and Jackson M B 2021 Recordings from neuron-HEK cell cocultures reveal the determinants of miniature excitatory postsynaptic currents *J. Gen. Physiol.* **153** e202012849
[40] Khanin R, Parnas H and Segel L 1994 Diffusion cannot govern the discharge of neurotransmitter in fast synapses *Biophys. J.* **67** 966–72
[41] Clements J D 1996 Transmitter timecourse in the synaptic cleft: its role in central synaptic function *Trends Neurosci.* **19** 163–71

[42] Stiles J R, Van Helden D, Bartol T M, Salpeter E E and Salpeter M M 1996 Miniature endplate current rise times <100 μs from improved dual recordings can be modeled with passive acetylcholine diffusion from a synaptic vesicle *Proc. Natl Acad. Sci. USA* **93** 5747–52

[43] Jackson M B 2007 In search of the fusion pore of exocytosis *Biophys. Chem.* **126** 201–8

[44] He L, Wu X-S, Mohan R and Wu L-G 2006 Two modes of fusion pore opening revealed by cell-attached recordings at a synapse *Nature* **444** 102–5

[45] Richards D A 2009 Vesicular release mode shapes the postsynaptic response at hippocampal synapses *J. Physiol.* **587** 5073–80

[46] Weber T, Zemelman B V, McNew J A, Westermann B, Gmachl M, Parlati F, Sollner T H and Rothman J E 1998 SNAREpins: minimal machinery for membrane fusion *Cell* **92** 759–72

[47] Jackson M B and Chapman E R 2006 Fusion pores and fusion machines in Ca^{2+}-triggered exocytosis *Annu. Rev. Biophys. Biomol. Struct.* **35** 135–60

[48] Karatekin E 2018 Toward a unified picture of the exocytotic fusion pore *FEBS Lett.* **592** 3563–85

[49] Shi L, Shen Q T, Kiel A, Wang J, Wang H W, Melia T J, Rothman J E and Pincet F 2012 SNARE proteins: one to fuse and three to keep the nascent fusion pore open *Science* **335** 1355–9

[50] Wu Z, Auclair S M, Bello O, Vennekate W, Dudzinski N R, Krishnakumar S S and Karatekin E 2016 Nanodisc-cell fusion: control of fusion pore nucleation and lifetimes by SNARE protein transmembrane domains *Sci. Rep.* **6** 27287

[51] Wu Z, Bello O D, Thiyagarajan S, Auclair S M, Vennekate W, Krishnakumar S S, O'Shaughnessy B and Karatekin E 2017 Dilation of fusion pores by crowding of SNARE proteins *eLife* **6** e22964

[52] Bao H, Das D, Courtney N A, Jiang Y, Briguglio J S, Lou X, Roston D, Cui Q, Chanda B and Chapman E R 2018 Dynamics and number of trans-SNARE complexes determine nascent fusion pore properties *Nature* **554** 260–3

[53] Fulop T, Radabaugh S and Smith C 2005 Activity-dependent differential transmitter release in mouse adrenal chromaffin cells *J. Neurosci.* **25** 7324–32

[54] MacDonald P E, Braun M, Galvanovskis J and Rorsman P 2006 Release of small transmitters through kiss-and-run fusion pores in rat pancreatic beta cells *Cell Metab.* **4** 283–90

[55] Delacruz J B, Sharma S, Rathore S S, Huang M, Lenz J S and Lindau M 2021 Fusion pores with low conductance are cation selective *Cell Rep.* **36** 109580

[56] Spruce A E, Breckenridge L J, Lee A K and Almers W 1990 Properties of the fusion pore that forms during exocytosis of a mast cell secretory granule *Neuron* **4** 643–54

[57] Bohannon K P, Bittner M A, Lawrence D A, Axelrod D and Holz R W 2017 Slow fusion pore expansion creates a unique reaction chamber for co-packaged cargo *J. Gen. Physiol.* **149** 921–34

[58] Monck J R, Alvarez de Toledo G and Fernandez J M 1990 Tension in secretory granule membranes causes extensive membrane transfer through the exocytotic fusion pore *Proc. Natl Acad. Sci. USA* **87** 7804–8

[59] Boal D 2002 *Mechanics of the Cell* (Cambridge: Cambridge University Press)

[60] Helfrich W 1973 Elastic properties of lipid bilayers: theory and possible experiments *Z. Naturforsch.* C **28** 693–703

[61] Chizmadzhev Y A, Cohen F S, Shcherbakov A and Zimmerberg J 1995 Membrane mechanics can account for fusion pore dilation in stages *Biophys. J.* **69** 2489–500
[62] Chernomordik L, Chanturiya A, Green J and Zimmerberg J 1995 The hemifusion intermediate and its conversion to complete fusion: regulation by membrane composition *Biophys. J.* **69** 922–9
[63] Haque M E and Lentz B R 2004 Roles of curvature and hydrophobic interstice energy in fusion: studies of lipid perturbant effects *Biochemistry* **43** 3507–17
[64] Zhang Z and Jackson M B 2010 Membrane bending energy and fusion pore kinetics in Ca(2 +)-triggered exocytosis *Biophys. J.* **98** 2524–34
[65] Kozlov M M, Leikin S L, Chernomordik L V, Markin V S and Chizmadzhev Y A 1989 Stalk mechanism of vesicle fusion. intermixing of aqueous contents *Eur. Biophys. J.* **17** 121–9
[66] Kreyszig E 1991 *Differential Geometry* (New York: Dover)
[67] Ito M and Sato T 2010 *In situ* observation of a soap-film catenoid—a simple educational physics experiment *Eur. J. Phys.* **31** 357–65
[68] Katsov K, Muller M and Schick M 2004 Field theoretic study of bilayer membrane fusion. I. Hemifusion mechanism *Biophys. J.* **87** 3277–90
[69] Siegel D P 2008 The Gaussian curvature elastic energy of intermediates in membrane fusion *Biophys. J.* **95** 5200–15
[70] Hu M, Briguglio J J and Deserno M 2012 Determining the Gaussian curvature modulus of lipid membranes in simulations *Biophys. J.* **102** 1403–10
[71] Kasson P M, Kelley N W, Singhal N, Vrljic M, Brunger A T and Pande V S 2006 Ensemble molecular dynamics yields submillisecond kinetics and intermediates of membrane fusion *Proc. Natl Acad. Sci. USA* **103** 11916–21
[72] Risselada H J and Grubmuller H 2012 How SNARE molecules mediate membrane fusion: recent insights from molecular simulations *Curr. Opin. Struct. Biol.* **22** 187–96
[73] Risselada H J, Kutzner C and Grubmuller H 2011 Caught in the act: visualization of SNARE-mediated fusion events in molecular detail *ChemBioChem* **12** 1049–55
[74] Sharma S and Lindau M 2018 Molecular mechanism of fusion pore formation driven by the neuronal SNARE complex *Proc. Natl Acad. Sci. USA* **115** 12751–6
[75] Hayashi T, McMahon H, Yamasaki S, Binz T, Hata Y, Sudhof T C and Niemann H 1994 Synaptic vesicle membrane fusion complex: action of clostridial neurotoxins on assembly *EMBO J.* **13** 5051–61
[76] Hanson P I, Roth R, Morisaki H, Jahn R and Heuser J E 1997 Structure and conformational changes in NSF and its membrane receptor complexes visualized by quick-freeze/deep-etch electron microscopy *Cell* **90** 523–35
[77] Sorensen J B, Wiederhold K, Muller E M, Milosevic I, Nagy G, de Groot B L, Grubmuller H and Fasshauer D 2006 Sequential N- to C-terminal SNARE complex assembly drives priming and fusion of secretory vesicles *EMBO J.* **25** 955–66
[78] Xu T, Rammner B, Margittai M, Artalejo A R, Neher E and Jahn R 1999 Inhibition of SNARE complex assembly differentially affects kinetic components of exocytosis *Cell* **99** 713–22
[79] Gao Y, Zorman S, Gundersen G, Xi Z, Ma L, Sirinakis G, Rothman J E and Zhang Y 2012 Single reconstituted neuronal SNARE complexes zipper in three distinct stages *Science* **337** 1340–3
[80] Melia T J, Weber T, McNew J A, Fisher L E, Johnston R J, Parlati F, Mahal L K, Sollner T H and Rothman J E 2002 Regulation of membrane fusion by the membrane-proximal coil of the t-SNARE during zippering of SNAREpins *J. Cell Biol.* **158** 929–40

[81] Ellena J F, Liang B, Wiktor M, Stein A, Cafiso D S, Jahn R and Tamm L K 2009 Dynamic structure of lipid-bound synaptobrevin suggests a nucleation-propagation mechanism for trans-SNARE complex formation *Proc. Natl Acad. Sci. USA* **106** 20306–11

[82] Stein A, Weber G, Wahl M C and Jahn R 2009 Helical extension of the neuronal SNARE complex into the membrane *Nature* **460** 525–8

[83] Jackson M B 2010 SNARE complex zipping as a driving force in the dilation of proteinaceous fusion pores *J. Membr. Biol.* **235** 89–100

[84] Kiessling V, Kreutzberger A J B, Liang B, Nyenhuis S B, Seelheim P, Castle J D, Cafiso D S and Tamm L K 2018 A molecular mechanism for calcium-mediated synaptotagmin-triggered exocytosis *Nat. Struct. Mol. Biol.* **25** 911–7

[85] Lindau M, Hall B A, Chetwynd A, Beckstein O and Sansom M S 2012 Coarse-grain simulations reveal movement of the synaptobrevin C-terminus in response to piconewton forces *Biophys. J.* **103** 959–69

[86] Sharma S and Lindau M 2018 The fusion pore, 60 years after the first cartoon *FEBS Lett.* **592** 3542–62

[87] Giraudo C G, Hu C, You D, Slovic A M, Mosharov E V, Sulzer D, Melia T J and Rothman J E 2005 SNAREs can promote complete fusion and hemifusion as alternative outcomes *J. Cell Biol.* **170** 249–60

[88] Mostafavi H, Thiyagarajan S, Stratton B S, Karatekin E, Warner J M, Rothman J E and O'Shaughnessy B 2017 Entropic forces drive self-organization and membrane fusion by SNARE proteins *Proc. Natl Acad. Sci. USA* **114** 5455–60

[89] Chernomordik L V and Kozlov M M 2003 Protein-lipid interplay in fusion and fission of biological membranes *Annu. Rev. Biochem.* **72** 175–207

[90] Jackson M B 2016 The hydrophobic effect in solute partitioning and interfacial tension *Sci. Rep.* **6** 19265

[91] Tanford C 1979 Interfacial free energy and the hydrophobic effect *Proc. Natl Acad. Sci. USA* **76** 4175–6

[92] Kozlovsky Y and Kozlov M M 2003 Membrane fission: model for intermediate structures *Biophys. J.* **85** 85–96

[93] Leikin S, Parsegian V A, Rau D C and Rand R P 1993 Hydration forces *Annu. Rev. Phys. Chem.* **44** 369–95

[94] Lentz B R 1994 Polymer-induced membrane fusion: potential mechanism and relation to cell fusion events *Chem. Phys. Lipids* **73** 91–106

[95] Kozlov M M and Chernomordik L V 2015 Membrane tension and membrane fusion *Curr. Opin. Struct. Biol.* **33** 61–7

[96] Chanturiya A, Chernomordik L V and Zimmerberg J 1997 Flickering fusion pores comparable with initial exocytotic pores occur in protein-free phospholipid bilayers *Proc. Natl Acad. Sci. USA* **94** 14423–8

[97] Bretou M *et al* 2014 Cdc42 controls the dilation of the exocytotic fusion pore by regulating membrane tension *Mol. Biol. Cell* **25** 3195–209

[98] Zhou Z, Misler S and Chow R H 1996 Rapid fluctuations in transmitter release from single vesicles in bovine adrenal chromaffin cells *Biophys. J.* **70** 1543–52

[99] Wang C-T, Grishanin R, Earles C A, Chang P-Y, Martin T F J, Chapman E R and Jackson M B 2001 Synaptotagmin modulation of fusion pore kinetics in regulated exocytosis of dense-core vesicles *Science* **294** 1111–5

[100] Chiang N, Hsiao Y T, Yang H J, Lin Y C, Lu J C and Wang C T 2014 Phosphomimetic mutation of cysteine string protein-alpha increases the rate of regulated exocytosis by modulating fusion pore dynamics in PC12 cells *PLoS One* **9** e99180
[101] Wang C T, Bai J, Chang P Y, Chapman E R and Jackson M B 2006 Synaptotagmin-Ca^{2+} triggers two sequential steps in regulated exocytosis in rat PC12 cells: fusion pore opening and fusion pore dilation *J. Physiol.* **570** 295–307
[102] Bai J, Wang C T, Richards D A, Jackson M B and Chapman E R 2004 Fusion pore dynamics are regulated by synaptotagmin*t-SNARE interactions *Neuron* **41** 929–42
[103] Jackson M B 1992 Stationary single channel analysis *Methods Enzymol.* **207** 729–46
[104] Almers W 1990 Exocytosis *Annu. Rev. Physiol.* **52** 607–24
[105] Lindau M and Almers W 1995 Structure and function of fusion pores in exocytosis and ectoplasmic membrane fusion *Curr. Opin. Cell Biol.* **7** 509–17
[106] Llobet A, Wu M and Lagnado L 2008 The mouth of a dense-core vesicle opens and closes in a concerted action regulated by calcium and amphiphysin *J. Cell Biol.* **182** 1017–28
[107] Hinshaw J E and Schmid S L 1995 Dynamin self-assembles into rings suggesting a mechanism for coated vesicle budding *Nature* **374** 190–2
[108] Bashkirov P V, Akimov S A, Evseev A I, Schmid S L, Zimmerberg J and Frolov V A 2008 GTPase cycle of dynamin is coupled to membrane squeeze and release, leading to spontaneous fission *Cell* **135** 1276–86
[109] Gonzalez-Jamett A M *et al* 2013 Dynamin-2 regulates fusion pore expansion and quantal release through a mechanism that involves actin dynamics in neuroendocrine chromaffin cells *PLoS One* **8** e70638
[110] Trouillon R and Ewing A G 2013 Amperometric measurements at cells support a role for dynamin in the dilation of the fusion pore during exocytosis *ChemPhysChem* **14** 2295–301
[111] Wu Q, Zhang Q, Liu B, Li Y, Wu X, Kuo S, Zheng L, Wang C, Zhu F and Zhou Z 2019 Dynamin 1 restrains vesicular release to a subquantal mode in mammalian adrenal chromaffin cells *J. Neurosci.* **39** 199–211
[112] Samasilp P, Chan S A and Smith C 2012 Activity-dependent fusion pore expansion regulated by a calcineurin-dependent dynamin-syndapin pathway in mouse adrenal chromaffin cells *J. Neurosci.* **32** 10438–47
[113] Neco P, Fernandez-Peruchena C, Navas S, Gutierrez L M, de Toledo G A and Ales E 2008 Myosin II contributes to fusion pore expansion during exocytosis *J. Biol. Chem.* **283** 10949–57
[114] Archer D A, Graham M E and Burgoyne R D 2002 Complexin regulates the closure of the fusion pore during regulated vesicle exocytosis *J. Biol. Chem.* **277** 18249–52
[115] Barclay J W 2008 Munc-18-1 regulates the initial release rate of exocytosis *Biophys. J.* **94** 1084–93
[116] Heuser J E and Reese T S 1981 Structural changes after transmitter release at the frog neuromuscular junction *J. Cell Biol.* **88** 564–80
[117] Torri-Tarelli F, Grohovaz F, Fesce R and Ceccarelli B 1985 Temporal coincidence between synaptic vesicle fusion and quantal secretion of acetylcholine *J. Cell Biol.* **101** 1386–99
[118] Watanabe S 2016 Flash-and-freeze: coordinating optogenetic stimulation with rapid freezing to visualize membrane dynamics at synapses with millisecond resolution *Front Synaptic Neurosci.* **8** 24
[119] Lollike K, Borregaard N and Lindau M 1998 Capacitance flickers and pseudoflickers of small granules, measured in the cell-attached configuration *Biophys. J.* **75** 53–9

[120] Takahashi N, Kishimoto T, Nemoto T, Kadowaki T and Kasai H 2002 Fusion pore dynamics and insulin granule exocytosis in the pancreatic islet *Science* **297** 1349–52
[121] Anantharam A, Bittner M A, Aikman R L, Stuenkel E L, Schmid S L, Axelrod D and Holz R W 2011 A new role for the dynamin GTPase in the regulation of fusion pore expansion *Mol. Biol. Cell* **22** 1907–18
[122] Wen P J *et al* 2016 Actin dynamics provides membrane tension to merge fusing vesicles into the plasma membrane *Nat. Commun.* **7** 12604
[123] Alvarez de Toledo G, Fernandez-Chacon R and Fernandez J M 1993 Release of secretory products during transient vesicle fusion *Nature* **363** 554–8
[124] Wang C T, Lu J C, Bai J, Chang P Y, Martin T F, Chapman E R and Jackson M B 2003 Different domains of synaptotagmin control the choice between kiss-and-run and full fusion *Nature* **424** 943–7
[125] Mellander L J, Trouillon R, Svensson M I and Ewing A G 2012 Amperometric post spike feet reveal most exocytosis is via extended kiss-and-run fusion *Sci. Rep.* **2** 907
[126] Perrais D, Kleppe I C, Taraska J W and Almers W 2004 Recapture after exocytosis causes differential retention of protein in granules of bovine chromaffin cells *J. Physiol.* **560** 413–28
[127] Chang C W, Hsiao Y T and Jackson M B 2021 Synaptophysin regulates fusion pores and exocytosis mode in chromaffin cells *J. Neurosci.* **41** 3563–78

Chapter 6

Mechanical regulation of exocytosis and endocytosis

Shan-Shan Lin, Jophin Joseph and Allen P Liu

Overview

Cells constantly encounter mechanical stresses present in, and generated by, their external environment. Such external stimuli include shear stress, compressive stress, and tensile stress. As the physical barrier between intra- and extracellular space, the plasma membrane serves as a border that preserves cell integrity and also possesses mechanosensory functions that guide downstream biochemical responses [1]. Apart from the mechanical stresses caused by external forces, cell behaviors such as division, spreading, and migration can also actively affect membrane shape and cell volume, resulting in alterations in the mechanical properties of the plasma membrane.

In this chapter, we will focus on the relationship between membrane tension and membrane trafficking, as well as their reciprocal regulation. Specifically, we will discuss how alterations in membrane tension resulting from external forces impact membrane trafficking, the molecular mechanisms underlying tension-regulated endo- and exocytosis, and how membrane tension homeostasis or asymmetry is facilitated by membrane trafficking.

6.1 Introduction

The concept that membrane trafficking is regulated by membrane tension can be dated back to morphometric studies of the mammalian urinary bladder in the late 1970s and early 1980s [2, 3]. The mammalian bladder undergoes a series of expansions and contractions which leads to a dramatic change in membrane surface area. Studies show that filled bladders have fewer cytosolic vesicles compared with contracted bladders and that bladder epithelium changes from a cuboidal to a squamous shape during bladder expansion, accompanied by fusion of cytoplasmic

doi:10.1088/978-0-7503-3771-7ch6

vesicles. In addition to bladder epithelium, cell types including cardiomyocytes, smooth muscle cells, vascular endothelial cells, adipocytes, and lung epithelial cells also experience constant changes in membrane tension [2–8].

To understand how cells respond to such mechanical stimuli, different approaches have been developed to manipulate membrane tension under physiologically relevant contexts [9, 10]. Using manipulations such as mechanical stretch and osmotic shock that mechanically deform cells to change plasma membrane tension, researchers have shown that membrane trafficking is stimulated in the tension-responsive cells [4–6, 11–13]. The secretion of growth factors, surfactant proteolipid, and hormones is enhanced when the osmolarity of the culture environment is changed [4] or cells are subjected to a stretching device that increases membrane tension [5, 6].

The membrane tension of a cell corresponds to the combination of in-plane membrane tension and cortical tension [14, 15]. When we consider the lipid bilayer to be a continuous 2D fluid film, stretching-compression elastic stress along the surface gives rise to in-plane surface tension [16]; whereas the structural rigidity and the actomyosin contractile force of the underlying actin meshwork that the plasma membrane is in close contact with contribute to cortical tension [17–19]. A caveat is that in plasma membranes, the lipid bilayer and the actin cortex act on one another. Thus, it is appropriate to consider their tension as a whole. The actomyosin contractile force at cell–substrate or cell–cell adhesion sites is another contributor to membrane tension. In adherent cells, when actin stress fibers anchor to focal adhesions, the contractile force generated by myosin can be transmitted through integrins [20]. Growing evidence shows that membrane trafficking is a key effector of membrane tension, which helps cells adapt to the external environment and maintain the homeostasis of the plasma membrane. Since the cell boundary is delimited by the plasma membrane, membrane trafficking in this context can be a key regulator of membrane tension, as exocytosis and endocytosis modulate the surface area of the plasma membrane [14, 21].

6.2 Membrane tension and membrane homeostasis

The surface-area regulation theory has been proposed to explain how membrane tension regulates membrane trafficking [21]. It suggests that when membrane tension changes due to externally imposed or cell-mediated shape changes, the tension perturbation results in surface-area adjustments that maintain membrane homeostasis. The resting membrane tension ranges from 0.003 mN m^{-1} in the growth cones of chick neurons to less than 1 mN m^{-1} in different cell types under physiological conditions [21]. Since a lipid bilayer has very low lateral elasticity and cannot be expanded by more than 3%–5% of its area without rupture [21], it naturally resists bending and has a preferred curvature due to compositional asymmetry. These properties of a lipid bilayer, especially the elastic limit, come into play when membrane tension is changed, and thus trigger membrane trafficking as a buffering mechanism to maintain cell membrane integrity [21]. Although membrane trafficking has been intensively studied in the context of the delivery of proteins and vesicle contents, it is important to note that along with its physiological roles, vesicle trafficking also brings or removes phospholipids that can expand or reduce the membrane surface area, thereby modulating membrane tension. When membrane tension increases, membrane lipids are supplied by endomembrane reserves through exocytosis. When membrane tension decreases, membrane lipids are retrieved via endocytosis. At a moderate tension of 0.02 mN m^{-1}, the endocytosis of a small 20 nm vesicle changes the elastic energy by ~4 kcal mol^{-1} and is expected to increase membrane strain. However, under higher tensions, the elastic energy change required to remove membrane would be prohibitive and must rely on external forces (i.e. the cytoskeleton).

One classic example of how cells maintain membrane homeostasis in response to membrane tension is the caveolae. Cells utilize caveolae as a reservoir for excess membrane and respond to increasing membrane tension by flattening these structures [22–24]. Caveolae are 50–60 nm wide flask-shaped plasma membrane pits composed of caveolins and cavins and regulated by caveolin-mediated endocytosis [25]. The discovery of caveolae can be traced back to the 1950s, when cytologists observed many submicroscopic structures in endothelial and epithelial cells using electron microscopy [26, 27]. The term 'caveola' was used to denote the cave-like or pocket-like pits that extend inward from the surface of a cell. It has been proposed that oligomerization of caveolin drives membrane bending and causes the formation of cup-like structures [28]. However, recent structural studies revealed that caveolin complexes may primarily function in stabilizing the flat membrane surfaces of the polyhedral structures rather than by imposing membrane curvature [29]. There are three different types of caveolin in mammalian cells, namely CAV1, CAV2, and CAV3. While CAV1 and CAV2 are abundantly expressed in non-muscle cells, CAV3 is expressed in skeletal muscle cells. All three caveolins are expressed in smooth muscle cells and cardiomyocytes [28]. Myotubes bearing dystrophy-associated CAV3 mutations showed abnormal response to mechanical stress due to a decrease in the number of caveolae [30]. Another essential component of caveolae is the cavin family of proteins. In mammals, there are four isoforms of cavins, which are Cavin1, Cavin2, Cavin3, and a muscle-specific isoform, Cavin4 [31]. Cavin1 is ubiquitously expressed and is indispensable for caveolae formation. Animals lacking cavin are viable but have no morphologically detectable caveolae [32]. In addition to cavin and caveolin, a curvature-generating Fes/CIP4 homology BAR (F-BAR) protein enriched at the necks of caveolae, Pacsin (also known as Syndapin), and an ATPase, EH domain-containing protein 2 (EHD2), are also important for caveolae formation and stabilization [33]. EHD2 oligomerizes around the caveolar neck to stabilize caveolae in an ATP-dependent manner. Removal of EHD2 promotes caveolae detachment and their intracellular trafficking [34]. Rosettes or clusters of caveolae are stabilized at low membrane tension and destabilized by high tension [8] (figure 6.1). It is known that the flattening of caveolae is integral to the process by which muscle cells protect themselves against membrane rupture due to high tension [23].

The density of caveolae varies in different cell types. In endothelial cells, muscle cells, and adipocytes, ~50% of the plasma membrane surface contains caveolae [28]. In tissues such as muscle fibers, the organization of caveolae has been found to not

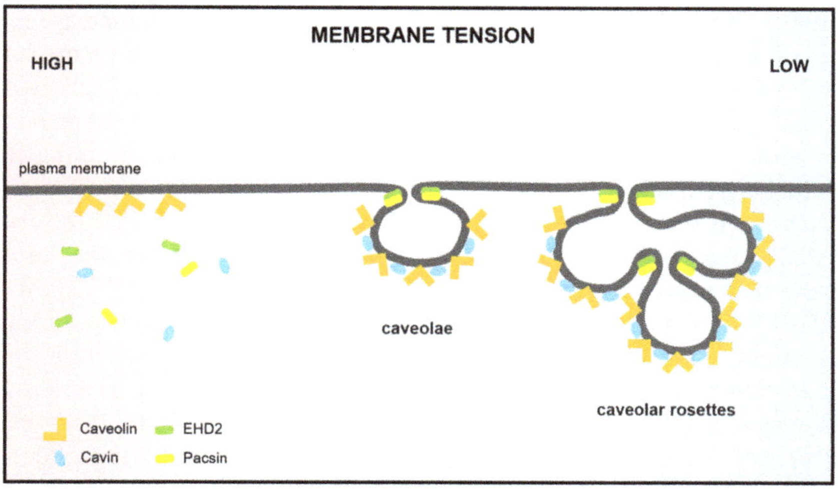

Figure 6.1. Caveolae as mechanosensors for membrane tension. Caveolar rosettes serve as membrane reservoirs at low membrane tensions. High-tension membrane conditions, such as hypo-osmotic shock and stretching, induce caveolae flattening, which causes detachment of the stabilizing proteins, including Pacsin and EHD2.

only consist of single pits but also higher-order structures with multiple lobular bulbs [35, 36]. These highly invaginated membrane domains play important roles as membrane tension regulators in cells that are frequently confronted by stretch and shear stress. It is believed that the invagination of membrane functions as a membrane reservoir that buffers changes in mechanical tension [37]. *In vivo* studies have demonstrated that increasing cardiac output that mimics the effects of exercise induces caveolae disassembly in lung endothelial cells as well as cardiac muscle in mice [38]. Mice lacking Cavin1 are deficient in caveolae in all tissues and had significantly lower exercise endurance [39]. Studies have shown that the number of caveolae in HeLa cells is decreased by hypo-osmotic shock and stretching [23]. After hypo-osmotic shock, flattened caveolae were observed using deep-etched electron microscopy (EM). In addition, lifetime imaging using fluorescence lifetime imaging microscopy showed decreased interaction between CAV1 and Cavin1 after hypo-osmotic shock. These observations showed that cells disassemble caveolae in response to membrane mechanical stress. Cells flatten caveolae to cope with hypo-osmotic shock so that a considerable amount of membrane can be released from the caveolar invagination to maintain membrane tension [37]. Although the hypo-osmotic condition that was used to induce 40%–50% of caveolar flattening in cultured HeLa cells was likely an extreme condition compared with situations *in vivo*, 3T3 mouse fibroblasts and T24 human bladder carcinoma cells showed different levels of caveolar invagination, including nearly flat structures [25, 40]. Shear stress and osmotic-stress-induced integrin signaling led to CAV1 phosphorylation at tyrosine residue 14 (pY14) [41]. CAV1Y14 phosphorylation inactivates the early growth response-1 (EGR1) transcription factor which decreases CAV1 and Cavin1 genetic expression via competition with specificity protein 1 (SP1) transcription factor that enhances target gene expression [42]. An increase in CAV1 and Cavin1 expression supports caveolae biogenesis and strengthens mechanosensation of the cell. Given the above, it is widely accepted that cells utilize caveolae in the plasma membrane as one of the buffer systems against rapid membrane tension changes, either by flattening caveolae (during tension increase) or forming caveolae rosettes (during tension decrease) [43].

A structure in yeast that is functionally similar to caveolae, called the eisosome, was initially discovered as large immobile protein assemblies at the plasma membrane that mark sites of endocytosis [44]. Using yeast protoplasts that lack the rigid cell wall, eisosomes disassembled as the surface area increased during hypo-osmotic stress [45]. In contrast, hyperosmotic shock that caused cell shrinkage resulted in membrane slackening and deepening of eisosomes [46]. The morphology of eisosomes in response to changes in plasma membrane tension directly relates to surface expression of nutrient transporters and thus helps to regulate nutrient transporter stability via endocytosis [46]. The modulation of eisosomes in response to an increase in membrane tension changes is also coupled with a temporary reduction of endocytosis and an increased rate of exocytosis to allow cells to add membrane to the plasma membrane [47]. Given these studies in yeasts and other studies in mammalian cells, it is likely that membrane tension regulation and its relationship to endocytosis, exocytosis, and osmotic response are highly conserved among eukaryotes.

6.3 Coupling of exo- and endocytosis

6.3.1 Synaptic transmission

Membrane homeostasis is essential for cell function. One well-known example is synaptic vesicle recycling. Synaptic vesicles in the neuronal terminals are frequently endocytosed following exocytosis [48]. During synaptic transmission, exocytosis of synaptic vesicles increases the surface area of axon terminals, whereas endocytosis counteracts their expansion [49]. The recycling process is critical in removing excess membrane from the plasma membrane and is proposed to maintain the optimal tension needed to enable synaptic vesicle fusion [49, 50].

The two major exocytosis modes that regulate vesicular content release are full-collapse fusion and kiss-and-run. The former releases synaptic vesicles by rapidly expanding the fusion pore and eventually leads to the flattening of the synaptic membrane into the plasma membrane, whereas the kiss-and-run mode bypasses vesicle collapse and releases vesicular contents through a narrow pore [51, 52]. Fusion is mediated by the soluble N-ethylmaleimide sensitive factor (NSF) attachment protein receptor (SNARE) complex, including vesicle SNAREs (v-SNAREs) and their counterparts in the target membrane, so-called t-SNAREs. The synaptic vesicle proteins, synaptobrevin/VAMP, and calcium-sensing synaptotagmins interact with the plasma membrane protein syntaxin and soluble NSF attachment proteins (SNAPs). Together, the SNARE complex mediates vesicle fusion in response to calcium entry upon neuron impulse [53, 54] (figure 6.2). Several endocytic modes have been demonstrated and classified. The recycling modes

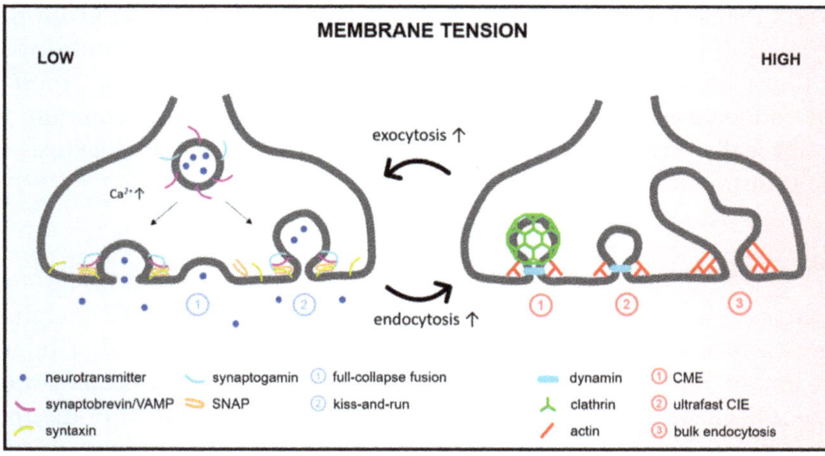

Figure 6.2. Membrane trafficking in the pre-synapse is mediated by membrane tension. SNARE complex-mediated release of synaptic vesicles increases surface area and reduces membrane tension in response to a neuronal impulse (left). Reduction in membrane tension in turn activates clathrin-dependent and independent endocytic routes that retrieve excess membrane and maintain membrane homeostasis (right).

include classical mechanisms, such as clathrin-mediated endocytosis (CME), and more recently established mechanisms that are independent of clathrin, such as bulk endocytosis and ultrafast endocytosis [48]. These mechanisms ensure that the plasticity of neuronal terminals is suitable for the different functions of neurotransmission. Compared with CME, which is well known for its function in internalizing receptors instead of mass lipid removal, the latter two mechanisms provide a more efficient way to remove excess membrane and play important roles in preventing exhaustion or overstimulation [49]. For instance, ultrafast endocytosis is characterized by rapid formation of membrane invagination without the involvement of clathrin. Neuronal synapses use ultrafast clathrin-independent endocytosis (CIE) mediated by actin and dynamin to recycle synaptic vesicles [49]. Actin is required to generate membrane curvature, whereas dynamin is required to pinch off the vesicles by membrane fission [55]. On the other hand, bulk endocytosis plays a critical role in responding to longer and stronger stimulation [48]. Bulk endocytosis is triggered by significant membrane addition due to high exocytotic loads and an increase in calcium at the neuron terminals. Like ultrafast endocytosis, bulk endocytosis retrieves the membrane through actin-dependent mechanisms, which mediates membrane bending and fission [51]. It has been shown that under a high level of stimulation, bulk endocytosis is upregulated in dynamin knockout synapses in which CME is dramatically impaired, demonstrating its ability to bypass CME during intense synaptic activity [56] (figure 6.2).

Overall, coupling between exo- and endocytosis plays a profound role in synaptic transmission. Membrane tension contributes to the rate, size, and morphology of membrane invagination morphologies which are critical for both endocytosis and exocytosis [43, 57, 58, 77, 78]. Rapid removal of excess membrane is necessary to restore the fusion sites and maintain membrane tension during rapid membrane turnover in the synapse. Moreover, experimental and modeling studies suggest that membrane tension contributes to membrane shape transition during ultrafast endocytosis [59]. Given this knowledge, a change in membrane tension may reciprocally facilitate the coupling between exo- and endocytosis during synaptic transmission.

6.3.2 Phagocytosis

Phagocytosis is another example of the use of exo- and endocytic coupling as a buffering system to maintain homeostasis. Phagocytosis is a vital cellular event for the innate immune response that is characterized by receptor-mediated engulfment of large particles [60]. During phagocytosis, the membranes beneath the particle grow over its surface, encircling the prey. Membranous protrusions, the so-called pseudopods, must gain extensive membrane to complete the engulfment. The pseudopod extension has two phases. In the first phase, the cell utilizes preexisting membrane to encircle the attached particle, and this is accompanied by actin polymerization that enforces the encirclement of the particle. The second phase occurs when the membrane reservoir is depleted, leading to an increase in membrane tension [60]. It has been found that the increased tension triggers exocytosis to provide an extra membrane for pseudopod extension, thereby facilitating the final step of particle uptake [61] (figure 6.3).

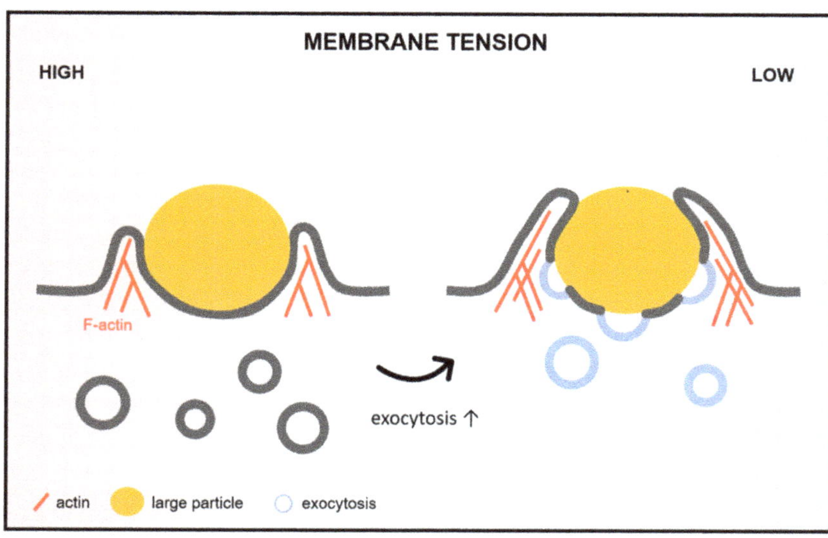

Figure 6.3. Phagocytosis is facilitated by local exocytosis that buffers a membrane tension increment. Membrane tension increases during pseudopod extension (left). The increase in membrane tension in turn triggers exocytosis that provides membrane for pseudopod extension and allows particle uptake (right).

6.3.3 Membrane sealing

The plasma membrane is challenged by various external stresses such as chemicals, pathogens, or mechanical stresses. Defects in membrane repair are linked to multiple neuronal and cardiac diseases, illustrating the importance of membrane sealing in the maintenance of membrane integrity [62, 63]. While membrane injuries at the nanometer scale can reseal spontaneously, for larger injuries, especially when the membrane is linked to the underlying cytoskeleton, the opposing force in the membrane cytoskeleton prevents spontaneous sealing. Since membrane tension limits sealing in this case, other mechanisms that reduce membrane tension, i.e. exocytosis, are needed. The idea that exocytosis contributes to the repair of wounded membrane has been supported by earlier studies [64, 65]. It has been found that laser-induced damage to the membrane of sea urchin eggs leads to a rapid burst of Ca^{2+}-dependent exocytosis that correlates with membrane sealing. Moreover, when membrane fusion is disrupted by neurotoxins, which selectively cleave SNARE protein complexes, sealing of the wound is inhibited [64]. On the other hand, the Ca^{2+}-binding protein, annexin A5, is also found to be involved in membrane sealing by forming a 2D array around the wound and promoting sealing via reducing membrane tension [66]. Conversely, lowering membrane tension by disrupting actin cortex enhances membrane sealing in low Ca^{2+} conditions in which exocytosis is inhibited [67]. Thus, lowering membrane tension is the key to successful membrane sealing.

6.3.4 Cell spreading

It is well known that motile functions rely on biochemical signaling to mediate the assembly of cytoskeletal networks. It is increasingly appreciated that physical forces provide a robust mechanism that coordinates biochemical reactions in order to govern cell behaviors. Membrane tension is becoming recognized as a key regulator of cell motility. Apart from the externally imposed mechanical stimuli that alter membrane tension, regulation of cell-mediated activities including migration, spreading, and growing is also highly intertwined with the mechanics of the plasma membrane. Cellular processes such as actin cytoskeletal reorganization, the formation of focal adhesions, cellular polarization, etc. can create spatiotemporal variations in membrane tension that affect membrane trafficking [68, 69]. Cell motility is a critical cellular process that impacts biological events spanning from embryogenesis, angiogenesis, immune response, and tissue repair to metastasis. It requires multifarious regulations that involve both biochemical and biophysical triggers, and they play critical roles in development and disease.

During spreading and migration, the cell undergoes a sequence of cytoskeletal rearrangement and membrane reorganization. Increasing evidence shows that membrane tension acts as a mechanical regulator for cell motility [70–75]. It has been shown that fast membrane recycling by endocytosis in combination with rapid exocytosis is utilized by fibroblasts to regulate membrane area during spreading [76]. When fibroblasts contact a surface, they adhere and spread in a stepwise manner [11, 79–81]. In the beginning, the cells expand the surface area rapidly at a uniform speed. During this phase, they utilize the membrane reservoir provided by folded

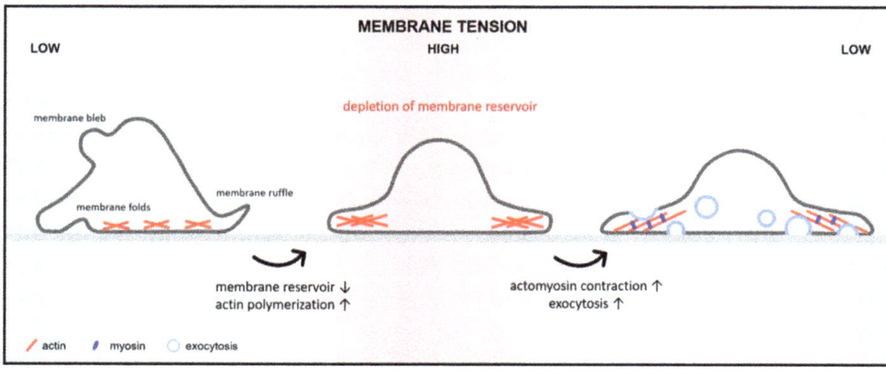

Figure 6.4. Exocytosis facilitates cell spreading. Unfolding and flattening of the membrane reservoir contained in blebs, ruffles, and folds provide extra membrane area at the beginning of cell–substrate attachment (left). As actin continues to polymerize and the membrane reservoir is depleted, membrane tension increases and triggers exocytosis (middle). Membrane tension gradually decreases as the membrane surface progressively expands (right). During this stage, actomyosin contraction increases and adhesions begin to mature.

regions, ruffles, and blebs to increase the membrane area as actin polymerizes, and there is only weak contraction between the cell and the substrate. After the membrane reservoir is used up, there is a sudden transition to the next phase. In the second phase, actin-driven protrusion slows down and actomyosin contraction starts to increase to facilitate focal adhesion formation, whereby the addition of ~40%–60% plasma membrane is supplied by exocytosis [79, 81]. The transition from the first phase to the second phase is later characterized by an abrupt increase in membrane tension due to depletion of the membrane reservoir [11] (figure 6.4). Using optical tweezers to perform a tether-pulling assay and measure the restoring force, Gauthier *et al* found that membrane tension increases twofold by the end of the first phase, increasing the exocytosis of lysosomes and glycosylphosphatidylinositol (GPI)-anchored protein vesicles. The effect of tension likely shifts the balance between exocytosis and endocytosis, buffering the membrane tension change [11, 81]. It is interesting to note that cells lacking CAV1 exhibit impaired cell polarity and directional migration but not impaired cell spreading [82]. However, since caveolae have been shown to contain GPI-anchored proteins, the involvement of caveolae during cell spreading requires further investigation [83]. Further, artificially increasing the membrane tension by hypo-osmotic treatment halts lamellipodia protrusion and activates exocytosis, suggesting that membrane tension is the trigger for the transition between the initial rapid spreading phase and the contractile spreading phase.

6.3.5 Cell migration

As the cell migrates, it has been shown that exocytosis is more prominent at the leading edge, whereas endocytosis mainly occurs in the cell rear [84]. During cell migration, the membrane of the recycling endosomes is incorporated into the leading edge that facilitates surface expansion through SNARE-mediated membrane fusion [85]. Exocytosis not only aids the expansion of the leading edge but also

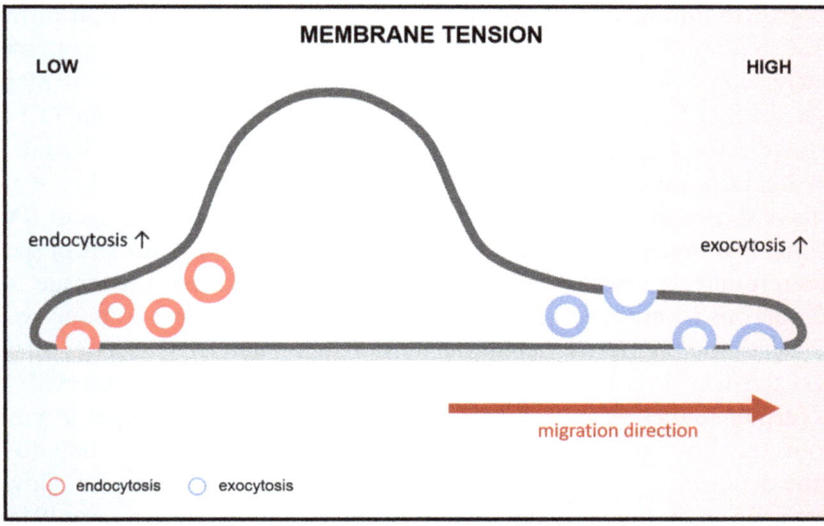

Figure 6.5. Front-to-rear tension gradient maintains asymmetric membrane trafficking and promotes directional cell migration. During directional cell migration, exocytosis is upregulated in the cell front to expand the membrane area and allow actin polymerization. In contrast, endocytosis is upregulated in the cell rear to remove membrane.

recycles and delivers integrins that allow adhesion formation at the cell front. This differential regulation is due to a front-to-rear gradient of membrane tension. Membrane tension at the leading edge is higher than that at the trailing edge, accompanied by a higher Rac activity and actin assembly at the cell front [75, 86]. Experiments demonstrate that when membrane tension is increased by micropipette aspiration, osmotic swelling, or stretching, Rac activity and the WASP family verprolin homologous protein (WAVE) complex recruitment at the leading edge are decreased, demonstrating that membrane tension is a limiting factor for leading-edge formation [74, 86, 88]. As protrusions form, membrane tension increases and rapidly propagates throughout the cell. The viscous friction due to adhesion between the membrane and the substrate also contributes to the tension gradient during cell migration [73]. At the cell rear, caveolae form in response to low membrane tension, which activates RhoA signaling, inducing contractility to promote cell translocation [88]. The positive feedback between the front–rear membrane tension differential and retraction at the cell rear reinforces directional cell migration. Theoretical modeling also highlights the contribution of membrane tension to symmetry breaking during cell spreading and migration [89]. Together, the coupling between the tension gradient and membrane trafficking maintains directional cell migration by limiting the formation of the leading edge to the cell front (figure 6.5).

6.3.6 Axon outgrowth and guidance

Similarly to its role in cell migration, exocytosis plays a role in other cell movements that require membrane expansion and directionality maintenance. One example is axon outgrowth and guidance during neural development. In addition to the well-established relationship between exocytosis and signaling in growth cone development, increasing evidence also suggests that the mechanical effect of membrane tension is critical for axon dynamics. A mechanical force of around 0.3–10 nN applied to cultured neurons breaks cell symmetry and stimulates axon initiation [90]. Axon elongation has also been reported to be regulated by membrane tension. Studies have shown that axons undergoing stretch exhibit elongation from 100 μm to 10 cm within two weeks, which is ten times faster than their typical movement speed [91]. It has been found that neuron growth speed is increased when the tensile force on cultured neurons is increased, either by glass needle or by hypo-osmotic treatment [92, 93], suggesting that rapid membrane tension increment triggers axon outgrowth. However, the way in which membrane tension impacts axon outgrowth *in vivo* requires further studies. A tension gradient arises that drives a retrograde membrane flow from the growth cone, where membrane tension is lower due to excess membrane deposition, to the soma [12]. Lower membrane tension also allows actin polymerization in filopodia at the peripheral region of the growth cone [94]. These findings demonstrate that membrane tension plays a role in axon outgrowth (figure 6.6(a)). Further, since the growth of actin filaments and microtubules is limited by the plasma membrane, axonal guidance requires an increase of membrane surface and involves asymmetric membrane transport [95, 96]. In cultured embryonic chick dorsal root ganglion neurons, expansion of the attraction side is mediated by delivery of VAMP7-positive endocytic vesicles through exocytosis, whereas exocytosis is inhibited in the repulsive side [97]. This asymmetric membrane trafficking is mediated by second messengers, such as cyclic nucleotides. The attractive signal, cAMP, stimulates microtubule growth and promotes transport of VAMP7 vesicles to the leading edge; conversely, cGMP in the repulsive side

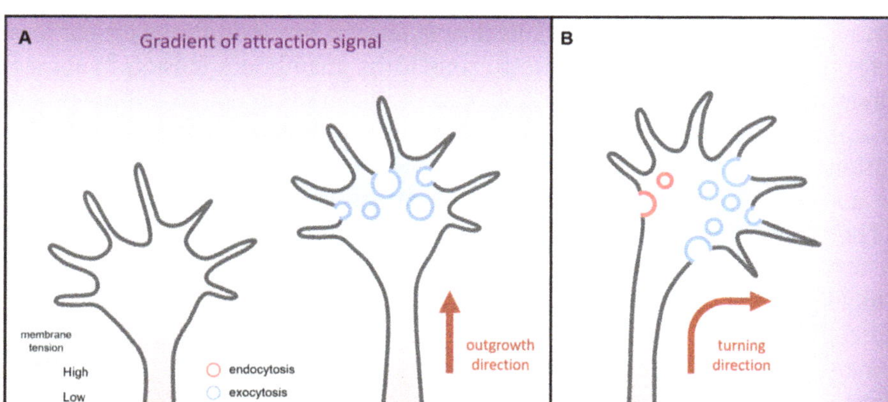

Figure 6.6. Membrane tension and asymmetric membrane trafficking promotes axon outgrowth and turning. An increase in membrane tension serves as a positive signal that induces exocytosis and facilitates axon outgrowth (left). During axon turning, exocytosis is activated at the attractive side and inhibited at the repulsive side to reduce membrane tension and facilitate membrane expansion, while endocytosis is activated at the repulsive side to retrieve excess membrane.

inhibits these events. Based on the significant role of membrane tension in cell migration, and the fact that membrane trafficking directly orchestrates axon extension, it has been proposed that exocytosis lowers membrane tension and facilitates axon guidance through actin polymerization [98] (figure 6.6(b)). Still, the way in which membrane tension modulates neuronal development *in vivo* requires further investigation.

6.4 Regulation of exocytosis by membrane tension

Exocytosis, the process through which an intracellular vesicle fuses with the plasma membrane and releases luminal contents, is a fundamental process that regulates various biological events. From immune responses to neuron transmission, exocytosis regulates the secretion of cytokines and neurotransmitters for cell communication, as well as the delivery of transmembrane proteins to plasma membrane for cell signaling and adaptation [11, 99, 100]. Exocytosis is a complicated process that comprises multiple stages. Vesicles are first transported along microtubules to the plasma membrane. Upon reaching the plasma membrane, the vesicles are tethered and docked to the plasma membrane. Following docking, the lipid bilayers of the vesicle and plasma membranes begin to merge. Membrane fusion is then initiated by formation of a nanometer-sized fusion pore and expansion of the fusion pore [101].

Growing evidence obtained from simulations and experiments shows that higher membrane tension is critical for successful fusion [16, 102, 103]. Molecular dynamics simulations show that tension in both of the opposing membranes should be high enough to promote fusion [103]. A recent study using model membranes equipped with a minimum set of SNARE fusion machinery further verified this concept [102]. The authors developed two *in vitro* strategies to manipulate membrane tension; one used a dilatable polydimethylsiloxane (PDMS) sheet to stretch patches of planar lipid bilayers; the other manipulated the membrane tension by adjusting the adhesion area between giant unilamellar vesicles (GUVs) and the substrate (figures 6.7(a) and (b)). In the latter approach, the substrate was coated with avidin and the GUVs were reconstituted using biotinylated phospholipids. When the concentration of Mg^{2+} ions increases, the adhesion area increases due to the bridging effect. Using these setups, they were able to monitor the fusion efficiency of large unilamellar vesicles (LUVs) in solution with lipid bilayers or GUVs. The results obtained using the first approach show that fusion efficiency was slightly increased when the tension reached 3.2 mN m^{-1} and then significantly increased when the membrane area was expanded to around ~1.4%–1.6%, corresponding to a membrane tension of ~4.0–4.4 mN m^{-1}. Since the fusion event was limited by the expanded area in the stretched patch and was restricted to a few percent, the author employed the second approach, which was able to adjust the tension over a larger range, i.e. ~0.1–5.0 mN m^{-1}. It was found that fusion efficiency increased slightly between ~0.6–2.6 mN m^{-1} and increased dramatically as the membrane tension reached ~4.0–5.0 mN m^{-1}. In conclusion, both assays suggest that fusion efficiency depends on membrane tension, and that a threshold tension has to be overcome to

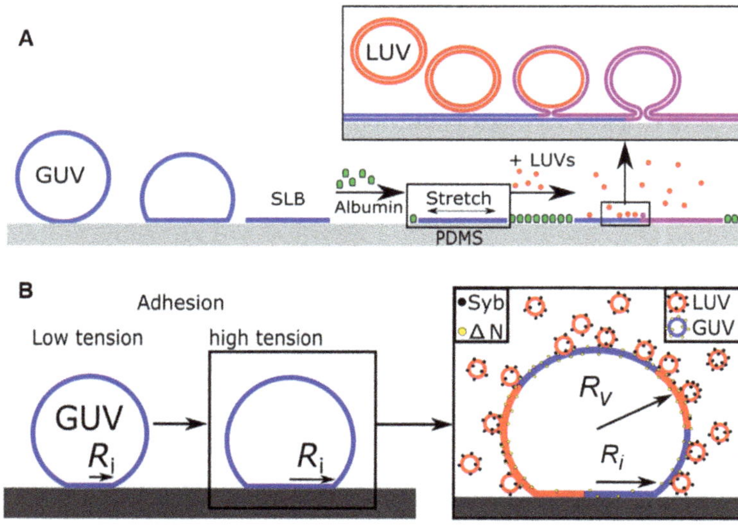

Figure 6.7. Schematic diagram of the two assays that manipulated membrane tension and its impact on membrane fusion. (a) Measurement of the tension-dependent fusion of LUVs (red) with pre-stretched bilayer (blue) GUVs. Membrane patches are formed by the spreading of GUVs. Albumin was added to passivate the surface, inhibiting LUV fusion on PDMS. The addition of LUVs to the sample resulted in the fusion of LUVs with single lipid bilayers (SLBs) in the presence of SNARE proteins. (b) Fusion of LUVs to adhered GUVs achieved by modifying the membrane tension via changing the adhesion between the GUV and the functionalized glass surface. The GUV membrane tension was calculated by evaluating the radius of the sessile GUV (*R* v) and the contact radius (*R i*) with the substrate. Reproduced under the terms of the CC BY 4.0. Copyright © 2017, Springer Nature [102].

promote fusion. It is noteworthy that a further increase in membrane tension does not further enhance fusion.

Experimental and theoretical studies have demonstrated that pore opening and expansion are the most energy-consuming processes during the late stage of fusion [104]. After pore opening, the vesicle forms an omega-shaped profile with a narrow neck connected to the plasma membrane. Shrinkage or enlargement of the omega-shaped structure is dependent on membrane tension [105]. It has been suggested that fusion proteins, such as the SNARE complex and its partners, drive the accumulation of local curvature stress and lower the energy barrier for fusion pore formation [106]. SNARE proteins bring the outer leaflets of the vesicle and the plasma membrane into contact by forming intertwined alpha helical bundles between the opposing membranes. Experimental and computational results have demonstrated that crowding of SNARE proteins generates entropic forces that facilitate pore expansion, and that the probability of pore expansion increases with SNARE density [105]. In addition, the cortical actin network also plays a direct role in mediating pore expansion and fusion. Studies have demonstrated that the actin network localizes to, and regulates, SNAP-25 clustering, which defines the final sites of exocytosis [107]. Furthermore, the inhibition or knockdown of Cdc42 reduces membrane tension by half and compromises fusion pore enlargement, resulting in impaired exocytosis in neuroendocrine cells [108]. Similarly, inhibiting the actin nucleation-promoting factor N-WASP also restricts pore expansion due to a reduction in actin polymerization [109]. Other studies have shown that an increase in myosin II activity correlates with the enlargement of fusion pore and exocytosis, suggesting that cortical actin provides membrane tension and promotes pore expansion and fusion [110].

In addition, several experiments have shed light on dynamin as a key regulator for fusion pore expansion via remodeling the cortical actin network. The activity of dynamin has been found to correlate to the rapidity of pore expansion [111, 112]. Inhibition of dynamin slows down fusion pore expansion, whereas an increase in dynamin activity causes rapid fusion pore widening, suggesting that expansion of fusion pores is bidirectionally regulated by dynamin. Inhibition of actin polymerization or dynamin-2 reduces fusion pore expansion in a non-additive manner, showing that dynamin and actin work in concert to mediate membrane tension and in turn regulate the expansion of fusion pores. It has also been proposed that dynamin restricts the pore size and counteracts F-actin and tension-dependent pore expansion [113, 114]. Studies combining super-resolution stimulated emission depletion imaging and force measurement have shown that the assembly of an actin network enhances membrane tension and promotes the release of vesicular contents [113, 115]. It has been found that the size of the fusion pore is reduced when membrane tension is lowered by either hyperosmotic treatment or the use of latrunculin A to disrupt filamentous actin. These studies have provided the missing evidence that shows how membrane tension regulates fusion pore size in living cells (figure 6.8).

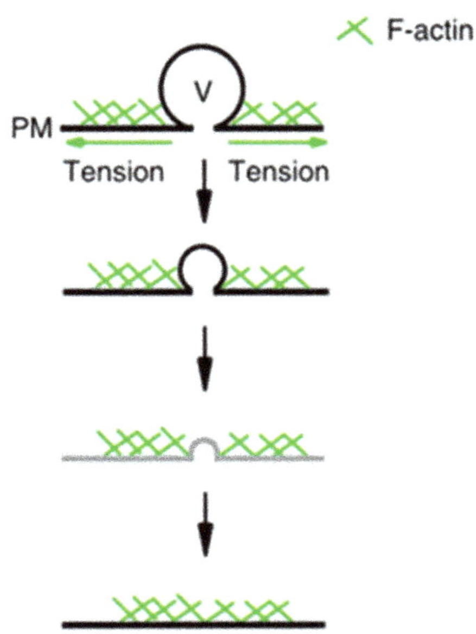

Figure 6.8. The actin network mediates membrane tension to facilitate vesicle fusion. F-actin shrinks the Ω-profile by enhancing membrane tension. Reproduced under the terms of the CC BY 4.0. Copyright © 2018, Springer Nature [115].

6.5 Regulation of endocytosis by membrane tension

Eukaryotic cells have evolved diverse endocytic pathways that internalize extracellular components. The canonical endocytic pathway, CME, involves the internalization of cargo by packaging it in 60–120 nm wide clathrin-coated vesicles [116]. Membrane tension plays a major role in regulating the reorganization of a flat membrane into a clathrin-coated bud [117–119]. It also affects the rate of nucleation and the lifetimes of endocytic events in different pathways by increasing the energy barrier for membrane budding and inhibiting curvature formation and membrane scission [7, 69, 120–122]. High membrane tension inhibits the transitions of flat membranes to open hemispherical invaginations (U shaped) and those of open hemispherical invaginations to a closed (Ω) topology [69, 123–125] (figure 6.9). Inhibition of these transitions leads to premature dissociation and/or stalling of vesicles in the membrane [69, 117, 121]. The interplay between membrane tension and the recruitment of endocytic coat proteins often controls morphology and the fate of the endocytic vesicles [123, 126]. The activities of actin cytoskeletons and membrane curvature generation proteins are needed to rescue membrane trafficking processes under high membrane tension conditions [69, 123, 126]. Coarse-grained modeling shows that lower membrane tension favors small-budded structures, whereas higher tension stabilizes large, flat clathrin plaques [127]. At physiologically relevant values of membrane tension (0.02 mN m^{-1}), the transition from a hemispherical open bud to an Ω-shaped closed bud occurs spontaneously via a snap-through instability [123, 124]. However, at high tension, a smooth transition from an open to closed bud morphology without snap-through instability is achieved by increasing clathrin-coat rigidity and the external force produced by actin polymerization [123]. This is consistent with experimental results which show that disrupting actin cytoskeleton by jasplakinolide causes stalling of clathrin buds in cells under hypo-osmotic shock [69]. In summary, membrane tension, endocytic coat proteins,

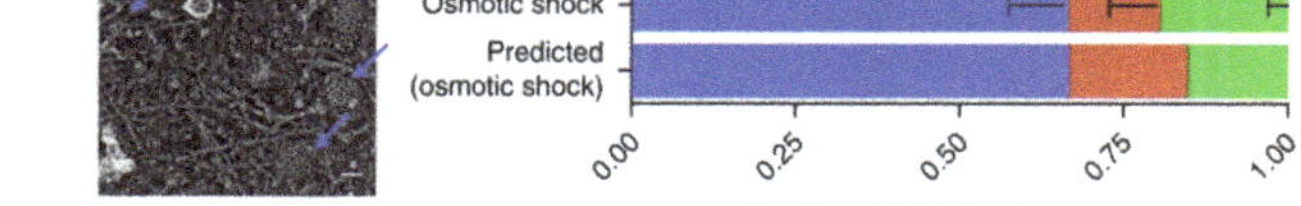

Figure 6.9. Membrane tension regulates a flat-to-curved transition. (a) Examples of clathrin-coated structures (CCSs) under normal and osmotic shock conditions. Flat structures are indicated by blue arrows. (b) Comparison of flat, domed, and pit-shaped CCSs under normal and osmotic shock conditions estimated using correlative EM images. Reproduced under the terms of the CC BY 4.0. Copyright © 2018, Springer Nature [125].

and membrane-bending proteins together mediate the transition of the clathrin coat to multiple invagination topologies [78].

Membrane tension also plays a direct role in the recruitment of membrane-bending proteins to CME nucleation sites. Membrane tension along with membrane rigidity control the critical density of the membrane-bending proteins required to form clathrin coats [124, 126]. In addition to actin, membrane-bending proteins such as Bin/Amphiphysin/Rvs (BAR) domain proteins and epsin drive the transition from open bud to closed bud over a range of membrane tensions [124]. Using coarse-grained molecular dynamics simulation, it has been found that high tensions (>1.5 mN m^{-1}) alter the assembly of membrane-associated BAR proteins by inhibiting protein oligomerization and interaction between the plasma membrane and the N-terminal amphipathic helix-containing BAR (N-BAR) domain [128]. It is noteworthy that curvature-dependent actin polymerization depends on the interaction between N-WASP and F-BAR proteins such as FBP17 and Toca-1, suggesting feedback regulation between membrane curvature and actin polymerization [129]. Interestingly, membrane-bending proteins appear to respond differently to membrane tension. For example, while high membrane tension reduces the binding of inverse BAR (I-BAR) proteins [15], the epsin N-terminal homology (ENTH) and AP180 N-terminal homology (ANTH) protein families have evolved to detect and counteract membrane tension variation by altering the properties of the membrane to achieve CME under different membrane tension conditions [130–132]. Our work has found that the recruitment of epsin, a member of the ENTH family, to CME nucleation sites increases with membrane tension. This elevated recruitment is mediated by the amphipathic insertion of the alpha helix in the N-terminus of epsin into the lipid bilayer, while the intrinsically disordered C-terminus domain stabilizes clathrin buds under high tension [131, 133]. CALM, which belongs to the ANTH family and is structurally similar to epsin, also stabilizes CME under high tension [132].

Cells also use multiple endocytic pathways that do not involve clathrin-coat formation, which are collectively known as CIE. CIE pathways can internalize plasma membrane at a high rate, thereby making them rapid responders to acute physical changes in cells [134, 135]. One of the major CIE pathways, the clathrin-Independent carriers (CLICs) and GPI-enriched endocytic compartments (GEECs), CLIC/GEEC (CG) pathway is involved in the rapid response to acute loss of membrane tension due to its ability to internalize plasma membrane rapidly [136, 137]. In this context, disruption of the CG pathway reduces membrane tension [136]. The CG pathway is activated when tension is increased due to cell attachment. By controlling the guanine-nucleotide exchange factor GBF1 in the plasma membrane, vinculin mediates the response of the CG pathway to membrane tension [136]. Conversely, membrane tension regulates the activity of the I-BAR protein IRSp53, which is necessary for tubule formation and membrane scission in the CG pathway [138]. In yeast cells, the target of the rapamycin complex TORC2 controls CIE pathways by regulating membrane tension homeostasis [139]. Disruption of TORC2 activity leads to elevated membrane tension, which inhibits the binding of adaptor proteins Sla2 and Ent1 to the actin cytoskeleton and hinders the recruitment of the N-BAR protein Rvs167 which is involved in vesicle fission to endocytic sites, leading to the downregulation of endocytosis in yeast. As a result, the CG pathway plays important role in membrane tension homeostasis by rapidly responding to tension change.

Macropinocytosis, another endocytic pathway, is also influenced by physical factors such as membrane tension [140, 141]. Acute losses of membrane tension caused by osmotic shock induce macropinocytosis by activating phospholipase D2 (PLD2), an enzyme that catalyzes the production of phosphatidic acid, leading to subsequent F-actin and PIP_2-enriched membrane ruffling in myoblasts [140]. Using GUV and super-resolution imaging of lipid nanodomains in myoblasts, it has been found that under normal tension, PLD2 is segregated from its substrate, phosphatidylcholine, and its activator, PI(4,5)P2. A reduction in membrane tension leads to a decrease in the lipid nanodomain and makes the substrate and activator more accessible to PLD2. Overall, these findings point to the vital role of both exo- and endocytic pathways in the rapid reorganization of the plasma membrane in response to physical stresses and membrane ruptures [142].

6.6 Outlook

Endocytosis and exocytosis have individually been studied in detail over the past half century [143]. Advancements in super-resolution microscopy and micromanipulation of cells have led to a more detailed understanding of exo- and endocytic structures in response to membrane tension change [135, 144]. Newer studies also emphasize the importance of CIE pathways and how they couple with exocytosis as a major response mechanism that acts against rapid changes in membrane tension [145]. Newly discovered endocytic pathways, such as fast endophilin-mediated endocytosis (FEME) and other ultrafast endocytosis pathways, are known to be active in synapses; they rapidly retrieve and recycle exocytosed membrane and synaptic vesicle components [146]. The major challenge to a mechanistic understanding of synaptic vesicle exocytotic–endocytotic coupling at synapses arises from the heterogeneity of synaptic transmission under physiological conditions. As a result of advances in super-resolution microscopy and ultra-structural analysis techniques, it is now known that synaptic vesicles are heterogeneous with respect to their size, content, motion, distribution, and release probability [147, 148]. Systematic analysis using flash-freezing EM and cryo-EM tomography combined with computational simulation may reveal the detailed biophysical mechanisms underlying exocytotic–endocytotic coupling at synapses.

During past decades, the function of caveolae in mechanosensation was revealed as imaging and micromanipulation approaches advanced. Caveolae can be regarded as mechanosensing complexes on the plasma membrane, as well as membrane reservoirs that buffer membrane tension fluctuations. The flattening of caveolae plays a critical role in maintaining membrane tension homeostasis. Several signaling transduction events might take place after caveolar flattening, due to its influence on membrane microdomains and lipid composition. Knowing the function of caveolae provides a clue as to why the density of caveolae varies in different cell types.

However, the effects of caveolae deformation and its downstream transduction events produced in response to distinct mechanical stimuli remain to be deciphered (e.g. local versus global; acute versus prolonged mechanical stress). In addition, more mathematical modeling can help to provide insights into the relationship between caveolae and membrane tension in higher-ordered caveolae-containing tissues.

Another important question is the contribution of membrane trafficking to biochemical signaling during membrane tension fluctuation. Since membrane trafficking redistributes membrane proteins and signaling lipids, downstream signaling may be activated. For example, it was recently reported that macropinocytosis is activated by mechanical stretching in muscle cells [141]. The newly formed macropinosome contains phosphatidic acid on its surface and can serve as a platform for mTOR activation. Thus, another outstanding question for future studies is to determine the extent to which membrane trafficking regulates membrane surface area by modulating the activities of biochemical signaling.

Finally, further investigations are required to discover how membrane tension coordinates biochemical events to produce integrated cell behavior and motility and how different exo- and endocytic pathways are precisely regulated. Does each membranous structure (i.e. blebs, ruffles, and caveolae) have a critical threshold of tension at which it collapses? Does exocytosis bring lipids that structurally fit best (i.e. cone, cylinder, or inverse cone) under different levels of tension increment? Furthermore, as we gain more insights into tension-induced balancing between endo- and exocytosis, our current knowledge will be improved by an understanding of how biochemical signaling is spatiotemporally mediated or involved in tension-induced membrane trafficking. Similarly, along with tension-induced membrane trafficking, second-order effects, such as cytoskeleton alteration, that accompany vesicle secretion may alter membrane–cytoskeleton interactions, changing membrane tension. Overall, the balancing of endo- and exocytosis upon membrane tension change may be a complicated process that involves both biochemical and biophysical inter-regulation. We expect such studies to uncover deeper insights that reveal how cell types that are regularly subjected to acute mechanical stimuli survive without compromising their structural and functional integrity. We are still at a nascent stage when it comes to developing a holistic understanding of membrane trafficking processes in the context of mechano-regulation. Such knowledge will redefine our understanding of cellular physiology.

References

[1] Le Roux A L, Quiroga X, Walani N, Arroyo M and Roca-Cusachs P 2019 The plasma membrane as a mechanochemical transducer *Phil. Trans. R. Soc.* B **374** 20180221

[2] Lewis S A and de Moura J L C 1982 Incorporation of cytoplasmic vesicles into apical membrane of mammalian urinary bladder epithelium *Nature* **297** 685–88

[3] Minsky B D and Chlapowski F J 1978 Morphometric analysis of the translocation of lumenal membrane between cytoplasm and cell surface of transitional epithelial cells during the expansion-contraction cycles of mammalian urinary bladder *J. Cell Biol.* **77** 685–97

[4] Edwards Y S, Sutherland L M, Power J H T, Nicholas T E and Murray A W 1998 Osmotic stress induces both secretion and apoptosis in rat alveolar type II cells *Am. J. Physiol. Lung Cell. Mol. Physiol.* **275** L670–78
[5] Xu J, Liu M, Liu J, Caniggia I and Post M 1996 Mechanical strain induces constitutive and regulated secretion of glycosaminoglycans and proteoglycans in fetal lung cells *J. Cell Sci.* **109** 1605–13
[6] Mourgeon E *et al* 2000 Mechanical stretch stimulates macrophage inflammatory protein-2 secretion from fetal rat lung cells *Am. J. Physiol. Lung Cell. Mol. Physiol.* **279** L699–706
[7] Apodaca G 2002 Modulation of membrane traffic by mechanical stimuli *Am. J. Physiol. Renal Physiol.* **282** F179–190
[8] Golani G, Ariotti N, Parton R G and Kozlov M M 2019 Membrane curvature and tension control the formation and collapse of caveolar superstructures *Dev. Cell* **48** 523–38.e4
[9] Brown T D 2000 Techniques for mechanical stimulation of cells *in vitro*: a review *J. Biomech.* **33** 3–14
[10] Davis C A *et al* 2015 Device-based *in vitro* techniques for mechanical stimulation of vascular cells: a review *J. Biomech. Eng.* **137**
[11] Gauthier N C, Fardin M A, Roca-Cusachs P and Sheetz M P 2011 Temporary increase in plasma membrane tension coordinates the activation of exocytosis and contraction during cell spreading *Proc. Natl Acad. Sci.* **108** 14467–72
[12] Dal J and Sheetz M P 1995 Axon membrane flows from the growth cone to the cell body *Cell* **83** 693–701
[13] Sadoshima J and Izumo S 1997 The cellular and molecular response of cardiac myocytes to mechanical stress *Annu. Rev. Physiol.* **59** 551–71
[14] Wang G and Galli T 2018 Reciprocal link between cell biomechanics and exocytosis *Traffic* **19** 741–49
[15] Diz-Muñoz A, Fletcher D A and Weiner O D 2013 Use the force: membrane tension as an organizer of cell shape and motility *Trends Cell Biol.* **23** 47–53
[16] Kozlov M M and Chernomordik L V 2015 Membrane tension and membrane fusion *Curr. Opin. Struct. Biol.* **33** 61–7
[17] Chugh P and Paluch E K 2018 The actin cortex at a glance *J. Cell Sci.* **131**
[18] Diz-Muñoz A, Weiner O D and Fletcher D A 2018 In pursuit of the mechanics that shape cell surfaces *Nat. Phys.* 648–52
[19] Sitarska E and Diz-Muñoz A 2020 Pay attention to membrane tension: Mechanobiology of the cell surface *Curr. Opin. Cell Biol.* **66** 11–8
[20] Rajalakshmi N, E M R and J T M 2009 Control of cell membrane tension by myosin-I *Proc. Natl Acad. Sci.* **106** 11972–1977
[21] Morris C E and Homann U 2001 Cell surface area regulation and membrane tension *J. Membr. Biol.* **179** 79–102
[22] Gervasio O L, Phillips W D, Cole L and Allen D G 2011 Caveolae respond to cell stretch and contribute to stretch-induced signaling *J. Cell Sci.* **124** 3581–90
[23] Sinha B *et al* 2011 Cells respond to mechanical stress by rapid disassembly of caveolae *Cell* **144** 402–13
[24] Kozera L, White E and Calaghan S 2009 Caveolae act as membrane reserves which limit mechanosensitive channel activation during swelling in the rat ventricular myocyte *PLoS One* **4** e8312

[25] Schlörmann W *et al* 2010 The shape of caveolae is omega-like after glutaraldehyde fixation and cup-like after cryofixation *Histochem. Cell Biol.* **133** 223–28
[26] Yamada E 1955 The fine structure of the gall bladder epithelium of the mouse *J. Biophys. Biochem. Cytol.* **1** 445–58
[27] Palade G E 1953 The fine structure of blood capillaries *J. Appl. Phys.* **24** 1424
[28] Parton R G and del Pozo M A 2013 Caveolae as plasma membrane sensors, protectors and organizers *Nat. Rev. Mol. Cell Biol.* **14** 98–112
[29] Porta J C *et al* 2022 Molecular architecture of the human caveolin-1 complex *Sci. Adv.* **8** eabn7232
[30] Dewulf M *et al* 2019 Dystrophy-associated caveolin-3 mutations reveal that caveolae couple IL6/STAT3 signaling with mechanosensing in human muscle cells *Nat. Commun.* **10** 1974
[31] Kovtun O, Tillu V A, Ariotti N, Parton R G and Collins B M 2015 Cavin family proteins and the assembly of caveolae *J. Cell Sci.* **128** 1269–78
[32] Liu L *et al* 2008 Deletion of Cavin/PTRF causes global loss of caveolae, dyslipidemia, and glucose intolerance *Cell Metab.* **8** 310–17
[33] del Pozo M A, Lolo F-N and Echarri A 2021 Caveolae: mechanosensing and mechano-transduction devices linking membrane trafficking to mechanoadaptation *Curr. Opin. Cell Biol.* **68** 113–23
[34] Matthaeus C and Taraska J W 2021 Energy and dynamics of caveolae trafficking *Front. Cell Dev. Biol.* **8** 614472
[35] Parton R G, Way M, Zorzi N and Stang E 1997 Caveolin-3 associates with developing T-tubules during muscle differentiation *J. Cell Biol.* **136** 137–54
[36] Thorn H *et al* 2003 Cell surface orifices of caveolae and localization of caveolin to the necks of caveolae in adipocytes *Mol. Biol.* **14** 3967–76
[37] Sens P and Turner M S 2006 Budded membrane microdomains as tension regulators *Phys. Rev.* E **73** 31918
[38] Cheng J P X *et al* 2015 Caveolae protect endothelial cells from membrane rupture during increased cardiac output *J. Cell Biol.* **211** 53–61
[39] Lo H P *et al* 2015 The caveolin–cavin system plays a conserved and critical role in mechanoprotection of skeletal muscle *J. Cell Biol.* **210** 833–49
[40] Nassoy P and Lamaze C 2012 Stressing caveolae new role in cell mechanics *Trends Cell Biol.* **22** 381–89
[41] Radel C and Rizzo V 2005 Integrin mechanotransduction stimulates caveolin-1 phosphorylation and recruitment of Csk to mediate actin reorganization *Am. J. Physiol. Heart Circul. Physiol.* **288** H936–945
[42] Joshi B *et al* 2012 Phosphocaveolin-1 is a mechanotransducer that induces caveola biogenesis via Egr1 transcriptional regulation *J. Cell Biol.* **199** 425–35
[43] Echarri A and Del Pozo M A 2015 Caveolae—mechanosensitive membrane invaginations linked to actin filaments *J. Cell Sci.* **128** 2747–58
[44] Walther T C *et al* 2006 Eisosomes mark static sites of endocytosis *Nature* **439** 998–1003
[45] Kabeche R, Howard L and Moseley J B 2015 Eisosomes provide membrane reservoirs for rapid expansion of the yeast plasma membrane *J. Cell Sci.*
[46] Appadurai D *et al* 2020 Plasma membrane tension regulates eisosome structure and function *Mol. Biol.* **31** 287–303
[47] Lemière J, Ren Y and Berro J 2021 Rapid adaptation of endocytosis, exocytosis, and eisosomes after an acute increase in membrane tension in yeast cells *eLife* **10** e62084

[48] Chanaday N L, Cousin M A, Milosevic I, Watanabe S and Morgan J R 2019 The synaptic vesicle cycle revisited: new insights into the modes and mechanisms *J. Neurosci.* **39** 8209
[49] Watanabe S and Boucrot E 2017 Fast and ultrafast endocytosis *Curr. Opin. Cell Biol.* **47** 64–71
[50] Chernomordik L V and Kozlov M M 2008 Mechanics of membrane fusion *Nat. Struct. Mol. Biol.* **15** 675–83
[51] Lou X 2018 Sensing exocytosis and triggering endocytosis at synapses: synaptic vesicle exocytosis–endocytosis coupling *Front. Cellular Neurosci.* **12** 66
[52] Wu L-G, Hamid E, Shin W and Chiang H-C 2014 Exocytosis and endocytosis: modes, functions, and coupling mechanisms *Annu. Rev. Physiol.* **76** 301–31
[53] Urbina F L and Gupton S L 2020 SNARE-mediated exocytosis in neuronal development *Front. Mol. Neurosci.* **13** 133
[54] Jahn R and Scheller R H 2006 SNAREs — engines for membrane fusion *Nat. Rev. Mol. Cell Biol.* **7** 631–43
[55] Watanabe S *et al* 2013 Ultrafast endocytosis at mouse hippocampal synapses *Nature* **504** 242–47
[56] Wu Y *et al* 2014 A dynamin 1-, dynamin 3- and clathrin-independent pathway of synaptic vesicle recycling mediated by bulk endocytosis *eLife* **3** e01621
[57] Baschieri F *et al* 2018 Frustrated endocytosis controls contractility-independent mechanotransduction at clathrin-coated structures *Nat. Commun.* **9** 3825
[58] Tan X, Heureaux J and Liu A P 2015 Cell spreading area regulates clathrin-coated pit dynamics on micropatterned substrate *Integr. Biol.* **7** 1033–43
[59] Shi Z and Baumgart T 2015 Membrane tension and peripheral protein density mediate membrane shape transitions *Nat. Commun.* **6** 1–8
[60] Botelho R J and Grinstein S 2011 Phagocytosis *Curr. Biol.* **21** R533–8
[61] Masters T A, Pontes B, Viasnoff V, Li Y and Gauthier N C 2013 Plasma membrane tension orchestrates membrane trafficking, cytoskeletal remodeling, and biochemical signaling during phagocytosis *Proc. Natl Acad. Sci. USA* **110** 11875–80
[62] Cooper S T and McNeil P L 2015 Membrane repair: mechanisms and pathophysiology *Physiol. Rev.* **95** 1205–40
[63] Blazek A D, Paleo B J and Weisleder N 2015 Plasma membrane repair: a central process for maintaining cellular homeostasis *Physiology (Bethesda)* **30** 438–48
[64] Bi G Q, Alderton J M and Steinhardt R A 1995 Calcium-regulated exocytosis is required for cell membrane resealing *J. Cell Biol.* **131** 1747–58
[65] Miyake K and McNeil P L 1995 Vesicle accumulation and exocytosis at sites of plasma membrane disruption *J. Cell Biol.* **131** 1737–45
[66] Bouter A *et al* 2011 Annexin-A5 assembled into two-dimensional arrays promotes cell membrane repair *Nat. Commun.* **2** 270
[67] Togo T, Krasieva T B and Steinhardt R A 2000 A decrease in membrane tension precedes successful cell-membrane repair *Mol. Biol. Cell* **11** 4339–46
[68] Ferguson J P *et al* 2016 Deciphering dynamics of clathrin-mediated endocytosis in a living organism *J. Cell Biol.* **214** 347–58
[69] Boulant S, Kural C, Zeeh J C, Ubelmann F and Kirchhausen T 2011 Actin dynamics counteract membrane tension during clathrin-mediated endocytosis *Nat. Cell Biol.* **13** 1124–31
[70] L B E *et al* 2011 Membrane tension regulates motility by controlling lamellipodium organization *Proc. Natl Acad. Sci.* **108** 11429–34

[71] Craig E M, Van Goor D, Forscher P and Mogilner A 2012 Membrane tension, myosin force, and actin turnover maintain actin treadmill in the nerve growth cone *Biophys. J.* **102** 1503–13
[72] Kinneret K 2011 Membrane tension leads the way *Proc. Natl Acad. Sci.* **108** 14379–80
[73] Schweitzer Y, Lieber A D, Keren K and Kozlov M M 2014 Theoretical analysis of membrane tension in moving cells *Biophys. J.* **106** 84–92
[74] Raucher D and Sheetz M P 2000 Cell spreading and lamellipodial extension rate is regulated by membrane tension *J. Cell Biol.* **148** 127–36
[75] Lieber A D, Schweitzer Y, Kozlov M M and Keren K 2015 Front-to-rear membrane tension gradient in rapidly moving cells *Biophys. J.* **108** 1599–603
[76] Gauthier N C, Rossier O M, Mathur A, Hone J C and Sheetz M P 2009 Plasma membrane area increases with spread area by exocytosis of a {GPI}-anchored protein compartment *Mol. Biol.* **20** 3261–72
[77] Irajizad E, Walani N, Veatch S L, Liu A P and Agrawal A 2017 Clathrin polymerization exhibits high mechano-geometric sensitivity *Soft Matter* **13** 1455–62
[78] Saleem M *et al* 2015 A balance between membrane elasticity and polymerization energy sets the shape of spherical clathrin coats *Nat. Commun.* **6** 6249
[79] Dubin-Thaler B J, Giannone G, Döbereiner H-G and Sheetz M P 2004 Nanometer analysis of cell spreading on matrix-coated surfaces reveals two distinct cell states and STEPs *Biophys. J.* **86** 1794–806
[80] Dubin-Thaler B J *et al* 2008 Quantification of cell edge velocities and traction forces reveals distinct motility modules during cell spreading *PLoS One* **3** e3735
[81] Gauthier N C, Rossier O M, Mathur A, Hone J C and Sheetz M P 2009 Plasma membrane area increases with spread area by exocytosis of a GPI-anchored protein compartment *Mol. Biol. Cell* **20** 3261–72
[82] Grande-García A *et al* 2007 Caveolin-1 regulates cell polarization and directional migration through Src kinase and Rho GTPases *J. Cell Biol.* **177** 683–94
[83] Satyajit M, Rothberg K G and Maxfield F R 1994 Sequestration of GPI-anchored proteins in caveolae triggered by cross-linking *Science* **264** 1948–51
[84] Keren K 2011 Cell motility: the integrating role of the plasma membrane *Eur. Biophys. J.* **40** 1013–27
[85] Veale K J, Offenhäuser C, Whittaker S P, Estrella R P and Murray R Z 2010 Recycling endosome membrane incorporation into the leading edge regulates lamellipodia formation and macrophage migration *Traffic* **11** 1370–379
[86] Houk A R *et al* 2012 Membrane tension maintains cell polarity by confining signals to the leading edge during neutrophil migration *Cell* **148** 175–88
[87] Labouesse M 2011 Rac GTPase signaling in mechanotransduction during embryonic morphogenesis *Small GTPases* **2** 305–09
[88] Hetmanski J H R *et al* 2019 Membrane tension orchestrates rear retraction in matrix--directed cell migration *Dev. Cell* **51** 460–75.e10
[89] Pierre S 2020 Stick–slip model for actin-driven cell protrusions, cell polarization, and crawling *Proc. Natl Acad. Sci.* **117** 24670–4678
[90] de Vincentiis S, Falconieri A, Scribano V, Ghignoli S and Raffa V 2020 Manipulation of axonal outgrowth via exogenous low forces *Int. J. Mol. Sci.* **21** 8009
[91] Pfister B J, Iwata A, Meaney D F and Smith D H 2004 Extreme stretch growth of integrated axons *J. Neurosci.* **24** 7978 LP–983

[92] Zheng J *et al* 1991 Tensile regulation of axonal elongation and initiation *J. Neurosci.* **11** 1117–25
[93] Lin C, Lamoureux P, Buxbaum R E and Heidemann S R 1995 Osmotic dilution stimulates axonal outgrowth by making axons more sensitive to tension *J. Biomech.* **28** 1429–38
[94] Craig E M, Van Goor D, Forscher P and Mogilner A 2012 Membrane tension, myosin force, and actin turnover maintain actin treadmill in the nerve growth cone *Biophys. J.* **102** 1503–13
[95] Cotrufo T *et al* 2011 A signaling mechanism coupling netrin-1/deleted in colorectal cancer chemoattraction to SNARE-mediated exocytosis in axonal growth cones *J. Neurosci.* **31** 14463–80
[96] Tojima T *et al* 2007 Attractive axon guidance involves asymmetric membrane transport and exocytosis in the growth cone *Nat. Neurosci.* **10** 58–66
[97] Akiyama H, Fukuda T, Tojima T, Nikolaev V O and Kamiguchi H 2016 Cyclic nucleotide control of microtubule dynamics for axon guidance *J. Neurosci.* **36** 5636–49
[98] Tojima T, Hines J H, Henley J R and Kamiguchi H 2011 Second messengers and membrane trafficking direct and organize growth cone steering *Nat. Rev. Neurosci.* **12** 191–203
[99] Mao F, Yang Y and Jiang H 2021 Endocytosis and exocytosis protect cells against severe membrane tension variations *Biophys. J.* **120** 5521–529
[100] Wu L-G, Hamid E, Shin W and Chiang H-C 2014 Exocytosis and endocytosis: modes, functions, and coupling mechanisms *Annu. Rev. Physiol.* **76** 301–31
[101] Han J, Pluhackova K and Böckmann R A 2017 The multifaceted role of SNARE proteins in membrane fusion *Front. Physiol.* **8** 5
[102] Kliesch T-T *et al* 2017 Membrane tension increases fusion efficiency of model membranes in the presence of SNAREs *Sci. Rep.* **7** 12070
[103] Shillcock J C and Lipowsky R 2005 Tension-induced fusion of bilayer membranes and vesicles *Nat. Mater.* **4** 225–28
[104] Cohen F S and Melikyan G B 2004 The energetics of membrane fusion from binding, through hemifusion, pore formation, and pore enlargement *J. Membrane Biol.* **199** 1–4
[105] Wu Z *et al* 2017 Dilation of fusion pores by crowding of SNARE proteins *eLife* **6** e22964
[106] McMahon H T, Kozlov M M and Martens S 2010 Membrane curvature in synaptic vesicle fusion and beyond *Cell* **140** 601–05
[107] Torregrosa-Hetland C J *et al* 2013 Cortical F-actin affects the localization and dynamics of SNAP-25 membrane clusters in chromaffin cells *Int. J. Biochem. Cell Biol.* **45** 583–92
[108] Bretou M *et al* 2014 Cdc42 controls the dilation of the exocytotic fusion pore by regulating membrane tension *Mol. Biol. Cell* **25** 3195–209
[109] Olivares M J *et al* 2014 Src kinases regulate de novo actin polymerization during exocytosis in neuroendocrine chromaffin cells *PLoS One* **9** e99001
[110] Ňeco P *et al* 2008 Myosin II contributes to fusion pore expansion during exocytosis *J. Biol. Chem.* **283** 10949–57
[111] González-Jamett A M *et al* 2013 Dynamin-2 regulates fusion pore expansion and quantal release through a mechanism that involves actin dynamics in neuroendocrine chromaffin cells *PLoS One* **8** e70638
[112] Anantharam A *et al* 2011 A new role for the dynamin GTPase in the regulation of fusion pore expansion *Mol. Biol. Cell* **22** 1907–18

[113] Shin W *et al* 2018 Visualization of membrane pore in live cells reveals a dynamic-pore theory governing fusion and endocytosis *Cell* **173** 934–45.e12
[114] Trexler A J, Sochacki K A and Taraska J W 2016 Imaging the recruitment and loss of proteins and lipids at single sites of calcium-triggered exocytosis *Mol. Biol. Cell* **27** 2423–34
[115] Wen P J *et al* 2016 Actin dynamics provides membrane tension to merge fusing vesicles into the plasma membrane *Nat. Commun.* **7** 12604
[116] Kaksonen M and Roux A A 2018 Mechanisms of clathrin-mediated endocytosis *Nat. Rev. Mol. Cell Biol.* **19** 313–26
[117] Ferguson J P *et al* 2017 Mechanoregulation of clathrin-mediated endocytosis *J. Cell Sci.* **130** 3631–636
[118] Lacy M M, Ma R, Ravindra N G and Berro J 2018 Molecular mechanisms of force production in clathrin-mediated endocytosis *FEBS Lett.* **592** 3586–605
[119] Joseph J G and Liu A P 2020 Mechanical regulation of endocytosis: new insights and recent advances *Adv. Biosyst.* **4** 1900278
[120] Dai J and Sheetz M P 1995 Regulation of endocytosis, exocytosis, and shape by membrane tension *Cold Spring Harbor Symp. Quant. Biol.* **60** 567–71
[121] Bucher D *et al* 2018 Clathrin-adaptor ratio and membrane tension regulate the flat-to-curved transition of the clathrin coat during endocytosis *Nat. Commun.* **9** 1109
[122] Wu X-S *et al* 2017 Membrane tension inhibits rapid and slow endocytosis in secretory cells *Biophys. J.* **113** 2406–14
[123] Hassinger J E, Oster G, Drubin D G and Rangamani P 2017 Design principles for robust vesiculation in clathrin-mediated endocytosis *Proc. Natl Acad. Sci.* **114** E1118–1127
[124] Walani N, Torres J and Agrawal A 2015 Endocytic proteins drive vesicle growth via instability in high membrane tension environment *Proc. Natl Acad. Sci.* **2015** 201418491
[125] Bucher D *et al* 2018 Clathrin-adaptor ratio and membrane tension regulate the flat-to-curved transition of the clathrin coat during endocytosis *Nat. Commun.* **9** 1109
[126] Shi Z and Baumgart T 2015 Membrane tension and peripheral protein density mediate membrane shape transitions *Nat. Commun.* **6** 1–8
[127] Cordella N, Lampo T J, Mehraeen S and Spakowitz A J 2014 Membrane fluctuations destabilize clathrin protein lattice order *Biophys. J.* **106** 1476–88
[128] Simunovic M and Voth G A 2015 Membrane tension controls the assembly of curvature-generating proteins *Nat. Commun.* **6** 1–8
[129] Takano K, Toyooka K and Suetsugu S 2008 EFC/F-BAR proteins and the N-WASP-WIP complex induce membrane curvature-dependent actin polymerization *EMBO J.* **27** 2817–28
[130] Jarsch I K, Daste F and Gallop J L 2016 Membrane curvature in cell biology: an integration of molecular mechanisms *J. Cell Biol.* **214** 375–87
[131] Joseph J G, Osorio C, Yee V, Agrawal A and Liu A P 2020 Complimentary action of structured and unstructured domains of epsin supports clathrin-mediated endocytosis at high tension *Commun. Biol.* **3** 1–16
[132] Willy N M *et al* 2021 CALM supports clathrin-coated vesicle completion upon membrane tension increase *Proc. Natl Acad. Sci. USA* **118** e2010438118
[133] Stachowiak J C, Hayden C C and Sasaki D Y 2010 Steric confinement of proteins on lipid membranes can drive curvature and tubulation *Proc. Natl Acad. Sci.* **107** 7781–786
[134] Sandvig K, Pust S, Skotland T and van Deurs B 2011 Clathrin-independent endocytosis: mechanisms and function *Curr. Opin. Cell Biol.* **23** 413–20

[135] Gauthier N C, Masters T A and Sheetz M P 2012 Mechanical feedback between membrane tension and dynamics *Trends Cell Biol.* **22** 527–35

[136] Thottacherry J J *et al* 2018 Mechanochemical feedback control of dynamin independent endocytosis modulates membrane tension in adherent cells *Nat. Commun.* **9** 4217

[137] Howes M T *et al* 2010 Clathrin-independent carriers form a high capacity endocytic sorting system at the leading edge of migrating cells *J. Cell Biol.* **190** 675–91

[138] Sathe M *et al* 2018 Small GTPases and BAR domain proteins regulate branched actin polymerisation for clathrin and dynamin-independent endocytosis *Nat. Commun.* **9** 1835

[139] Riggi M *et al* 2019 TORC2 controls endocytosis through plasma membrane tension *J. Cell Biol.* **218** 2265–76

[140] Loh J *et al* 2019 An acute decrease in plasma membrane tension induces macropinocytosis via PLD2 activation *J. Cell Sci.* **132** jcs232579

[141] Lin S-S and Liu Y-W 2019 Mechanical stretch induces mTOR recruitment and activation at the phosphatidic acid-enriched macropinosome in muscle cell *Front. Cell Dev. Biol.* **7** 78

[142] Donaldson J, Poratshliom N and Cohen L 2009 Clathrin-independent endocytosis: a unique platform for cell signaling and remodeling *Cell. Signalling* **21** 1–6

[143] McMahon H T and Boucrot E 2011 Molecular mechanism and physiological functions of clathrin-mediated endocytosis *Nat. Rev. Mol. Cell Biol.* **12** 517–33

[144] Djakbarova U, Madraki Y, Chan E T and Kural C 2021 Dynamic interplay between cell membrane tension and clathrin-mediated endocytosis *Biol. Cell* **113** 344–73

[145] Thottacherry J J, Sathe M, Prabhakara C and Mayor S 2019 Spoiled for choice: diverse endocytic pathways function at the cell surface *Annu. Rev. Cell Dev. Biol.* **35** 55–84

[146] Llobet A *et al* 2011 Endophilin drives the fast mode of vesicle retrieval in a ribbon synapse *J. Neurosci.* **31** 8512–519

[147] Neher E 2015 Merits and limitations of vesicle Pool models in view of heterogeneous populations of synaptic vesicles *Neuron* **87** 1131–42

[148] Joensuu M *et al* 2016 Subdiffractional tracking of internalized molecules reveals heterogeneous motion states of synaptic vesicles *J. Cell Biol.* **215** 277–92

IOP Publishing

Exocytosis: From Molecules to Cells

Arun Anantharam and Jefferson Knight

Chapter 7

A dynamin perspective on endocytosis and coupling to exocytosis

Nidhi Kundu and Jenny E Hinshaw

Endocytosis is a cellular process that internalizes material including nutrients, hormones, proteins, and some viruses. Endocytosis also plays a role in keeping a cell's surface area and tension constant by the retrieval of membrane. This is especially critical in neuronal cells, where stimulation leads to excess membrane at the cell surface from the fusion of numerous synaptic vesicles. The main driver of most endocytosis pathways is dynamin, a large GTPase that plays a direct role in internalizing vesicles by wrapping around the necks of invaginating pits and undergoing a GTP-induced conformational change that leads to fission. There are three isoforms of dynamin, and only mutations in dynamin 2 are associated with neuropathies. The details of dynamin's mechanism has been unraveled by numerous *in vitro* fission assays and structural studies. Additional dynamin-binding proteins also localize and contribute to endocytosis. Several of these dynamin partners, as well as dynamin, are involved in coupling endocytosis to exocytosis.

7.1 Introduction

Membrane fusion and fission in exocytosis and endocytosis, respectively, share the same fundamental process, the fusion of two membranes to generate a single bilayer (figure 7.1). During endocytosis, extracellular material and membrane receptors are internalized through vesiculation events at the plasma membrane. The large GTPase dynamin is the mechanochemical driver for most endocytotic events, including clathrin-mediated endocytosis (CME), a major endocytotic pathway that is characterized by a clathrin coat on the budding vesicle (Jimah and Hinshaw 2019, Mettlen *et al* 2018). CME is a universal endocytotic cellular pathway; in synapses, clathrin-mediated budding, either from the plasma membrane or endosomes, contributes to the recycling of the plasma membrane for continuous synaptic vesicle production

doi:10.1088/978-0-7503-3771-7ch7

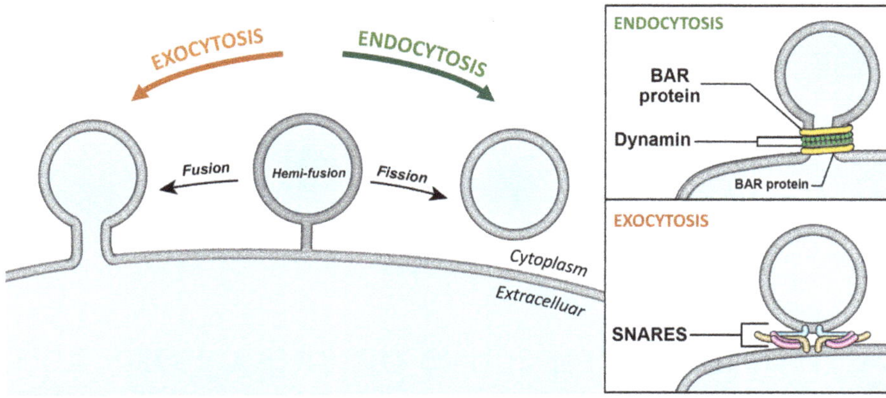

Figure 7.1. Schematic diagram of membrane fusion and fission through a hemi-fusion state. During endocytosis, dynamin and BAR-domain proteins play a pivotal role in pinching off vesicles from the plasma membrane, while SNARE proteins facilitate the fusion of vesicles to the plasma membrane during exocytosis. Both endocytosis and exocytosis are believed to go through a hemi-fusion state prior to fission and fusion, respectively.

(Milosevic 2018). However, most synaptic membrane retrieval pathways do not require a clathrin coat, although dynamin is necessary (Gan and Watanabe 2018, López-Hernández *et al* 2022). During endocytosis, dynamin assembles as a helix around the necks of budding vesicles, including clathrin-coated vesicles, and constricts and pinches off the vesicles upon guanine triphosphate (GTP) hydrolysis (Antonny *et al* 2016). The recruitment of dynamin to necks is facilitated by Bin/Amphiphysin/Rvs (BAR)-domain proteins (e.g. endophilin and amphiphysin) (Meinecke *et al* 2013, Simunovic *et al* 2019). BAR-domain proteins are characterized by their banana-shaped dimers that self-assemble into helices around the membrane with a distinct diameter (Simunovic *et al* 2019). During vesiculation, BAR-domain proteins help to stabilize an ideal diameter of the budding vesicle neck for dynamin to wrap around. Upon assembly, dynamin's GTPase activity is stimulated ten to a hundred times, leading to membrane constriction and eventually membrane fission (Antonny *et al* 2016). In addition to its well-established role in endocytosis, there is growing evidence that in some cases dynamin also plays a role in exocytosis. Whether dynamin behaves in a similar or different manner in the two pathways is just beginning to be unraveled.

7.2 Dynamin's role in endocytosis

Dynamin was initially shown to be involved in endocytosis through genetic screening of *Drosophila* (Koenig and Ikeda 1989). Specifically, a temperature-sensitive mutant of the *Drosophila* dynamin homolog, *shibire*, blocked endocytosis at the non-permissive temperature. As a result, coated pits accumulated at the plasma membrane and a dense collar was observed at the base of coated pits in the synapse (Koenig and Ikeda 1989). *Shibire* was subsequently identified as a homolog of dynamin (Chen *et al* 1991, Van der Bliek and Meyerowrtz 1991) and purified dynamin was shown to assemble into rings and helices similar to the dense collar seen at the necks of invaginating pits in the synapse (Hinshaw and Schmid 1995). Dynamin 1 knockout (KO) mice also display severe defects in endocytosis of synaptic membrane, which result in the accumulation of clathrin-coated pits at the presynaptic neuron (Ferguson *et al* 2007, Lou *et al* 2008)

Structurally, dynamin can be subdivided into five domains: GTPase, bundle signaling element (BSE), stalk, pleckstrin homology (PH), and the proline/arginine-rich domain (PRD) (figure 7.4(a)). The stalk and BSE domains are involved in dynamin assembly and GTP-induced conformational changes, the PH domain interacts with lipids, and the C-terminal PRD contains several SH3 binding motifs that dynamin partners can bind to (i.e. amphiphysin, endophilin, and SNX9). Over the years, dynamin mutants used in cell-based assays, biochemical analysis, and structural studies have further dissected the functional properties of each domain (Faelber *et al* 2012, Schmid and Frolov 2011). In general, dynamin mutants that inhibit endocytosis are defective in four areas: targeting to the plasma membrane (the PRD), GTP hydrolysis (the GTPase domain), oligomeric assembly (the stalk and BSE domains), and lipid binding (the PH domain) (Faelber *et al* 2012, Schmid and Frolov 2011). One particular mutant in the GTPase domain (K44A), which inhibits GTPase activity and blocks endocytosis, has been used extensively to determine whether a cargo of interest enters cells by a dynamin-dependent pathway, such as CME (Damke *et al* 1994, 2001). Structural biologists have taken advantage of assembly mutants and GTP hydrolysis-deficient mutants to capture dynamin in stable transition states (Chappie *et al* 2011, Faelber *et al* 2011, Ford *et al* 2011, Kong *et al* 2018, Reubold *et al* 2015, Sundborger *et al* 2014).

Lipid composition at endocytotic sites also plays a role in dynamin-mediated fission. Dynamin's PH domain preferentially binds to negatively charged phosphatidylinositol lipids with a preference for PI(4,5)P_2 at endocytotic sites (Klein *et al* 1998, Salim *et al* 1996, Zheng *et al* 1996). The affinity to lipids is significantly greater for PH domain dimers than for monomers (Klein *et al* 1998). Thus, the helical assembly of dynamin at the necks of budding vesicles clusters PI(4,5)P_2 in this region (Bethoney *et al* 2009). PI(4,5)P_2 also recruits AP2 (adapter protein 2), a major component of the clathrin coat, to endocytotic sites (Kelly *et al* 2014). The binding of AP2 to the membrane then triggers clathrin to assemble into a cage that forms an invaginating bud prior to vesiculation (Kelly *et al* 2014). Additional studies have indicated that the lipid composition at endocytotic sites is dynamic. For example, a P(4)I phosphatase, INNP4, is activated at highly curved membrane sites, converting PI(3,4)P_2 to PI(3)P (Daste *et al* 2017). This transient lipid modification coincides with SNX9 recruitment and actin polymerization prior to membrane fission during CME (Daste *et al* 2017). In addition, after fission, the conversion of PI(4,5)P_2 to PI(4)P by synaptojanin promotes the release of the clathrin coat (Dickson and Hille 2019, Saheki and De Camilli 2012).

7.2.1 Dynamin isoforms

Mammals contain three dynamin isoforms that share 79% of their sequence identity; the greatest variability is found in the PRD (Ferguson and De Camilli 2012). The isoforms are expressed preferentially in tissue: dynamin 1 is primarily found in the brain, dynamin 2 is ubiquitously expressed, and dynamin 3 is located in the brain, lung, and testis (Ferguson and De Camilli 2012). Dynamin 1 is essential for the recycling of synaptic vesicles, while dynamin 2 is present in all cell types for routine endocytosis. However, in non-neuronal cells, dynamin 1, which is normally quiescent, can be activated by inhibiting the kinase GSK3β (Reis *et al* 2015). GSK3β phosphorylates Ser774 in the PRD of dynamin 1, a residue not conserved in dynamin 2 (Reis *et al* 2015). In neuronal cells, dynamin 1 is also negatively regulated by phosphorylation. In this case, two serine residues in dynamin 1 are constitutively phosphorylated by Cdk5, S774, and S778 (Armbruster *et al* 2013, Tan *et al* 2003). When calcium signaling occurs, dynamin 1 is dephosphorylated by the calcium-dependent phosphatase calcineurin and subsequently recruited to sites of endocytosis (Lai *et al* 1999, Liu *et al* 1994). Dynamin 3 may also be regulated by phosphorylation in a similar manner to dynamin 1 (Larsen *et al* 2004) and KO mouse studies suggest it plays a supportive role in synaptic membrane recycling (Raimondi *et al* 2011) in addition to its conventional role in the postsynapse (Gray *et al* 2003).

The variability of the isoforms is directly associated with the regulation of endocytosis in different tissue types. For example, dynamin 1 has a higher propensity to bend membranes than dynamin 2, due to an amino-acid change in variable loop 3 of the PH domain (Liu *et al* 2011). Dynamin 2 is characterized as a curvature sensor that depends on BAR-domain proteins to help generate high curvature at the necks of coated pits (Liu *et al* 2011). The increase in the bending efficiency of dynamin 1 is directly related to the stringent control and rapid endocytosis necessary in synapses. In non-neuronal cells, dynamin 2 may have evolved to slow the process of endocytosis in order to ensure that proper clathrin-coated pit formation and cargo recruitment take place (Liu *et al* 2011).

The most divergent region between dynamin 1 and 2 is in the PRD, a region responsible for targeting dynamin toward sites of endocytosis through multiple SH3 binding motifs located there. There are ~30 proteins with SH3 domains that bind to dynamin's PRD (Krishnan *et al* 2015). In non-neuronal cells, replacing dynamin 2 PRD with dynamin 1 PRD drastically reduced the recruitment of dynamin to coated pits and inhibited endocytosis (Bhave *et al* 2020, Liu *et al* 2011). This is consistent with data which shows that dynamin 2 has a longer residence time than dynamin 1 at CME sites (Rappoport and Simon 2003). Future studies are needed to determine whether specific dynamin partners or affinities, with their signature SH3 domains, are responsible for differences in isoform recruitment.

7.2.2 Dynamin 2 and neuropathies

Mutations in dynamin 2 are linked to neurological diseases including Charcot–Marie–Tooth (CMT) disease, centronuclear myopathy (CNM), and hereditary spastic paraplegia (HSP) (Bitoun *et al* 2005, Chin *et al* 2015, Faelber *et al* 2013, González-Jamett *et al* 2017, Kenniston and Lemmon 2010, McNiven 2005, Sambuughin *et al* 2015, Tassin *et al* 2021). CMT disease results in defects in motor and sensory neurons, CNM causes congenital muscular disorders, and HSP affects the upper motor neurons and their long axonal projections, resulting in muscle weakness and stiffness (Murala *et al* 2021, Sambuughin *et al* 2015, Zhao *et al* 2018). Most of the CMT and CNM mutations cluster in and around the PH domain, with CMT mutations located primarily on the lipid-binding side of the PH domain, while CNM mutations are grouped in the PH/stalk interface, potentially affecting dynamin assembly properties (Chin *et al* 2015, Kenniston and Lemmon 2010, Reubold *et al* 2015). Evidence suggests that CMT and CNM are loss- and gain-of-function disorders, respectively. CMT mutations inhibit CME, while CNM mutations have little effect on CME (Chin *et al* 2015, Kenniston and Lemmon 2010, Sidiropoulos *et al* 2012). Most CMT mutations block PI(4,5)P_2 binding and lipid-induced GTPase activity (Kenniston and Lemmon 2010, Reubold *et al* 2015, Züchner *et al* 2005), however there is at least one CMT mutant (555ΔDEE557) that exhibits gain of function, with increased assembly and GTPase activity (Tassin *et al* 2021). CNM mutations lead to increased helical assembly, membrane binding, GTPase activity, and fission (Chin *et al* 2015, Kenniston and Lemmon 2010, Wang *et al* 2010). CNM and CMT mouse models present muscle weakness and atrophy (Durieux *et al* 2010, Massana Muñoz *et al* 2019). For CNM, this phenotype can be reversed by decreasing the levels of dynamin 2 with intramuscular injections of shRNA Dnm2 (the dynamin 2 gene) (Buono *et al* 2018). In myoblast cell lines, CNM mutations do not impair CME but lead to T-tubule fragmentation, suggesting that CNM mutations are hyperactive and result in unregulated gain of function (Chin *et al* 2015). The CNM mutations also disrupt the interaction between dynamin and actin in muscle cells, resulting in reduced actin polymerization and subsequent defects in intracellular trafficking (González-Jamett *et al* 2017). A novel mutation in dynamin 2 causes HSP and is located in the BSE domain (R719W) (Sambuughin *et al* 2015). The way in which this HSP mutant effects dynamin's mechanochemical properties remains unknown.

7.2.3 Dynamin endocytotic partners

Although dynamin alone has been shown to mediate membrane fission, additional membrane-remodeling proteins, including amphiphysin, endophilin, and SNX9, are essential for efficient endocytosis. These BAR-domain proteins dimerize into a banana shape that further oligomerizes into helices. Endophilin, amphiphysin, and SNX9 can each tubulate liposomes to generate protein-decorated tubes in the range of tube diameters generated by dynamin (20–100 nm) (Farsad *et al* 2001, Shin *et al* 2008, Sundborger *et al* 2011, Takei *et al* 1999). Similar tubes are seen *in vivo* in dynamin knockout cells that display tubular networks extending from the plasma membrane, commonly capped with clathrin coats (Ferguson *et al* 2007). Immunofluorescence studies with dynamin KO cells localize endophilin, amphiphysin, SNX9, and SNX18 in these tubular networks, suggesting that BAR-domain proteins are capable of preforming the necks of budding vesicles in cells (Ferguson *et al* 2007, 2009, Lou *et al* 2008, Park *et al* 2013). *In vitro* assays further show that amphiphysin enhances dynamin-mediated membrane fission, while endophilin inhibits fission (Farsad *et al* 2001, Meinecke *et al* 2013, Mettlen *et al* 2018, Takei *et al* 1999). The results of cell-based assays reveal that amphiphysin is essential for CME (Meinecke *et al* 2013), and, to a lesser extent, SNX9 is also required for efficient CME in certain cell types (Soulet *et al* 2005). Endophilin knockout studies suggest endophilin is not involved in the fission of the clathrin-coated pits, but plays a role in uncoating the clathrin coat from newly formed vesicles through the recruitment of synaptojanin, a PI(4,5)P_2 phosphatase (Milosevic *et al* 2011). A defect in uncoating clathrin-coated vesicles was also observed in *Drosophila* cells when loss-of-function mutants of endophilin and synaptojanin were co-expressed (Verstreken *et al* 2003). Furthermore, overexpression of endophilin in cells can slow dynamin-mediated membrane fission and transferrin uptake (Hohendahl *et al* 2017). SNX9 plays a role in sensing lipid composition and curvature, specifically PI(4,5)P_2, PI(3,4)P_2, and PI(3)P lipids at curved membranes through its two membrane-interacting domains, BAR and phox homology (PX) (Daste *et al* 2017, Schöneberg *et al* 2017). In addition, SNX9 recruits dynamin to the membrane (Lundmark and Carlsson 2004) and also localizes actin at the sites of clathrin-independent endocytotic events in cells (Mettlen *et al* 2018, Yarar *et al* 2007). In summary, evidence suggests that all three BAR-domain proteins play a role in generating high-curvature necks, while SNX9 and amphiphysin directly recruit dynamin to the necks (Lundmark and Carlsson 2004, Meinecke *et al* 2013). Amphiphysin and SNX9 then enhance dynamin's fission reaction, while endophilin augments the clathrin-coat uncoating that occurs after fission (Mettlen *et al* 2018).

Two additional BAR-domain proteins, FCHo and syndapin (also referred to as Pacsin), also interact with dynamin at endocytotic sites, again through SH3 binding motifs. FCHo and syndapin contain F-BAR domains that favor more shallow membrane curvature and tubulate membranes with wider diameters (40–200 nm) (Itoh *et al* 2005). FCHo localizes in clathrin lattices in the early stages of CME and interacts directly with AP2 and Eps15 (a scaffolding adapter protein) (Henne *et al* 2010, Taylor *et al* 2011). FCHo and Eps15 stabilize AP2 at endocytotic sites, allowing endocytosis to progress with bound cargo (Ma *et al* 2016). In addition to FCHo and Eps15, several other scaffolding proteins are involved in initiating clathrin-mediated endocytosis (Eps15R, intersectin 1–2, and CALM) (Kaksonen and Roux 2018, Maritzen 2022). An amphiphathic helix in CALM and one in epsin (another endocytotic protein) also play a role in generating membrane curvature at endocytotic sites (Ford *et al* 2002, Miller *et al* 2015, Saleem *et al* 2015). FCHo2, syndapin, SNX9, amphiphysin, and dynamin 2 are located at the edges of clathrin structures (Sochacki *et al* 2017).The concentration of FCHo decreases during the transition from flat to highly curved clathrin lattices, which is consistent with the view that FCHo binds to shallow curvature and plays a role in the growth of the clathrin-coated pit (Sochacki *et al* 2017). Dynamin 2 increases with the maturation of the highly curved neck of the budding vesicles, while amphiphisin, SNX9, and syndapin are rarely found on flat structures but concentrate at the edges of highly curved clathrin-coated pits, priming the neck for dynamin-mediated fission (Sochacki *et al* 2017).

In the synapse, syndapin 1 binds dephosphorylated dynamin 1 after stimulation, and evidence suggests that this interaction increases the efficiency of endocytosis (Andersson *et al* 2008, Anggono *et al* 2006, Armbruster *et al* 2013, Quan and Robinson 2013). Syndapin 1 specifically recruits the dynamin splice variant dynamin1xA to the plasma membrane resulting in a pool of dynamin primed for ultrafast endocytosis (Imoto *et al* 2021). Indeed, inhibiting the syndapin 1/dynamin 1 interaction blocks ultrafast endocytosis (Imoto *et al* 2021). Evidence also suggests that blocking syndapin's interaction with dynamin prevents bulk endocytosis (Clayton *et al* 2009) and the formation of synaptic vesicles from the large bulk endocytotic vesicles (Gan and Watanabe 2018, Kittelmann *et al* 2013). However, dynamin 1,3 double KO mice and dynamin 1,2,3 triple KO fibroblasts showed no defect in bulk endocytosis, suggesting that bulk endocytosis is clathrin- and dynamin-independent (Park *et al* 2013, Wu *et al* 2014b). The apparent contradiction may be attributed to different methods, synapse origins, or model systems.

Actin also contributes to membrane invagination in several endocytotic pathways, including CME, where Arp2/3 is the main actin nucleator (Chakrabarti *et al* 2021, Kaksonen and Roux 2018). In yeast, which curiously lacks dynamin, actin, as well as type 1 myosin, are essential and play a dynamic role in clathrin-coated vesicle formation (Kaksonen and Roux 2018, Pedersen and Drubin 2019, Skruzny *et al* 2015, Sun *et al* 2006). In mammalian cells, actin dependence varies with different cellular environments and is directly linked to membrane tension (Kaksonen and Roux 2018, Lamaze *et al* 1997). Increasing membrane tension at the plasma membrane increases the dependence on actin in generating clathrin-coated vesicles (Boulant *et al* 2011, Kaksonen and Roux 2018, Kaplan *et al* 2022). This is further illustrated during mitosis, in which CME is blocked in the early mitotic stages due to an increase in membrane tension and a redirection of actin to the cortex, decreasing actin at CME sites (Kaur *et al* 2014).

7.2.4 Clathrin-independent, dynamin-dependent endocytotic pathways

Clathrin-independent pathways play a major role in synaptic membrane retrieval and in most cases are appreciably faster than CME (Gan and Watanabe 2018, Kononenko *et al* 2014). Depletion of clathrin-coat proteins in model systems extensively inhibits endocytosis in non-neuronal cells but surprisingly has only a slight effect in neuronal cells, suggesting that synaptic membrane recycling involves clathrin-independent pathways (Kononenko *et al* 2014, Koo *et al* 2015, Sato *et al* 2009). This result contradicts previous experiments that showed that CME plays a critical role in synaptic endocytosis; however, many of the earlier experiments were performed at room temperature, and several clathrin-independent pathways, including ultrafast endocytosis, are blocked at room temperature, which may in turn upregulate CME (Watanabe *et al* 2014).

In the synapse, several clathrin-independent pathways have been characterized, including kiss-and-run (10 ms), ultrafast (30–1000 ms), fast (1–2 s), and slow bulk endocytosis (minutes) (Gan and Watanabe 2018, Kononenko *et al* 2014). With the possible exception of bulk endocytosis, all these pathways require dynamin. For ultrafast endocytosis, a single stimulus rapidly triggers endocytosis that is dependent on dynamin and actin (Delvendahl *et al* 2016, Watanabe *et al* 2013). This was observed by both membrane capacitance methods at physiological temperatures (Delvendahl *et al* 2016) and by freezing samples within 50–100 ms after simulation and examining them using electron microscopy (Watanabe *et al* 2013). During ultrafast endocytosis, endosome-like vesicles bud from the plasma membrane in milliseconds followed by the generation of synaptic vesicles (SVs) from the endosome in seconds via a clathrin-mediated process (Kononenko *et al* 2014, Watanabe *et al* 2014).

Another newly defined clathrin-independent pathway, called fast-endophilin-mediated endocytosis (FEME), is driven by the binding of specific cargo to receptors and the recruitment of endophilin (Boucrot *et al* 2015, Chan Wah Hak *et al* 2018, Renard *et al* 2015) as well as numerous other BAR-domain proteins (Ferreira *et al* 2021). Both bacterial Shiga and cholera toxins induce and are internalized by this pathway (Renard *et al* 2015). This pathway takes seconds compared to minutes for the traditional clathrin-mediated endocytotic pathway. The speed of endocytosis is likely due to the amplified recruitment of BAR-domain proteins, which induce high membrane curvature that favors dynamin assembly, and the time saved by not generating a clathrin coat. In addition, the FEME pathway involves modifying the membrane to favor endophilin binding. Specifically, the 5’-phosphatase SHIP2 converts PIP_3 to $PI(3,4)P_2$, a lipid favored by endophilin (Chan Wah Hak *et al* 2018).

In neuroendocrine chromaffin cells, a novel calcium-triggered clathrin-independent endocytotic pathway exists that generates large endocytotic vesicles (0.2–1.5 μm) (Shin *et al* 2021). Within this pathway, the endocytotic membrane undergoes a transition from flat to Λ-shaped, Λ- to Ω-shaped, and Ω- to O-shaped vesicles, in contrast to the conventional mechanism of flat-to-round transformation of the membrane (Shin *et al* 2021). This chromaffin endocytotic pathway is dynamin- and actin-dependent, is mediated by calcium influx, and leads to slow, fast, ultrafast, compensatory, overshoot, and bulk endocytosis (Shin *et al* 2021).

7.3 Mechanochemical properties of dynamin leading to membrane fission

7.3.1 Dynamin assembly and fission assays

Dynamin is considered to be the force-generating molecule solely responsible for membrane fission. *In vitro* tubulation and structural studies of dynamin have been at the forefront of understanding the mechanism of dynamin-mediated membrane fission during endocytosis.

The interaction between dynamin and the membrane has been mimicked by numerous assays which demonstrated that dynamin self-assembles as a helix around lipid tubules that further constrict the tubules upon GTP addition and undergo fission upon GTP hydrolysis. However, a direct connection between GTP hydrolysis

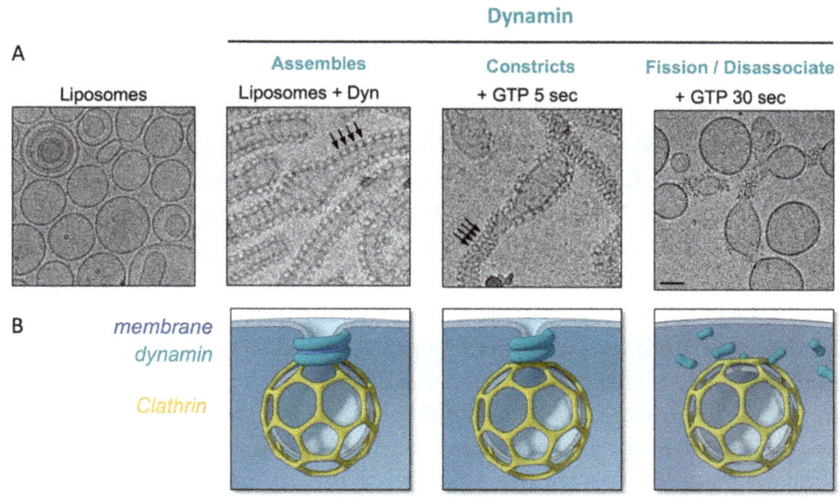

Figure 7.2. Dynamin rapidly constricts the lipid bilayer, leading to fission. (a) Cryo-EM images of liposomes before and after the addition of dynamin. Dynamin assembles on the outside of liposomes as a helix with an outer diameter of ~50 nm. Upon GTP addition, the tubes initially constrict to ~40 nm and then dynamin disassociates from the membrane, leading to membrane fission. *Scale bar, 50 nm.* (b) Cartoon diagram of dynamin (teal) during assembly, constriction, and fission of the necks of clathrin-coated pits (orange) on the plasma membrane (blue).

and the final fission step, in which the two membranes fuse, is still under debate (Galli *et al* 2017). Early electron microscopy studies revealed that dynamin is capable of self-assembly into helices (Hinshaw and Schmid 1995) and the generation of protein-decorated lipid tubes with a diameter of ~50 nm that constrict to ~40 nm, twist into supercoils, and fragment upon GTP addition (Danino *et al* 2004, Sweitzer and Hinshaw 1998) (figures 7.2 and 7.3(a)). Dynamin-decorated lipid tubes were generated by the addition of purified dynamin with liposomes and imaged by electron microscopy before and after the addition of nucleotides (Chappie *et al* 2011, Chen *et al* 2004, Kong *et al* 2018, Sundborger *et al* 2014, Sweitzer and Hinshaw 1998, Zhang and Hinshaw 2001) (figure 7.2). In addition, EM studies of synaptosomes treated with GTPγS revealed long dynamin-decorated tubules with similar diameters extending from the plasma membrane (Takei *et al* 1995). In another assay, liposomes incubated with dynamin and clathrin-coat proteins resulted in long decorated dynamin tubes that often were capped with clathrin-coated buds (Takei *et al* 1999). The percentage of clathrin caps was significantly increased when the BAR protein amphiphysin was added (Takei *et al* 1999). Further studies showed that amphiphysin increased dynamin binding to large liposomes and subsequently increased dynamin's GTPase activity, leading to more fission events and the formation of small vesicles (Yoshida *et al* 2004). In addition, high-speed atomic force microscopy revealed that amphiphysin increased the number of constriction sites on dynamin/amphiphysin-decorated lipid tubes, again resulting in more fission events and smaller vesicles (Takeda *et al* 2018).

Various studies have suggested that in addition to constriction by dynamin, other parameters such as membrane tension and rigidity are required for the final fission event (Morlot *et al* 2012, Roux *et al* 2006). Discovery of the required membrane properties was achieved by real-time fission assays using fluorescent light microscopy and conductance measurements. The role of tension in fission was demonstrated by an *in vitro* assay in which dynamin-decorated lipid tubes were tethered at both ends (Roux *et al* 2006). The addition of GTP to the dynamin tubes led to twisting, the generation of torque and tension, and ultimately breakage (fission) of the tubes, but only when both ends of the tubes were anchored (Morlot *et al* 2012, Roux *et al* 2006). To measure torque during membrane constriction, another assay was used that measured the rotation of beads attached to dynamin tubes after GTP addition (Morlot *et al* 2010, 2012, Roux *et al* 2006). Further studies that used optical tweezers to pull fluorescently labeled lipid tubes from a stabilized giant unilamellar vesicle (GUV), followed by the addition of dynamin and GTP, similarly revealed that fission depends on membrane tension (Morlot *et al* 2012) (figure 7.3(b)). In addition, fission preferentially occurred at the interface between the large GUV and the tube, where there is high membrane curvature, a model system that resembles the necks of budding vesicles on the relatively flat plasma membrane (Morlot *et al* 2012). Overall, dynamin-mediated membrane constriction leads to an increase in torque and elastic energy at the edge of the dynamin coat, which reduces the overall energy barrier to carrying out membrane fission (Morlot *et al* 2012). This study also points out that the constriction rate is significantly faster (milliseconds) than the fission event (seconds), which is dependent on membrane tension (Morlot *et al* 2012). Constriction within milliseconds was also observed by time-resolved cryo-EM, in which dynamin tubes and GTP were sprayed onto electron micrograph grids and frozen in milliseconds (Dandey *et al* 2020). GUV assays also showed that BAR-domain proteins alone (e.g. endophilin) assembled on pulled tubes can trigger scission due to increased membrane tension and friction produced by the protein scaffold (Renard *et al* 2015, Simunovic *et al* 2017).

Supported bilayers with excess membrane reservoir (SUPER) templates provide another unique way of studying vesiculation *in vitro* (Pucadyil and Schmid 2008) (figure 7.3(c)). In this system, multiple coats of lipid bilayers are attached to silica beads. Real-time studies with this assay using fluorescence microscopy have suggested that the addition of dynamin 1 alone was able to generate membrane tubules with an outer diameter of 45.5 ± 5.0 nm from the SUPER templates. When dynamin 2 was added to the SUPER templates, significantly fewer tubes were observed due to a single amino-acid difference in the PH domain between dynamin 1 and dynamin 2 (Dyn1Y600 to Dyn2L600) (Liu *et al* 2011). The addition of GTP to the long preformed dynamin tubes did not lead to fission but disassembly of the protein. However, when GTP was present in the SUPER template assay prior to the addition of dynamin, long tube formation did not occur, but rather various small vesicles (74 ± 33 nm) were released from the template into the solution. Under these conditions, small dynamic assemblies of dynamin appeared on the membrane prior to membrane fission (Pucadyil and Schmid 2008). These results showed that dynamin can undergo fission without tethering or externally induced membrane

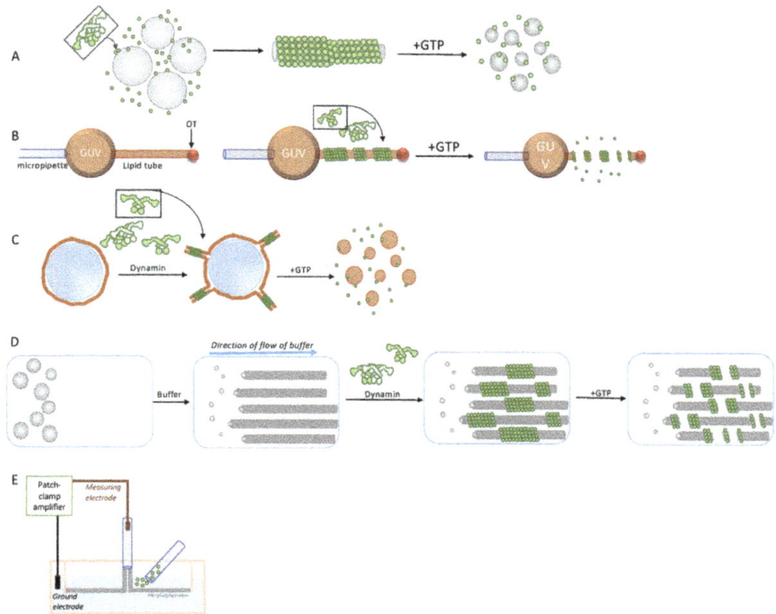

Figure 7.3. Membrane fission assays. (a) *Liposome:* purified dynamin (green) incubated with 1,2- dioleoyl-sn-glycero-3-phospho-L-serine (DOPS) liposomes forms dynamin-decorated tubes through the assembly of higher-order helical oligomers. In the presence of GTP, dynamin undergoes structural rearrangements and constricts the membrane, which is followed by fission. (b) *GUV:* a micropipette on the left (in blue) holds a giant unilamellar vesicle (GUV) and maintains membrane tension. Using optical tweezers (red circle), a nanotube is pulled from the right-hand side of the GUV. Dynamin is released into the system to assemble and further constrict the lipid tube. The addition of GTP leads to membrane fission. (c) *Supported bilayers with excess membrane reservoir (SUPER) template:* fluorescently labeled liposomes, whose composition is equivalent to the plasma membrane, are absorbed onto silica beads (blue). Dynamin tubulates these SUPER templates and fission can be observed using real-time fluorescence microscopy when GTP is added. (d) *SMrT:* a fluorescent labeled lipid is placed on a PEGylated coverslip and membrane tubes are generated by buffer flowing over the surface. The addition of dynamin and GTP leads to fission. (e) *Nanotube conductance:* lipid nanotubes are pulled from a lipid bilayer reservoir by a pipette (top) and dynamin is added to the reservoir. Nanotube conductance is then determined using a patch-clamp amplifier.

tension; however, membrane fission took significantly longer (close to a minute) compared to the milliseconds measured in tethered assays (Pucadyil and Schmid 2008).

Dynamin-mediated fission was further studied using a supported membrane tube (SMrT) assay system that generated multiple membrane tubes on a surface using a flow cell system (Dar *et al* 2015) (figure 7.3(d)). In this system, the lipid mixture was spotted onto a passivated glass coverslip, followed by hydration and the formation of membrane tubes by a flow of buffer. This method of preparing flow-induced membrane tubes (SMrT templates) is advantageous compared to other methods due to its simplicity and the attachment of tubes to the surface, which allows for the analysis of membrane-curvature-sensing properties (Dar *et al* 2015). When dynamin was assembled on the SMrT templates of $PI(4,5)P_2$ followed by the addition of GTP, the preassembled dynamin tubes underwent scission at multiple sites. Under simulated conditions that were closer to physiological, tube constriction and scission were also observed when GTP was added to an SMrT template solution prior to dynamin addition (Dar *et al* 2015). Further measurements revealed that the membrane tubes reached a critical diameter of 14.6 nm, with an inner lumen of 4.6 nm, just prior to membrane fission (Dar *et al* 2015).

Dynamin-mediated constriction was also studied by measuring the ionic permeability of lipid tubes (Bashkirov *et al* 2008). For this assay, tubes were pulled from a lipid reservoir by a pipette and then dynamin was added to the reservoir, allowing dynamin to decorate the pulled-out tube (figure 7.3(e)). The conductance between the pipette and the reservoir was measured during dynamin assembly with and without GTP. From these measurements, the lumen of the tube was calculated; in the presence of GTP, the diameter of the lumen reached ~2 nm. As observed with the SUPER template assay, fission did not occur with the long tubes, but only occurred with short dynamin helices after rounds of GTPase hydrolysis (Bashkirov *et al* 2008). To dissect the GTP hydrolysis state responsible for the fission, dynamin was crosslinked to mimic a GTP transition state (Mattila *et al* 2015). When the crosslinked protein was used in the SUPER template assay, fission was undetectable; however, an ionic permeability assay revealed no conductance, suggesting that in a GTP transition state, dynamin can stabilize a lipid hemi-fission intermediate (Mattila *et al* 2015).

Utilizing the fission assays described above, researchers have explored the role of the PH domain in membrane fission. For example, when the PH domain was removed from dynamin, the SUPER templates and SMrT assays revealed that the fission rate decreased tenfold (Dar and Pucadyil 2017). Further evidence suggesting that PH domains act as catalysts for membrane fission was provided by Förster resonance energy transfer (FRET) assays that established the PH domain inserts into the membrane in a GTP-hydrolysis-dependent manner (Mehrotra *et al* 2014). Within the PH domain, the I533A mutation in variable loop 1 (VL1) blocked membrane insertion in the FRET assay and inhibited fission *in vitro* as well as endocytosis in cells (Mehrotra *et al* 2014, Ramachandran *et al* 2009).

7.3.2 Dynamin structure

Atomic models of dynamin in different nucleotide states obtained from x-ray crystallography structures of soluble dynamin and cryo-EM volumes of assembled dynamin have provided clues about dynamin's mechanochemical properties (Chappie *et al* 2010, 2011, Faelber *et al* 2011, Ford *et al* 2011, Kong *et al* 2018, Reubold *et al* 2015, Sundborger *et al* 2014) (figure 7.4). The architecture of dynamin consists of a GTPase domain connected to a stalk domain by a BSE three-helix bundle (figures 7.4(a) and (b)). The PH domain is on the other side of the stalk, where it interacts with the membrane. The PRD is located near the GTPase domain, but due to its inherent disorder it lacks any structural information. Purified dynamin preferentially exists as a tetramer in physiological solutions (Binns *et al* 1999, Muhlberg *et al* 1997, Ramachandran *et al* 2007), where the tetramer consists of two

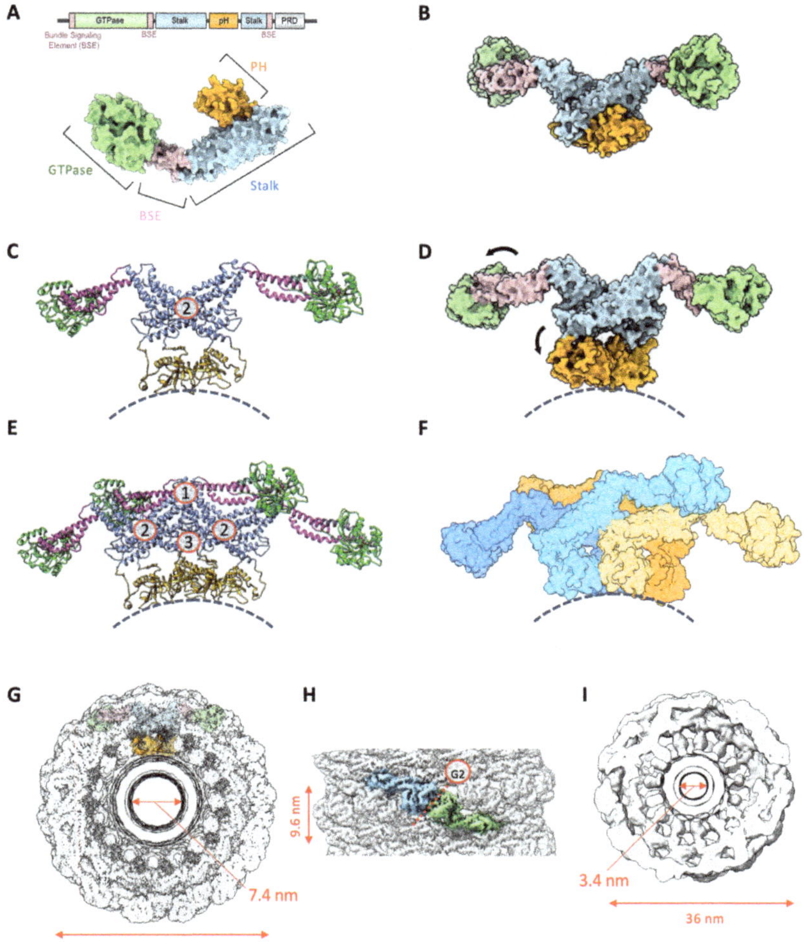

Figure 7.4. Structural organization of dynamin. (a) Linear representation of dynamin domains and the three-dimensional structure of a dynamin monomer with its domains colored: the GTPase domain is shown in green, the BSE in pink, the stalk in blue, the PH domain in orange, and the PRD in gray (from the Protein Data Bank (PDB): 3SNH). (b) Space-filling x-ray crystal structure of the dynamin dimer (PDB: 3SNH) with the PH and GTPase domains tucked inward. (c and d) Cryo-EM atomic models of the dynamin dimer bound to lipid, highlighting interface 2 (red circle in C), which is responsible for dimer assembly, the PH domains (orange) extending down (towards the membrane), and GTPase domains (green) extending outward (see the arrows in D; modified from PDB: 6DLU). (e) Atomic model of a dynamin tetramer with the dimer–dimer interfaces 1, 2, and 3 highlighted; modified from PDB: 6DLU. (f) Tetramer organization of dynamin, consisting of two dimers (blue and orange; modified from the PDB: 6DLU). (g) Cryo-EM map of ΔPRD-dynamin assembled around a lipid tube in the GMPPCP (Guanylyl 5'- (beta,gamma-methylenediphosphonate), a nonhydrolyzable GTP analogue) constricted state with an inner lumen of 7.4 nm, an outer diameter of 40 nm, and a pitch of 9.6 nm (a one-start helix). The dynamin dimer atomic model is shown in the map and colored as in D (EMD-7957). (h) Side view of cryo-EM ΔPRD-dynamin helix illustrating the G2 interface (dashed red line) between the rungs of the helix. (i) Cryo-EM map of ΔPRD-dynamin in the GTP super-constricted state with an inner lumen of 3.4 nm, an outer diameter of 36 nm, and a pitch of 20.2 nm (a two-start helix) (EMD-7958). Dashed curved lines indicate the membrane location in (c)–(f).

equivalent dimers (Reubold *et al* 2015). The dimer is stabilized by a large interface between stalk domains called interface 2 (figures 7.4(c) and (d)) (Faelber *et al* 2011, Ford *et al* 2011). Two additional interfaces (1 and 3; figure 7.4(e)) drive dynamin to assemble into tetramers and higher-order structures (Kong *et al* 2018, Reubold *et al* 2015). This is consistent with early biochemical data that demonstrated that the residues R361 and R399, located in interface 3, play a crucial role in dynamin assembly (Ramachandran *et al* 2007).

The crystal structure of the PH domain (Ferguson *et al* 1994, Timm *et al* 1994) has three variable loops extending from a β-sandwich toward the membrane where they preferentially bind to $PI(4,5)P_2$ with an affinity that increases with the oligomerization of the PH domains (Klein *et al* 1998, Lemmon and Ferguson 2000). Specifically, variable loop 1 contains hydrophobic residues that are inserted into the lipid bilayer (Ramachandran and Schmid 2008). Mutations in this loop (I533A, M534C, K535M) block endocytosis (Achiriloaie *et al* 1999, Ramachandran *et al* 2009) and the I533A mutant has been shown to also block membrane constriction, as observed in a conductance assay (figure 7.3(e)) (Shnyrova *et al* 2013). Theoretical modeling suggests that shallow insertion of VL1 residues into the lipids facilitates fission (Shnyrova *et al* 2013) and coarse-grained simulation predicts that the PH domain, once inserted, can facilitate a hemi-fission state of the lipid bilayer (Mattila *et al* 2015). The importance of the variable loops is manifested by mutations in this region that are associated with CMT disease (Tassin *et al* 2021, Züchner *et al* 2005). On the opposite side of the PH domain, an amphipathic α-helix interacts with the stalk, where it plays a role in the assembled state of dynamin (Reubold *et al* 2015). The associated CNM mutations in this helix disrupt the PH-stalk interface, enhance helical assembly, and ultimately uncouple dynamin assembly from lipid binding (Chin *et al* 2015, Kenniston and Lemmon 2010, Reubold *et al* 2015, Wang *et al* 2010).

Cryo-EM studies further show the architecture of dynamin bound to the membrane. In solution, the PH domain is tucked up under the stalk (Faelber *et al* 2011, Reubold *et al* 2015) (figure 7.4(b)) and when dynamin binds to the membrane, the PH domains swing down to interact with negatively charged lipids (figures 7.4(c) and (d)), allowing the protein to self-assemble into a helix around the necks of budding vesicles (Kong *et al* 2018, Sundborger *et al* 2014). Cryo-EM structures reveal how dynamin–lipid tubes constrict in the presence of GTP: first to an inner lumen of 7.4 nm upon GTP binding (the constricted state) (figure 7.4(g)) and then to 3–4 nm upon GTP hydrolysis (the super-constricted state) (figure 7.4(i)). The transition from the constricted state to the super-constricted state is aided by the transformation from a one-start (9.6 nm pitch) to a two-start (20.2 nm pitch) helix (Kong *et al* 2018, Sundborger *et al* 2014). An inner lumen of 3–4 nm followed by disassembly of the protein scaffold may be sufficient for spontaneous bilayer fusion (Kozlovsky and Kozlov 2003). Although evidence, including cryo-EM images of dynamin tubes, suggests that dynamin is capable of stabilizing a hemi-fission state (Mattila *et al* 2015), the nuances of how dynamin domains and lipid head-groups interact at the exact point of bilayer fusion remain unknown.

In the cytoplasm, the soluble or non-assembled state of dynamin has an exceptionally low GTPase activity (~1 molecule hydrolyzed per minute) (Mohanakrishnan *et al* 2017, Praefcke and McMahon 2004). This activity is stimulated by 10–100 times upon assembly on the membrane (Stowell *et al* 1999, Warnock *et al* 1996). Structural studies have begun to uncover how dynamin's GTP hydrolysis translates to its mechanical properties and ultimately membrane fission. In the assembled state, GTPase domains extend outward, allowing a GTPase domain dimer interface (interface G2) to form between the rungs of the helix (figure 7.4(h)) (Chappie *et al* 2011, Kong *et al* 2018), which resembles the G-domain dimers observed in x-ray crystallography structures (Chappie *et al* 2010, 2011). *Trans* interactions between the GTPase domains in the G2 interface stabilize nucleotide binding, resulting in stimulated GTPase activity upon assembly (Chappie *et al* 2010). In addition, during the GTPase cycle the BSE undergoes a large downward swing (69° rotation) from an extended position in the GTP bound state to a downward position close to the GTPase domain in the guanosine diphosphate (GDP) or transition states (Chappie *et al* 2010, 2011). The swing of the BSE is called the power stroke, and based on molecular modeling studies the power stroke leads to membrane constriction through a ratchet mechanism (Ganichkin *et al* 2021). Surprisingly, very little, if any, swing of the BSE is observed in dynamin–lipid tubes in the transient super-constricted post-hydrolysis state (Kong *et al* 2018), suggesting that the power stroke destabilizes the assembled dynamin polymer. Indeed, further rounds of GTP hydrolysis lead to disassembly of the protein from the lipid bilayer, which is speculated to lead to fission in the cell (Antonny *et al* 2016).

7.4 Coupling endocytosis and exocytosis

A balance between exocytosis and endocytosis is crucial in all cells to maintain cell shape and membrane homeostasis and tension. In neuronal cells, several endocytotic pathways (described above) are coupled to exocytosis, including fast clathrin-independent pathways and slow full-collapse fusion followed by CME (Fesce *et al* 1994, Wu *et al* 2014a) (figure 7.5). The most direct coupling of exocytosis and endocytosis is observed in the fast kiss-and-run pathway, in which secretory vesicles fuse with the membranes for a short time, long enough to release the vesicular contents into extracellular space, followed by quick closure of the fusion pore with the help of dynamin to generate a synaptic vesicle (Wen *et al* 2017). The vesicles are rapidly replenished and filled with neurotransmitters to perform another round of exocytosis (Wen *et al* 2017) (figure 7.5). Although this kiss-and-run may not be a dominant pathway in most synapses, evidence suggests it is prevalent in photoreceptors (Wen *et al* 2017) and chromaffin cells (Shin *et al* 2021).

Unlike kiss-and-run, full-collapse fusion of secretory vesicles in the active zone is commonly followed by endocytosis in the periactive zone. Endocytosis in the periactive zone consists of slow clathrin-mediated or ultrafast clathrin-independent pathways (Gan and Watanabe 2018, Harata *et al* 2006) (figure 7.5). In both cases, synaptic vesicle proteins diffuse rapidly from the active zone after fusion and end up

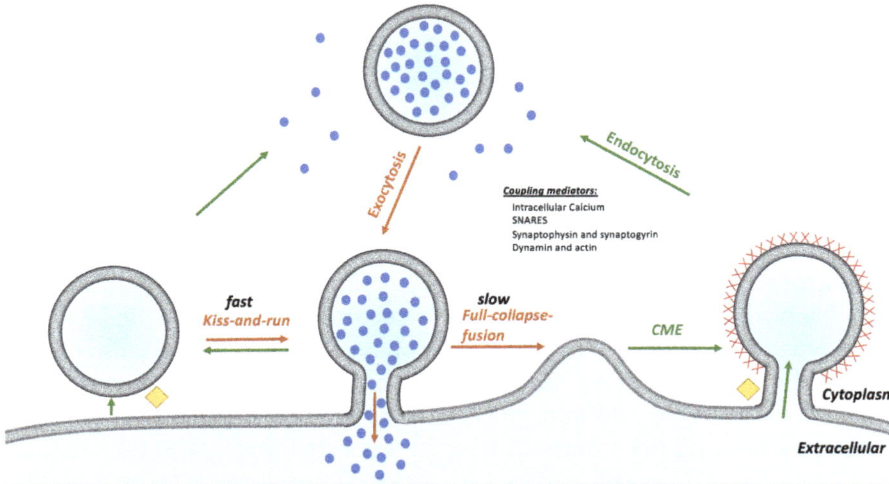

Figure 7.5. Coupling endocytotis and exocytosis. The fusion of synaptic vesicles with the plasma membrane occurs through two exocytotic pathways (orange): fast kiss-and-run and slow full-collapse fusion mode. For kiss-and-run, exocytosis is coupled to endocytosis (green) by fast retrieval of the plasma membrane via a vesicle, while the shape of the vesicle is maintained. The slow full-collapse pathway may follow the canonical clathrin-mediated endocytosis approach (CME, green) that retrieves the synaptic membrane by forming clathrin (shown in red) -coated pits followed by membrane fission by dynamin (yellow diamond). Common factors that mediate the coupling of endocytosis and exocytosis are intracellular calcium ions, the SNARE complex and associated proteins (synaptophysin and synaptotagmin), dynamin, and actin.

clustered in the periactive zone, where they are retrieved by CME or clathrin-independent endocytosis (Gimber *et al* 2015, López-Hernández *et al* 2022).

Several factors regulate and couple endocytosis with exocytosis, including calcium, synaptic vesicle proteins, dynamin, and actin. In the case of exocytosis, an increase in the local Ca^{2+} concentration triggers the fusion of synaptic vesicles in the active zone at the presynaptic neuron (Maritzen and Haucke 2018, Wu *et al* 2009). An influx of calcium via voltage-gated channels activates calcium sensors such as synaptotagmin and calmodulin (which play complementary roles during exocytosis), facilitates clathrin-independent and clathrin-dependent endocytosis (Hosoi *et al* 2009, Wu *et al* 2009), and stimulates endocytosis through dephosphorylation of endocytotic proteins by the calcium/calmodulin-dependent phosphatase calcineurin (Sun *et al* 2010). The endocytotic proteins dephosphorylated by this pathway, commonly referred to as dephosphins, include dynamin, synaptojanin, AP180, amphiphysin, and phosphatidylinositol 3-kinase (Bauerfeind *et al* 1997, Clayton *et al* 2007, Robinson *et al* 1993). A possible mechanism for calcium-accelerated endocytosis may derive from the enhancement of interactions between the dephosphorylated proteins in the periactive zone (Cano and Tabares 2016).

Three SNARE proteins, synatopbrevin, SNAP-25, and syntaxin, have been shown to play a dual role in maintaining exo- and endocytotic functions (Xu *et al* 2013). Disruption of the SNARE proteins by either cleavage or peptide inhibitors halted both fast and slow endocytosis in the calyx of Held nerve terminal (Xu *et al* 2013).

Furthermore, two of the most abundant synaptic vesicle proteins, synaptobrevin-2 and synaptophysin (Wittig *et al* 2021), play a role in coupling exocytosis and CME endocytosis (Cousin 2021). Following fusion, synaptophysin is believed to facilitate the clearance of synaptobrevin-2 from the active zone and form a complex with synaptobrevin-2 and endocytotic proteins (AP180, CALM, AP2) in the periactive zone (Cousin 2021). The recruitment of the endocytotic proteins is predicted to enhance clathrin assembly prior to endocytosis (Cousin 2021). The role of synaptophysin in exocytosis was further clarified in another study, which showed that synaptophysin and a related protein, synaptogyrin, play an important function in controlling the fate of fusion pores between full fusion and kiss-and-run (Chang *et al* 2021).

Dynamin plays a critical role in both endocytosis and exocytosis events by controlling the internal membrane source and fusion pore dynamics. Inhibiting endocytosis with dynamin mutants, inhibitors, or by knocking out the gene, blocks exocytosis by depleting the membrane source needed to regenerate synaptic vesicles (Ferguson *et al* 2007, Gan and Watanabe 2018, Koenig and Ikeda 1989, Newton *et al* 2006). The dual role of dynamin has been also highlighted in studies using chromaffin cells in which dynamin functions in membrane internalization via endocytosis and regulates fusion pore size (Anantharam *et al* 2011, Deisl *et al* 2021, González-Jamett *et al* 2013, Jackson *et al* 2015, Shin *et al* 2018). Several studies have suggested that dynamin enhances fusion pore expansion (Anantharam *et al* 2011, González-Jamett *et al* 2013, Jackson *et al* 2015), while another study suggested that dynamin plays a direct role in fusion pore constriction (Shin *et al* 2018). Further studies are needed to resolve this apparent contradiction, which may be a result of the method used to block or enhance dynamin's GTPase activity and/or assembly properties.

The clathrin adapter protein AP2 also has a dual role in exo- and endocytosis. For example, in hippocampal neurons, AP2 plays a role in the clathrin-independent retrieval of a few synaptic vesicle proteins (López-Hernández *et al* 2022). Although retrieval of all synaptic vesicle proteins was normal in cells depleted of clathrin, a subset of synaptic vesicle proteins failed to recycle in cells depleted of AP2 (López-Hernández *et al* 2022). This suggests the sorting ability of AP2 is essential in order to recycle some synaptic vesicle proteins even in the absence of clathrin.

Actin is a crucial component in the regulation of synaptic vesicle recycling (Wu and Chan 2022). For clathrin-mediated endocytosis, the actin cytoskeleton functions throughout the cycle from membrane invagination, which forms the budding vesicle, to vesicle release in the cytoplasm (Qualmann *et al* 2000). It is also well established that there is a correlation between F-actin and membrane tension such that greater membrane tension leads to an increase in the dependence on actin in generating clathrin-coated vesicles (Boulant *et al* 2011, Kaksonen and Roux 2018, Kaplan *et al* 2022). Removal of actin by knockout methods also inhibits fast clathrin-independent endocytosis as well as slow and bulk endocytosis (Wu *et al* 2016). Furthermore, addition of the actin-polymerization inhibitor latrunculin A blocks ultrafast endocytosis (Delvendahl *et al* 2016). For exocytosis, actin is proposed to be involved in several steps of the pathway for synaptic vesicle fusion. First, F-actin, along with syndapin, may play a role in clustering synaptic vesicles, although this effect may be due to syndapin alone (Milovanovic *et al* 2018, Wu and Chan 2022). Second, depletion of actin slows the regeneration of the readily releasable pool necessary for reliable synaptic transmission (Wu *et al* 2016). Third, actin-dependent membrane tension is also crucial for fusion pore expansion during exocytosis at the synapse (González-Jamett *et al* 2013, Shin *et al* 2018). Finally, the actin-polymerization inhibitor latrunculin A inhibits fusion pore shrinkage (Wen *et al* 2016).

7.5 Summary

From multiple fission assays and structural studies, a consensus model has emerged that describes how dynamin drives membrane fission during endocytosis. First, dynamin is recruited to the necks of budding vesicles in concert with BAR-domain proteins and preferentially binds to the negatively charged lipid with the highest affinity to PI(4,5)P_2. Upon GTP hydrolysis, short dynamin helical assemblies further constrict the membrane relatively quickly (milliseconds), followed by fission of the membrane in seconds after multiple rounds of GTP hydrolysis. Several modeling studies have also indicated that during the fission event, the PH domain plays an active role in disrupting the membrane. Dynamin's role in synaptic vesicle recycling also highlights the coupling of endocytosis with exocytosis in neuronal cells. We look forward to future studies that will dissect exo- and endocytosis coupling in non-neuronal cells.

References

Achiriloaie M, Barylko B and Albanesi J P 1999 Essential role of the dynamin pleckstrin homology domain in receptor-mediated endocytosis *Mol. Cell. Biol.* **19** 1410–15

Anantharam A, Bittner M A, Aikman R L, Stuenkel E L, Schmid S L, Axelrod D and Holz R W 2011 A new role for the dynamin GTPase in the regulation of fusion pore expansion *Mol. Biol. Cell* **22** 1907–18

Andersson F, Jakobsson J, Löw P, Shupliakov O and Brodin L 2008 Perturbation of syndapin/PACSIN impairs synaptic vesicle recycling evoked by intense stimulation *J. Neurosci.* **28** 3925–33

Anggono V, Smillie K J, Graham M E, Valova V A, Cousin M A and Robinson P J 2006 Syndapin I is the phosphorylation-regulated dynamin I partner in synaptic vesicle endocytosis *Nat. Neurosci.* **9** 752–60

Antonny B, Burd C, De Camilli P, Chen E, Daumke O, Faelber K, Ford M, Frolov V A, Frost A and Hinshaw J E 2016 Membrane fission by dynamin: what we know and what we need to know *EMBO J.* **35** 2270–84

Armbruster M, Messa M, Ferguson S M, De Camilli P and Ryan T A 2013 Dynamin phosphorylation controls optimization of endocytosis for brief action potential bursts *eLife* **2** e00845

Bashkirov P V, Akimov S A, Evseev A I, Schmid S L, Zimmerberg J and Frolov V A 2008 GTPase cycle of dynamin is coupled to membrane squeeze and release, leading to spontaneous fission *Cell* **135** 1276–86

Bauerfeind R, Takei K and De Camilli P 1997 Amphiphysin I is associated with coated endocytic intermediates and undergoes stimulation-dependent dephosphorylation in nerve terminals *J. Biol. Chem.* **272** 30984–92

Bethoney K A, King M C, Hinshaw J E, Ostap E M and Lemmon M A 2009 A possible effector role for the pleckstrin homology (PH) domain of dynamin *Proc. Natl Acad. Sci.* **106** 13359–64

Bhave M, Mettlen M, Wang X and Schmid S L 2020 Early and nonredundant functions of dynamin isoforms in clathrin-mediated endocytosis *Mol. Biol. Cell* **31** 2035–47

Binns D D, Barylko B, Grichine N, Atkinson M A, Helms M K, Jameson D M, Eccleston J F and Albanesi J P 1999 Correlation between self-association modes and GTPase activation of dynamin *J. Protein Chem.* **18** 277–90

Bitoun M, Maugenre S, Jeannet P-Y, Lacene E, Ferrer X, Laforêt P, Martin J-J, Laporte J, Lochmüller H and Beggs A H 2005 Mutations in dynamin 2 cause dominant centronuclear myopathy *Nat. Genet.* **37** 1207–9

Boucrot E, Ferreira A, Almeida-Souza L, Debard S, Vallis Y, Howard G, Bertot L, Sauvonnet N and McMahon H T 2015 Endophilin marks and controls a clathrin-independent endocytic pathway *Nature* **517** 460–65

Boulant S, Kural C, Zeeh J-C, Ubelmann F and Kirchhausen T 2011 Actin dynamics counteract membrane tension during clathrin-mediated endocytosis *Nat. Cell Biol.* **13** 1124–31

Buono S, Ross J A, Tasfaout H, Levy Y, Kretz C, Tayefeh L, Matson J, Guo S, Kessler P and Monia B P 2018 Reducing dynamin 2 (DNM2) rescues DNM2-related dominant centronuclear myopathy *Proc. Natl Acad. Sci.* **115** 11066–71

Cano R and Tabares L 2016 The active and periactive zone organization and the functional properties of small and large synapses *Front. Synapt. Neurosci.* **8** 12

Chakrabarti R, Lee M and Higgs H N 2021 Multiple roles for actin in secretory and endocytic pathways *Curr. Biol.* **31** R603–18

Chan Wah Hak L, Khan S, Di Meglio I, Law A-L, Lucken-Ardjomande Häsler S, Quintaneiro L M, Ferreira A, Krause M, McMahon H T and Boucrot E 2018 FBP17 and CIP4 recruit SHIP2 and lamellipodin to prime the plasma membrane for fast endophilin-mediated endocytosis *Nat. Cell Biol.* **20** 1023–31

Chang C-W, Hsiao Y-T and Jackson M B 2021 Synaptophysin regulates fusion pores and exocytosis mode in chromaffin cells *J. Neurosci.* **41** 3563–78

Chappie J S, Acharya S, Leonard M, Schmid S L and Dyda F 2010 G domain dimerization controls dynamin's assembly-stimulated GTPase activity *Nature* **465** 435–40

Chappie J S, Mears J A, Fang S, Leonard M, Schmid S L, Milligan R A, Hinshaw J E and Dyda F 2011 A pseudoatomic model of the dynamin polymer identifies a hydrolysis-dependent powerstroke *Cell* **147** 209–22

Chen M S, Obar R A, Schroeder C C, Austin T W, Poodry C A, Wadsworth S C and Vallee R B 1991 Multiple forms of dynamin are encoded by shibire, a *Drosophila* gene involved in endocytosis *Nature* **351** 583–86

Chen Y-J, Zhang P, Egelman E H and Hinshaw J E 2004 The stalk region of dynamin drives the constriction of dynamin tubes *Nat. Struct. Mol. Biol.* **11** 574–75

Chin Y-H, Lee A, Kan H-W, Laiman J, Chuang M-C, Hsieh S-T and Liu Y-W 2015 Dynamin-2 mutations associated with centronuclear myopathy are hypermorphic and lead to T-tubule fragmentation *Hum. Mol. Genet.* **24** 5542–54

Clayton E L, Anggono V, Smillie K J, Chau N, Robinson P J and Cousin M A 2009 The phospho-dependent dynamin–syndapin interaction triggers activity-dependent bulk endocytosis of synaptic vesicles *J. Neurosci.* **29** 7706–17

Clayton E L, Evans G J and Cousin M A 2007 Activity-dependent control of bulk endocytosis by protein dephosphorylation in central nerve terminals *J. Physiol.* **585** 687–91

Cousin M A 2021 Synaptophysin-dependent synaptobrevin-2 trafficking at the presynapse-Mechanism and function *J. Neurochem.* **159** 78–89

Damke H, Baba T, Warnock D E and Schmid S L 1994 Induction of mutant dynamin specifically blocks endocytic coated vesicle formation *J. Cell Biol.* **127** 915–34

Damke H, Binns D D, Ueda H, Schmid S L and Baba T 2001 Dynamin GTPase domain mutants block endocytic vesicle formation at morphologically distinct stages *Mol. Biol. Cell* **12** 2578–89

Dandey V P, Budell W C, Wei H, Bobe D, Maruthi K, Kopylov M, Eng E T, Kahn P A, Hinshaw J E and Kundu N 2020 Time-resolved cryo-EM using spotiton *Nat. Methods* **17** 897–900

Danino D, Moon K-H and Hinshaw J E 2004 Rapid constriction of lipid bilayers by the mechanochemical enzyme dynamin *J. Struct. Biol.* **147** 259–67

Dar S, Kamerkar S C and Pucadyil T J 2015 A high-throughput platform for real-time analysis of membrane fission reactions reveals dynamin function *Nat. Cell Biol.* **17** 1588–96

Dar S and Pucadyil T J 2017 The pleckstrin-homology domain of dynamin is dispensable for membrane constriction and fission *Molecular Biology of the Cell* **28** 152–60

Daste F, Walrant A, Holst M R, Gadsby J R, Mason J, Lee J-E, Brook D, Mettlen M, Larsson E and Lee S F 2017 Control of actin polymerization via the coincidence of phosphoinositides and high membrane curvature *J. Cell Biol.* **216** 3745–65

Deisl C, Hilgemann D W, Syeda R and Fine M 2021 TMEM16F and dynamins control expansive plasma membrane reservoirs *Nat. Commun.* **12** 1–11

Delvendahl I, Vyleta N P, von Gersdorff H and Hallermann S 2016 Fast, temperature-sensitive and clathrin-independent endocytosis at central synapses *Neuron* **90** 492–98

Dickson E J and Hille B 2019 Understanding phosphoinositides: rare, dynamic, and essential membrane phospholipids *Biochem. J.* **476** 1–23

Durieux A-C, Vignaud A, Prudhon B, Viou M T, Beuvin M, Vassilopoulos S, Fraysse B, Ferry A, Lainé J and Romero N B 2010 A centronuclear myopathy-dynamin 2 mutation impairs skeletal muscle structure and function in mice *Hum. Mol. Genet.* **19** 4820–36

Faelber K, Gao S, Held M, Posor Y, Haucke V, Noé F and Daumke O 2013 Oligomerization of dynamin superfamily proteins in health and disease *Prog. Mol. Biol. Transl. Sci.* **117** 411–43

Faelber K, Held M, Gao S, Posor Y, Haucke V, Noé F and Daumke O 2012 Structural insights into dynamin-mediated membrane fission *Structure* **20** 1621–28

Faelber K, Posor Y, Gao S, Held M, Roske Y, Schulze D, Haucke V, Noé F and Daumke O 2011 Crystal structure of nucleotide-free dynamin *Nature* **477** 556–60

Farsad K, Ringstad N, Takei K, Floyd S R, Rose K and De Camilli P 2001 Generation of high curvature membranes mediated by direct endophilin bilayer interactions *J. Cell Biol.* **155** 193–200

Ferguson K M, Lemmon M A, Schlessinger J and Sigler P B 1994 Crystal structure at 2.2 Å resolution of the pleckstrin homology domain from human dynamin *Cell* **79** 199–209

Ferguson S, Raimondi A, Paradise S, Shen H, Mesaki K, Ferguson A, Destaing O, Ko G, Takasaki J and Cremona O 2009 Coordinated actions of actin and BAR proteins upstream of dynamin at endocytic clathrin-coated pits *Dev. Cell* **17** 811–22

Ferguson S M, Brasnjo G, Hayashi M, Wölfel M, Collesi C, Giovedi S, Raimondi A, Gong L-W, Ariel P and Paradise S 2007 A selective activity-dependent requirement for dynamin 1 in synaptic vesicle endocytosis *Science* **316** 570–74

Ferguson S M and De Camilli P 2012 Dynamin, a membrane-remodelling GTPase *Nat. Rev. Mol. Cell Biol.* **13** 75–88

Ferreira A, Casamento A, Carrillo Roas S, Halff E F, Panambalana J, Subramaniam S, Schützenhofer K, Chan Wah Hak L, McGourty K and Thalassinos K 2021 Cdk5 and GSK3β inhibit fast endophilin-mediated endocytosis *Nat. Commun.* **12** 1–16

Fesce R, Grohovaz F, Valtorta F and Meldolesi J 1994 Neurotransmitter release: fusion or 'kiss-and-run'? *Trends Cell Biol.* **4** 1–4

Ford M G, Jenni S and Nunnari J 2011 The crystal structure of dynamin *Nature* **477** 561–66

Ford M G, Mills I G, Peter B J, Vallis Y, Praefcke G J, Evans P R and McMahon H T 2002 Curvature of clathrin-coated pits driven by epsin *Nature* **419** 361–66

Galli V, Sebastian R, Moutel S, Ecard J, Perez F and Roux A 2017 Uncoupling of dynamin polymerization and GTPase activity revealed by the conformation-specific nanobody dynab *eLife* **6** e25197

Gan Q and Watanabe S 2018 Synaptic vesicle endocytosis in different model systems *Fron. Cellular Neurosci.* **12** 171

Ganichkin O M, Vancraenenbroeck R, Rosenblum G, Hofmann H, Mikhailov A S, Daumke O and Noel J K 2021 Quantification and demonstration of the collective constriction-by-ratchet mechanism in the dynamin molecular motor *Proc. Natl Acad. Sci.* 118

Gimber N, Tadeus G, Maritzen T, Schmoranzer J and Haucke V 2015 Diffusional spread and confinement of newly exocytosed synaptic vesicle proteins *Nat. Commun.* **6** 1–11

González-Jamett A M, Baez-Matus X, Olivares M J, Hinostroza F, Guerra-Fernández M J, Vasquez-Navarrete J, Bui M T, Guicheney P, Romero N B and Bevilacqua J A 2017 Dynamin-2 mutations linked to centronuclear myopathy impair actin-dependent trafficking in muscle cells *Sci. Rep.* **7** 1–16

González-Jamett A M, Momboisse F, Guerra M J, Ory S, Báez-Matus X, Barraza N, Calco V, Houy S, Couve E and Neely A 2013 Dynamin-2 regulates fusion pore expansion and quantal release through a mechanism that involves actin dynamics in neuroendocrine chromaffin cells *PLoS One* **8** e70638

Gray N W, Fourgeaud L, Huang B, Chen J, Cao H, Oswald B J, Hémar A and McNiven M A 2003 Dynamin 3 is a component of the postsynapse, where it interacts with mGluR5 and Homer *Curr. Biol.* **13** 510–15

Harata N C, Aravanis A M and Tsien R W 2006 Kiss-and-run and full-collapse fusion as modes of exo-endocytosis in neurosecretion *J. Neurochem.* **97** 1546–70

Henne W M, Boucrot E, Meinecke M, Evergren E, Vallis Y, Mittal R and McMahon H T 2010 FCHo proteins are nucleators of clathrin-mediated endocytosis *Science* **328** 1281–84

Hinshaw J E and Schmid S L 1995 Dynamin self-assembles into rings suggesting a mechanism for coated vesicle budding *Nature* **374** 190–92

Hohendahl A, Talledge N, Galli V, Shen P S, Humbert F, De Camilli P, Frost A and Roux A 2017 Structural inhibition of dynamin-mediated membrane fission by endophilin *eLife* **6** e26856

Hosoi N, Holt M and Sakaba T 2009 Calcium dependence of exo-and endocytotic coupling at a glutamatergic synapse *Neuron* **63** 216–29

Imoto Y, Raychaudhuri S, Fenske P, Sandoval E, Itoh K, Blumrich E-M, Mamer L, Zarebidaki F, Söhl-Kielczynski B and Trimbuch T 2021 Dynamin is primed at endocytic sites for ultrafast endocytosis *BioRxiv*:2021.02.15.431332

Itoh T, Erdmann K S, Roux A, Habermann B, Werner H and De Camilli P 2005 Dynamin and the actin cytoskeleton cooperatively regulate plasma membrane invagination by BAR and F-BAR proteins *Dev. Cell* **9** 791–804

Jackson J, Papadopulos A, Meunier F, McCluskey A, Robinson P and Keating D 2015 Small molecules demonstrate the role of dynamin as a bi-directional regulator of the exocytosis fusion pore and vesicle release *Mol. Psychiatry* **20** 810–19

Jimah J R and Hinshaw J E 2019 Structural insights into the mechanism of dynamin superfamily proteins *Trends Cell Biol.* **29** 257–73

Kaksonen M and Roux A 2018 Mechanisms of clathrin-mediated endocytosis *Nat. Rev. Mol. Cell Biol.* **19** 313–26

Kaplan C, Kenny S J, Chen X, Schöneberg J, Sitarska E, Diz-Muñoz A, Akamatsu M, Xu K and Drubin D G 2022 Load adaptation by endocytic actin networks *Mol. Biol. Cell* E21-11-0589

Kaur S, Fielding A B, Gassner G, Carter N J and Royle S J 2014 An unmet actin requirement explains the mitotic inhibition of clathrin-mediated endocytosis *eLife* **3** e00829

Kelly B T, Graham S C, Liska N, Dannhauser P N, Höning S, Ungewickell E J and Owen D J 2014 AP2 controls clathrin polymerization with a membrane-activated switch *Science* **345** 459–63

Kenniston J A and Lemmon M A 2010 Dynamin GTPase regulation is altered by PH domain mutations found in centronuclear myopathy patients *EMBO J.* **29** 3054–67

Kittelmann M, Liewald J F, Hegermann J, Schultheis C, Brauner M, Costa W S, Wabnig S, Eimer S and Gottschalk A 2013 *In vivo* synaptic recovery following optogenetic hyper-stimulation *Proc. Natl Acad. Sci.* **110** E3007–16

Klein D E, Lee A, Frank D W, Marks M S and Lemmon M A 1998 The pleckstrin homology domains of dynamin isoforms require oligomerization for high affinity phosphoinositide binding *J. Biol. Chem.* **273** 27725–33

Koenig J and Ikeda K 1989 Disappearance and reformation of synaptic vesicle membrane upon transmitter release observed under reversible blockage of membrane retrieval *J. Neurosci.* **9** 3844–60

Kong L, Sochacki K A, Wang H, Fang S, Canagarajah B, Kehr A D, Rice W J, Strub M-P, Taraska J W and Hinshaw J E 2018 Cryo-EM of the dynamin polymer assembled on lipid membrane *Nature* **560** 258–62

Kononenko N L, Puchkov D, Classen G A, Walter A M, Pechstein A, Sawade L, Kaempf N, Trimbuch T, Lorenz D and Rosenmund C 2014 Clathrin/AP-2 mediate synaptic vesicle reformation from endosome-like vacuoles but are not essential for membrane retrieval at central synapses *Neuron* **82** 981–88

Koo S J, Kochlamazashvili G, Rost B, Puchkov D, Gimber N, Lehmann M, Tadeus G, Schmoranzer J, Rosenmund C and Haucke V 2015 Vesicular synaptobrevin/VAMP2 levels guarded by AP180 control efficient neurotransmission *Neuron* **88** 330–44

Kozlovsky Y and Kozlov M M 2003 Membrane fission: model for intermediate structures *Biophys. J.* **85** 85–96

Krishnan S, Collett M and Robinson P J 2015 SH3 domains differentially stimulate distinct dynamin I assembly modes and G domain activity *PLoS One* **10** e0144609

Lai M M, Hong J J, Ruggiero A M, Burnett P E, Slepnev V I, De Camilli P and Snyder S H 1999 The calcineurin-dynamin 1 complex as a calcium sensor for synaptic vesicle endocytosis *J. Biol. Chem.* **274** 25963–66

Lamaze C, Fujimoto L M, Yin H L and Schmid S L 1997 The actin cytoskeleton is required for receptor-mediated endocytosis in mammalian cells *J. Biol. Chem.* **272** 20332–35

Larsen M R, Graham M E, Robinson P J and Roepstorff P 2004 Improved detection of hydrophilic phosphopeptides using graphite powder microcolumns and mass spectrometry: evidence for *in vivo* doubly phosphorylated dynamin I and dynamin III *Mol. Cellular Proteom.* **3** 456–65

LEMMON M A and FERGUSON K M 2000 Signal-dependent membrane targeting by pleckstrin homology (PH) domains *Biochem. J.* **350** 1–18

Liu J-P, Sim A T and Robinson P J 1994 Calcineurin inhibition of dynamin I GTPase activity coupled to nerve terminal depolarization *Science* **265** 970–73

Liu Y-W, Neumann S, Ramachandran R, Ferguson S M, Pucadyil T J and Schmid S L 2011 Differential curvature sensing and generating activities of dynamin isoforms provide opportunities for tissue-specific regulation *Proc. Natl Acad. Sci.* **108** E234–42

López-Hernández T, Takenaka K-i, Mori Y, Kongpracha P, Nagamori S, Haucke V and Takamori S 2022 Clathrin-independent endocytic retrieval of SV proteins mediated by the clathrin adaptor AP-2 at mammalian central synapses *eLife* **11** e71198

Lou X, Paradise S, Ferguson S M and De Camilli P 2008 Selective saturation of slow endocytosis at a giant glutamatergic central synapse lacking dynamin 1 *Proc. Natl Acad. Sci.* **105** 17555–60

Lundmark R and Carlsson S R 2004 Regulated membrane recruitment of dynamin-2 mediated by sorting nexin 9 *J. Biol. Chem.* **279** 42694–702

Ma L, Umasankar P K, Wrobel A G, Lymar A, McCoy A J, Holkar S S, Jha A, Pradhan-Sundd T, Watkins S C and Owen D J 2016 Transient Fcho1/2· Eps15/R· AP-2 nanoclusters prime the AP-2 clathrin adaptor for cargo binding *Dev. Cell* **37** 428–43

Maritzen T 2022 Endocytic proteins: an expanding repertoire of presynaptic functions *Curr. Opin. Neurobiol.* **73** 102519

Maritzen T and Haucke V 2018 Coupling of exocytosis and endocytosis at the presynaptic active zone *Neurosci. Res.* **127** 45–52

Massana Muñoz X, Buono S, Koebel P, Laporte J and Cowling B S 2019 Different *in vivo* impacts of dynamin 2 mutations implicated in Charcot–Marie–Tooth neuropathy or centronuclear myopathy *Hum. Mol. Genet.* **28** 4067–77

Mattila J-P, Shnyrova A V, Sundborger A C, Hortelano E R, Fuhrmans M, Neumann S, Müller M, Hinshaw J E, Schmid S L and Frolov V A 2015 A hemi-fission intermediate links two mechanistically distinct stages of membrane fission *Nature* **524** 109–13

McNiven M A 2005 Dynamin in disease *Nat. Genet.* **37** 215–16

Mehrotra N, Nichols J and Ramachandran R 2014 Alternate pleckstrin homology domain orientations regulate dynamin-catalyzed membrane fission *Mol. Biol. Cell* **25** 879–90

Meinecke M, Boucrot E, Camdere G, Hon W-C, Mittal R and McMahon H T 2013 Cooperative recruitment of dynamin and BIN/amphiphysin/Rvs (BAR) domain-containing proteins leads to GTP-dependent membrane scission*[S]♦ *J. Biol. Chem.* **288** 6651–61

Mettlen M, Chen P-H, Srinivasan S, Danuser G and Schmid S L 2018 Regulation of clathrin-mediated endocytosis *Annu. Rev. Biochem.* **87** 871–96

Miller S E, Mathiasen S, Bright N A, Pierre F, Kelly B T, Kladt N, Schauss A, Merrifield C J, Stamou D and Höning S 2015 CALM regulates clathrin-coated vesicle size and maturation by directly sensing and driving membrane curvature *Dev. Cell* **33** 163–75

Milosevic I 2018 Revisiting the role of clathrin-mediated endoytosis in synaptic vesicle recycling *Front. Cellular Neurosci.* **12** 27

Milosevic I, Giovedi S, Lou X, Raimondi A, Collesi C, Shen H, Paradise S, O'Toole E, Ferguson S and Cremona O 2011 Recruitment of endophilin to clathrin-coated pit necks is required for efficient vesicle uncoating after fission *Neuron* **72** 587–601

Milovanovic D, Wu Y, Bian X and De Camilli P 2018 A liquid phase of synapsin and lipid vesicles *Science* **361** 604–7

Mohanakrishnan A, Tran T V M, Kumar M, Chen H, Posner B A and Schmid S L 2017 A highly-sensitive high throughput assay for dynamin's basal GTPase activity *PLoS One* **12** e0185639

Morlot S, Galli V, Klein M, Chiaruttini N, Manzi J, Humbert F, Dinis L, Lenz M, Cappello G and Roux A 2012 Membrane shape at the edge of the dynamin helix sets location and duration of the fission reaction *Cell* **151** 619–29

Morlot S, Lenz M, Prost J, Joanny J-F and Roux A 2010 Deformation of dynamin helices damped by membrane friction *Biophys. J.* **99** 3580–88

Muhlberg A B, Warnock D E and Schmid S L 1997 Domain structure and intramolecular regulation of dynamin GTPase *EMBO J.* **16** 6676–83

Murala S, Nagarajan E and Bollu P C 2021 Hereditary spastic paraplegia *Neurol. Sci.* **42** 883–94

Newton A J, Kirchhausen T and Murthy V N 2006 Inhibition of dynamin completely blocks compensatory synaptic vesicle endocytosis *Proc. Natl Acad. Sci.* **103** 17955–60

Park R J, Shen H, Liu L, Liu X, Ferguson S M and De Camilli P 2013 Dynamin triple knockout cells reveal off target effects of commonly used dynamin inhibitors *J. Cell Sci.* **126** 5305–12

Pedersen R T and Drubin D G 2019 Type I myosins anchor actin assembly to the plasma membrane during clathrin-mediated endocytosis *J. Cell Biol.* **218** 1138–47

Praefcke G J and McMahon H T 2004 The dynamin superfamily: universal membrane tubulation and fission molecules? *Nat. Rev. Mol. Cell Biol.* **5** 133–47

Pucadyil T J and Schmid S L 2008 Real-time visualization of dynamin-catalyzed membrane fission and vesicle release *Cell* **135** 1263–75

Qualmann B, Kessels M M and Kelly R B 2000 Molecular links between endocytosis and the actin cytoskeleton *J. Cell Biol.* **150** F111–6

Quan A and Robinson P J 2013 Syndapin–a membrane remodelling and endocytic F-BAR protein *FEBS J.* **280** 5198–212

Raimondi A, Ferguson S M, Lou X, Armbruster M, Paradise S, Giovedi S, Messa M, Kono N, Takasaki J and Cappello V 2011 Overlapping role of dynamin isoforms in synaptic vesicle endocytosis *Neuron* **70** 1100–14

Ramachandran R, Pucadyil T J, Liu Y-W, Acharya S, Leonard M, Lukiyanchuk V and Schmid S L 2009 Membrane insertion of the pleckstrin homology domain variable loop 1 is critical for dynamin-catalyzed vesicle scission *Mol. Biol. Cell* **20** 4630–39

Ramachandran R and Schmid S L 2008 Real-time detection reveals that effectors couple dynamin's GTP-dependent conformational changes to the membrane *EMBO J.* **27** 27–37

Ramachandran R, Surka M, Chappie J S, Fowler D M, Foss T R, Song B D and Schmid S L 2007 The dynamin middle domain is critical for tetramerization and higher-order self-assembly *EMBO J.* **26** 559–66

Rappoport J Z and Simon S M 2003 Real-time analysis of clathrin-mediated endocytosis during cell migration *J. Cell Sci.* **116** 847–55

Reis C R, Chen P H, Srinivasan S, Aguet F, Mettlen M and Schmid S L 2015 Crosstalk between Akt/GSK 3β signaling and dynamin-1 regulates clathrin-mediated endocytosis *EMBO J.* **34** 2132–46

Renard H-F, Simunovic M, Lemière J, Boucrot E, Garcia-Castillo M D, Arumugam S, Chambon V, Lamaze C, Wunder C and Kenworthy A K 2015 Endophilin-A2 functions in membrane scission in clathrin-independent endocytosis *Nature* **517** 493–96

Reubold T F, Faelber K, Plattner N, Posor Y, Ketel K, Curth U, Schlegel J, Anand R, Manstein D J and Noé F 2015 Crystal structure of the dynamin tetramer *Nature* **525** 404–8

Robinson P J, Sontag J-M, Liu J-P, Fykse E M, Slaughter C, McMahontt H and Südhof T C 1993 Dynamin GTPase regulated by protein kinase C phosphorylation in nerve terminals *Nature* **365** 163–66

Roux A, Uyhazi K, Frost A and De Camilli P 2006 GTP-dependent twisting of dynamin implicates constriction and tension in membrane fission *Nature* **441** 528–31

Saheki Y and De Camilli P 2012 Synaptic vesicle endocytosis *Cold Spring Harb. Perspect. Biol.* **4** a005645

Saleem M, Morlot S, Hohendahl A, Manzi J, Lenz M and Roux A 2015 A balance between membrane elasticity and polymerization energy sets the shape of spherical clathrin coats *Nat. Commun.* **6** 1–10

Salim K, Bottomley M J, Querfurth E, Zvelebil M, Gout I, Scaife R, Margolis R, Gigg R, Smith C and Driscoll P 1996 Distinct specificity in the recognition of phosphoinositides by the pleckstrin homology domains of dynamin and Bruton's tyrosine kinase *EMBO J.* **15** 6241–50

Sambuughin N, Goldfarb L G, Sivtseva T M, Davydova T K, Vladimirtsev V A, Osakovskiy V L, Danilova A P, Nikitina R S, Ylakhova A N and Diachkovskaya M P 2015 Adult-onset autosomal dominant spastic paraplegia linked to a GTPase-effector domain mutation of dynamin 2 *BMC Neuro.* **15** 1–12

Sato K, Ernstrom G G, Watanabe S, Weimer R M, Chen C-H, Sato M, Siddiqui A, Jorgensen E M and Grant B D 2009 Differential requirements for clathrin in receptor-mediated endocytosis and maintenance of synaptic vesicle pools *Proc. Natl Acad. Sci.* **106** 1139–44

Schmid S L and Frolov V A 2011 Dynamin: functional design of a membrane fission catalyst *Annu. Rev. Cell Dev. Biol.* **27** 79–105

Schöneberg J, Lehmann M, Ullrich A, Posor Y, Lo W-T, Lichtner G, Schmoranzer J, Haucke V and Noé F 2017 Lipid-mediated PX-BAR domain recruitment couples local membrane constriction to endocytic vesicle fission *Nat. Commun.* **8** 1–17

Shin N, Ahn N, Chang-Ileto B, Park J, Takei K, Ahn S-G, Kim S-A, Di Paolo G and Chang S 2008 SNX9 regulates tubular invagination of the plasma membrane through interaction with actin cytoskeleton and dynamin 2 *J. Cell Sci.* **121** 1252–63

Shin W, Ge L, Arpino G, Villarreal S A, Hamid E, Liu H, Zhao W-D, Wen P J, Chiang H-C and Wu L-G 2018 Visualization of membrane pore in live cells reveals a dynamic-pore theory governing fusion and endocytosis *Cell* **173** 934–45. e912

Shin W, Wei L, Arpino G, Ge L, Guo X, Chan C Y, Hamid E, Shupliakov O, Bleck C K and Wu L-G 2021 Preformed Ω-profile closure and kiss-and-run mediate endocytosis and diverse endocytic modes in neuroendocrine chromaffin cells *Neuron* **109** 3119–34. e3115

Shnyrova A V, Bashkirov P V, Akimov S A, Pucadyil T J, Zimmerberg J, Schmid S L and Frolov V A 2013 Geometric catalysis of membrane fission driven by flexible dynamin rings *Science* **339** 1433–36

Sidiropoulos P N, Miehe M, Bock T, Tinelli E, Oertli C I, Kuner R, Meijer D, Wollscheid B, Niemann A and Suter U 2012 Dynamin 2 mutations in Charcot–Marie–Tooth neuropathy highlight the importance of clathrin-mediated endocytosis in myelination *Brain* **135** 1395–411

Simunovic M, Evergren E, Callan-Jones A and Bassereau P 2019 Curving cells inside and out: roles of BAR domain proteins in membrane shaping and its cellular implications *Annu. Rev. Cell Dev. Biol.* **35** 111–29

Simunovic M, Manneville J-B, Renard H-F, Evergren E, Raghunathan K, Bhatia D, Kenworthy A K, Voth G A, Prost J and McMahon H T 2017 Friction mediates scission of tubular membranes scaffolded by BAR proteins *Cell* **170** 172–84. e111

Skruzny M, Desfosses A, Prinz S, Dodonova S O, Gieras A, Uetrecht C, Jakobi A J, Abella M, Hagen W J and Schulz J 2015 An organized co-assembly of clathrin adaptors is essential for endocytosis *Dev. Cell* **33** 150–62

Sochacki K A, Dickey A M, Strub M-P and Taraska J W 2017 Endocytic proteins are partitioned at the edge of the clathrin lattice in mammalian cells *Nat. Cell Biol.* **19** 352–61

Soulet F, Yarar D, Leonard M and Schmid S L 2005 SNX9 regulates dynamin assembly and is required for efficient clathrin-mediated endocytosis *Mol. Biol. Cell* **16** 2058–67

Stowell M H, Marks B, Wigge P and McMahon H T 1999 Nucleotide-dependent conformational changes in dynamin: evidence for a mechanochemical molecular spring *Nat. Cell Biol.* **1** 27–32

Sun T, Wu X-S, Xu J, McNeil B D, Pang Z P, Yang W, Bai L, Qadri S, Molkentin J D and Yue D T 2010 The role of calcium/calmodulin-activated calcineurin in rapid and slow endocytosis at central synapses *J. Neurosci.* **30** 11838–47

Sun Y, Martin A C and Drubin D G 2006 Endocytic internalization in budding yeast requires coordinated actin nucleation and myosin motor activity *Dev. Cell* **11** 33–46

Sundborger A, Soderblom C, Vorontsova O, Evergren E, Hinshaw J E and Shupliakov O 2011 An endophilin–dynamin complex promotes budding of clathrin-coated vesicles during synaptic vesicle recycling *J. Cell Sci.* **124** 133–43

Sundborger A C, Fang S, Heymann J A, Ray P, Chappie J S and Hinshaw J E 2014 A dynamin mutant defines a superconstricted prefission state *Cell Rep.* **8** 734–42

Sweitzer S M and Hinshaw J E 1998 Dynamin undergoes a GTP-dependent conformational change causing vesiculation *Cell* **93** 1021–29

Takeda T, Kozai T, Yang H, Ishikuro D, Seyama K, Kumagai Y, Abe T, Yamada H, Uchihashi T and Ando T 2018 Dynamic clustering of dynamin-amphiphysin helices regulates membrane constriction and fission coupled with GTP hydrolysis *eLife* **7** e30246

Takei K, McPherson P S, Schmid S L and Camilli P D 1995 Tubular membrane invaginations coated by dynamin rings are induced by GTP-γS in nerve terminals *Nature* **374** 186–90

Takei K, Slepnev V I, Haucke V and De Camilli P 1999 Functional partnership between amphiphysin and dynamin in clathrin-mediated endocytosis *Nat. Cell Biol.* **1** 33–9

Tan T C, Valova V A, Malladi C S, Graham M E, Berven L A, Jupp O J, Hansra G, McClure S J, Sarcevic B and Boadle R A 2003 Cdk5 is essential for synaptic vesicle endocytosis *Nat. Cell Biol.* **5** 701–10

Tassin T C, Barylko B, Hedde P N, Chen Y, Binns D D, James N G, Mueller J D, Jameson D M, Taussig R and Albanesi J P 2021 Gain-of-function properties of a dynamin 2 mutant implicated in Charcot–Marie–Tooth disease *Front. Cellular Neurosci.* **15** 745940

Taylor M J, Perrais D and Merrifield C J 2011 A high precision survey of the molecular dynamics of mammalian clathrin-mediated endocytosis *PLoS Biol.* **9** e1000604

Timm D, Salim K, Gout I, Guruprasad L, Waterfield M and Blundell T 1994 Crystal structure of the pleckstrin homology domain from dynamin *Nat. Struct. Biol.* **1** 782–88

Van der Bliek A M and Meyerowrtz E M 1991 Dynamin-like protein encoded by the *Drosophila* shibire gene associated with vesicular traffic *Nature* **351** 411–14

Verstreken P, Koh T-W, Schulze K L, Zhai R G, Hiesinger P R, Zhou Y, Mehta S Q, Cao Y, Roos J and Bellen H J 2003 Synaptojanin is recruited by endophilin to promote synaptic vesicle uncoating *Neuron* **40** 733–48

Wang L, Barylko B, Byers C, Ross J A, Jameson D M and Albanesi J P 2010 Dynamin 2 mutants linked to centronuclear myopathies form abnormally stable polymers *J. Biol. Chem.* **285** 22753–57

Warnock D E, Hinshaw J E and Schmid S L 1996 Dynamin self-assembly stimulates its GTPase activity *J. Biol. Chem.* **271** 22310–14

Watanabe S, Rost B R, Camacho-Pérez M, Davis M W, Söhl-Kielczynski B, Rosenmund C and Jorgensen E M 2013 Ultrafast endocytosis at mouse hippocampal synapses *Nature* **504** 242–47

Watanabe S, Trimbuch T, Camacho-Pérez M, Rost B R, Brokowski B, Söhl-Kielczynski B, Felies A, Davis M W, Rosenmund C and Jorgensen E M 2014 Clathrin regenerates synaptic vesicles from endosomes *Nature* **515** 228–33

Wen P J, Grenklo S, Arpino G, Tan X, Liao H-S, Heureaux J, Peng S-Y, Chiang H-C, Hamid E and Zhao W-D 2016 Actin dynamics provides membrane tension to merge fusing vesicles into the plasma membrane *Nat. Commun.* **7** 1–14

Wen X, Saltzgaber G W and Thoreson W B 2017 Kiss-and-run is a significant contributor to synaptic exocytosis and endocytosis in photoreceptors *Front. Cellular Neurosci.* **11** 286

Wittig S, Ganzella M, Barth M, Kostmann S, Riedel D, Pérez-Lara Á, Jahn R and Schmidt C 2021 Cross-linking mass spectrometry uncovers protein interactions and functional assemblies in synaptic vesicle membranes *Nat. Commun.* **12** 1–14

Wu L-G and Chan C Y 2022 Multiple roles of actin in exo- and endocytosis *Front. Synaptic Neurosci.* **14** 841704

Wu L-G, Hamid E, Shin W and Chiang H-C 2014a Exocytosis and endocytosis: modes, functions, and coupling mechanisms *Annu. Rev. Physiol.* **76** 301–31

Wu X-S, Lee S H, Sheng J, Zhang Z, Zhao W-D, Wang D, Jin Y, Charnay P, Ervasti J M and Wu L-G 2016 Actin is crucial for all kinetically distinguishable forms of endocytosis at synapses *Neuron* **92** 1020–35

Wu X-S, McNeil B D, Xu J, Fan J, Xue L, Melicoff E, Adachi R, Bai L and Wu L-G 2009 Ca 2+ and calmodulin initiate all forms of endocytosis during depolarization at a nerve terminal *Nat. Neurosci.* **12** 1003–10

Wu Y, T O'Toole E, Girard M, Ritter B, Messa M, Liu X, McPherson P S, Ferguson S M and De Camilli P 2014b A dynamin 1-, dynamin 3-and clathrin-independent pathway of synaptic vesicle recycling mediated by bulk endocytosis *eLife* **3** e01621

Xu J, Luo F, Zhang Z, Xue L, Wu X-S, Chiang H-C, Shin W and Wu L-G 2013 SNARE proteins synaptobrevin, SNAP-25, and syntaxin are involved in rapid and slow endocytosis at synapses *Cell Rep.* **3** 1414–21

Yarar D, Waterman-Storer C M and Schmid S L 2007 SNX9 couples actin assembly to phosphoinositide signals and is required for membrane remodeling during endocytosis *Dev. Cell* **13** 43–56

Yoshida Y, Kinuta M, Abe T, Liang S, Araki K, Cremona O, Di Paolo G, Moriyama Y, Yasuda T and De Camilli P 2004 The stimulatory action of amphiphysin on dynamin function is dependent on lipid bilayer curvature *EMBO J.* **23** 3483–91

Zhang P and Hinshaw J E 2001 Three-dimensional reconstruction of dynamin in the constricted state *Nat. Cell Biol.* **3** 922–26

Zhao M, Maani N and Dowling J J 2018 Dynamin 2 (DNM2) as cause of, and modifier for, human neuromuscular disease *Neurotherapeutics* **15** 966–75

Zheng J, Cahill S M, Lemmon M A, Fushman D, Schlessinger J and Cowburn D 1996 *Identification of the Binding Site for Acidic Phospholipids on the PH Domain of Dynamin: Implications for Stimulation of GTPase Activity* (Amsterdam: Elsevier) pp 14–21

Züchner S, Noureddine M, Kennerson M, Verhoeven K, Claeys K, Jonghe P D, Merory J, Oliveira S A, Speer M C and Stenger J E 2005 Mutations in the pleckstrin homology domain of dynamin 2 cause dominant intermediate Charcot–Marie–Tooth disease *Nat. Genet.* **37** 289–94

Part II

Biophysical insights from synthetic and reconstituted systems

IOP Publishing

Exocytosis: From Molecules to Cells

Arun Anantharam and Jefferson Knight

Chapter 8

Understanding the molecular mechanism of fusion pores through reconstruction

Sarah Dallo and Huan Bao

Vesicle fusion influences all aspects of life and is thus under explicit control to ensure cargo release to the right place at the right time. Fusion pores are the first crucial intermediates formed to connect the two opposing membranes during fusion. Pore properties play a critical role in determining the outcome of vesicle fusion and the level of cargo release. Despite extensive studies, the molecular mechanism that determines pore size and lifetime remains unclear. Recent efforts using bottom-up reconstruction approaches have provided novel insights into this mystery. Here, we summarize this progress and discuss future directions of research that may advance mechanistic understandings of fusion pores.

8.1 Introduction

Eukaryotic cells are compartmentalized by membranes and thus need to communicate with one another. Vesicular transport along the secretory pathway represents one such essential mechanism for cell-to-cell communication [1, 2]. Through a process originally called 'membrane fusion' and later termed 'exocytosis' [3], secretory vesicles discharge their content of biochemical signals into the extracellular space, resulting in the flow and computation of information among cells at close or distant sites. A landmark example is the quantum release at synapses, in which Fatt and Katz provided the very first evidence that the secretion of neurotransmitters occurs in the form of discrete packages in frog muscle fibers [4]. A few years later, electron microscopy (EM) studies showed that these packages are small membrane-bound vesicles in the presynaptic nerve terminals [5]. During synaptic vesicle exocytosis, a series of critical bilayer structural changes culminate in the fusion of two opposing membranes and the release of neurotransmitters (figure 8.1(a)). Detailed EM analyses of these secretion events has revealed a crucial intermediate state that exhibits omega-shaped structures called fusion pores (figure 8.1(b)) [6],

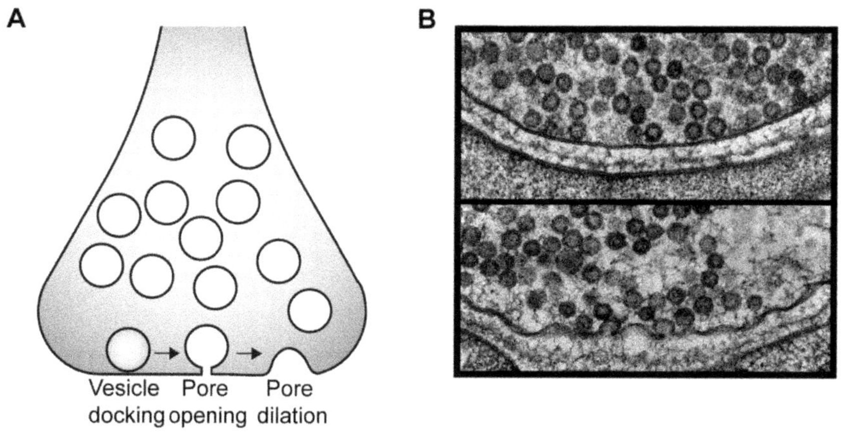

Figure 8.1. Fusion pore: the first crucial intermediate state in vesicle exocytosis (a) illustration of fusion pores in a neuron; (b) EM micrographs of fusion pore formation at the synapse before (upper) and after stimulation (bottom). Copyright 1981 Heuser *et al.* Originally published by The Rockefeller University Press. Figures were adapted from reference [6] with permission.

which establish the initial connection between vesicles and the plasma membrane to allow the escape of neurotransmitters and hormones [7].

The formation of a fusion pore results in a current flow between the cell exterior and the lumen of a secretory vesicle [8]. The conductance of this current flow can be detected using whole-cell patch-clamp electrophysiology approaches with microsecond time resolution [9, 10]. The data thus obtained demonstrate that fusion pores are remarkably dynamic and exhibit several interesting features [11, 12]. The initial pore diameters range from 0.5 to 2 nm and the pores may further dilate or reversibly open and close in a kiss-and-run fashion [10, 13–16]. These features are vital in order to control the level of cargo secretion, and their defects are associated with a multitude of diseases [17–20]. It has been proposed that the specific engagement of reactive molecules on the vesicle and the plasma membrane forms the fusion pore and merges the two opposing membranes [10]. This hypothesis inspired decades of extensive research to identify the reactive molecules and to delineate the underlying molecular mechanism that is responsible for the exquisite speed and precision of exocytosis.

8.2 SNARE-mediated membrane fusion

Biochemical reconstitution is essential for the identification of these reactive molecules, the core fusion machinery, i.e. soluble *N*-ethylmaleimide-sensitive factor attachment protein receptors. Through the cell-free reconstruction of protein transport across the Golgi stack, the Rothman laboratory uncovered two critical soluble proteins needed for membrane fusion. The first protein isolated was an AAA ATPase that is inhibited by mild treatment with the cysteine-alkylating reagent *N*-ethylmaleimide (NEM), and was hence named NEM-sensitive factor (NSF) [21]. The second protein was identified as a result of its ability to bind NSF to Golgi membranes and was thus named soluble NSF attachment protein (SNAP) [22, 23]. It was then found that NSF and SNAP could assemble with detergent-solubilized integral membrane components from brain extracts into a protein complex with a sedimentation coefficient of 20S [24]. This 20S complex could also be disassembled by a process that requires adenosine triphosphate (ATP) hydrolysis. Taking advantage of this inherent assembly and disassembly property, Rothman and coworkers isolated the membrane components, the SNAP receptors (SNAREs) from the 20S complex immobilized on a solid matrix. Interestingly, the SNAREs derived from whole brains are proteins that had previously been cloned and localized in synapses [25]. Two of these proteins, syntaxin (syx, isoforms 1A and 1B) and SNAP-25, reside on the presynaptic plasma membrane. The other protein, synaptobrevin II (syb2), also termed vesicle-associated membrane protein (VAMP2), is confined to synaptic vesicles. In reconstituted systems, the pairing of syb2 with syx and SNAP-25 alone could catalyze the fusion of lipid bilayers [26]. Thus, the SNARE hypothesis was coined that the assembly of a vesicle SNARE (v-SNARE) with its cognate target membrane SNAREs (t-SNAREs) would form a SNAREpin to drive membrane fusion [26–28]. The assembly specificity of SNARE complexes is encoded in their inherent physicochemical properties; only complementary v- and t-SNAREs can pair between bilayers, maintaining the identity of membrane-bound compartments. At the same time, other groups demonstrated that clostridial neurotoxins inhibit synaptic vesicle secretion by cleaving the SNARE proteins [29–33], further supporting the SNARE hypothesis.

Simultaneously, the Schekman laboratory took a completely independent genetic approach and identified that the homologs of NSF, SNAP, and SNARE proteins play essential roles in mediating vesicular transport and membrane trafficking in yeast [34]. In addition, a few other regulatory proteins stood out from these studies as they exhibited the ability to modulate the function of SNAREs [35]. Thus, the SNARE complex is subjected to extensive control by a panel of regulators to ensure the precision of the secretory pathway and is conserved in eukaryotic cells as evolutionarily distant as neurons and yeast [36].

The molecular mechanism of the SNARE hypothesis was further elucidated by structural and biophysical studies [37, 38]. SNARE proteins all contain conserved stretches of 60–70 residues termed SNARE motifs. Although individual SNARE motifs are largely unstructured, four cognate SNARE motifs spontaneously assemble into parallel four-helix bundles that are composed of mostly leucine zipper-like layers. The zippering of these layers forms the *trans*-SNARE complex between two opposing membranes and generates the force required to overcome the energy barrier associated with membrane fusion (figures 8.2(a) and (b)). The completion of membrane fusion converts the *trans*-SNARE complex into a *cis*-SNARE complex, in which all SNARE proteins are anchored in the same membrane (figure 8.2(c)). This conformational transition from the *trans* to *cis* complex is further dissected by single-molecule biophysical approaches [39]. The data suggest that v- and t-SNAREs zipper into the *trans* complex via a stepwise process that pulls the two membranes together and eventually open a fusion pore to allow cargo release. The N-terminal half of the SNARE motif assembles slowly to

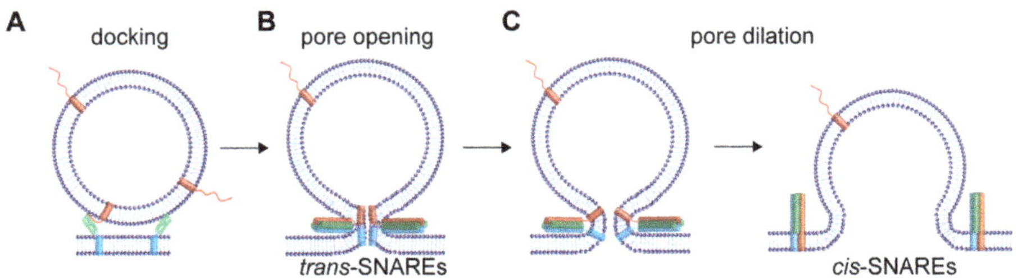

Figure 8.2. Illustration of SNARE-mediated vesicle fusion: (a) the binding of v- to t-SNAREs promotes vesicle docking; (b) formation of the *trans*-SNARE complex results in pore opening; (c) transition from the *trans* to *cis* complex causes pore dilation and bilayer merging.

form a partially zippered intermediate state [39–41] that can be clamped by complexin to prevent the zippering of the C-terminal half [42–44]. In contrast, Sec1/Munc18-family proteins markedly change the kinetics of this stepwise assembly reaction and chaperone the full zippering of the SNARE complex [40, 45, 46]. Therefore, the partially assembled *trans*-SNARE complex is a critical intermediate that allows the action of regulators to control the property and fate of fusion pores.

8.3 The mystery of fusion pores

Despite these remarkable mechanistic insights into the SNARE proteins, a number of mysteries regarding the operating principles of fusion pores remain unresolved [10, 47, 48]. First, the structure of fusion pores is heavily debated. Specifically, it is unknown whether the pore is purely lipidic or proteinaceous in the initial open state (figures 8.3(a) and (b)). Based on free energy calculations in theoretical models, the fusion of opposing membranes can proceed via a hemifusion intermediate state in the absence of any proteins [49–51], providing that the two membranes are close enough (figure 8.3(a)). In this model, the fusion pore is purely lipidic and the function of the SNARE proteins is to bring the two membranes into close proximity via the zippering process. As such, the transmembrane domains (TMDs) of SNAREs merely serve as membrane anchors and do not have any functional roles in the formation of fusion pores. Supporting this model, bilayer fusion can occur with lipid-tethered SNAREs lacking TMDs [52]. However, it has been demonstrated that replacing the TMD of syb2 with lipid anchors or TMDs from other proteins eliminated cargo release in both cell-free and cell-based assays [53–55]. Therefore, the SNARE TMDs are important for fusion pore formation and their role in the lipidic fusion pore model remains to be elucidated.

In contrast, the proteinaceous fusion pore model suggests that the pore is formed by the TMDs of SNAREs as a gap junction-like structure (figure 8.3(b)) [10, 47, 48]. This model was based on the observation that the conductance of the initial fusion pore in mast cells is in the range of ~230 pS[8], which is very similar to that of an ion channel. The other similar ion channel formed between two membranes is the gap junction that connects adjacent cells. Supporting this model, site-specific Trp-substitutions on the TMDs of SNAREs restrained catecholamine release in PC12 cells [54, 56]. However, these data have other possible interpretations, such as the

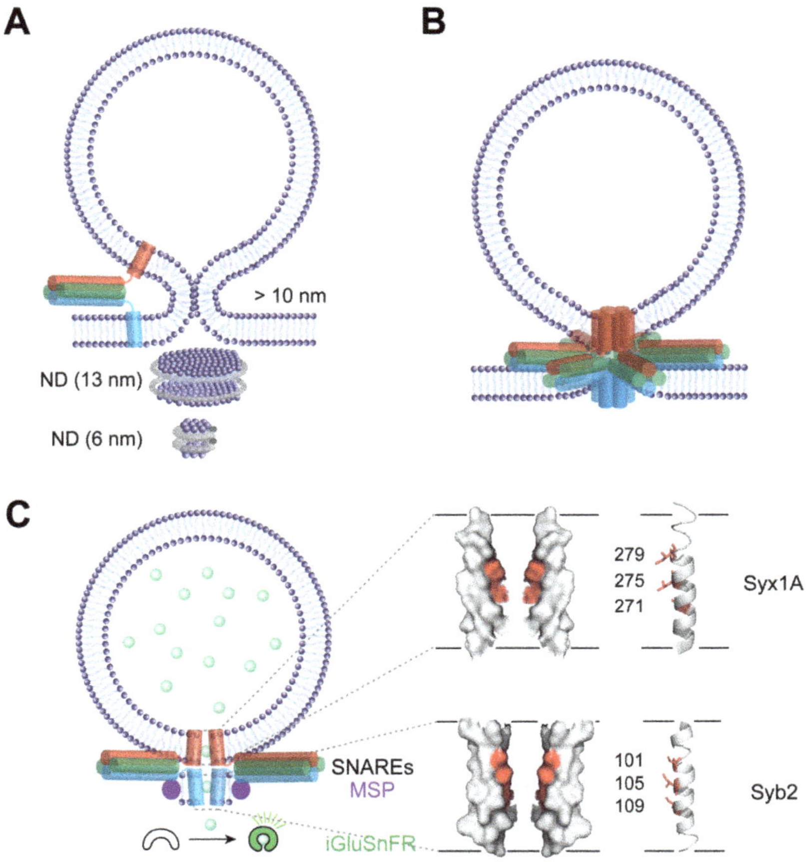

Figure 8.3. Probing the composition of fusion pores using nanodiscs: (a) nanodiscs with well-defined diameters enable the interrogation of the lipidic fusion pore model; (b) model of a gap junction-like fusion pore formed by multiple SNARE TMDs; (c) a hybrid fusion pore reconstituted in nanodiscs. *Left*, fusion between nanodiscs and vesicles results in pore opening and glutamate release, monitored using iGluSnFR. *Right*, the residues of SNARE TMDs lining the fusion pore formed between nanodiscs and vesicles are highlighted in red.

altered interaction of the SNARE complex with cofactors in the membrane. To further examine this possibility, it would be necessary to develop new methods to isolate the fusion pore in the early open state for structural and biochemical interrogations [13, 48].

Another mystery is the molecular mechanism that determines pore characteristics [12, 57]. Fusion pores in cells exhibit a range of diameters from 0.5 to 3 nm [58, 59]. At the early stage of a fusion event, the pore is quite dynamic and can flicker between open and closed states with distinct lifetimes. It is thus unclear how the same zippering process of the SNARE proteins can lead to completely different fusion pores with unique physical and chemical properties. One intriguing mechanism is that the interaction between regulatory proteins and the SNARE proteins encodes fundamental features of individual pores. As more than a dozen regulatory proteins are involved in exocytosis, reconstitution approaches are key to decoding the underlying molecular mechanism.

The last mystery is the conformational state of the *trans*-SNARE complex at the stage of initial pore opening. In light of the crystal structure obtained using cytosolic fragments of SNARE proteins, it has been suggested that the formation of fusion pores requires the full assembly of the SNARE motif. However, SNAREs cleaved by botulinum neurotoxin A (BoNT/A) could still mediate glutamate release with altered kinetics in neurons [60, 61], suggesting that partially assembled SNAREs might be able to catalyze the formation of fusion pores. Therefore, the structure of a *trans*-SNARE complex at the initial pore-opening stage remains largely unknown. This problem arises because the *trans*-SNARE complex is a transient intermediate and is rapidly converted to the *cis*-SNARE complex after fusion. As a consequence, it is difficult to isolate the *trans* complex for mechanistic interrogation. In the following sections, we summarize recent progress in new cell-free systems for dissecting nascent fusion pores and *trans*-SNARE complexes achieved by us and others and also discuss future directions and challenges on this front. Since there is extensive literature detailing various aspects of fusion pores that cannot be comprehensively covered here, we refer the readers to several excellent recent reviews on this subject for further reading [10, 62, 63].

8.4 Interrogation of fusion pores using liposomes

The biochemical reconstitution of pore formation was initially performed in liposome-based membrane fusion assays. Here, v- and t-SNAREs were incorporated into two separate vesicles encapsulating a pair of reporters that resulted in fluorescence readouts upon pore opening [64–66]. Reporters of various sizes were used to infer the properties of fusion pores. Using TIRF microscopy, individual pore formation events could be dissected at single-vesicle level with ~100 ms time resolution. These experiments demonstrated that synaptotagmin-1 (syt1) and complexin could expand pore sizes and possibly also stabilize fusion pores, thereby promoting the release of large cargoes. In addition, transient pore-opening and -closing dynamics were captured using reporters with high sensitivity [67]. To mimic exocytosis in cells, these assays could also be adapted to monitor fusion between vesicles and supported lipid bilayers [68, 69]. Together, these experiments provided evidence that SNAREs and regulatory proteins play important roles in controlling pore properties.

However, it is difficult to quantitatively determine pore sizes and dynamics through single-vesicle imaging, because reconstituted fusion pores in liposomes can dilate within a short time window. To resolve this problem, electrophysiology approaches with higher temporal resolutions were used to study the fusion pore formed between liposomes and supported lipid bilayers [70]. In these systems, the liposomes either contained a salt gradient or other nanopores. The fusion of the liposome with the supported lipid bilayers created a transient current passing through the pore. One caveat here is that it became impossible to track the properties of single pores once the liposome fully merged onto the bilayer. Thus, a pressing need exists to create tools that can trap nascent fusion pores in the open state for a longer period of time.

8.5 Isolation of nascent fusion pores using nanodiscs

To isolate nascent fusion pores, the Rothman group developed a nanodisc-based reconstitution assay [53]. Nanodiscs are nanoscale lipid bilayers encased by membrane scaffold proteins (MSPs); they can be tailored to have diameters from 6 to 50 nm [71–74]. When a nanodisc bearing v-SNARE fuses with a vesicle bearing t-SNAREs, the rigid framework of MSPs restrains the size and expansion of fusion pores. Considering that each bilayer is 4–5 nm thick and the diameter of the α-helix of MSP is ~1.2 nm, 13 nm nanodiscs would only allow fusion pores to open to 1–3 nm (figure 8.3(c)), thereby stabilizing the pore in the initial open state.

In addition, nanodiscs can enclose a well-defined number of proteins in lipid bilayers, providing a means to study the stoichiometric relationship between membrane fusion and SNAREs. It has been demonstrated that a single SNARE complex is sufficient to mediate the fusion of lipid bilayers, but three are required to keep the fusion pore open for cargo release [53]. It is proposed that the radial force from multiple SNARE zippering is essential to prevent the nascent pore from resealing. Interestingly, the minimum number of SNARE copies required to drive cargo release also depends on the presence of regulatory proteins. For example, two SNARE complexes were enough to allow cargo release in the presence of synaptotagmin-1 (syt1) [55]. This observation is most likely due to the ability of syt1 to chaperone the assembly of the *trans*-SNARE complex, and thus the formation of fusion pores [75, 76].

The isolation of nascent fusion pores in nanodiscs paved the way for a series of structure–function relationship studies [55]. In particular, the wide range of available nanodisc sizes makes the nanodisc a perfect tool with which to interrogate the pore composition. As alluded to earlier, there are two existing models for this. In one model, the pore is purely lipidic, formed via a hemifusion stalk structure [49, 51]. In the other model, the fusion pore is formed by the TMD of the SNARE proteins in a gap junction-like structure [53–55]. One distinct feature that separates the two models is the pore dimension. If the pore is formed by pure lipids, it would only be reconstituted in nanodiscs that have a diameter larger than the thickness of two lipid bilayers, which is ~10 nm (figure 8.3(a)). As such, fusion pores would not be formed in nanodiscs that have diameters of 6 nm. Using the biochemical reconstitution of glutamate release, the formation of fusion pores can readily be monitored. In this system, t-SNARE vesicles were loaded with glutamate and allowed to fuse with v-SNARE nanodiscs. The glutamate released via fusion pores was quantified using a highly sensitive fluorescent reporter, iGluSnFR [77, 78]. Using this assay, we found that both 6 and 13 nm nanodiscs could form fusion pores with robust glutamate release efficiencies [55]. Based on these data, we propose that the nascent fusion pore is not purely lipidic, and therefore might be a proteinaceous pore formed by the TMDs of SNAREs.

To further test this possibility, we performed chemical labeling experiments to probe the accessibility of the SNARE TMDs during pore formation, taking advantage of the fact that nanodiscs could keep the pore open for over 1 h. This idea was inspired by studies from the Karlin group in mapping the water-accessible crevices of GPCRs and ion channels [79, 80]. In this assay, the residues on the TMDs of SNAREs would only be labeled if they were transiently exposed to the solvent during the opening of fusion pores. Indeed, we found that several specific residues on one face of SNARE TMDs were labeled (figure 8.3(c), right), so they might line the lumen face of fusion pores. However, as only two copies of SNAREs are needed for robust cargo release in the presence of syt1, the pore must be formed as a hybrid structure by both lipids and SNARE TMDs (figure 8.3(c)). These conclusions are further supported by molecular dynamics simulation studies showing that the TMDs and lipids together form a permeable channel sufficiently large to allow the passage of neurotransmitters [10, 81, 82].

8.6 Dissection of pore properties using electric recording

Although nanodiscs are capable of isolating individual nascent fusion pores, it remains challenging to capture pore dynamics at the single-molecule level. Previous single-vesicle imaging assays were limited by low time resolutions (10–100 ms) and were only able to track the pore for a few seconds. These limitations precluded quantitative studies of individual pore-opening and -closing events using single-vesicle imaging methods. To bypass these limitations, we and others turned to electrical recordings that could offer a time resolution of 20 μs [83, 86]. This high temporal resolution was previously employed to probe the fusion of vesicles with supported lipid bilayers and enabled earlier attempts to capture pore dynamics in reconstituted systems [70].

Using patch-clamp recordings, Wu *et al* characterized the fusion pore formed between nanodiscs bearing v-SNAREs and cell membranes displaying t-SNAREs [83, 84]. In this nanodisc–cell fusion assay, pore-opening events could readily be detected, but the isolation of single pores was challenging due to high background signals. In addition, it was difficult to interrogate the conformational state of the *trans*-SNARE complex that mediates pore formation, as the pore is no longer accessible for manipulation after opening. Nevertheless, the two studies by Wu *et al* were the first to demonstrate the superiority of electric recording for monitoring pore dynamics with microsecond time resolution *in vitro*.

Because of the challenge of the nanodisc–cell fusion assay, we attempted to use planar lipid bilayer (PLB) electrophysiology for the reconstitution of single fusion pores (figure 8.4) [85]. Fusion between nanodiscs and planar lipid bilayers (ND-PLBs) results in the formation of single pores that can readily be detected by monitoring the current passing through them (figure 8.4(a)). The pore sizes and dynamics can be determined by electrophysiology approaches and the time resolution can readily achieve 200 μs. Using the ND-PLB system, we found that both pore size and stability were affected by the number of SNARE complexes. For example, three copies of SNAREs only led to a 1.1 nm rapidly flickering pore, whereas five copies give rise to a stable 2.2 nm pore (figure 8.4(b)). Thus, we proposed that the number of SNAREs recruited to drive fusion determines the pore properties in cells.

Furthermore, the open accessibility of ND-PLB enabled us to probe the degree of SNARE zippering in the *trans* complex at the stage of initial pore opening. As the

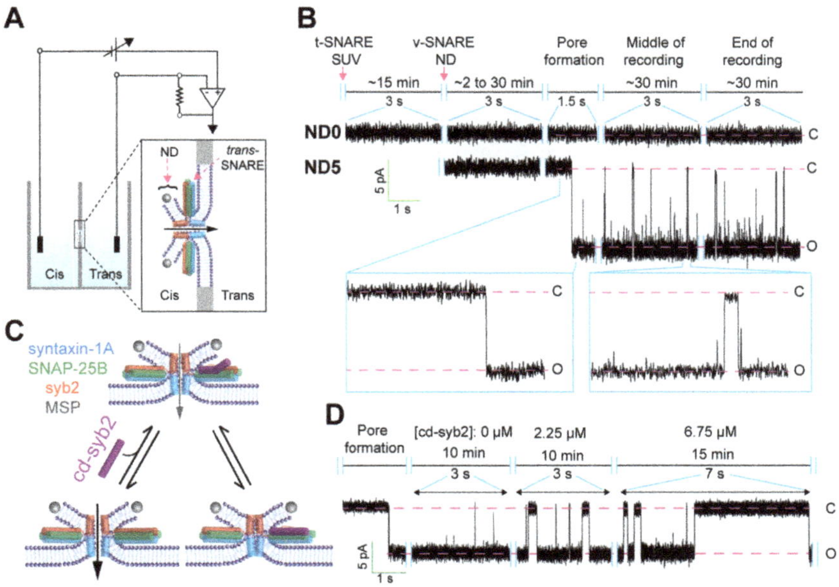

Figure 8.4. Dynamics and number of *trans*-SNARE complexes determine nascent fusion pore properties. (a) Illustration of the ND-PLB system. Fusion pores formed between nanodiscs and PLB were dissected using electric recordings with high temporal resolution. (b) Representative traces of reconstituted fusion pores using ND-PLB. ND0, empty nanodiscs; ND5, nanodiscs bearing an average of five v-SNAREs. (c) Illustration of the impact of cd-syb2 on reconstituted fusion pores. (d) Representative traces of pore closure caused by the addition of cd-syb2. Huan Bao *et al*, 2018, Copyright 2018, Macmillan Publishers Limited, part of Springer Nature. All rights reserved. Figures were adapted from [85] with permission of Springer.

formation of fusion pores is dependent on the pairing between v- and t-SNAREs, reagents that disrupt the *trans*-SNARE complex were able to close the pore. Thus, we could determine whether the SNARE motif of the *trans* complex is fully or partially zippered by examining the impact of conformational specific reagents on pore dynamics. The fully zippered SNARE motif is remarkably stable and is resistant to sodium dodecyl sulfate and neurotoxin cleavage in reconstituted systems. In cells, its unwinding requires the ATP-fueled disassembly machinery composed of NSF and SNAP. In contrast, partially zippered SNAREs are much less stable and sensitive to neurotoxin cleavage and can be disrupted by peptides corresponding to the N-terminal half of the SNARE motif [87]. In the ND-PLB system, the addition of the cytosolic fragment of v-SNARE (cd-syb2) closed the pore in a concentration-dependent manner (figures 8.4(c) and (d)) [85]. These results suggest that cd-syb2 binds the *trans*-SNARE complex and displaces the v-SNARE nanodisc from t-SNAREs. Although many possible explanations exist, the simplest is that the *trans* complex is partially zippered and exposes the C-terminal half of t-SNAREs to bind cd-syb2. Otherwise, the fully assembled SNARE complex should be resistant to the cd-syb2 triggered disassembly. Single-molecule FRET (smFRET) characterizations of the *trans*-SNARE complex reconstituted in two opposing 13 nm nanodiscs consistently confirmed that the N-terminal residues of v- and t-SNAREs are in close proximity, whereas the C-terminal residues are still far apart from each other [86]. Overall, we propose that the *trans*-SNARE complex is only partially zippered in the ND-PLB system, yet able to mediate the formation of fusion pores. The dynamics of these *trans* complexes are the underlying mechanism for the flickering behavior of nascent fusion pores in cells. In line with this hypothesis, molecular dynamics simulation studies revealed that the zippering to layer six of the SNARE motif provided sufficient energy to open a ~1–3 nm pore [82]. In addition, previous studies have shown that synaptic vesicle exocytosis in neurons exhibits distinct sensitivity to different neurotoxin treatments [60, 61]. These findings could be explained if the SNARE motif is not fully zippered in the *trans*-SNARE complex that mediates neurotransmitter release *in vivo*, so that the neurotoxin cleavage sites at the C-termini are still accessible.

8.7 Reconstitution of expanded fusion pores

Recent engineering efforts in MSPs have substantially extended the size range of nanodiscs [74], which has allowed us to study the mechanism of pore dilation. Using large nanodiscs bearing both v-SNARE and syt1, we could observe Ca^{2+}-stimulated pore expansion [86]. These expanded fusion pores became resistant to disassembly by cd-syb2, but were still sensitive to the recycling machinery of NSF and α-SNAP. Therefore, the SNARE motif is chaperoned by syt1 to an irreversible state that results in the dilation of fusion pores. In smFRET experiments, the C-terminal residues of the SNARE motif in the *trans* complex reconstituted in nanodiscs were also brought into close proximity in the presence of syt1. In contrast, pores remained sensitive to cd-syb2 and the SNARE motif was still partially zippered in the absence of syt1. With the continuous development of nanodiscs and other nanolipoprotein particles [84, 88, 89], we should be able to readily delineate the relationship between pore sizes and the level of SNARE zippering.

8.8 Future directions

Further insights into fusion pores require structural information. The main challenge here is to generate homogeneous samples for single-particle cryo-EM studies. On this front, there are three main difficulties. First, the average number of SNAREs in nanodiscs is primarily controlled by using different amounts of MSPs during reconstitution. This strategy often led to a binomial distribution of protein stoichiometry within each nanodisc and thus could not provide the homogeneity required for structural studies. Second, the orientation of SNAREs in nanodiscs is random, which further increases the heterogeneity of reconstituted fusion pores. Third, the yield of MSPs suitable for making large nanodiscs is ~20 times less than that for small MSPs. As such, it is technically challenging to obtain a sufficient amount of expanded fusion pores. To overcome these difficulties, future work will need to engineer simple and straightforward platforms that can tighten the control of the stoichiometry and orientation of SNARE proteins in nanodiscs. Recently, DNA origami technology has enabled the construction of membrane systems with programmable geometry arrangements [90, 91]. These DNA scaffolds can

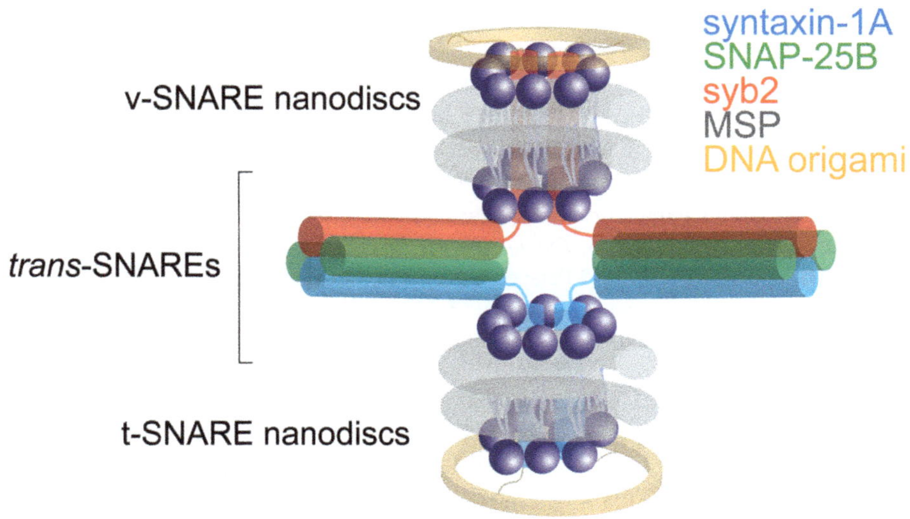

Figure 8.5. Reconstitution of *trans*-SNARE complexes and fusion pores using DNA-origami templated nanodiscs. DNA origami would allow precise control of the number of SNARE copies and their orientation, supporting the reconstitution of homogenous fusion pores, which are useful in structural studies.

rigorously orientate a defined number of membrane proteins reconstituted in nanodiscs with various diameters. Therefore, we believe that DNA-origami-assisted nanodisc reconstitution will profoundly advance the mechanistic understanding of fusion pores in the coming years (figure 8.5).

Acknowledgments

This work was made possible by the support of the Scripps Research Institute and the NIH Director's New Innovator Award (DP2GM140920 to B.H.).

References

[1] Sudhof T C and Rothman J E 2009 Membrane fusion: grappling with SNARE and SM proteins *Science* **323** 474–77
[2] Wickner W and Schekman R 2008 Membrane fusion *Nat. Struct. Mol. Biol.* **15** 658–64
[3] Palade G 1975 Intracellular aspects of the process of protein synthesis *Science* **189** 867
[4] Fatt P and Katz B 1950 Some observations on biological noise *Nature* **166** 597–98
[5] De Robertis E D and Bennett H S 1955 Some features of the submicroscopic morphology of synapses in frog and earthworm *J. Biophys. Biochem. Cytol.* **1** 47–58
[6] Heuser J E and Reese T S 1981 Structural changes after transmitter release at the frog neuromuscular junction *J. Cell Biol.* **88** 564–80
[7] Rothman J E 2014 The principle of membrane fusion in the cell (Nobel lecture) *Angew. Chem. Int. Ed. Engl.* **53** 12676–94
[8] Breckenridge L J and Almers W 1987 Currents through the fusion pore that forms during exocytosis of a secretory vesicle *Nature* **328** 814–17
[9] Fernandez J M, Neher E and Gomperts B D 1984 Capacitance measurements reveal stepwise fusion events in degranulating mast cells *Nature* **312** 453–55

[10] Sharma S and Lindau M 2018 The fusion pore, 60 years after the first cartoon *FEBS Lett.* **592** 3542–62

[11] Spruce A E, Breckenridge L J, Lee A K and Almers W 1990 Properties of the fusion pore that forms during exocytosis of a mast cell secretory vesicle *Neuron* **4** 643–54

[12] Klyachko V A and Jackson M B 2002 Capacitance steps and fusion pores of small and large-dense-core vesicles in nerve terminals *Nature* **418** 89–92

[13] Jackson M B and Chapman E R 2006 Fusion pores and fusion machines in Ca^{2+}-triggered exocytosis *Annu. Rev. Biophys. Biomol. Struct.* **35** 135–60

[14] Alabi A A and Tsien R W 2013 Perspectives on kiss-and-run: role in exocytosis, endocytosis, and neurotransmission *Annu Rev Physiol* **75** 393–422

[15] Zhao W D *et al* 2016 Hemi-fused structure mediates and controls fusion and fission in live cells *Nature* **534** 548–52

[16] Shin W *et al* 2018 Visualization of membrane pore in live cells reveals a dynamic-pore theory governing fusion and endocytosis *Cell* **173** 934–45 e912

[17] Fulop T, Radabaugh S and Smith C 2005 Activity-dependent differential transmitter release in mouse adrenal chromaffin cells *J. Neurosci.* **25** 7324–32

[18] Richards D A 2009 Vesicular release mode shapes the postsynaptic response at hippocampal synapses *J. Physiol.* **587** 5073–80

[19] Collins S C *et al* 2016 Increased expression of the diabetes gene SOX4 reduces insulin secretion by impaired fusion pore expansion *Diabetes* **65** 1952–61

[20] Logan T, Bendor J, Toupin C, Thorn K and Edwards R H 2017 Alpha-synuclein promotes dilation of the exocytotic fusion pore *Nat. Neurosci.* **20** 681–89

[21] Block M R, Glick B S, Wilcox C A, Wieland F T and Rothman J E 1988 Purification of an *N*-ethylmaleimide-sensitive protein catalyzing vesicular transport *Proc. Natl Acad. Sci. USA* **85** 7852–56

[22] Clary D O and Rothman J E 1990 Purification of 3 related peripheral membrane-proteins needed for vesicular transport *J. Biol. Chem.* **265** 10109–17

[23] Clary D O, Griff I C and Rothman J E 1990 Snaps, a family of Nsf attachment proteins involved in intracellular membrane-fusion in animals and yeast *Cell* **61** 709–21

[24] Wilson D W, Whiteheart S W, Wiedmann M, Brunner M and Rothman J E 1992 A multisubunit particle implicated in membrane-fusion *J. Cell Biol.* **117** 531–38

[25] Sollner T, Bennett M K, Whiteheart S W, Scheller R H and Rothman J E 1993 A protein assembly-disassembly pathway *in vitro* that may correspond to sequential steps of synaptic vesicle docking, activation, and fusion *Cell* **75** 409–18

[26] Weber T *et al* 1998 SNAREpins: minimal machinery for membrane fusion *Cell* **92** 759–72

[27] Parlati F *et al* 2000 Topological restriction of SNARE-dependent membrane fusion *Nature* **407** 194–98

[28] McNew J A *et al* 2000 Compartmental specificity of cellular membrane fusion encoded in SNARE proteins *Nature* **407** 153–59

[29] Blasi J *et al* 1993 Botulinum neurotoxin a selectively cleaves the synaptic protein SNAP-25 *Nature* **365** 160–63

[30] Blasi J *et al* 1993 Botulinum neurotoxin C1 blocks neurotransmitter release by means of cleaving HPC-1/syntaxin *EMBO J.* **12** 4821–28

[31] Schiavo G *et al* 1992 Tetanus and botulinum-B neurotoxins block neurotransmitter release by proteolytic cleavage of synaptobrevin *Nature* **359** 832–35

[32] Schiavo G, Matteoli M and Montecucco C 2000 Neurotoxins affecting neuroexocytosis *Physiol. Rev.* **80** 717–66

[33] Montecucco C, Schiavo G and Pantano S 2005 SNARE complexes and neuroexocytosis: how many, how close? *Trends Biochem. Sci.* **30** 367–72

[34] Novick P, Field C and Schekman R 1980 Identification of 23 complementation groups required for post-translational events in the yeast secretory pathway *Cell* **21** 205–15

[35] Schekman R and Novick P 2004 23 genes, 23 years later *Cell* **116** S13–15 11 p following S19

[36] Ferro-Novick S and Jahn R 1994 Vesicle fusion from yeast to man *Nature* **370** 191–93

[37] Sutton R B, Fasshauer D, Jahn R and Brunger A T 1998 Crystal structure of a SNARE complex involved in synaptic exocytosis at 2.4 A resolution *Nature* **395** 347–53

[38] Stein A, Weber G, Wahl M C and Jahn R 2009 Helical extension of the neuronal SNARE complex into the membrane *Nature* **460** 525–U105

[39] Gao Y *et al* 2012 Single reconstituted neuronal SNARE complexes zipper in three distinct stages *Science* **337** 1340–3

[40] Jiao J *et al* 2018 Munc18-1 catalyzes neuronal SNARE assembly by templating SNARE association *eLife* **7** e41771

[41] Zhang Y and Hughson F M 2021 Chaperoning SNARE Folding and Assembly *Annu. Rev. Biochem.* **90** 581–603

[42] Giraudo C G *et al* 2009 Alternative zippering as an on-off switch for SNARE-mediated fusion *Science* **323** 512–16

[43] Maximov A, Tang J, Yang X F, Pang Z P P and Sudhof T C 2009 Complexin controls the force transfer from SNARE complexes to membranes in fusion *Science* **323** 516–21

[44] Shon M J, Kim H and Yoon T Y 2018 Focused clamping of a single neuronal SNARE complex by complexin under high mechanical tension *Nat. Commun.* **9** 3639

[45] Baker R W *et al* 2015 Membrane fusion a direct role for the secl/Munc18-family protein Vps33 as a template for SNARE assembly *Science* **349** 1111–14

[46] Baker R W and Hughson F M 2016 Chaperoning SNARE assembly and disassembly *Nat. Rev. Mol. Cell Biol.* **17** 465–79

[47] Lindau M and Almers W 1995 Structure and function of fusion pores in exocytosis and ectoplasmic membrane fusion *Curr. Opin. Cell Biol.* **7** 509–17

[48] Jackson M B and Chapman E R 2008 The fusion pores of Ca^{2+}-triggered exocytosis *Nat. Struct. Mol. Biol.* **15** 684–89

[49] Kozlovsky Y and Kozlov M M 2002 Stalk model of membrane fusion: solution of energy crisis *Biophys. J.* **82** 882–95

[50] Chernomordik L V and Kozlov M M 2005 Membrane hemifusion: crossing a chasm in two leaps *Cell* **123** 375–82

[51] Chernomordik L V and Kozlov M M 2008 Mechanics of membrane fusion *Nat. Struct. Mol. Biol.* **15** 675–83

[52] McNew J A *et al* 2000 Close is not enough: SNARE-dependent membrane fusion requires an active mechanism that transduces force to membrane anchors *J. Cell Biol.* **150** 105–17

[53] Shi L *et al* 2012 SNARE proteins: one to fuse and three to keep the nascent fusion pore open *Science* **335** 1355–59

[54] Chang C W *et al* 2015 A structural role for the synaptobrevin 2 transmembrane domain in dense-core vesicle fusion pores *J. Neurosci.* **35** 5772–80

[55] Bao H *et al* 2016 Exocytotic fusion pores are composed of both lipids and proteins *Nat. Struct. Mol. Biol.* **23** 67–73

[56] Han X, Wang C T, Bai J, Chapman E R and Jackson M B 2004 Transmembrane segments of syntaxin line the fusion pore of Ca^{2+}-triggered exocytosis *Science* **304** 289–92

[57] Albillos A *et al* 1997 The exocytotic event in chromaffin cells revealed by patch amperometry *Nature* **389** 509–12

[58] Henkel A W, Meiri H, Horstmann H, Lindau M and Almers W 2000 Rhythmic opening and closing of vesicles during constitutive exo- and endocytosis in chromaffin cells *EMBO J.* **19** 84–93

[59] Staal R G W, Mosharov E V and Sulzer D 2004 Dopamine neurons release transmitter via a flickering fusion pore *Nat. Neurosci.* **7** 341–46

[60] Hua S Y and Charlton M P 1999 Activity-dependent changes in partial VAMP complexes during neurotransmitter release *Nat. Neurosci.* **2** 1078–83

[61] Sakaba T, Stein A, Jahn R and Neher E 2005 Distinct kinetic changes in neurotransmitter release after SNARE protein cleavage *Science* **309** 491–94

[62] Karatekin E 2018 Toward a unified picture of the exocytotic fusion pore *FEBS Lett.* **592** 3563–85

[63] Chang C W, Chiang C W and Jackson M B 2017 Fusion pores and their control of neurotransmitter and hormone release *J. Gen. Physiol.* **149** 301–22

[64] Diao J *et al* 2010 A single-vesicle content mixing assay for SNARE-mediated membrane fusion *Nat. Commun.* **1** 54

[65] Diao J *et al* 2012 A single vesicle-vesicle fusion assay for *in vitro* studies of SNAREs and accessory proteins *Nat. Protoc.* **7** 921–34

[66] Lai Y *et al* 2013 Fusion pore formation and expansion induced by Ca^{2+} and synaptotagmin 1 *Proc. Natl Acad. Sci. USA* **110** 1333–38

[67] Gong B *et al* 2015 High Affinity host-guest FRET pair for single-vesicle content-mixing assay: observation of flickering fusion events *J. Am. Chem. Soc.* **137** 8908–11

[68] Ramakrishnan S *et al* 2018 High-throughput monitoring of single vesicle fusion using freestanding membranes and automated analysis *Langmuir* **34** 5849–59

[69] Kreutzberger A J B *et al* 2019 *In vitro* fusion of single synaptic and dense core vesicles reproduces key physiological properties *Nat. Commun.* **10** 3904

[70] Zimmerberg J 1993 Simultaneous electrical and optical measurements of individual membrane fusion events during exocytosis *Methods Enzymol.* **221** 99–112

[71] Bayburt T H, Grinkova Y V and Sligar S G 2002 Self-assembly of discoidal phospholipid bilayer nanoparticles with membrane scaffold proteins *Nano Lett.* **2** 853–56

[72] Denisov I G, Grinkova Y V, Lazarides A A and Sligar S G 2004 Directed self-assembly of monodisperse phospholipid bilayer nanodiscs with controlled size *J. Am. Chem. Soc.* **126** 3477–87

[73] Hagn F, Etzkorn M, Raschle T and Wagner G 2013 Optimized phospholipid bilayer nanodiscs facilitate high-resolution structure determination of membrane proteins *J. Am. Chem. Soc.* **135** 1919–25

[74] Nasr M L *et al* 2017 Covalently circularized nanodiscs for studying membrane proteins and viral entry *Nat. Methods* **14** 49–52

[75] Bhalla A, Chicka M C, Tucker W C and Chapman E R 2006 Ca^{2+}-synaptotagmin directly regulates t-SNARE function during reconstituted membrane fusion *Nat. Struct. Mol. Biol.* **13** 323–30

[76] Zhou Q J *et al* 2017 The primed SNARE-complexin-synaptotagmin complex for neuronal exocytosis *Nature* **548** 420

[77] Marvin J S *et al* 2013 An optimized fluorescent probe for visualizing glutamate neurotransmission *Nat. Methods* **10** 162–70
[78] Marvin J S *et al* 2018 Stability, affinity, and chromatic variants of the glutamate sensor iGluSnFR *Nat. Methods* **15** 936–39
[79] Javitch J A, Fu D Y, Chen J Y and Karlin A 1995 Mapping the binding-site crevice of the dopamine D2 receptor by the substituted-cysteine accessibility method *Neuron* **14** 825–31
[80] Karlin A and Akabas M H 1998 Substituted-cysteine accessibility method *Methods Enzymol.* **293** 123–45
[81] Sharma S and Lindau M 2016 The mystery of the fusion pore *Nat. Struct. Mol. Biol.* **23** 5–6
[82] Sharma S and Lindau M 2018 Molecular mechanism of fusion pore formation driven by the neuronal SNARE complex *Proc. Natl Acad. Sci. USA* **115** 12751–56
[83] Wu Z *et al* 2016 Nanodisc-cell fusion: control of fusion pore nucleation and lifetimes by SNARE protein transmembrane domains *Sci. Rep.* **6** 27287
[84] Wu Z Y *et al* 2017 Dilation of fusion pores by crowding of SNARE proteins *eLife* **6** e22964
[85] Bao H *et al* 2018 Dynamics and number of trans-SNARE complexes determine nascent fusion pore properties *Nature* **554** 260–63
[86] Das D, Bao H, Courtney K C, Wu L and Chapman E R 2020 Resolving kinetic intermediates during the regulated assembly and disassembly of fusion pores *Nat. Commun.* **11** 231
[87] Melia T J *et al* 2002 Regulation of membrane fusion by the membrane-proximal coil of the t-SNARE during zippering of SNAREpins *J. Cell Biol.* **158** 929–40
[88] Bello O D, Auclair S M, Rothman J E and Krishnakumar S S 2016 Using ApoE nanolipoprotein particles to analyze SNARE-induced fusion pores *Langmuir* **32** 3015–23
[89] Zhang S *et al* 2021 One-step construction of circularized nanodiscs using SpyCatcher-SpyTag *Nat. Commun.* **12** 5451
[90] Zhao Z *et al* 2018 DNA-corralled nanodiscs for the structural and functional characterization of membrane proteins and viral entry *J. Am. Chem. Soc.* **140** 10639–43
[91] Zhang Z and Chapman E R 2020 Programmable nanodisc patterning by DNA origami *Nano Lett.* **20** 6032–37

IOP Publishing

Exocytosis: From Molecules to Cells

Arun Anantharam and Jefferson Knight

Chapter 9

Pore-spanning membranes: a tool for studying neuronal fusion

Peter Mühlenbrock and Claudia Steinem

In vitro single-vesicle fusion assays are an invaluable tool for the analysis of neuronal fusion mediated by soluble *N*-ethylmaleimide-sensitive-factor attachment receptors (SNAREs). A number of different assays have been developed, which provide access to the intermediate states of fusion as well as fusion kinetics. Despite this progress in assay development, the different membrane geometries and architectures appear to influence the obtained results and conclusions. In this chapter, we provide a short overview of the current *in vitro* single-vesicle fusion assays based on fluorescence microscopy; we also introduce and discuss an assay that we have developed in recent years, which is based on pore-spanning membranes (PSMs). These PSMs, spanning micrometer-sized pores in silicon, may overcome some of the drawbacks associated with the other membrane architectures that are used to mimic the target presynaptic membrane. We describe the preparation of the PSMs with reconstituted proteins, the fluorescence readout used to detect single-vesicle fusion events, and how the intermediate fusion states can be distinguished. We further discuss the impact of synaptotagmin-1 on the fusion process and how content-release detection provides further information about fusion pore formation.

9.1 Introduction

Exocytosis is the hallmark of neuronal signal transduction; it occurs when a synaptic vesicle loaded with neurotransmitters fuses with the presynaptic membrane of synaptic boutons to release neurotransmitters into the synaptic cleft. This fusion process is energetically costly, mainly due to the hydration repulsion of the two membranes that need to come close together. To overcome this barrier, specific proteins are required, namely SNAREs. In neuronal fusion, three different SNAREs have been identified as the fusion machinery, namely full-length syntaxin 1A (full-length syx 1A), synaptosom-associated protein of 25 kDa (SNAP-25), and synaptobrevin 2 (syb 2) [1]. Syb 2 is

doi:10.1088/978-0-7503-3771-7ch9

a single-span transmembrane protein that is anchored to the vesicular membrane (v-SNARE), while the presynaptic target membrane contains the other two SNAREs, full-length syx 1A and SNAP-25 (t-SNAREs). Full-length syx 1A is a single-span transmembrane protein that harbors an additional regulatory Habc domain [2]. In contrast, SNAP-25 is anchored to the membrane via palmitoyl side chains covalently bound to cysteine residues of the protein. The three proteins contain a total of four so-called SNARE motifs. Based on various structural information and modeling, Liang *et al* [3] suggested a model that illustrates the possible progression of the engagement of *trans-* and *cis*-SNARE complexes at the site of fusion (figure 9.1). Prior to fusion, the characteristic and highly conserved cytosolic SNARE motifs are largely unstructured (figure 9.1(I)), but transform into α-helices if they interact with each other and form a very stable tetrameric coiled-coil complex (figure 9.1(II)) [1, 4–7]. This structure is known as the *trans*-SNARE complex. Within the coiled-coil structure, there are 16 layers of mainly hydrophobic residues facing inwards, and one hydrophilic zero layer with three glutamine residues (Q) and one arginine (R) residue [8–10]. This zero layer led to the categorization of SNAREs into four classes, namely Q_a, Q_b, Q_c, and R-SNAREs. It is hypothesized that the transformation from the unstructured forms into the tight four-helix bundle (the *trans*-SNARE complex) is initiated at the *N*-termini of the SNARE domains and proceeds to the *C*-termini in a zipper-like fashion [11]. This process pulls the opposing membranes close together, leading to membrane fusion [12, 13]. Upon membrane mergence, both transmembrane helices of syx 1A and syb 2 are located inside the same lipid bilayer, and the SNARE complex changes its conformation from a high-energy *trans*-configuration into a low-energy *cis*-configuration (figure 9.1(III)) [14].

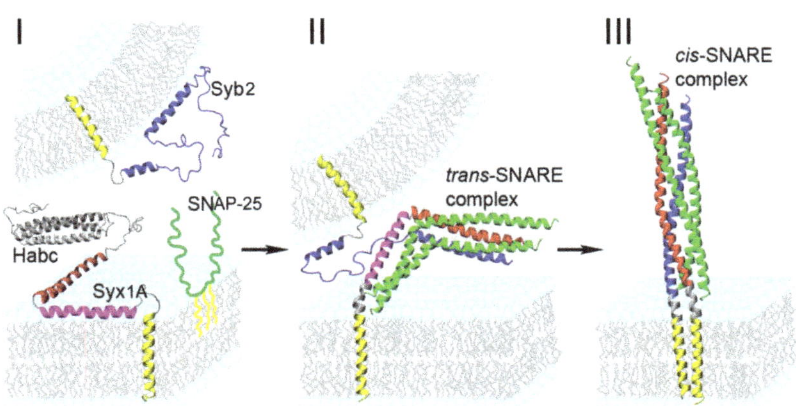

Figure 9.1. SNARE structures postulated to form during the process of membrane fusion. (I) Prior to fusion, SNAP-25 (unknown structure) and syntaxin 1A (syx 1A, Nuclear magnetic resonance (NMR) structure in micelles including the regulatory Habc domain [17], validated by fluorescence interference contrast (FLIC) microscopy [3]) are located in the target membrane, and synaptobrevin 2 (syb 2, NMR structure in micelles [18]) is in the vesicular membrane. (II) If the two membranes come into close proximity, a four-helix bundle is formed (*trans*-SNARE complex, the model of the ΔN49 acceptor complex determined by FLIC microscopy [3, 11, 18]). (III) After the two membranes have merged, the *cis*-SNARE complex (x-ray structure and FLIC microscopy [3, 11]) is formed, where the two transmembrane helices are within one membrane. Copyright (2013) National Academy of Sciences. Modified according to Liang *et al* [3].

SNARE-mediated fusion, which has a number of regulatory/accessory proteins, such as muncs, synaptotagmin [15], and complexin [16], is a highly efficient process. To unravel the molecular details of the fusion process without considering the large number of accessory proteins involved in its regulation, *in vitro* assays have been developed to dissect the individual steps of the process.

9.2 *In vitro* fusion assays

Bulk fusion assays. Historically, Weber *et al* [19] were the first to observe SNARE-mediated membrane fusion *in vitro.* Based on a Förster resonance energy transfer (FRET) pair (nitrobenzoxadiazole (NBD)/rhodamine), they observed lipid mixing of two vesicle populations with reconstituted syb 2 and syx 1A/SNAP-25, which is called the minimal fusion machinery. Syx 1A is a protein lacking the regulatory Habc domain that is not required for fusion [2] and recombinantly expressed SNAP-25 lacks the palmitoyl side chains. Fusion of the two vesicle populations leads to a dilution of the FRET pair concentration and hence a decrease in FRET efficiency. To be able to distinguish between full fusion and hemifusion, in which only the outer leaflets of the bilayers mix, dithionite was applied, which is incapable of penetrating the lipid bilayer but quenches the NBD fluorescence intensity. Thus, a change in FRET efficiency under these conditions can be correlated to inner leaflet mixing [20, 21]. These assays only provide information about lipid mixing, but do not illuminate the process of fusion pore formation, which is a prerequisite for the transfer of the vesicle's content into the other compartment. It turns out that inner lipid leaflet mixing does not necessarily lead to content mixing [22], and bursting/leakage of the vesicles cannot readily be distinguished from fusion [23–25]. Moreover, some information cannot be read out from bulk fusion assays, e.g. fusion intermediates or kinetic aspects such as docking lifetimes. Therefore, fusion assays based on single vesicles have been established to resolve individual fusion events and thus provide access to fusion intermediates, their formation, and their kinetics [25, 26].

Single-vesicle fusion assays based on surface-adhered vesicles. For this type of assay, unilamellar vesicles with sizes of about 100 nm and doped with t-SNAREs are immobilized on a flat surface, and v-SNARE doped vesicles of a similar size are added to the solution. As the reaction occurs at the surface, individual vesicle–vesicle fusion events can be readily monitored by fluorescence microscopy in a time-resolved manner (figure 9.2(a)) [27–32]. Fluorescence intensity time traces are read out within a region of interest (ROI) and report on the fusion state of the vesicles. It is possible to distinguish between hemi-fused intermediates as well as full fusion with fusion pore formation. However, even though such a single-vesicle assay has certain advantages over bulk fusion assays, one major drawback remains the same as in bulk fusion assays, i.e. highly curved vesicles serve as the target membrane. This geometry does not resemble the natural situation, in which a synaptic vesicle that has a size of 40–50 nm fuses with a planar presynaptic membrane. Moreover, several studies have demonstrated that the high vesicle curvature affects the fusion behavior [33, 34].

Single-vesicle fusion assays based on supported lipid bilayers. To address the membrane geometry issue, fusion assays based on planar supported lipid bilayers

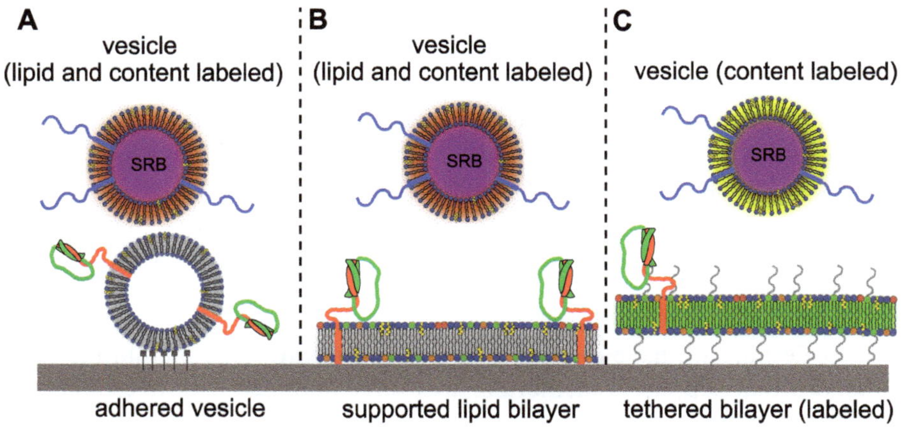

Figure 9.2. Schematics of different single-vesicle fusion assays generally found in research literature. Fusion of syb-2-containing vesicles with: (A) t-SNARE-doped liposomes that are adhered to a planar surface; (B) t-SNARE-doped planar membranes directly in contact with the support; and (C) t-SNARE-doped planar membranes that are separated from the support by a polyethylene glycol (PEG) linker. The vesicle sizes are typically in the range of 50–100 nm. Note that the orientation of the reconstituted proteins in the vesicles is random. Those that point to the inner lumen of the vesicles are not accessible to the fusion process.

(SLBs) have been developed. In the first attempts, vesicles doped with t-SNAREs were directly spread and fused onto glass substrates to produce SLBs, and the fusion of v-SNARE-doped small unilamellar vesicles (SUVs, diameters of about 50 nm) was monitored by means of total internal reflection fluorescence (TIRF) microscopy (figure 9.2(b)). However, the direct contact between the planar membrane and the support greatly hampered protein mobility [35]. Moreover, it turned out that fusion was SNAP-25 independent [35, 36] or Ca^{2+}-dependent, even though the Ca^{2+}-sensor synaptotagmin-1 was not present [37]. In a subsequent study, Wang *et al* [38] analyzed lipid mixing and content release and found that the fusing vesicles preferentially burst, releasing their contents into the bulk solution rather than transferring them across the SLB. The explanation given for their findings was that the lack of space underneath the target membrane was insufficient for the uptake of the vesicular content.

To generate additional space between the membrane and the support, Karatekin *et al* [39, 40] introduced an approximately 4 nm long polyethylene glycol (PEG) linker in the vesicles that were spread and fused on glass substrates (figure 9.2(c)), resulting in tethered bilayers with a random orientation of t-SNAREs. This approach not only partially detached the membrane from the support but also increased the mobility of the SNAREs in the membrane. By labeling the target membrane, the quality of the SLB preparation was controlled and SNARE-mediated fusion of SUVs was monitored by means of lipid mixing. Another approach that increased the mobility of the SNAREs within the SLB was developed by Tamm and coworkers. They first produced a protein-free monolayer on glass substrates deposited by a Langmuir–Blodgett transfer, onto which t-SNARE-doped vesicles were fused [41]. With this setup in hand, they were able to address different factors that influence the fusion process, such as lipid composition, asymmetric lipid distribution, vesicle size, or the type of reconstituted t-SNARE acceptor complex [42–45]. Even though the membrane-support space was still very small, successful content transfer across the SLB was demonstrated by showing that a two-dimensional diffusion kinetics better describes the observed efflux of fluorescent molecules from the vesicle than a three-dimensional one [44].

9.3 Setting up a single-vesicle assay based on pore-spanning membranes

In the last decade, we have developed an alternative artificial membrane to SLBs with the idea that supported membrane areas are combined with freestanding ones to generate a chip-based system which provides mechanically robust membranes that are stable in the long term [46, 47], while the membrane components are laterally mobile [48, 49]. In addition, there is sufficient space to harbor the incoming lipid material from the fusing vesicle [50] and transfer the vesicle's content into the underlying aqueous compartment. For a functional fusion assay based on pore-spanning membranes (PSMs), we envisioned the different setups depicted in figure 9.3. A quasi-planar membrane that harbors the t-SNAREs is produced on the porous substrate, while vesicles doped with the v-SNARE syb 2 are added (figure 9.3(a)). This arrangement resembles the geometric membrane situation at the presynaptic site. Dependent on the readout procedure, either both membranes (the vesicular and target membranes) are labeled with a lipid dye (lipid mixing, figure 9.3(a/b)) or the vesicles are filled with a water-soluble dye (content release, figure 9.3(c)).

Generation of pore-spanning membranes (PSMs). To produce PSMs with the ability to reconstitute proteins as required for a fusion assay to be built up, a method was developed based on spreading giant unilamellar vesicles (GUVs) on functionalized porous substrates. The general procedure is depicted in figure 9.4. It uses Si/Si_3N_4 or Si/SiO_2 substrates with micrometer-sized pores (typically in the range of 1–5 μm) arranged in a hexagonal array and a typical surface porosity of 35%–40% (SEM image I, figure 9.4). Titanium is first evaporated onto the substrate to serve as an adhesive layer, followed by gold (about 40 nm). It is crucial to evaporate the gold layer only on top of the pore rims (SEM image II, figure 9.4) in order to ensure that the top surface is chemically distinct from the inner pore walls and that the membranes *span* the pores. These freshly coated substrates are functionalized with a self-assembled monolayer (SAM) of 6-mercapto-1-hexanol (6-MH), which renders the surface hydrophilic and chemically distinguishes the upper part of the porous substrate from the inner pore walls. To form PSMs, a GUV-containing solution is added on top of the substrates and incubated for about 30 min. If the surface is clean and the SAM has been appropriately formed, GUVs adhere to the hydrophilic surface and an increase in contact area subsequently results in the spontaneous rupture of individual GUVs and the formation of PSMs [48, 51–54]. GUVs with sizes larger than 10 μm are used instead of small or LUVs with typical sizes in the 100 nm regime, as the latter would diffuse into the micrometer-sized cavities and thus no freestanding membranes would form. As the membrane patch is derived

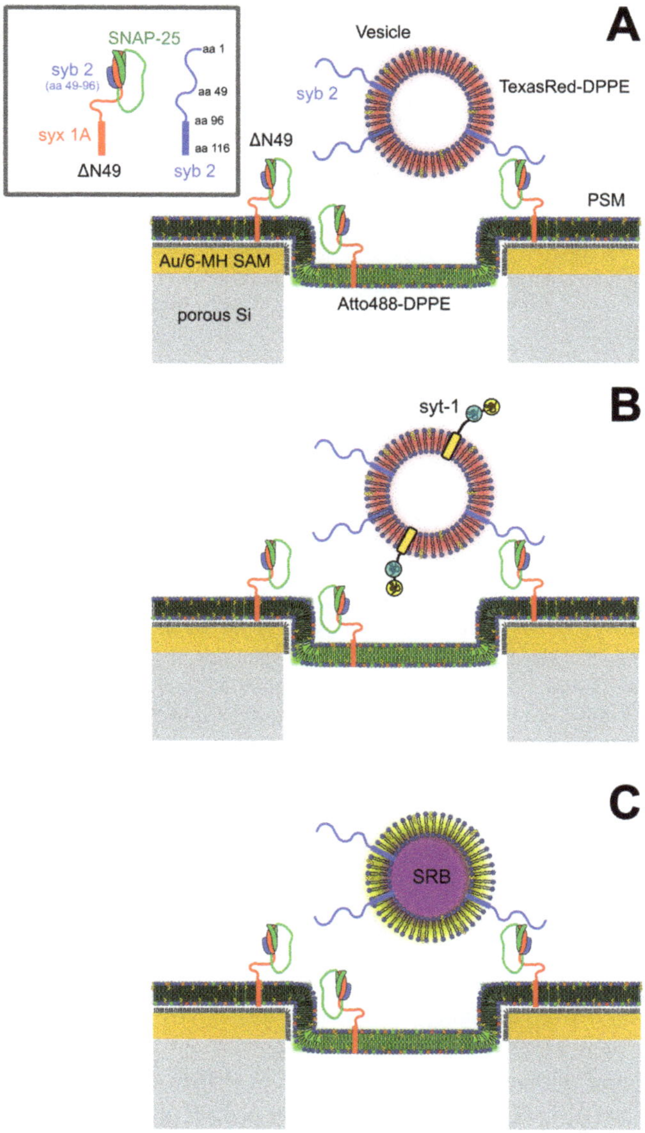

Figure 9.3. Schematic illustration of the SNARE-mediated fusion assay based on pore-spanning membranes (PSMs): (A) lipid-mixing assay with syb 2 in the vesicle and the ΔN49 complex in the PSM; (B) lipid-mixing assay with reconstituted syb 2 and syt-1 in the vesicle and the ΔN49 complex in the PSM; (C) content-release assay with syb 2 in the vesicle and the ΔN49 acceptor complex in the PSM. Large unilamellar vesicles (LUVs) with diameters of about 250 nm are used. For the PSMs, different pore sizes ranging between 1 and 5 μm in diameter are used.

from a single GUV, its size is approximately that of the area of a single GUV, and the patches are typically either round or heart-shaped, depending on the rupture point of the adhered GUV [55]. The membrane is symmetric in its composition and floats on a thin aqueous layer.

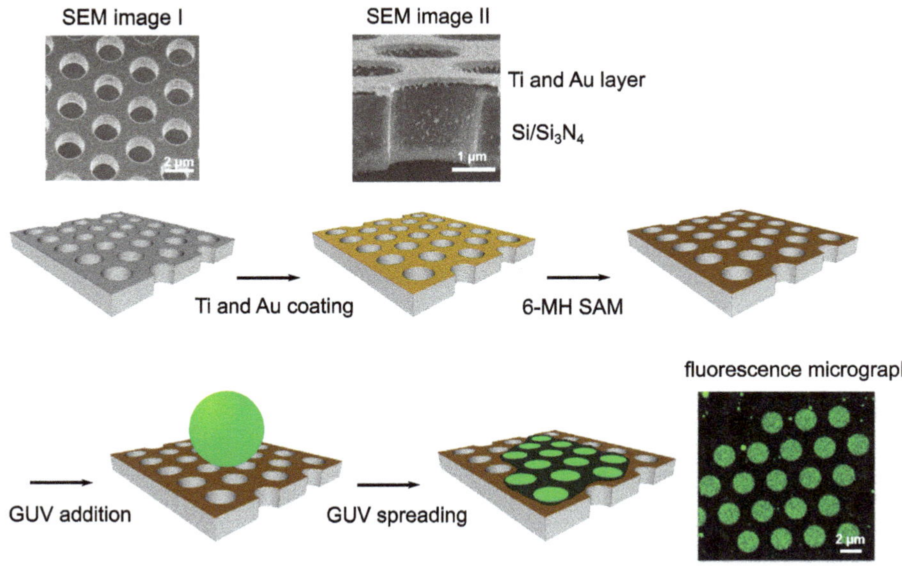

Figure 9.4. Schematic illustration of the preparation of PSMs along with scanning electron microscopy (SEM) images of the substrates and a fluorescence micrograph of the PSMs. The porous Si/Si_3N_4 substrate is orthogonally coated with a thin layer of titanium and a ~40 nm-thick gold layer. The addition of *n*-propanolic 6-mercapto-1-hexanol (6-MH) solution leads to the formation of a self-assembled monolayer (SAM) that enables the spread of individual GUVs on top of the substrate to form PSMs.

A fluorescence micrograph of PSMs containing a lipid label obtained by spreading GUVs on a silicon substrate with 5 µm-sized pores is depicted in figure 9.4. The micrograph shows regions of high fluorescence intensity, which are membranes that span the pores (freestanding PSMs, f-PSMs), while dark areas are visible in between, which are the supported parts (supported PSMs, s-PSMs). Since gold surfaces quench fluorescent emission over short distances of up to 15 nm [56, 57], only the f-PSMs are visible in the fluorescence micrographs even though a continuous lipid bilayer is formed on the surface, which was verified by fluorescence recovery after photobleaching (FRAP) experiments [58].

Reconstitution of the membrane components in PSMs. As individual GUVs spread on the surface, the composition of the GUV is expected to be directly transferred to the PSMs. While the lipid composition of the GUVs can readily be controlled within a certain regime, the reconstitution of proteins is, in general, a great challenge and can be a real bottleneck for the generation of fully functional PSMs. For the fusion experiments depicted in figure 9.3, the t-SNAREs syx 1A (aa 183–288, lacking the regulatory Habc domain) and SNAP-25a ('a' indicates that all cysteine residues are mutated to serine residues, aa 1–206) are reconstituted into the PSMs. In our studies, we made use of the so-called ΔN49 acceptor complex (t-SNARE) [59], which is a preassembled syx 1A/SNAP25a/syb 2 fragment (aa 49-96)-complex (1:1:1). As outlined in figure 9.1, SNARE complex formation depends on the interactions between syb 2 and the t-SNAREs syx 1A and SNAP-25 resulting in a 1:1:1 stoichiometry. However, in *in vitro* systems a 2:1 syx 1A/SNAP-25 complex is spontaneously formed on the target membrane, which kinetically slows down the binding of syb 2 [59]. To prevent the formation of this 2:1 'dead end' complex and thus accelerate the fusion kinetics *in vitro*, the ΔN49 acceptor complex became the t-SNARE complex of choice. Only in recent years did Kreutzberger *et al* [42] develop new procedures to assemble membrane-bound 1:1 SNARE acceptor complexes that produce fast and efficient fusion without the need of the syb 2 fragment (aa 49–96) *in vitro*.

To generate GUVs harboring the t-SNARE complex, small unilamellar vesicles containing the ΔN49 acceptor complex are first produced using a standard procedure [59], which are then used to generate GUVs via electroformation [54]. It is expected that the ΔN49 acceptor complex is randomly oriented in the GUVs. These GUVs are then spread onto the porous substrate. Quantification of the protein reconstitution efficiency revealed that there is a great variation ranging from almost 0%–100% in the GUVs, with a median value of 26% ± 24% [60]. This large variation can be attributed to the electroformation process used to generate GUVs [61, 62]. Even though one can statistically analyze the influence of the number of SNAREs in the PSMs on the extracted fusion parameters [60], there is still a certain uncertainty and it would be desirable to have better control over the reconstitution efficiency.

The success of protein reconstitution into PSMs (i.e. whether the transfer from the GUV to the PSM occurs) can be verified by fluorescence microscopy. We used different fluorescently labeled peptides and proteins and started with a simple synthetic syntaxin derivative, namely the transmembrane domain (TMD) of syx 1A that was fluorescently labeled with Atto647N (Atto647N–syx 1A-TMD). In the next step, we labeled syx 1A with Alexa488 via a cysteine residue (Alexa488–syx 1A). Both fluorescent proteins were visible in the fluorescence micrographs of PSMs, indicating their successful reconstitution [54]. Consequently, as a final test, the fully assembled ΔN49 acceptor complex was fluorescently visualized in PSMs. To achieve this, we used a labeled syb 2 fragment (residues 49–96 S79C labeled with Atto488–maleimide) to assemble the complex. This labeling procedure has the advantage that only the fusion active heterodimer composed of syx 1A and SNAP-25a (1:1) is visualized. If the label were attached to either syx 1A or SNAP-25a, the fusion inactive complex composed of syx 1A and SNAP-25a (2:1) would also be visualized [59].

In figure 9.5(b1), a fluorescence micrograph obtained with the fluorescently labeled ΔN49 acceptor complex is depicted. It shows a homogeneous fluorescence intensity, indicating a homogeneous distribution of the ΔN49 acceptor complex. However, this kind of distribution is not always observed, as shown in figure 9.5(b2). In this case, the ΔN49 complex is localized at the pore edges, which might be caused by the bend structure of the membrane (figure 9.3) [53, 63, 64].

To initiate the highly dynamic fusion process, it is not only required that t-SNAREs are functionally reconstituted into the PSMs, but it is also necessary for the t-SNAREs to be mobile in the plane of the membrane. Hence, we investigated the diffusion behavior of these components within PSMs. The diffusion constants of Atto647N–syx 1A-TMD and Alexa488–syx 1A in the f-PSM were determined by means of fluorescence correlation spectroscopy. For Atto647N–syx 1A-TMD a diffusion constant D_{TMD} (f-PSM) = 3.4 ± 0.2 μm^2 s^{-1} was found, whereas for Alexa488–syx 1A $D_{\mathrm{syx\ 1A}}$ (f-PSM) = 2.3 ± 0.2 μm^2 s^{-1} was determined. These values are in good

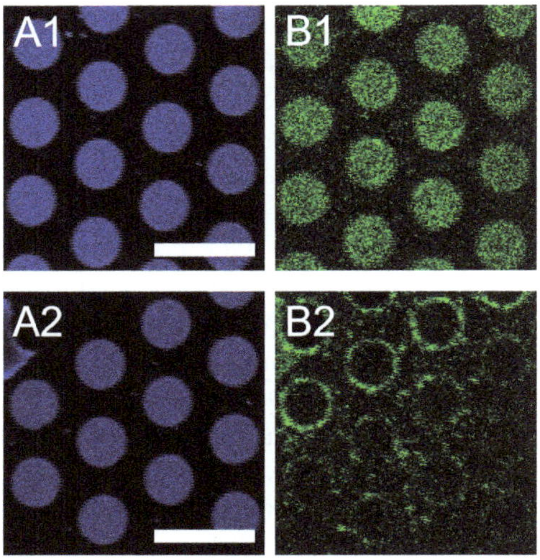

Figure 9.5. Fluorescence micrographs: (A1/A2) Atto390–1,2-dipalmitoyl-sn-glycero-3-phosphoethanolamine (DPPE)-labeled PSMs and (B1/B2) reconstituted ΔN49–Atto488 complex. While a homogeneous distribution of the ΔN49–Atto488 complex is observed in (B1), in (B2) the fluorescence of the ΔN49–Atto488 complex is preferentially localized at the pore edges. Scale bars: 10 μm. Reproduced under the terms of the CC BY 4.0. Copyright 2020, Springer. Adapted from Mühlenbrock *et al* [66] with the permission of Springer.

agreement with the diffusion constants found for syx 1A in GUVs [65], and they clearly demonstrate that both syx 1A derivatives are mobile within the f-PSMs [54].

For the s-PSMs, a direct measurement of the diffusion constants is not feasible by fluorescence correlation spectroscopy or by FRAP experiments, as the fluorescence intensity of the fluorophores on the gold-coated surfaces is quenched. To access the diffusion behavior of the membrane components in the s-PSMs, we instead performed so-called indirect FRAP experiments. In such an experiment, the fluorescence of one individual f-PSM is bleached and the recovery of the fluorescence in the f-PSMs is recorded. This fluorescence recovery is dominated by the diffusion of molecules from the s-PSMs surrounding the one f-PSM. We applied this technique to determine the diffusion coefficient of syx 1A-TMD, obtaining a value of 1.1 $\mu m^2\ s^{-1}$ [58]. For the ΔN49 acceptor complex we obtained a diffusion constant of $D_{\Delta N49}$ (s-PSM) = 1.0 ± 0.5 $\mu m^2\ s^{-1}$. Both values indicate that the mobility of the proteins in the s-PSMs is lower than in the f-PSMs. While the proteins are still mobile, the reduced diffusion constants are likely caused by interactions with the underlying solid support.

9.4 Single-vesicle fusion assay—lipid mixing

Once the PSMs with the reconstituted ΔN49 acceptor complex are established, syb 2 (aa 1–116)-doped vesicles can be added to observe single-vesicle fusion events. Depending on the readout parameter, we use either lipid-labeled vesicles (figure 9.3(a), lipid mixing) or vesicles filled with a water-soluble dye (figure 9.3(c), content release; see section 9.6). For the lipid-mixing assay, the PSMs were labeled with Atto488–DPPE (green channel) and the vesicles with Texas

Red–DPPE (red channel). With a dual-color fluorescence approach using an upright confocal microscope, two fluorescent time traces can be read out simultaneously, giving access to the docking and fusing behavior of the vesicle (figure 9.6). Notably, syb 2-containing vesicles only dock to the PSMs if the functional ΔN49 acceptor complex is present. If the ΔN49 acceptor complex is blocked with a soluble syb 2 fragment (aa 1–96) no docking is observed [60, 67].

In general, we distinguished between vesicles docked to the f-PSMs and those that docked to the s-PSMs. We found that vesicles docked to the f-PSMs were mobile, while vesicles docked to the s-PSMs were instantaneously immobilized. Vesicles that docked to the edges of the f-PSMs and are thus partially in contact to the pore rim are also immobile (figure 9.6(a)). We used this mobility criterion to distinguish between f-PSM docked vesicles and s-PSM docked vesicles. This criterion led to the result that 99% of the vesicles are s-PSM docked ones. They either dock directly to the pore-rim area or first dock to the f-PSM, but are subsequently immobilized at the edges of the pores, so that they also belong to immobilized s-PSM docked vesicles. Only very rarely do vesicles dock and fuse with the f-PSM. Ramakrishnan *et al* [64] also observed vesicles immobilized at the pore edges and they discussed their finding

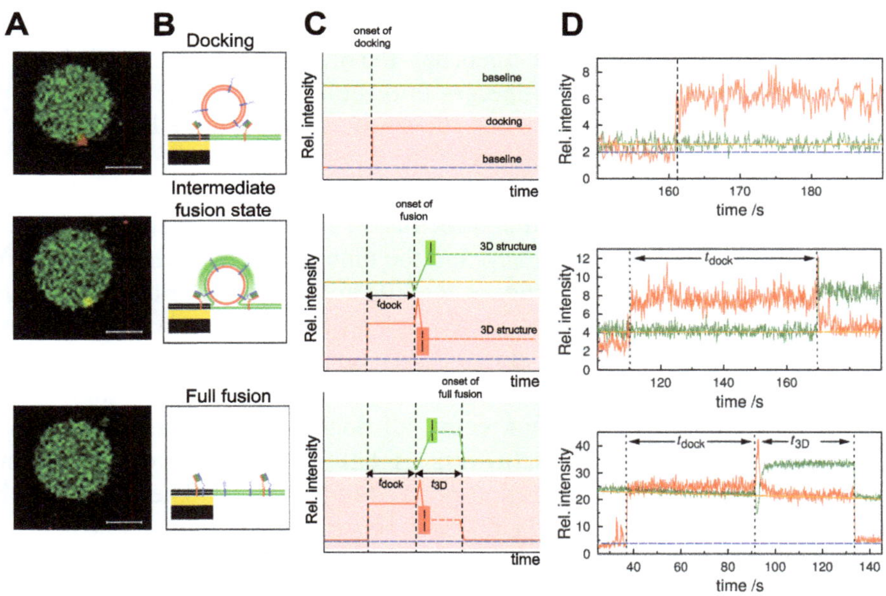

Figure 9.6. Classification of fusion events according to the two-color fluorescence time traces: (A) fluorescence micrographs of a PSM (green) and a vesicle (red) that fuses with the PSM. Scale bars: 2 μm. (B) Schematics of the fusion states. (C) Idealized fluorescence time traces of the red and green fluorescence intensities with the corresponding baselines in orange and blue (dashed lines). The dashed black lines indicate the time points of docking, the onset of fusion, and full fusion with the corresponding time constants t_{dock} and t_{3D}. t_{dock} is defined as the time period between the onset of docking of the vesicle and the onset of fusion; t_{3D} is the time period during which the vesicle remains in an intermediate state until it proceeds to a fully fused state. The dark shaded boxes indicate that the fluorescent intensities of the intermediate fusion states vary as a result of the different vesicle sizes located in a defined region of interest and the position of vesicle fusion on the PSMs (s-PSM or f-PSM). (D) Experimental time traces of the red and green fluorescent intensities.

in terms of immobile t-SNARE aggregates in this region. This might be a plausible explanation, given the fact that we observed the ΔN49 acceptor complex to be preferentially localized in the membrane region at the pore edges (see figure 9.5). Second, the overall mobility of the membrane components is lower in s-PSMs than in f-PSMs, adding to the observed immobility of the docked vesicles on the s-PSMs. The diffusion coefficients of the lipids and peptides are lower by a factor of two to four in the s-PSMs compared to those in f-PSMs [54, 58, 60]. Third, once docked, conformal contact between the vesicle and the membrane on the gold-covered support can further immobilize the vesicles, as there is a large van der Waals attraction between the gold surface and the membrane [68].

Figure 9.6 depicts the expected and measured intensity time traces. The intensities observed in the red channel are rather straightforward. If the vesicle (red) docks, an increase in fluorescence intensity is observed that decreases to an in-between level if an intermediate three-dimensional fusion state remains; it returns to the baseline level if the vesicle has fully merged into the PSM. The intensities observed in the green channel are a bit more complicated. If the vesicle docks to the PSM (green), no change in the green fluorescence intensity is observed. However, if an intermediate state is formed in which the outer leaflets of the two membranes merge, an increase in the green fluorescence is observed, as the Atto488–DPPE lipids move further away from the gold surface that quenches the fluorescence. Only if the three-dimensional vesicle structure fully merges into the PSM does the Atto488–DPPE dye return closer to the gold surface and the lipids diffuse into the PSM, so that the green fluorescence intensity also returns to the baseline level.

To obtain statistically meaningful data about the fusion process, all vesicles docked onto the PSMs are detected and the fusion efficiency as well as the kinetics, i.e. the docking time t_{dock} (figure 9.6(c)) and the time required for the vesicle to fully fuse into the planar PSM (t_{3D}) (figure 9.6(c)), are extracted. For the statistical analysis, we only take those vesicles into consideration that have been immobilized, i.e. those classified as vesicles docked to s-PSMs (either directly on the support or at the edges of the pores, where vesicles become immobilized even if they are first mobile on the f-PSM part). We first count all docked vesicles and then take the fractions of (i) only docked vesicles, (ii) vesicles that progress to fusion but remain in an intermediate state, and (iii) those that proceed to full fusion. The classification of the fusion events is deduced from the fluorescence time traces (figure 9.6(b)–(d)). As the lipid composition of the PSMs as well as that of vesicles can be nicely controlled, information about the influence of different lipid components on the fusion efficiency can be readily gathered. Fusion efficiency is defined as the fraction of all detected docked vesicles that progress to fusion, remaining in either an intermediate fusion state or in a fully fused state. As an example, we investigated the influence of the lipid PtdInsP(4,5)P_2 on fusion efficiency using this assay. Proteo-LUVs composed of DOPC/POPE/POPS/cholesterol/Texas Red–DPPE (50:19:10:20:1) doped with syb 2 (p/l = 1:500) were added to PSMs composed of 1,2-dioleoyl-sn-glycero-3-phosphocholine (DOPC)/1-palmitoyl-2-oleoyl-sn-glycero-3-phosphocholine (POPE)/1-palmitoyl-2-oleoyl-sn-glycero-3-phospho-L-serine (POPS)/cholesterol/PtdInsP(4,5)P_2/Atto488–DPPE (50-χ(PtdInsP(4,5)P_2)/19/10/20/χ(PtdInsP(4,5)P_2)/1) and doped with the ΔN49 acceptor complex (p/l = 1:500). We found that the amount of PtdInsP(4,5)P_2 greatly influenced the fusion efficiency. It reached a maximum at 2 mol% PtdInsP(4,5)P_2, where the fusion efficiency was 92 ± 4%, as opposed to a fusion efficiency of 47% in the absence of PtdInsP(4,5)P_2 (figure 9.7(a)).

Numerous recent studies have demonstrated that PtdInsP(4,5)P_2 plays an important role in the formation of nano- and mesoscale domains of the t-SNARE syx 1A [54, 69]. It has been shown that nanodomains of syx 1A with diameters of ~70 nm are present in inverted sheets of PC12 cell-derived plasma membranes. PtdInsP(4,5)P_2 was found to be the predominant inner leaflet lipid in these syx 1A domains, accounting for ~82% of the total surface area [70, 71]. Thus, we interpret our findings as showing that PtdInsP(4,5)P_2-induced clustering of the t-SNARE complex is a prerequisite for docking and efficient fusion.

In addition to fusion efficiencies, kinetics data can be extracted. First, the lifetime of the docking state t_{dock} can be determined as exemplarily shown in figure 9.7(b). To interpret the data, a general model proposed by Floyd *et al* [72] was employed. In this model, the rate-limiting step from docking to the onset of fusion is not defined as a one-step transition but as a series of N hidden transitions with a single rate constant k_1 for each transition (equation (9.1)):

$$\text{pdf}(t_{\text{dock}}) = \frac{k_1^{N} \cdot t_{\text{dock}}^{N-1}}{\Gamma(N)} \cdot \exp(-k_1 \cdot t_{\text{dock}}), \tag{9.1}$$

with $\Gamma(N)$ being the gamma function. Assuming a single rate constant k_1 for all intermediate steps is reasonable, as only the slowest step in the entire reaction process significantly contributes to pdf(t_{dock}) [72]. For $N = 4$ (which showed the lowest reduced χ^2-values), $\bar{t}_{\text{dock}} = 48 \pm 2$ s was determined. Regardless of the PtdInsP(4,5)P_2 concentration, all docking lifetimes were in the 10 s regime, in agreement with previous results of vesicle fusion with PSMs [54, 58]. These docking times are certainly much slower than those found in nature. However, one has to keep in mind that the experiments were performed with the ΔN49 acceptor complex, in which the displacement of the stabilizing syb 2 fragment by the entire SNARE-motif of syb 2 needs to occur to produce the fusion active *trans*-SNARE complex. Pobatti *et al* [59] showed that this displacement takes place within tens of seconds and might thus be the rate-limiting step in our fusion system. We are aware of the fact that other groups found much smaller docking times, even though they also used the ΔN49 acceptor complex. However, they used a different membrane system, which might partly explain this discrepancy [41].

If intermediate fusion states are identified, their lifetimes $t_{3\text{D}}$ can be extracted from the fluorescence intensity traces (figure 9.6(c)) and a mono-exponential function (equation (9.2)) can be fitted to the data:

$$N(t_{3\text{D}}) = N_0 \cdot \exp(-k_2 \cdot t_{3\text{D}}). \tag{9.2}$$

N_0 is the total number of intermediate fusion states that proceed to full fusion and k_2 is the rate constant of the decay of the intermediate fusion state. The average lifetime of the intermediate fusion states were in the range of 4–7 s and had no significant dependency on the PtdInsP(4,5)P_2 concentration.

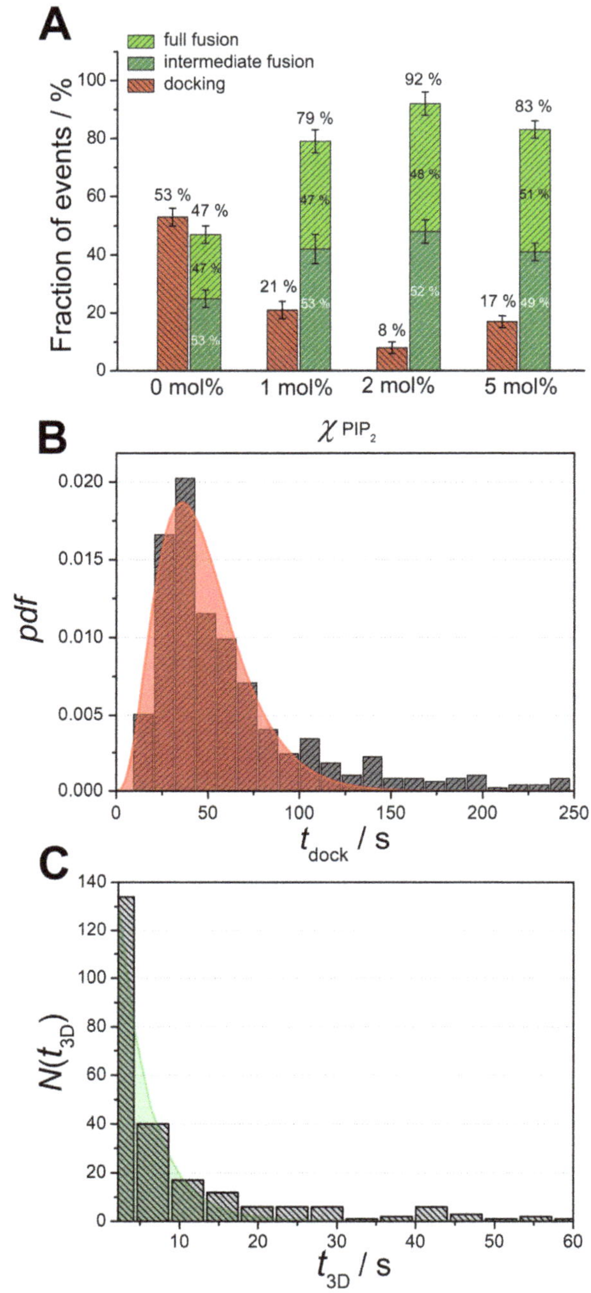

Figure 9.7. Statistical analysis of: (A) the fraction of events (the sum of intermediate and full fusion refers to the fusion efficiency) as a function of PtdInsP(4,5)P_2 content in the PSMs. (B) Lifetime of the docking-state distribution t_{dock} and the result of fitting equation (9.1) to the data (red). (C) Distribution of the lifetime of the intermediate fusion states t_{3D} together with the fit (green) providing a $\bar{t}_{3D}$ = 4.0 ± 0.2 s. The data sets were obtained for a PSM containing 5 mol% PtdIns(4,5)P_2. Adapted with permission from [73]. Copyright (2021) American Chemical Society.

9.5 Impact of synaptotagmin-1 on fusion efficiency and kinetics

In addition to the minimal fusion machinery, there are additional proteins involved that control the fusion process. *In vivo*, neuronal fusion is Ca^{2+} dependent, for which a sensitive Ca^{2+}-sensor is required. Synaptotagmin-1 (syt-1) has been identified as the main candidate for this function [74–76]. Syt-1 is a transmembrane protein harboring two C2 domains with Ca^{2+} binding sites. Most studies in the context of syt-1 have been performed with only the soluble C2A and C2B domains [77–80] and it has been shown that the Ca^{2+}-triggered fusion process in the presence of syt-1 depends strongly on negatively charged phosphatidylserine (PS) lipids and the amount of PtdInsP(4,5)P_2 in the plasma membrane [81]. Despite a large number of studies, we still lack a unified picture of the interactions and function of full-length syt-1 in neuronal fusion. In addition to the binding of PtdInsP(4,5)P_2 [70, 82] and the cross-linking of membranes [83, 84], membrane bending [77] has been discussed, as well as the interaction of syt-1 with complexin [85–87] and oligomerization leading to a ring-like assembly [64, 88–90].

To shed some light on the influence of syt-1 on the fusion process in our fusion assay (figure 9.3(b)), we decided to insert full-length syt-1 harboring the trans-membrane domain. Syt-1 was co-reconstituted with syb 2 in the vesicles, which required an adapted protocol to ensure that the full-length syt-1 was functional [73]. We used 2 mol% of PtdIns(4,5)P_2 in the PSMs, as this concentration showed, the largest fusion efficiency without syt-1 (figure 9.7). However, if syt-1 was co-reconstituted in the vesicles, a significantly reduced fusion efficiency of only 46% ± 5% was found in the absence of Ca^{2+} (figure 9.8). Among the 46% of the vesicles that fused with the PSM, 74% proceeded to full fusion, whereas 26% remained in an intermediate fusion state. One could interpret this finding according to the results of Chapman and coworkers [75], who described a syt-1 interaction with the SNAREs, which thus directly interfered with the assembly of the SNARE complex. However, according to Park *et al* [91], direct syt-1 regulation of SNARE zippering is potentially irrelevant under physiological salt conditions, and several *in vitro* and *in vivo* studies have rather suggested that syt-1 binds PtdIns(4,5)P_2 prior to SNARE complex formation to dock vesicles [70, 71, 92, 93]. This interpretation agrees better with our findings, since in the presence of negatively charged PS and PtdIns(4,5)P_2 but in the absence of SNAREs, we observed a tethering of the vesicles to the planar membranes, which means that these vesicles are counted as quasi-docked. However, as SNAREs are not present, these vesicles cannot proceed to fusion [73]. We observe a similar effect if Ca^{2+} is added: the fusion efficiency (38%) is even more reduced than in the Ca^{2+}-free condition (46%). In addition to this observation, one characteristic difference can be observed: among the 38% of fusing LUVs, almost all vesicles (99%) proceeded to full fusion. This was never observed in any of the other conditions and is probably a result of the strongly enhanced syt-1–membrane interaction in the presence of Ca^{2+}, PtdIns(4,5)P_2, and PS [73]. Others also found that there is an increased strong affinity between the C2AB domains and acidic phospholipids in their Ca^{2+}-bound state [70].

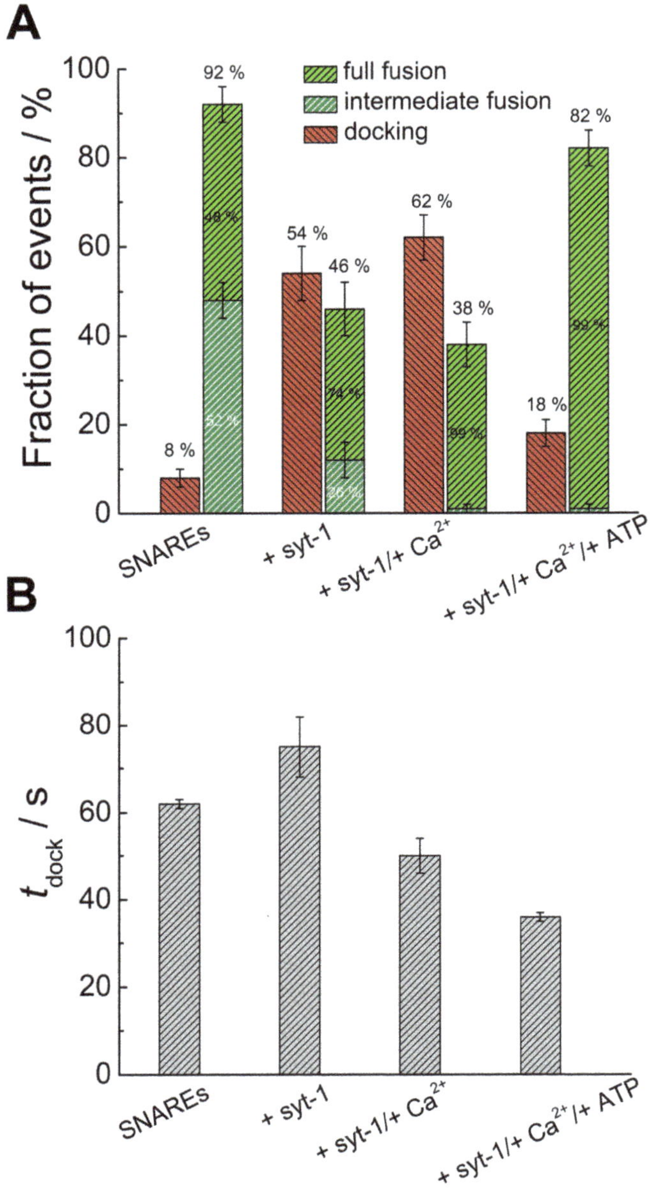

Figure 9.8. (A) Statistical analysis of the fraction of events involving proteo-LUVs (DOPC/POPE/POPS/cholesterol/Texas Red–DPPE, 50:19:10:20:1) doped with only syb 2 (p/l = 1:500) (SNAREs) or syb 2 and syt-1 (p/l = 1:1000) (+ syt-1) docked to PSMs (DOPC/POPE/POPS/Chol/PtdIns(4,5)P_2/Atto488–DPPE, 48:19:10:20:2:1) doped with the ΔN49 acceptor complex (p/l = 1:500). 100 μM of $CaCl_2$ (+ syt-1/+ Ca^{2+}) or 100 μM $CaCl_2$ and 5 mM of adenosine triphosphate (ATP) (+ syt-1/+ Ca^{2+}/+ ATP) were added. (B) Docking times as a function of the different conditions. Adapted with permission from [73]. Copyright (2021) American Chemical Society.

Another aspect that was investigated by our PSM-based fusion assay was the influence of the *cis*- (binding of syt-1 to the vesicle membrane) and *trans*-binding (binding of syt-1 to the target membrane) of syt-1 on fusion efficiency. Park *et al* [92, 94] found that *cis*-binding inactivates the fusion process, while *trans*-binding activates it. We added the negatively charged lipid PS to both membranes (PSM and vesicle), which resembles the natural situation and in principle allows the *cis*- and *trans*-binding of syt-1. Park *et al* [94] further showed that ATP and other polyphosphates at physiological concentrations reduce binding to acidic phospholipids and thus diminish or even cancel the inactivating *cis*-binding without affecting the *trans*-interaction. To show whether this was the case in our *in vitro* assay, we added ATP together with Ca^{2+} to the system, which indeed resulted in a significantly increased fusion efficiency of 82 ± 4%, a value nearly reaching the fusion efficiency observed in the absence of syt-1 (92%). The entire fusion system was pushed towards full fusion (99%), which was similar to what we observed for the +syt-1/+Ca^{2+} system.

For the kinetics analysis, we only evaluated the docking times, as the intermediate states were quasi absent in presence of syt-1. We evaluated the docking times for the different conditions (figure 9.8(b)) using the model of Floyd *et al* [72], again with $N = 4$. Even though the docking times were still in the 10 s regime, t_{dock} was slightly larger in the presence of syt-1 and the absence of Ca^{2+} than t_{dock} obtained in the absence of syt-1. In the presence of Ca^{2+}, the time period from first docking to a fused state became smaller, probably as a result of the deeper insertion of the C2AB domains of syt-1 into the target membrane and a resulting conformational change of the SNAREs [95, 96]. This effect became even more pronounced in the presence of both Ca^{2+} and ATP, as the *cis*-interaction between syt-1 and the membrane, which is strongly dependent on the ionic strength, can be completely abolished in the presence of polyvalent ions [94]. Moreover, fusion kinetics is further enhanced by the ATP-induced shift from the *cis*-interaction to the *trans*-interaction between the target and the vesicle membrane [92, 94].

9.6 Single-vesicle fusion assay—content release

Even though the lipid-mixing assay reports on docking, intermediate fusion states, and full fusion, information about fusion pore formation is missing. This information can be gathered by a content-release assay. In this approach, syb 2-doped LUVs are filled with a water-soluble dye and fused to lipid-labeled PSMs harboring the ΔN49 acceptor complex (figure 9.3(c)). The LUVs are filled with a water-soluble sulforhodamine dye (sulforhodamine B (SRB)) at self-quenching concentrations. Upon release of the dye, a fluorescence enhancement is expected. By simultaneously reading out the fluorescences of the vesicle and the target membrane, additional insight into the interplay between fusion pore formation and lipid mixing can be gathered.

In contrast to assays based on solid-supported membranes, the porous substrates provide sufficient space not only for the incoming lipid material but also for the content to be released underneath the planar membrane. Under certain conditions, this process can be visualized (figure 9.9(a1)). The fluorescent time traces

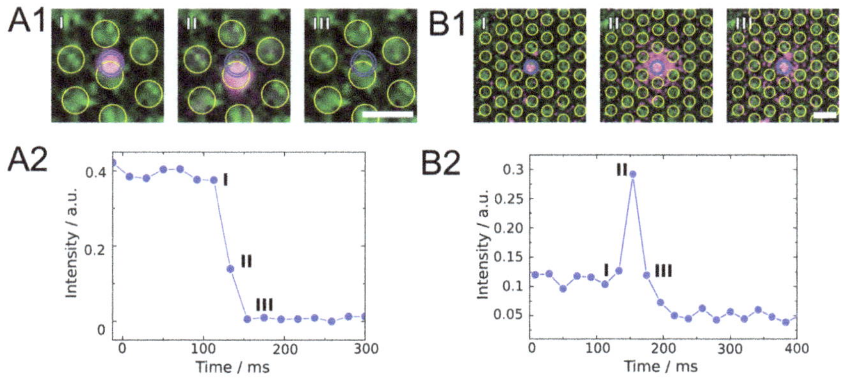

Figure 9.9. Fluorescence micrographs of: (A1) a fusing vesicle (magenta) that transfers its content across the PSM (green, underlying pores encircled in yellow) into the second aqueous compartment and (B1) a bursting vesicle. Scale bars: 2 μm. The fluorescence time traces in (A2/B2) show the corresponding time-dependent fluorescence intensity changes in the region of interest depicted in (A1/B1). If the vesicle releases its content into the underlying pores, the fluorescence intensity decreases upon fusion at the position where the vesicle is docked (blue circle), while a fluorescence peak is observed if it bursts and the fluorescent dye is released atop the membrane. Reprinted from [60], Copyright (2020), with permission from the Biophysical Society.

(figure 9.9(A2)) obtained at the site of vesicle docking demonstrate that the content is released underneath the membrane, which is indicative of fusion pore formation. This process can be distinguished from that of vesicle rupture (figure 9.9(B1)). In this case, the content is released into the bulk solution directly above the membrane/docking site, which is discernable as a spike in the SRB fluorescence intensity time trace (figure 9.9(B2)) at the position of vesicle docking. For our system, we found that only 0.003% of all events were assigned to burst vesicles.

Similar to the lipid-mixing assay with a dual-color fluorescence readout, the content release (SRB fluorescence) and the diffusion of lipids (Atto655–DPPE) from the PSM into the vesicular structure can be extracted for each individual fusion event. Using these two fluorescence time traces, different fusion states and fusion intermediates can be identified and quantified [60].

We found that most of the docked vesicles released their content completely (75% of fusing vesicles). We further found that hemifusion without fusion pore formation was negligible. If only lipid mixing was used to observe fusion, the observed intermediate state was assigned to a hemifusion state when the vesicle's fluorescence intensity dropped and the PSM fluorescence increased, reaching a constant value above the baseline. However, with the additional content-release information, we can claim that a hemifusion diaphragm is not formed without a fusion pore, as content release is observed in all cases. Apparently, the intermediate state is defined by the quick opening and closing of the fusion pore, rather than a stable hemifusion state.

The system further allowed us to investigate the kinetics of fusion pore opening and lipid diffusion from the s-PSM into the fusing vesicle. We determined the time span $\Delta\tau_{3D}$ between Atto655–DPPE influx into the fusing vesicle and the onset of the collapse of the three-dimensional vesicular structure. The cumulative distribution function of $\Delta\tau_{3D}$ suggested two vesicle populations, one that starts to collapse fast ($k_1 = 0.43 \pm 0.01$ s^{-1}) and one that remains longer in the three-dimensional structure ($k_2 = 0.043 \pm 0.004$ s^{-1}). This observation agrees with the results we obtained using the lipid-mixing assay, where we also proposed two vesicle populations [58].

We were also able to determine the lag time Δt between Atto655–DPPE diffusion into the three-dimensional structure and fusion pore formation within the given time resolution of 20 ms. A value of $\Delta t = 42 \pm 11$ ms means that the influx of Atto655–DPPE into the three-dimensional vesicular structure occurs slightly prior to fusion pore formation, but is basically a quasi-simultaneous process, in agreement with previous reports [97, 98]. Additionally, we quantified the half-life time of the process of content release, which occurred within 48 ± 7 ms. This very short time period between fusion pore opening and complete content release is likely a result of the free space underneath the f-PSM that can take up the incoming vesicular content. In the case of single-vesicle fusion assays based on supported lipid bilayers, the observed content release required up to several seconds [39, 43, 44].

9.7 Conclusions

While several single-vesicle fusion assays have been developed over recent decades, PSMs have only been added to this repertoire in the last couple of years. In contrast to vesicle–vesicle fusion assays, PSMs provide a quasi-planar membrane geometry that nicely mimics the natural situation. This holds also true for lipid bilayers attached to a support. However, PSMs appear to be advantageous over supported lipid bilayers and tethered bilayers, as they harbor freestanding bilayer parts and a large second aqueous compartment that allows take-up of the incoming lipid material and the content of the fusing vesicle. Our results have clearly shown that it is feasible to observe individual vesicle fusion events and that a lipid-mixing assay using a dual-color approach enables one to disentangle the different stages of the vesicle during the fusion process. Measuring the release of a water-soluble dye in the underlying compartments provides additional important information about fusion pore formation, which is not accessible from lipid-mixing assays. Fusion pore opening is the important step during the fusion process occurring at the synapse, which transfers the neurotransmitter into the synaptic cleft.

Starting with the simple minimal fusion machinery, the complexity of the membrane system can be stepwise increased, as has been shown by the addition of different lipid components such as PtdInsP(4,5)P_2 and full-length synaptotagmin-1; further components can be investigated in the future. Moreover, the fusion assay based on PSMs is certainly not limited to neuronal SNARE-mediated fusion, and future experiments will demonstrate how the system can be adapted to other fusion-relevant processes.

Acknowledgements

We thank all students that have contributed to this project over the last few years. We thank the Deutsche Forschungsgemeinschaft (SFB 803, project B04) for financial support.

References

[1] Jahn R and Südhof T C 1999 Membrane fusion and exocytosis *Annu. Rev. Biochem.* **68** 863–911

[2] Zhou P, Pang Z P, Yang X, Zhang Y, Rosenmund C, Bacaj T and Südhof T C 2013 Syntaxin-1 N-peptide and Habc-domain perform distinct essential functions in synaptic vesicle fusion *EMBO J.* **32** 159–71

[3] Liang B Y, Kiessling V and Tamm L K 2013 Prefusion structure of syntaxin-1A suggests pathway for folding into neuronal trans-SNARE complex fusion intermediate *Proc. Natl. Acad. Sci. USA* **110** 19384–9

[4] Sutton B R, Fasshauer D, Jahn R and Brunger A T 1998 Crystal structure of a SNARE complex involved in synaptic exocytosis at 2.4 Å resolution *Nature* **395** 347–53

[5] Fasshauer D, Bruns D, Shen B, Jahn R and Brunger A T 1997 A structural change occurs upon binding of syntaxin to SNAP-25 *J. Biol. Chem.* **272** 4582–90

[6] Fasshauer D, Antonin W, Subramaniam V and Jahn R 2002 SNARE assembly and disassembly exhibit a pronounced hysteresis *Nat. Struct. Biol.* **9** 144–51

[7] Fasshauer D, Otto H, Eliason W K, Jahn R and Brunger A T 1997 Structural changes are associated with soluble *N*-ethylmaleimide-sensitive fusion protein attachment protein receptor complex formation *J. Biol. Chem.* **272** 28036–41

[8] Fasshauer D, Sutton R B, Brunger A T and Jahn R 1998 Conserved structural features of the synaptic fusion complex: SNARE proteins reclassified as Q- and R-SNAREs *Proc. Natl. Acad. Sci. USA* **95** 15781–6

[9] Kloepper T H, Kienle C N and Fasshauer D 2007 An elaborate classification of SNARE proteins sheds light on the conservation of the eukaryotic endomembrane system *Mol. Biol. Cell* **18** 3463–71

[10] Weimbs T, Low S H, Chapin S J, Mostov K E, Bucher P and Hofmann K 1997 A conserved domain is present in different families of vesicular fusion proteins: a new superfamily *Proc. Natl. Acad. Sci. USA* **94** 3046–51

[11] Stein A, Weber G, Wahl M C and Jahn R 2009 Helical extension of the neuronal SNARE complex into the membrane *Nature* **460** 525–8

[12] van den Bogaart G, Lang T and Jahn R 2013 Microdomains of SNARE proteins in the plasma membrane *Curr. Top. Membr.* **72** 193–230

[13] Li F, Pincet F, Perez E, Eng W S, Melia T J, Rothman J E and Tareste D 2007 Energetics and dynamics of SNAREpin folding across lipid bilayers *Nat. Struct. Mol. Biol.* **14** 890–6

[14] Jahn R and Fasshauer D 2012 Molecular machines governing exocytosis of synaptic vesicles *Nature* **490** 201–7

[15] Munson M 2015 Synaptic-vesicle fusion: a need for speed *Nat. Struct. Mol. Biol.* **22** 509–11

[16] Lai Y, Diao J, Cipriano D J, Zhang Y, Pfuetzner R A, Padolina M S and Brunger A T 2014 Complexin inhibits spontaneous release and synchronizes Ca^{2+}-triggered synaptic vesicle fusion by distinct mechanisms *eLife* **3** e03756

[17] Fernandez I, Ubach J, Dulubova I, Zhang X Y, Südhof T C and Rizo J 1998 Three-dimensional structure of an evolutionarily conserved N-terminal domain of syntaxin 1A *Cell* **94** 841–9

[18] Ellena J F, Liang B Y, Wiktor M, Stein A, Cafiso D S, Jahn R and Tamm L K 2009 Dynamic structure of lipid-bound synaptobrevin suggests a nucleation-propagation mechanism for trans-SNARE complex formation *Proc. Natl Acad. Sci. USA* **106** 20306–11

[19] Weber T, Zemelman B V, McNew J A, Westermann B, Gmachl M, Parlati F, Söllner T H and Rothman J E 1998 SNAREpins: minimal machinery for membrane fusion *Cell* **92** 759–72

[20] Xu Y B, Zhang F, Su Z L, McNew J A and Shin Y K 2005 Hemifusion in SNARE-mediated membrane fusion *Nat. Struct. Mol. Biol.* **12** 417–22

[21] Hernandez J M, Stein A, Behrmann E, Riedel D, Cypionka A, Farsi Z, Walla P J, Raunser S and Jahn R 2012 Membrane fusion intermediates via directional and full assembly of the SNARE complex *Science* **336** 1581–4

[22] Chan Y H M, van Lengerich B and Boxer S G 2009 Effects of linker sequences on vesicle fusion mediated by lipid-anchored DNA oligonucleotides *Proc. Natl Acad. Sci. USA* **106** 979–84

[23] Yu H, Rathore S S, Lopez J A, Davis E M, James D E, Martin J L and Shen J 2013 Comparative studies of Munc18c and Munc18-1 reveal conserved and divergent mechanisms of Sec1/Munc18 proteins *Proc. Natl Acad. Sci. USA* **110** E3271–E3280

[24] Ma C, Su L J, Seven A B, Xu Y B and Rizo J 2013 Reconstitution of the vital functions of Munc18 and Munc13 in neurotransmitter release *Science* **339** 421–5

[25] Brunger A T, Cipriano D J and Diao J 2015 Towards reconstitution of membrane fusion mediated by SNAREs and other synaptic proteins *Crit. Rev. Biochem. Mol. Biol.* **50** 231–41

[26] Lira R B and Dimova R 2019 *Multiresponsive Behavior of Biomembranes and Giant Vesicles* ed R Lipowsky (New York: Academic) pp 229–70

[27] Kyoung M, Srivastava A, Zhang Y X, Diao J J, Vrljic M, Grob P, Nogales E, Chu S and Brunger A T 2011 *In vitro* system capable of differentiating fast Ca^{2+}-triggered content mixing from lipid exchange for mechanistic studies of neurotransmitter release *Proc. Natl Acad. Sci. USA* **108** E304–E13

[28] Kyoung M J, Zhang Y X, Diao J J, Chu S and Brunger A T 2013 Studying calcium-triggered vesicle fusion in a single vesicle-vesicle content and lipid-mixing system *Nat. Protocols* **8** 1–16

[29] Yoon T-Y, Okumus B, Zhang F, Shin Y-K and Ha T 2006 Multiple intermediates in SNARE-induced membrane fusion *Proc. Natl Acad. Sci. USA* **103** 19731–6

[30] Diao J *et al* 2012 Synaptic proteins promote calcium-triggered fast transition from point contact to full fusion *eLife* **1** e00109

[31] Gong B, Choi B-K, Kim J-Y, Shetty D, Ko Y H, Selvapalam N, Lee N K and Kim K 2015 High affinity host-guest FRET pair for single-vesicle content-mixing assay: observation of flickering fusion events *J. Am. Chem. Soc.* **137** 8908–11

[32] Lai Y, Diao J, Liu Y, Ishitsuka Y, Su Z, Schulten K, Ha T and Shin Y-K 2013 Fusion pore formation and expansion induced by Ca^{2+} and synaptotagmin 1 *Proc. Natl Acad. Sci. USA* **110** 1333–8

[33] Hernandez J M, Kreutzberger A J B, Kiessling V, Tamm L K and Jahn R 2014 Variable cooperativity in SNARE-mediated membrane fusion *Proc. Natl Acad. Sci. USA* **111** 12037–42

[34] Malinin V S and Lentz B R 2004 Energetics of vesicle fusion intermediates: comparison of calculations with observed effects of osmotic and curvature stresses *Biophys. J.* **86** 2951–64

[35] Bowen M E, Weninger K, Brunger A T and Chu S 2004 Single molecule observation of liposome-bilayer fusion thermally induced by soluble *N*-ethyl maleimide sensitive-factor attachment protein receptors (SNAREs) *Biophys. J.* **87** 3569–84

[36] Liu T, Tucker W C, Bhalla A, Chapman E R and Weisshaar J C 2005 SNARE-driven, 25-millisecond vesicle fusion *in vitro Biophys. J.* **89** 2458–72

[37] Fix M, Melia T J, Jaiswal J K, Rappoport J Z, You D, Söllner T H, Rothman J E and Simon S M 2004 Imaging single membrane fusion events mediated by SNARE proteins *Proc. Natl Acad. Sci. USA* **101** 7311–6

[38] Wang T, Smith E A, Chapman E R and Weisshaar J C 2009 Lipid mixing and content release in single-vesicle, SNARE-driven fusion assay with 1–5 ms resolution *Biophys. J.* **96** 4122–31

[39] Karatekin E, Di Giovanni J, Iborra C, Coleman J, O'Shaughnessy B, Seagar M and Rothman J E 2010 A fast, single-vesicle fusion assay mimics physiological SNARE requirements *Proc. Natl Acad. Sci. USA* **107** 3517–21

[40] Karatekin E and Rothman J E 2012 Fusion of single proteoliposomes with planar, cushioned bilayers in microfluidic flow cells *Nat. Protocols* **7** 903–20

[41] Domanska M K, Kiessling V, Stein A, Fasshauer D and Tamm L K 2009 Single vesicle millisecond fusion kinetics reveals number of SNARE complexes optimal for fast SNARE-mediated membrane fusion *J. Biol. Chem.* **284** 32158–66

[42] Kreutzberger A J B, Liang B, Kiessling V and Tamm L K 2016 Assembly and comparison of plasma membrane SNARE acceptor complexes *Biophys. J.* **110** 2147–50

[43] Kreutzberger A J B, Kiessling V, Liang B, Yang S T, Castle J D and Tamm L K 2017 Asymmetric phosphatidylethanolamine distribution controls fusion pore lifetime and probability *Biophys. J.* **113** 1912–5

[44] Kreutzberger A J B, Kiessling V and Tamm L K 2015 High cholesterol obviates a prolonged hemifusion intermediate in fast SNARE-mediated membrane fusion *Biophys. J.* **109** 319–29

[45] Kiessling V, Liang B, Kreutzberger A J B and Tamm L K 2017 Planar supported membranes with mobile SNARE proteins and quantitative fluorescence microscopy assays to study synaptic vesicle fusion *Frontiers Mol. Neurosci.* **10** 72

[46] Römer W, Lam Y H, Fischer D, Watts A, Fischer W B, Göring P, Wehrspohn R B, Gösele U and Steinem C 2004 Channel activity of a viral transmembrane peptide in micro-BLMs: Vpu_{1-32} from HIV-1 *J. Am. Chem. Soc.* **126** 16267–74

[47] Römer W and Steinem C 2004 Impedance analysis and single-channel recordings on nano-black lipid membranes based on porous alumina *Biophys. J.* **86** 955–65

[48] Spindler S, Sibold J, Gholami Mahmoodabadi R, Steinem C and Sandoghdar V 2018 High-speed microscopy of diffusion in pore-spanning lipid membranes *Nano Lett.* **18** 5262–71

[49] Weiskopf D, Schmitt E K, Klühr M H, Dertinger S K and Steinem C 2007 Micro-BLMs on highly ordered porous silicon substrates: rupture process and lateral mobility *Langmuir* **23** 9134–9

[50] Höfer I and Steinem C 2011 A membrane fusion assay based on pore-spanning lipid bilayers *Soft Matter* **7** 1644

[51] Kocun M, Lazzara T D, Steinem C and Janshoff A 2011 Preparation of solvent-free, pore-spanning lipid bilayers: modeling the low tension of plasma membranes *Langmuir* **27** 7672–80

[52] Mey I, Steinem C and Janshoff A 2012 Biomimetic functionalization of porous substrates: towards model systems for cellular membranes *J. Mater. Chem.* **22** 19348

[53] Schütte O M, Mey I, Enderlein J, Savić F, Geil B, Janshoff A and Steinem C 2017 Size and mobility of lipid domains tuned by geometrical constraints *Proc. Natl Acad. Sci. USA* **114** E6064–E6071

[54] Schwenen L L G, Hubrich R, Milovanovic D, Geil B, Yang J, Kros A, Jahn R and Steinem C 2015 Resolving single membrane fusion events on planar pore-spanning membranes *Sci. Rep.* **5** 12006

[55] Hamai C, Cremer P S and Musser S M 2007 Single giant vesicle rupture events reveal multiple mechanisms of glass-supported bilayer formation *Biophys. J.* **92** 1988–99

[56] Chi Y S, Byon H R, Lee B S, Kong B, Choi H C and Choi I S 2008 Polymeric rulers: distance-dependent emission behaviors of fluorophores on flat gold surfaces and bioassay platforms using plasmonic fluorescence enhancement *Adv. Func. Mater* **18** 3395–402

[57] Acuna G P *et al* 2012 Distance dependence of single-fluorophore quenching by gold nanoparticles studied on DNA origami *ACS Nano* **6** 3189–95

[58] Kuhlmann J W, Junius M, Diederichsen U and Steinem C 2017 SNARE-mediated single-vesicle fusion events with supported and freestanding lipid membranes *Biophys. J.* **112** 2348–56

[59] Pobbati A V, Stein A and Fasshauer D 2006 N- to C-terminal SNARE complex assembly promotes rapid membrane fusion *Science* **313** 673–6

[60] Mühlenbrock P, Herwig K, Vuong L, Mey I and Steinem C 2020 Fusion pore formation observed during SNARE-mediated vesicle fusion with pore-spanning membranes *Biophys. J.* **119** 151–61

[61] Garten M, Aimon S, Bassereau P and Toombes G E S 2015 Reconstitution of a trans-membrane protein, the voltage-gated ion channel, KvAP, into giant unilamellar vesicles for microscopy and patch clamp studies *J. Vis. Exp* **95** 52281

[62] Girard P, Pecreaux J, Lenoir G, Falson P, Rigaud J L and Bassereau P 2004 A new method for the reconstitution of membrane proteins into giant unilamellar vesicles *Biophys. J.* **87** 419–29

[63] Böcker M, Muschter S, Schmitt E K, Steinem C and Schäffer T E 2009 Imaging and patterning of pore-suspending membranes with scanning ion conductance microscopy *Langmuir* **25** 3022–8

[64] Ramakrishnan S, Bera M, Coleman J, Krishnakumar S S, Pincet F and Rothman J E 2019 Synaptotagmin oligomers are necessary and can be sufficient to form a Ca^{2+}-sensitive fusion clamp *FEBS Lett.* **593** 154–62

[65] Bacia K, Schuette C G, Kahya N, Jahn R and Schwille P 2004 SNAREs prefer liquid-disordered over “raft” (liquid-ordered) domains when reconstituted into giant unilamellar vesicles *J. Biol. Chem.* **279** 37951–5

[66] Mühlenbrock P, Sari M and Steinem C 2021 *In vitro* single vesicle fusion assays based on pore-spanning membranes: merits and drawbacks *Eur. Biophys. J.* **50** 239–52

[67] Hubrich R, Park Y, Mey I, Jahn R and Steinem C 2019 SNARE-mediated fusion of single chromaffin granules with pore-spanning membranes *Biophys. J.* **116** 308–18

[68] Jing H Y, Sinha S, Sachar H S and Das S 2019 Interactions of gold and silica nanoparticles with plasma membranes get distinguished by the van der Waals forces: implications for drug delivery, imaging, and theranostics *Coll. Surf. B-Biointerfaces* **177** 433–9

[69] Milovanovic D and Jahn R 2015 Organization and dynamics of SNARE proteins in the presynaptic membrane *Front. Physiol* **6** 89

[70] Honigmann A *et al* 2013 Phosphatidylinositol 4,5-bisphosphate clusters act as molecular beacons for vesicle recruitment *Nat. Struct. Mol. Biol.* **20** 679–86

[71] van den Bogaart G *et al* 2011 Membrane protein sequestering by ionic protein–lipid interactions *Nature* **479** 552–5

[72] Floyd D L, Harrison S C and van Oijen A M 2010 Analysis of kinetic intermediates in single-particle dwell-time distributions *Biophys. J.* **99** 360–6

[73] Dietz J, Oelkers M, Hubrich R, Pérez-Lara A, Jahn R, Steinem C and Janshoff A 2022 Forces, kinetics, and fusion efficiency altered by the full-length synaptotagmin-1-PI(4,5)P_2 interaction in constrained geometries *Nano Lett.* **22** 1449–55

[74] Bowers M R and Reist N E 2020 Synaptotagmin: mechanisms of an electrostatic switch *Neurosci. Lett.* **722** 134834

[75] Chicka M C, Hui E F, Liu H S and Chapman E R 2008 Synaptotagmin arrests the SNARE complex before triggering fast, efficient membrane fusion in response to Ca^{2+} *Nat. Struct. Mol. Biol.* **15** 827–35

[76] Fernandez-Chacon R, Konigstorfer A, Gerber S H, Garcia J, Matos M F, Stevens C F, Brose N, Rizo J, Rosenmund C and Südhof T C 2001 Synaptotagmin I functions as a calcium regulator of release probability *Nature* **410** 41–9

[77] Hui E F, Johnson C P, Yao J, Dunning F M and Chapman E R 2009 Synaptotagmin-mediated bending of the target membrane Is a critical step in Ca^{2+}-regulated fusion *Cell* **138** 709–21

[78] Martens S, Kozlov M M and McMahon H T 2007 How synaptotagmin promotes membrane fusion *Science* **316** 1205–8

[79] Stein A, Radhakrishnan A, Riedel D, Fasshauer D and Jahn R 2007 Synaptotagmin activates membrane fusion through a Ca^{2+}-dependent trans interaction with phospholipids *Nat. Struct. Mol. Biol.* **14** 904–11

[80] Xue M S, Ma C, Craig T K, Rosenmund C and Rizo J 2008 The Janus-faced nature of the C2B domain is fundamental for synaptotagmin-1 function *Nat. Struct. Mol. Biol.* **15** 1160–8

[81] Kreutzberger A J B, Kiessling V, Liang B, Seelheim P, Jakhanwal S, Jahn R, Castle J D and Tamm L K 2017 Reconstitution of calcium-mediated exocytosis of dense-core vesicles *Science Adv.* **3** e1603208

[82] van den Bogaart G, Meyenberg K, Diederichsen U and Jahn R 2012 Phosphatidylinositol 4,5-bisphosphate increases Ca^{2+} affinity of synaptotagmin-1 by 40-fold *J. Biol. Chem.* **287** 16447–53

[83] Chang S W, Trimbuch T and Rosenmund C 2018 Synaptotagmin-1 drives synchronous Ca^{2+}-triggered fusion by C2B-domain-mediated synaptic-vesicle-membrane attachment *Nat. Neurosci.* **21** 33–40

[84] Nyenhuis S B, Thapa A and Cafiso D S 2019 Phosphatidylinositol 4,5 bisphosphate controls the cis and trans interactions of synaptotagmin 1 *Biophys. J.* **117** 247–57

[85] Krishnakumar S S *et al* 2011 A conformational switch in complexin is required for synaptotagmin to trigger synaptic fusion *Nat. Struct. Mol. Biol.* **18** 934–U98

[86] Ramakrishnan S, Bera M, Coleman J, Rothman J E and Krishnakumar S S 2020 Synergistic roles of synaptotagmin-1 and complexin in calcium-regulated neuronal exocytosis *eLife* **9** e54506

[87] Trimbuch T and Rosenmund C 2016 Should I stop or should I go? The role of complexin in neurotransmitter release *Nat. Rev. Neurosci.* **17** 118–25

[88] Tagliatti E, Bello O D, Mendonca P R F, Kotzadimitriou D, Nicholson E, Coleman J, Timofeeva Y, Rothman J E, Krishnakumar S S and Volynski K E 2020 Synaptotagmin 1 oligomers clamp and regulate different modes of neurotransmitter release *Proc. Natl Acad. Sci. USA* **117** 3819–27

[89] Wang J, Li F, Bello O D, Sindelar C V, Pincet F, Krishnakumar S S and Rothman J E 2017 Circular oligomerization is an intrinsic property of synaptotagmin *eLife* **6** e27441

[90] Zanetti M N, Bello O D, Wang J, Coleman J, Cai Y Y, Sindelar C V, Rothman J E and Krishnakumar S S 2016 Ring-like oligomers of synaptotagmins and related C2 domain proteins *eLife* **5** e17262

[91] Park Y and Ryu J K 2018 Models of synaptotagmin-1 to trigger Ca^{2+}-dependent vesicle fusion *FEBS Lett.* **592** 3480-92

[92] Park Y *et al* 2015 Synaptotagmin-1 binds to PIP_2-containing membrane but not to SNAREs at physiological ionic strength *Nat. Struct. Mol. Biol.* **22** 815–23

[93] Chen Y, Wang Y H, Zheng Y, Li M, Wang B, Wang Q W, Fu C L, Liu Y N, Li X and Yao J 2021 Synaptotagmin-1 interacts with PI(4,5)P2 to initiate synaptic vesicle docking in hippocampal neurons *Cell Rep.* **34** 108842

[94] Park Y, Hernandez J M, van den Bogaart G, Ahmed S, Holt M, Riedel D and Jahn R 2012 Controlling synaptotagmin activity by electrostatic screening *Nat. Struct. Mol. Biol.* **19** 991–7

[95] Bai J H, Tucker W C and Chapman E R 2004 PIP_2 increases the speed of response of synaptotagmin and steers its membrane-penetration activity toward the plasma membrane *Nat. Struct. Mol. Biol.* **11** 36–44

[96] Kiessling V, Kreutzberger A J B, Liang B, Nyenhuis S B, Seelheim P, Castle J D, Cafiso D S and Tamm L K 2018 A molecular mechanism for calcium-mediated synaptotagmin-triggered exocytosis *Nat. Struct. Mol. Biol.* **25** 911–7

[97] Ramakrishnan S, Gohlke A, Li F, Coleman J, Xu W, Rothman J E and Pincet F 2018 High-throughput monitoring of single vesicle fusion using freestanding membranes and automated analysis *Langmuir* **34** 5849–59

[98] Stratton B S, Warner J M, Wu Z, Nikolaus J, Wei G, Wagnon E, Baddeley D, Karatekin E and O'Shaughnessy B 2016 Cholesterol increases the openness of SNARE-mediated flickering fusion pores *Biophys. J.* **110** 1538–50

IOP Publishing

Exocytosis: From Molecules to Cells

Arun Anantharam and Jefferson Knight

Chapter 10

Effects of anesthetics on membrane fusion and exocytosis

Robert E Coffman and Dixon J Woodbury

We review reported molecular targets of anesthetics that impact synaptic transmission, focusing specifically on the exocytotic event. Because many general anesthetics have potencies predicted by their solubility in lipids (the Meyer–Overton rule), we focus primarily on molecular mechanisms that may account for this observation. Most research has focused on two targets for the interactiaon of anesthetics at the cellular level: lipids and proteins in the cell membrane. Early research focused exclusively on lipid membranes, but eventually a joint lipid and protein hypothesis was developed. Some studies have shown that anesthetics alter lipid phases, but generally, these changes occur at doses greater than physiologically relevant doses. Other studies have shown the direct effects of a specific anesthetic on a specific membrane protein in neurons. Such results cannot be explained by the Meyer–Overton rule. A protein-centric hypothesis consistent with the Meyer–Overton rule is that anesthetics partition into the hydrophobic core of lipid membranes, where they interfere with lipid–protein interactions either directly or indirectly. This interference alters normal protein function and reduces synaptic transmission. An additional hypothesis is that anesthetics also directly alter vesicle-membrane fusion. This is supported by the observation that alcohol alters vesicle-membrane fusion in a SNARE-driven and protein-free system [1, 2]. This later hypothesis adds to the list of ways anesthetics may directly and specifically alter the exocytotic step of synaptic transmission.

10.1 Introduction

In this chapter, we present the effects of anesthetics on the exocytotic release of neurotransmitters (NTs). In our discussion of anesthetics, we include alcohols (e.g. ethanol) and other small compounds that produce unconsciousness, but only briefly consider molecules that bind directly to specific proteins in the pathway for neurotransmission. We begin with a background and overview of anesthetics,

doi:10.1088/978-0-7503-3771-7ch10

then summarize their effects on exocytosis, and end by discussing an additional way that anesthetics can alter exocytosis by acting directly on the fusing membranes.

10.1.1 Historical background

Alcohol has been in use since before recorded history as a drink, intoxicant, and disinfectant [3, 4]. Alcohol intoxication is known to decrease anxiety and motor skills at lower doses and in larger doses is known to depress the central nervous system as well as to cause unconsciousness and amnesia. Due to these properties, alcohol can be considered an anesthetic. Alcohol is not used as a general anesthetic because doses that lead to unconsciousness can also be fatal. It should be noted that alcohols are chemically much simpler than many general anesthetics and have several distinct physiological effects. General anesthetics were first used during surgical procedures in 1846 at the Massachusetts General Hospital and are currently used for surgery, rendering a patient unconscious and unresponsive to painful stimuli; they are amnestic and analgesic. Currently, the molecular mechanisms by which general anesthetics and alcohol produce unconsciousness are not well understood. This lack of knowledge stems from the fact that the structure(s) of the brain responsible for consciousness are not well defined, so no molecular studies can be performed directly on it (them). As an alternative, the simpler and generally accepted hypothesis is that consciousness requires synaptic transmission and any process that disrupts transmission can result in unconsciousness [5].

Before proceeding into the details of the possible molecular mechanisms by which anesthetics impact exocytosis, it is valuable to briefly consider what synaptic disruption might do to various cognitive processes, including consciousness [5, 6]. The following paragraph, including figure 10.1, is adapted from the research notes of the father of one of the authors, Professor of Physiology J Walter Woodbury (1923–2017).

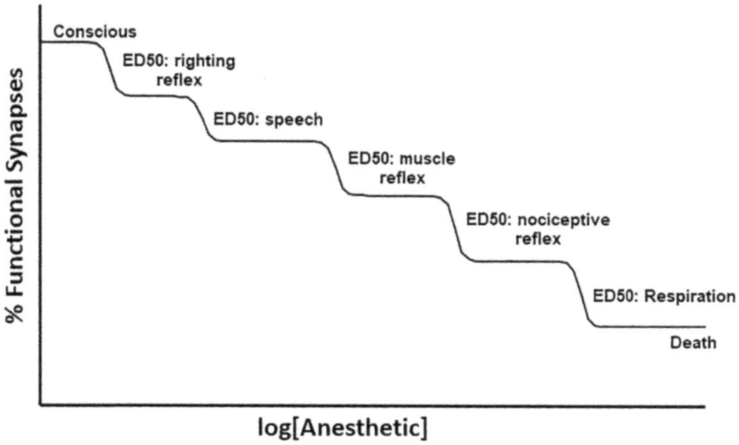

Figure 10.1. Hypothetical plot of the loss of different brain processes as a function of the dose of a specific anesthetic. ED_{50} indicates the dose at which 50% of organisms tested lose the specific function. The sequence shown is just one of many possibilities and would depend on the anesthetic as well as genetic and other factors. Consciousness might be lost at any dose, depending on the specific conditions.

We assume two well-established observations: (1) there are clusters of neurons in the brain responsible for processes such as balance, respiration, speech, pain perception, thinking, and presumably also consciousness, and (2) different types of neuron (e.g. cholinergic, glutamatergic, and gamma-aminobutyric acid (GABA)-ergic) respond differently to different anesthetics at different doses. From this foundation, we can conclude that for any anesthetic, there is an effective dose (ED_{50}) at which the clusters of neurons responsible for one of these processes no longer perform their task in 50% of organisms tested. Additionally, the ED_{50} would be expected to be different for different clusters of neurons carrying out different processes with diverse and unique blends of neuron types. A plot of anesthetic dose versus functioning synapses, or, more accurately, functioning clusters of neurons, is shown in figure 10.1. Consciousness would be lost at some specific dose, which could coincide with the loss of one or more other cognitive functions. For example, the drug MK-801 causes unconsciousness, but is not clinically used as an anesthetic because it also causes psychotic behavior and neurodegeneration, which is not desirable in a general anesthetic.

This model of anesthetic action, namely that molecular alterations at synapses compromise the nerve clusters responsible for consciousness, provides a framework for interpreting biochemical and biophysical experiments on molecular changes associated with anesthetics. Over the last hundred years, such experiments have focused on two general approaches. As described by Dr Bertil Hille, we refer to them as 'general perturbation' and 'specific receptor' [7] (see also [8]). These approaches are discussed below.

The 'general perturbations' approach states that anesthetics act by altering some 'phase' in neurons that compromises the molecular components essential for synaptic transmission. This 'phase' is assumed to be the lipid membrane, based on observations first made over a century ago. In 1899, Hans Horst Meyer and two years later, Charles Ernest Overton independently noticed that the potency of an anesthetic is correlated to its partition coefficient in olive oil compared to that in water. This correlation is known as the Meyer–Overton rule (figure 10.2); it suggests the simple hypothesis that anesthetics act on, and within, the cell membrane—the only part of the cell that is similar to oil.

On the other hand, the 'specific receptors' approach is that anesthetics bind to specific proteins, causing altered structure and loss of neurotransmission. The list of proteins identified as having altered function following the addition of different anesthetics is long and includes many described in the next section. A current hypothesis blends both mechanisms and states that general anesthetics affect synaptic transmission by disrupting lipids that affect proteins and by direct action on proteins, and that all of these effects combined contribute to anesthesia.

10.2 Neurotransmission

Before proceeding further with other possible mechanisms and sites of action for anesthetics, it is valuable to consider the key neuronal process of neurotransmission, since some aspect of the general process of neurotransmission must be the target of

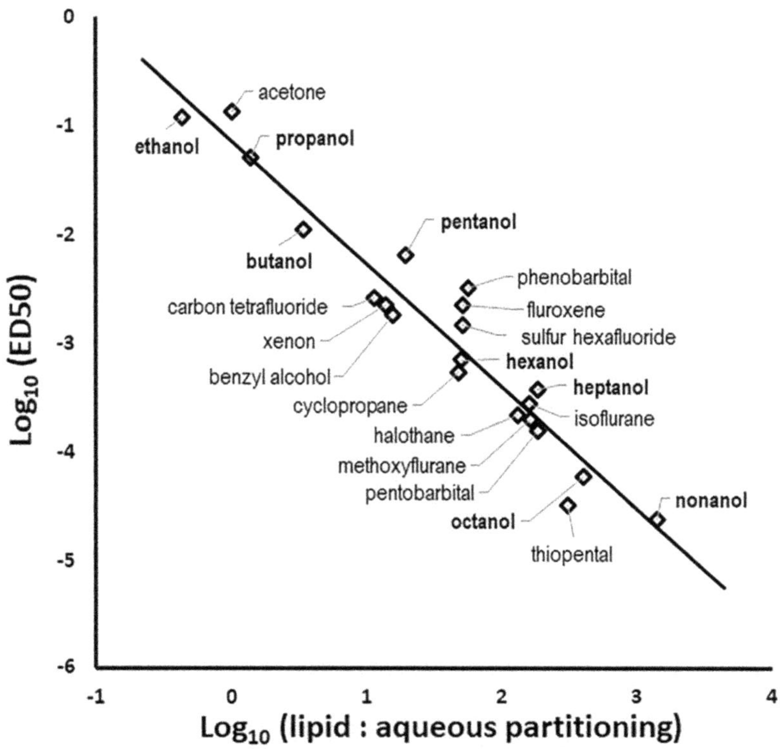

Figure 10.2. Meyer–Overton rule: there is a strong correlation between the potency (ED_{50}) of an anesthetic and its oil or lipid to water partition coefficient. A linear relation is observed when the log of the anesthetic dose (ED_{50}) is plotted against the log of its relative partitioning in oil. N-alcohols are bolded. Adapted from Janoff *et al* 1981 [99].

anesthetics. Neurotransmitter release (see figure 10.3) begins with an action potential traveling down the axon via voltage-gated Na^+ and K^+ channels. At the synaptic terminal the action potential opens voltage-gated Ca^{++} channels. Ca^{++} flows into the terminal, where it binds synaptotagmin on the synaptic vesicle membrane, leading to vesicle fusion. Before Ca^{++} influx takes place, neurotransmitter-filled synaptic vesicles are tethered near the release site ready to be docked by assorted proteins (e.g. RIM (Rab3-interacting molecule), RIM-BP (RIM-binding protein), Rab and Munc13 [9]) or tightly docked at the synaptic terminal under the control of *s*oluble *N*SF *a*ttachment protein *re*ceptors (SNAREs) and other accessory proteins [10]. The SNAREs are synaptobrevin (also known as vesicle-associated membrane protein (VAMP)) in the synaptic vesicle membrane and syntaxin and synaptosomal-associated protein, 25 kDa (SNAP-25) in the (presynaptic) nerve cell membrane. Docked vesicles are held close to the plasma membrane by these proteins, ready for fusion. The binding of Ca^{++} to synaptotagmin alters its interaction with SNARE proteins and membranes, triggering the fusion of vesicles with the plasma membrane and the dumping of their NT contents (e.g. glutamate, gamma-aminobutyric acid (GABA), dopamine, and acetycholine (ACh)) into the synaptic cleft [11]. The NTs

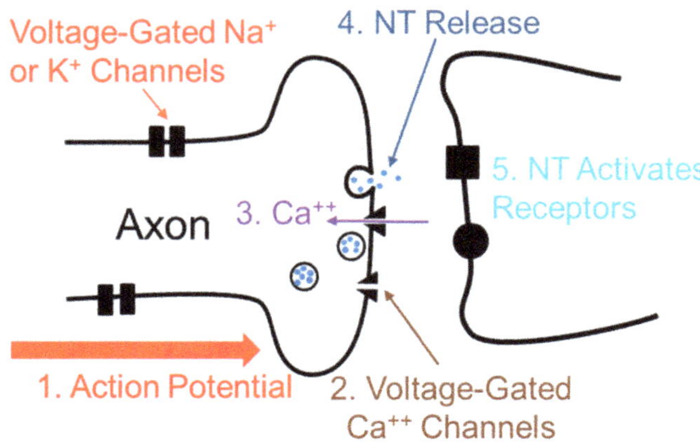

Figure 10.3. NT release: (1) action potential propagation by voltage-gated Na^+ channels; (2) opening of voltage-gated Ca^{++} channels; (3) influx of Ca^{++}; (4) NT release; (5) activation of receptors on postsynaptic membrane.

diffuse across the synaptic cleft, where they bind to postsynaptic receptors, initiating signaling in the postsynaptic cell.

The fusion of vesicles with the plasma membrane is the last event leading to NT release and is under the control of synaptotagmin and the SNARE proteins. Ultimately, release follows the opening of a pore or pathway out of the vesicle due to the fusion of vesicles to cell membranes and is thought to proceed in several steps (figure 10.4) [12]: (1) loose interaction between the two membranes as the vesicle approaches the plasma membrane; (2) water molecules and membrane-bound proteins move aside, and electrostatic repulsion of headgroups is overcome to allow the two membranes to enter a tightly docked state; (3) the formation of a lipid stalk, which is thought to be mediated by the splay of a lipid tail; (4) the hemi-fused state; and (5) the opening and expansion of the fusion pore [13–15].

The transition from step three to step five has two debated pathways. One pathway is called a hemi-fused state (step (4) in figure 10.4), where the stalk expands laterally, bringing the inner leaflets into closer proximity until they form a single bilayer separating the two compartments, and then a pore opens in this bilayer. The other pathway is that a pore opens very shortly after stalk formation without the need for lateral expansion and formation of a single bilayer separating the two compartments. There is evidence for both pathways [16]. Regardless, once the pore is open, it allows the vesicle contents (if they are small enough) to leak out until either the pore closes back up (known as the 'kiss and run' hypothesis) or the pore expands until complete fusion and total release of the vesicle contents is achieved (full fusion).

In addition to any postsynaptic or earlier presynaptic step, anesthetics may target any of these steps (figures 10.3 and 10.4) to reduce NT release and produce unconsciousness.

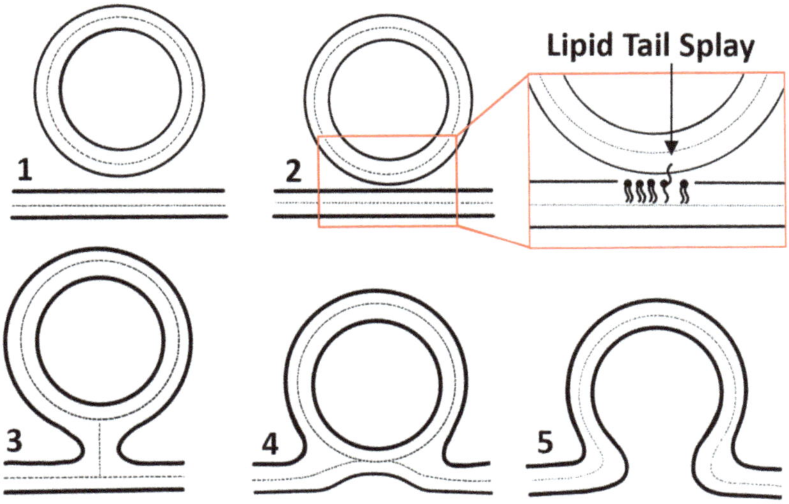

Figure 10.4. The process of vesicle fusion: the vesicle approaches the plasma membrane (1 and 2); a lipid tail splays (box), initiating stalk formation (3); the stalk expands laterally, forming the hemi-fused state (4); the fusion pore opens and expands (5).

10.2.1 Anesthetics change neurotransmitter release from cells

It is known that anesthetics can change the release of neurotransmitters from cells. For example, it has been shown that isoflurane and propofol inhibit glutamate and GABA release from isolated rat cerebrocortical nerve terminals [17] and that ethanol can increase or decrease glutamate and GABA release, depending on the experimental setup [18]. However, such correlations do not reveal where, along the multi-step process from action potential to NT release, the anesthetic is acting. Additionally, it is possible, and perhaps likely, that there are multiple sites of action. In cellular systems, where many components are required to interact harmoniously, it is difficult to design experiments to determine exactly where and how anesthetics alter neurotransmitter release.

Encouragingly, there have been many studies of the possible mechanism behind these changes. The Meyer–Overton rule provides strong evidence that anesthetics act by altering cell membranes in some way. Therefore, many early experiments focused on identifying a specific property of membrane lipids that correlated with anesthetic potency. As discussed in the section 'Changes in membrane properties' (10.2.2), no simple membrane property has been found that fits well with this expectation. With membrane properties generally discounted, researchers focused on anesthetic effects on membrane-spanning proteins in order to explain the observed potency. Recently, researchers have identified multiple sites of possible action including synaptic receptors, voltage-gated ion channels, and interactions with the protein machinery responsible for exocytosis [19, 20] and other proteins involved in vesicular trafficking and tethering, such as the Rab proteins [21]. For reviews of the mechanisms of anesthesia on proteins, see references [6, 22, 23].

Multiple studies of the anesthetic effect on membrane-spanning proteins have found that anesthetics do change protein properties, but in different ways. For example, two postsynaptic inhibitory receptors, $GABA_A$ and glycine, are potentiated by several anesthetics and pentanol [23, 24], whereas the postsynaptic excitatory receptors α-amino-3-hydroxy-5-methyl-4-isoxazolepropionic acid (AMPA), *N*-methyl-D-aspartic acid (NMDA), and nicotinic acetylcholine (nACh) receptors are generally inhibited by anesthetics [23, 25]. One example of an anesthetic-like molecule with a known protein binding site is MK-801 [26]. It has been shown through x-ray crystallography and molecular dynamics simulations to interact physically with the NMDA receptor [27]. There is also extensive evidence that general anesthetics and alcohols bind to specific sites on pentameric ligand-gated ion channels (pLGIC) [6]. This family of channels includes nicotinic acetylcholine receptors, $GABA_A$ receptors, glycine receptors, and 5-HT_3 (serotonin) receptors. Other groups have shown that anesthetics inhibit two important families of channels: (1) the presynaptic voltage-gated Na^+ channels involved in the propagation of the action potential and (2) voltage-gated Ca^{++} channels that are responsible for the influx of Ca^{++} at the axon terminal [28–31]. It has also been found that anesthetics can potentiate or inhibit different types of K^+ channel in the pre- and postsynaptic membranes [32]. These potassium-channel families are involved in regulating the resting membrane potential (leak channels) and repolarization of the membrane potential during the action potential (voltage-gated potassium channels). However, not all channels in these families of protein are affected by anesthetics.

It is clear that synaptic communication release can be blocked by altering presynaptic ion channels and postsynaptic receptors, but it can be more difficult to design experiments that detect the effects of anesthetics on the fusion machinery. In PC12 cells and rat cortical synaptosomes, catecholamine release is increased in the presence of propofol compared to controls using ionomycin to induce Ca^{++} influx. Propofol is known to affect voltage-gated Ca^{++} channels, so when those channels are bypassed, the authors suggest that 'propofol enhances the release machinery' in both preparations [19]. Other researchers tested anesthetic sensitivity of the fruit fly *Drosophila melanogaster*, using mutations to syntaxin-1A (a SNARE protein) following isoflurane treatment. They found that deleting 14 amino acids in the protein provided resistance to isoflurane, while a two-amino-acid substitution in the protein made the flies hypersensitive to the general anesthetic [20]. Experiments like these show that the exocytotic process is also a target of anesthetics [22].

10.2.2 Changes in membrane properties

Since there is such a wide variety of membrane-spanning proteins and anesthetic structures involved, the hypothesis that membrane proteins alone explain the Meyer–Overton rule seems unlikely. This leaves open the possibility that a membrane-specific hypothesis might still be part of the molecular mechanism responsible for the action of anesthetics.

For years, scientists have looked for a measurable membrane property that was always affected in the same way by relevant doses of anesthetics, but they have not convincingly found one that explains the observed effects on the nervous system. What is especially difficult to reconcile in a membrane property hypothesis is that the stereoisomers of some general anesthetics affect the measured membrane property to the same degree, but they do not have the same anesthetic potency [33]. For example, *S*(+)-isoflurane is 53% more potent than *R*(−)-isoflurane in rats [34] but partitions equally into lipid bilayers [35]. The ED_{50} of *R*(+)-etomidate was ~16 times lower (more potent) than *S*(−)-etomidate when tested using the righting reflex in tadpoles *Rana temporaria*, but the melting temperature of dipalmitoyl-L-alpha-phosphatidylcholine (DPPC) vesicles was equally changed by both enantiomers [36]. Some membrane properties that have been measured in response to anesthetics include: membrane thickness, fluidity, lipid tail order, area per lipid molecule, and phase transition/melting point. Some of these will be explored in the next paragraphs, along with example anesthetics and their effects on these properties.

One of the first membrane properties that was thought to correlate with general anesthetic potency was increased membrane thickness [37], achieved either by increasing lipid tail order or by physically pushing the two leaflets of a membrane apart. For example, Haydon and colleagues found through capacitance measurements of a lipid membrane that the incorporation of the sequence *n*-pentane through *n*-decane increases membrane thickness [38] (these alkanes are anesthetically potent). Interestingly, alkanes as well as their alcohols have a cutoff effect. As shown in figure 10.2, lipid solubility and potency increase with increasing chain length, but at some length (around nine carbons) the potency begins to decrease and eventually disappears [39–42]. This cutoff is hypothesized to occur because the chain lengths approach the lengths of the lipid tail chains, reducing their effect on protein–lipid interactions [43, 44]. Regardless, not many years later, another group using x-ray and neutron diffraction found that the incorporation of nitrous oxide, halothane, and cyclopropane (which are general anesthetics, with the exception of nitrous oxide, which has anesthetic properties) into a lipid membrane do not change membrane thickness [45]. This suggests that membrane thickness is not a common bilayer property perturbed by simple anesthetics.

Another lipid property considered was membrane fluidity; changes in membrane fluidity can be measured as changes in the transition temperature of lipid phases or the ability of lipids to diffuse in the membrane. These lipid phases correspond to the average orientation of lipid tails. For example, when most of the lipid tails are perpendicular to the plane of the bilayer, it is in the gel phase, and fluidity is low. When the tails are more disorganized, the bilayer is in the liquid phase, and fluidity is high (figure 10.5). It is well established that several anesthetics increase membrane fluidity [46–51], for example, ethanol and isoflurane [52, 53]. However, the purified stereoisomers of isoflurane have been shown to have different anesthetic potencies [34, 54], even though they change fluidity by the same magnitude [55].

Lipid rafts are membrane features related to fluidity. Lipid rafts form when dissimilar lipids separate into distinct clusters or rafts (figure 10.5). These clusters may dissolve or form in response to different conditions, including membrane proteins, detergents, or temperature. Allen *et al* have postulated that rafts may modulate NT release [56] and Machta *et al* showed a correlation between the

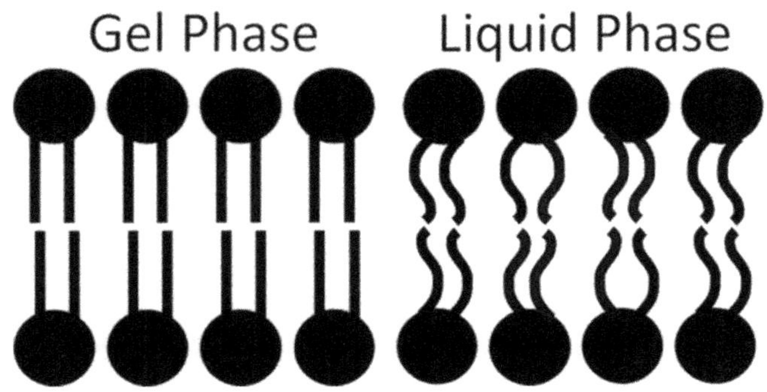

Figure 10.5. Representation of lipids in two phases. Lipid rafts occur when two phases coexist in the same membrane.

melting temperature of membrane domains and certain anesthetic properties of alcohols [57]. Heimburg and colleagues have taken this one step further; they measured how anesthetics alter the conductance of ions through 'lipid channels' [58, 59]. They show that near a phase transition, where two domains exist, lipid membranes become leaky and that anesthetics alter this leakiness. They also point out that real membranes often have phase transitions near enough to environmental temperatures that the effects of anesthetics cannot be dismissed [60]. Lipid channels presumably occur at the boundaries of phase domains (although lipid channels have also been seen in fusing vesicles without lipid-phase domains [61]). The alteration of membrane domains or rafts by anesthetics continues to be an important membrane property to be considered in any discussion of the mechanisms of general anesthetics [62, 63].

Other researchers also looked at molecules that were predicted to have anesthetic potencies based on the Meyer–Overton rule and found that some were anesthetically potent, while others caused excitation or convulsions or were 'nonanesthetic' [64, 65]. Although some of these disparities might be explained by the model shown in figure 10.1, many researchers believe that it is changes in membrane proteins—specifically ion channels—and not lipids that best explain anesthetic effects. While many other membrane properties have been measured, most follow a similar trend and will not be discussed further here. A blended hypothesis is that anesthetics alter membrane protein interactions. These are more difficult to measure and will be discussed next.

10.3 Current hypotheses of general anesthesia

A common hypothesis of general anesthetics is that they can affect protein function in various ways. One of the ways mentioned above is that a few of these molecules have binding sites on specific proteins [6, 66]. Another similar hypothesis is that these molecules displace lipids from a lipid binding site within the protein to alter protein conformation [67]. A third possibility is that anesthetics disrupt lipid rafts, as explained above [68]. A fourth, more general idea is that anesthetics partition into and perturb the center of the membrane in such a way as to create a lateral pressure

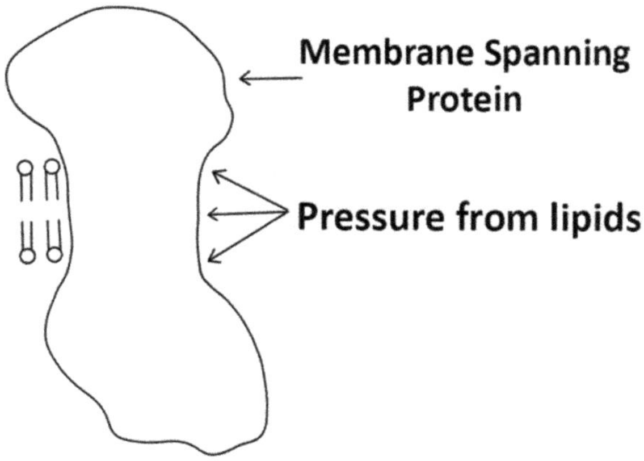

Figure 10.6. Schematic showing how lipids exert pressure on a protein that spans the membrane.

that physically alters the membrane protein structure (and hence function). This last idea is called the lateral pressure profile.

10.3.1 Lateral pressure profile

The lateral pressure hypothesis can explain many anesthetic effects. This difficult-to-measure membrane property describes the relative lateral pressure (perpendicular to the membrane normal) that membrane lipids place on membrane-spanning proteins (figure 10.6). This possible mechanism was discussed theoretically by Robert Cantor in 1997 [69]. Since then, researchers have been looking for a way to measure this property experimentally and have had some success. Fluorescence anisotropy measurements of two fluorescent probes that reside at different depths in a membrane suggest that isoflurane alters the lateral pressure profile [70]. In addition, through an experiment that combined ^{2}H-NMR (deuterium-nuclear magnetic resonance) spectroscopy and dissociation of the tetrameric KcsA (K+ channel from *Streptomyces lividans*) potassium channel, researchers found that several short-chain alcohols caused increased dissociation of the tetramer and pointed to potential changes in the lateral pressure profile as being responsible [71].

Another method that has had success and provides finer resolution is the use of computer simulations; these predict that the lateral pressure profile is changed by chloroform, halothane, diethyl ether, enflurane, methanol, and ethanol [72–75]. This hypothesis is attractive since there are so many different membrane-spanning proteins that are affected by anesthetics and because many of the proposed binding sites for anesthetics have low affinity and are within membrane-spanning domains [6, 76–78]. The lateral pressure hypothesis continues to successfully describe many aspects of anesthetics.

10.4 Membrane–membrane interactions

Many experiments have shown that membrane fusion is sensitive to membrane properties, which can be affected by lipid composition, temperature, and other small molecules, including alcohols and general anesthetics. For example, several *in vitro* fusion assays have shown that the incorporation of phosphatidylethanolamine (PE) increases fusion rates [79, 80]. PE is a lipid that induces negative curvature in a membrane; it is the second most abundant phospholipid [81]. Another example is cholesterol, which facilitates fusion in several ways [82], e.g. by clustering syntaxin in the plasma membrane [83], lowering the energy of stalk formation [84, 85], and increasing efficiency of fusion [86–89]. The enhancing effect of cholesterol on exocytosis has also been shown in chromaffin and blood platelets using amperometry [90–92]. In most of these studies, with the exception of the protein-free system of Lee *et al* [89], it is not possible to distinguish whether the alteration of the membrane's composition and its properties directly alters membrane fusion (e.g. by lowering the activation energy required to fuse two membranes), or whether the change alters protein properties which then affect the fusion process, as is usually assumed.

10.4.1 Anesthetics and membrane fusion

As seen in the above discussion, there are many targets for which anesthetics may alter neurotransmitter release. A few specific targets include the voltage-gated Na^+ and Ca^{++} channels, synaptotagmin, SNARE proteins, lipid tail splay, or other direct actions on membrane lipids. Many of these targets have been discussed here or elsewhere [6, 18, 21, 23, 25, 29, 32, 33]. In this section, we focus only on the potential effects anesthetics have on lipids that lead to vesicle-membrane fusion.

Using a vesicle-membrane fusion assay containing no proteins, Paxman *et al* [2] reported that the effects of alcohol (methanol through n-butanol) on membrane fusion are dependent on the side of the membrane to which the alcohol is applied. They showed that when 0.4% v/v ethanol is applied to the same side of the membrane as the vesicles (termed the *cis* side), fusion rates increased to 170% of the control rate. When ethanol and other short-chain alcohols are applied to the opposite (*trans*) side at the same dose, fusion rates decreased to 45%–80%. The inhibitory ethanolic dose required to reduce fusion by 50% (IC_{50}) was 0.20% v/v or 34 mM [2]. For reference, a blood alcohol content of 100 mM is considered to be the upper nonlethal limit [93], so the observed inhibition by *trans* ethanol is well within the physiological range. However, the physiological relevance of these results is questionable, since fusion was driven by an osmotic gradient, instead of the SNARE proteins that are known to drive fusion in all cellular systems [94, 95]. In support of the membrane effects reported [2], we also report an effect similar to cis addition when fusion is driven by SNARE proteins [1].

It is intriguing that opposite effects were seen, depending on which side of the membrane alcohol was added. In figure 10.7, five distinct membrane locations are identified, any of which could be the site of action. In *cis* addition, alcohol can interact with the vesicle membrane (site 1), alter the stability of the stalk (site 2), change the dehydration kinetics of the two approaching membranes (site 3), or alter the target membrane (site 4). Previous studies suggest that *cis* excitation is due to alcohol decreasing the tension needed for vesicle lysis [96] and accelerating vesicle distention and rupture rates (site 1) [97], as well as increasing the probability of stalk

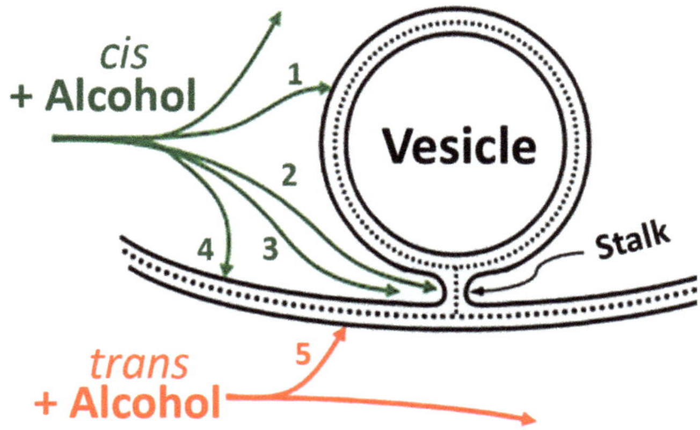

Figure 10.7. Potential membrane sites of action: (1) alcohol interacting with the vesicle membrane; (2) alcohol interacting with the highly curved stalk; (3) alcohol influencing dehydration as the membranes approach each other; (4) alcohol interacting with the target membrane from the *cis* chamber; (5) alcohol interacting with the target membrane from the *trans* chamber.

formation (site 2 or 3) [98]. Sites 4 and 5 in figure 10.7 show that in either *cis* or *trans* addition, alcohol interacts with the target membrane. Given this view, the authors predicted that excitation of fusion is due to effects on the vesicle membrane, which dominates over the inhibition induced in the target membrane. To test the authors' prediction, symmetric addition of alcohol should excite fusion at rates similar to *cis* addition, because alcohol's interaction with sites 1–3 should overcome any interaction with sites 4 and 5. This is a prediction waiting to be tested.

10.5 Summary and conclusions

In connecting the unconscious state produced by anesthetics to exocytosis (or any specific molecular event), a number of assumptions must be made. The largest of these is the problem that the location of the part of the brain that produces consciousness is not known, so the clusters of neurons that presumably perform this function cannot be studied directly. This is in contrast to other functions of the brain that have known locations (e.g. speech formation, which is associated with Brocha's region and the regulation of respiration, which is controlled by clusters of neurons in the medulla). This shortcoming in interpreting results for the molecular action of anesthetics is generally dismissed by the reasonable assumption that specific molecular events reduce synaptic transmission, and that at some level of reduction (the ED_{50}) the functioning of the clusters of neurons that produce consciousness ceases (figure 10.1). This assumption allows one to study the effects of anesthetics on any process contributing to synaptic transmission and interpret it as potentially producing unconsciousness. This has led to a plethora of studies identifying many potential targets in the pre- and postsynaptic cells as the site of action of one or more anesthetics (figure 10.3).

Because many general anesthetics have potencies predicted by the Meyer–Overton rule (figure 10.2), much effort has been made to connect a consistent action of these anesthetics to general disruption of synaptic transmission at the cellular level. This has led to two general targets for the interaction of anesthetics at the cellular level: lipids or proteins in the cell membrane [7]. Most studies of the anesthetic effects on lipids have focused on alterations in lipid phases (figure 10.5). Some studies have described the direct effects of specific anesthetics on membrane proteins in neurons. However, such results fail to explain the molecular basis for the Meyer–Overton rule that applies to many, but not all, anesthetics.

A harmonizing hypothesis, which is protein-centric, is that many anesthetics partition into the hydrophobic core of the lipid membrane, where they increase the lateral pressure of the lipids on membrane proteins and compromise protein function, therefore reducing synaptic transmission (figure 10.6). A lipid-centric hypothesis is suggested by the observation that short-chain alcohols can directly alter vesicle-membrane fusion in a protein-free system (figure 10.7) [1, 2]. This adds to the list of possible ways in which anesthetics alter membrane lipids to directly decrease the exocytotic step of synaptic transmission and NT release.

Acknowledgments

All authors declare they have no conflict of interest. This work was supported by Brigham Young University. Thanks to Jared Leblow and Emilia Coffman who redesigned several of the figures.

References

[1] Coffman R E, Kraichely K N, Kreutzberger A J B, Kiessling V, Tamm L K and Woodbury D J 2022 Drunken lipid membranes, not drunken SNARE proteins, promote fusion in a model of neurotransmitter release *Front. Mol. Neurosci.* **15** 1022756

[2] Paxman J, Hunt B, Hallan D, Zarbock S R and Woodbury D J 2017 Drunken membranes: short-chain alcohols alter fusion of liposomes to planar lipid bilayers *Biophys. J.* **112** 121–32

[3] Vallee B L 1998 Alcohol in the western world *Sci. Am.* **278** 80–5

[4] Vallee B L 1994 *Toward a Molecular Basis of Alcohol Use and Abuse* ed B Jansson *et al* (Basel: Birkhäuser Basel) 1–8 pp

[5] Woodbury J W, D'Arrigo J S and Eyring H 1975 Physiological mechanism of general anesthesia: synaptic blockade *Mol. Mech. Anesth.* **1** 53–91

[6] Hemmings H C, Riegelhaupt P M, Kelz M B, Solt K, Eckenhoff R G, Orser B A and Goldstein P A 2019 Towards a xomprehensive understanding of anesthetic mechanisms of action: a decade of discovery *Trends Pharmacol. Sci.* **40** 464–81

[7] Hille B 1980 Theories of anesthesia: general perturbations versus specific receptors *Mol. Mech. Anesth.* **2** 1–5

[8] Franks N P and Lieb W R 1994 Molecular and cellular mechanisms of general anaesthesia *Nature* **367** 607

[9] Südhof T C 2013 Neurotransmitter release: the last millisecond in the life of a synaptic vesicle *Neuron* **80** 675–90

[10] Zhang Y and Hughson F M 2021 Chaperoning SNARE folding and assembly *Annu. Rev. Biochem.* **90** 581–603

[11] Brunger A T, Choi U B, Lai Y, Leitz J, White K I and Zhou Q 2019 The pre-synaptic fusion machinery *Curr. Opin. Struct. Biol.* **54** 179–88

[12] Witkowska A, Heinz L P, Grubmüller H and Jahn R 2021 Tight docking of membranes before fusion represents a metastable state with unique properties *Nat. Commun.* **12** 3606
[13] Risselada H J and Grubmüller H 2012 How SNARE molecules mediate membrane fusion: recent insights from molecular simulations *Curr. Opin. Struct. Biol.* **22** 187–96
[14] Larsson P and Kasson P M 2013 Lipid tail protrusion in simulations predicts fusogenic activity of influenza fusion peptide mutants and conformational models *PLoS Comput. Biol.* **9** e1002950
[15] Scheidt H A, Kolocaj K, Veje Kristensen J, Huster D and Langosch D 2018 Transmembrane helix induces membrane fusion through lipid binding and splay *J. Phys. Chem. Lett.* **9** 3181–6
[16] Sharma S and Lindau M 2018 The fusion pore, 60 years after the first cartoon *FEBS Lett.* **592** 3542–62
[17] Westphalen R I and Hemmings H C 2003 Selective depression by general anesthetics of glutamate versus GABA release from isolated cortical nerve terminals *J. Pharmacol. Exper. Therapeutics* **304** 1188
[18] Siggins G R, Roberto M and Nie Z 2005 The tipsy terminal: presynaptic effects of ethanol *Pharmacol. Therapeutics* **107** 80–98
[19] Han L, Fuqua S, Li Q, Zhu L, Hao X, Li A, Gupta S, Sandhu R, Lonart G and Sugita S 2016 Propofol-induced inhibition of catecholamine release is reversed by maintaining calcium influx *Anesthesiology* **124** 878–84
[20] Zalucki O H, Menon H, Kottler B, Faville R, Day R, Bademosi A T, Lavidis N, Karunanithi S and van Swinderen B 2015 Syntaxin1A-mediated resistance and hypersensitivity to isoflurane in *Drosophila melanogaster Anesthesiology* **122** 1060–74
[21] Barclay Jeff W, Graham Margaret E, Edwards Mark R, Johnson James R, Morgan A and Burgoyne Robert D 2010 Presynaptic targets for acute ethanol sensitivity *Biochem. Soc. Trans.* **38** 172–6
[22] Das J 2020 SNARE complex–associated proteins and alcohol *Alcoholism: Clin. Exper. Res.* **44** 7–18
[23] Son Y 2010 Molecular mechanisms of general anesthesia *Korean J. Anesthesiol.* **59** 3–8
[24] Lin L H, Chen L L, Zirrolli J A and Harris R A 1992 General anesthetics potentiate gamma-aminobutyric acid actions on gamma-aminobutyric acidA receptors expressed by *Xenopus* oocytes: lack of involvement of intracellular calcium *J. Pharmacol. Exp. Ther.* **263** 569 https://jpet.aspetjournals.org/content/263/2/569
[25] Sonner J M and Cantor R S 2013 Molecular mechanisms of drug action: an emerging view *Annu. Rev. Biophys.* **42** 143–67
[26] Kovacic P and Somanathan R 2010 Clinical physiology and mechanism of dizocilpine (MK-801): electron transfer, radicals, redox metabolites and bioactivity *Oxid. Med. Cell Longev.* **3** 13–22
[27] Song X, Jensen M Ø, Jogini V, Stein R A, Lee C-H, McHaourab H S, Shaw D E and Gouaux E 2018 Mechanism of NMDA receptor channel block by MK-801 and memantine *Nature* **556** 515–9
[28] Herold K F, Andersen O S and Hemmings H C Jr. 2017 Divergent effects of anesthetics on lipid bilayer properties and sodium channel function *Eur. Biophys. J.* **46** 617–26
[29] Orestes P and Todorovic S M 2010 Are neuronal voltage-gated calcium channels valid cellular targets for general anesthetics? *Channels (Austin)* **4** 518–22
[30] Baumgart J P, Zhou Z-Y, Hara M, Cook D C, Hoppa M B, Ryan T A and Hemmings H C 2015 Isoflurane inhibits synaptic vesicle exocytosis through reduced Ca^{2+} influx, not Ca^{2+}-exocytosis coupling *Proc. Natl Acad. Sci.* **112** 11959–64

[31] Denomme N, Hull J M and Mashour G A 2019 Role of voltage-gated sodium channels in the mechanism of ether-induced unconsciousness *Pharmacol. Rev.* **71** 450

[32] Li Y, Xu J, Xu Y, Zhao X-Y, Liu Y, Wang J, Wang G-M, Lv Y-T, Tang Q-Y and Zhang Z 2018 Regulatory effect of general anesthetics on activity of potassium channels *Neurosci. Bull.* **34** 887–900

[33] Chau P L 2010 New insights into the molecular mechanisms of general anaesthetics *Br. J. Pharmacol.* **161** 288–307

[34] Lysko G S, Robinson J L, Casto R and Ferrone R A 1994 The stereospecific effects of isoflurane isomers *in vivo Eur. J. Pharmacol.* **263** 25–9

[35] Dickinson R, Franks N P and Lieb W R 1994 Can the stereoselective effects of the anesthetic isoflurane be accounted for by lipid solubility? *Biophys. J.* **66** 2019–23

[36] Tomlin Sarah L, Jenkins A, Lieb William R and Franks Nicholas P 1998 Stereoselective effects of etomidate optical isomers on gamma-aminobutyric acid type a receptors and animals *Anesthesiology* **88** 708–17

[37] Mullins L J 1954 Some physical mechanisms in narcosis *Chemical Reviews* **54** 289–323

[38] Haydon D A, Hendry B M, Levinson S R and Requena J 1977 The molecular mechanisms of anaesthesia *Nature* **268** 356–8

[39] Mitchell D C, Lawrence J T R and Litman B J 1996 Primary alcohols modulate the activation of the G protein-coupled receptor rhodopsin by a lipid-mediated mechanism *J. Biol. Chem.* **271** 19033–6

[40] Pringle M J, Brown K B and Miller K W 1981 Can the lipid theories of anesthesia account for the cutoff in anesthetic potency in homologous series of alcohols? *Mol. Pharmacol.* **19** 49–55 https://molpharm.aspetjournals.org/content/19/1/49

[41] Godden E L, Harris R A and Dunwiddie T V 2001 Correlation between molecular volume and effects of n-alcohols on human neuronal nicotinic acetylcholine receptors expressed in xenopus oocytes *J. Pharmacol. Exp. Ther.* **296** 716–22 https://jpet.aspetjournals.org/content/296/3/716.long

[42] Kubo I, Muroi H and Kubo A 1995 Structural functions of antimicrobial long-chain alcohols and phenols *Bioorg. Med. Chem.* **3** 873–80

[43] Balgavý P and Devínsky F 1996 Cut-off effects in biological activities of surfactants *Adv. Colloid Interface Sci.* **66** 23–63

[44] Ingólfsson Helgi I and Andersen Olaf S 2011 Alcohol's effects on lipid bilayer properties *Biophys. J.* **101** 847–55

[45] Franks N P and Lieb W R 1979 The structure of lipid bilayers and the effects of general anaesthetics: an x-ray and neutron diffraction study *J. Mol. Biol.* **133** 469–500

[46] Trudell J R, Hubbell W L and Cohen E N 1973 The effect of two inhalation anesthetics of the order of spin-labeled phospholipid vesicles *Biochim. Biophys. Acta Biomemb.* **291** 321–7

[47] Ueda I M D P D, Shieh Donald D P D and Eyring H P D 1974 Anesthetic interaction with a model cell membrane: expansion, phase transition, and melting of the lecithin monolayer *Anesthesiology* **41** 217–25

[48] Miller K W and Pang K-Y Y 1976 General anaesthetics can selectively perturb lipid bilayer membranes *Nature* **263** 253–5

[49] Rosenberg P H, Jansson S E and Gripenberg J 1977 Effects of halothane, thiopental, and lidocaine on fluidity of synaptic plasma membranes and artificial phospholipid membranes *Anesthesiology* **46** 322–6

[50] Trudell J R 1977 A unitary theory of anesthesia based on lateral phase separations in nerve membranes [proceedings] *Biophys. J.* **18** 358–9

[51] Ueda I, Hirakawa M, Arakawa K and Kamaya H 1986 Do anesthetics fluidize membranes? *Anesthesiology* **64** 67–71

[52] You K M, Rosenfield C L and Knipple D C 2003 Ethanol tolerance in the yeast *Saccharomyces cerevisiae* is dependent on cellular oleic acid content *Appl. Environ. Microbiol.* **69** 1499–503

[53] Ono J, Fushimi S, Suzuki S, Ameno K, Kinoshita H, Shirakami G and Kabayama K 2018 Effect of the volatile anesthetic agent isoflurane on lateral diffusion of cell membrane proteins *FEBS Open Bio.* **8** 1127–34

[54] Dickinson R P D, White I M B B S, Lieb William R P D and Franks Nicholas P P D 2000 Stereoselective loss of righting reflex in rats by isoflurane *Anesthesiology* **93** 837–43

[55] Franks N P and Lieb W R 1991 Stereospecific effects of inhalational general anesthetic optical isomers on nerve ion channels *Science* **254** 427

[56] Allen J A, Halverson-Tamboli R A and Rasenick M M 2007 Lipid raft microdomains and neurotransmitter signalling *Nat. Rev. Neurosci.* **8** 128–40

[57] Machta B B, Gray E, Nouri M, McCarthy N L C, Gray E M, Miller A L, Brooks N J and Veatch S L 2016 Conditions that stabilize membrane domains also antagonize *n*-alcohol anesthesia *Biophys. J.* **111** 537–45

[58] Wodzinska K, Blicher A and Heimburg T 2009 The thermodynamics of lipid ion channel formation in the absence and presence of anesthetics. BLM experiments and simulations *Soft Matter* **5** 3319–30

[59] Blicher A, Wodzinska K, Fidorra M, Winterhalter M and Heimburg T 2009 The temperature dependence of lipid membrane permeability, its quantized nature, and the influence of anesthetics *Biophys. J.* **96** 4581–91

[60] Heimburg T 2019 Phase transitions in biological membranes *Thermodynamics and Biophysics of Biomedical Nanosystems: Applications and Practical Considerations* ed C Demetzos and N Pippa (Singapore: Springer Singapore) pp 39–61

[61] Woodbury D J 1989 Pure lipid vesicles can induce channel-like conductances in planar bilayers *J. Membr. Biol.* **109** 145–50

[62] Weinrich M and Worcester D L 2018 The actions of volatile anesthetics: a new perspective *Acta Crystall. D* **74** 1169–77

[63] Jin W, Zucker M and Pralle A 2021 Membrane nanodomains homeostasis during propofol anesthesia as function of dosage and temperature *Biochim. Biophys. Acta Biomembr.* **1863** 183511

[64] Fang Z, Sonner J, Laster M J, Ionescu P, Kandel L, Koblin D D, Eger Ii E I and Halsey M J 1996 Anesthetic and convulsant properties of aromatic compounds and cycloalkanes: implications for mechanisms of narcosis *Anesth. Analg.* **83** 1097–104

[65] Koblin D D, Chortkoff B S, Laster M J, Eger E I, Halsey M J and Ionescu P 1994 Polyhalogenated and perfluorinated compounds that disobey the Meyer–Overton hypothesis *Anesth. Analg.* **79** 1043–8

[66] Kelz M B and Mashour G A 2019 The biology of general anesthesia from paramecium to primate *Curr. Biol.* **29** R1199–R210

[67] Nury H, Van Renterghem C, Weng Y, Tran A, Baaden M, Dufresne V, Changeux J-P, Sonner J M, Delarue M and Corringer P-J 2011 X-ray structures of general anaesthetics bound to a pentameric ligand-gated ion channel *Nature* **469** 428–31

[68] Pavel M A, Petersen E N, Wang H, Lerner R A and Hansen S B 2020 Studies on the mechanism of general anesthesia *Proc. Natl Acad. Sci.* **117** 13757–66
[69] Cantor R S 1997 The lateral pressure profile in membranes: a physical mechanism of general anesthesia *Biochemistry* **36** 2339–44
[70] Nelson S C, Neeley S K, Melonakos E D, Bell J D and Busath D D 2012 Fluorescence anisotropy of diphenylhexatriene and its cationic trimethylamino derivative in liquid dipalmitoylphosphatidylcholine liposomes: opposing responses to isoflurane *BMC Biophys.* **5** 5
[71] van den Brink-van der Laan E, Chupin V, Killian J A and de Kruijff B 2004 Small alcohols destabilize the KcsA tetramer via their effect on the membrane lateral pressure *Biochemistry* **43** 5937–42
[72] Fábián B, Sega M, Voloshin V P, Medvedev N N and Jedlovszky P 2017 Lateral pressure profile and free volume properties in phospholipid membranes containing anesthetics *J. Phys. Chem.* B **121** 2814–24
[73] Hantal G, Fabian B, Sega M, Jojart B and Jedlovszky P 2019 Effect of general anesthetics on the properties of lipid membranes of various compositions *Biochim. Biophys. Acta Biomembr.* **1861** 594–609
[74] Terama E, Ollila O H S, Salonen E, Rowat A C, Trandum C, Westh P, Patra M, Karttunen M and Vattulainen I 2008 Influence of ethanol on lipid membranes: from lateral pressure profiles to dynamics and partitioning *J. Phys. Chem.* B **112** 4131–9
[75] Patra M, Salonen E, Terama E, Vattulainen I, Faller R, Lee B W, Holopainen J and Karttunen M 2006 Under the influence of alcohol: the effect of ethanol and methanol on lipid bilayers *Biophys. J.* **90** 1121–35
[76] Brannigan G, LeBard D N, Hénin J, Eckenhoff R G and Klein M L 2010 Multiple binding sites for the general anesthetic isoflurane identified in the nicotinic acetylcholine receptor transmembrane domain *Proc. Natl Acad. Sci.* **107** 14122
[77] Olsen R W and Li G-D 2011 GABAA receptors as molecular targets of general anesthetics: identification of binding sites provides clues to allosteric modulation *Can. J. Anesth.* **58** 206–15
[78] Chiara D C, Jounaidi Y, Zhou X, Savechenkov P Y, Bruzik K S, Miller K W and Cohen J B 2016 General anesthetic binding sites in human $\alpha4\beta3\delta$ γ-aminobutyric acid type a receptors (GABAARs) *J. Biol. Chem.* **291** 26529–39
[79] Kreutzberger A J B, Kiessling V, Liang B, Yang S T, Castle J D and Tamm L K 2017 Asymmetric phosphatidylethanolamine distribution controls fusion pore lifetime and probability *Biophys. J.* **113** 1912–5
[80] Chernomordik L V, Kozlov M M, Melikyan G B, Abidor I G, Markin V S and Chizmadzhev Y A 1985 The shape of lipid molecules and monolayer membrane-fusion *Biochim. Biophys.* **812** 643–55
[81] Takamori S *et al* 2006 Molecular anatomy of a trafficking organelle *Cell* **127** 831–46
[82] Yang S T, Kreutzberger A J B, Lee J, Kiessling V and Tamm L K 2016 The role of cholesterol in membrane fusion *Chem. Phys. Lipids* **199** 136–43
[83] Lang T, Bruns D, Wenzel D, Riedel D, Holroyd P, Thiele C and Jahn R 2001 SNAREs are concentrated in cholesterol-dependent clusters that define docking and fusion sites for exocytosis *EMBO J.* **20** 2202–13
[84] Churchward M A, Rogasevskaia T, Höfgen J, Bau J and Coorssen J R 2005 Cholesterol facilitates the native mechanism of Ca^{2+}-triggered membrane fusion *J. Cell Sci.* **118** 4833

[85] Churchward M A, Rogasevskaia T, Brandman D M, Khosravani H, Nava P, Atkinson J K and Coorssen J R 2008 Specific lipids supply critical negative spontaneous curvature—an essential component of native Ca^{2+}-triggered membrane fusion *Biophys. J.* **94** 3976–86
[86] Tong J, Borbat P P, Freed J H and Shin Y-K 2009 A scissors mechanism for stimulation of SNARE-mediated lipid mixing by cholesterol *Proc. Natl Acad. Sci.* **106** 5141
[87] Kreutzberger Alex J B, Kiessling V and Tamm Lukas K 2015 High cholesterol obviates a prolonged hemifusion intermediate in fast SNARE-mediated membrane fusion *Biophys. J.* **109** 319–29
[88] Stratton B S, Warner J M, Wu Z, Nikolaus J, Wei G, Wagnon E, Baddeley D, Karatekin E and O'Shaughnessy B 2016 Cholesterol increases the openness of SNARE-mediated flickering fusion pores *Biophys. J.* **110** 1538–50
[89] Lee D E, Lew M G and Woodbury D J 2013 Vesicle fusion to planar membranes is enhanced by cholesterol and low temperature *Chem. Phys. Lipids* **166** 45–54
[90] Ge S, White J G and Haynes C L 2010 Critical role of membrane cholesterol in exocytosis revealed by single platelet study *ACS Chem. Biol.* **5** 819–28
[91] Wang N, Kwan C, Gong X, de Chaves E P, Tse A and Tse F W 2010 Influence of cholesterol on catecholamine release from the fusion pore of large dense core chromaffin granules *J. Neurosci.* **30** 3904
[92] Koseoglu S, Love S A and Haynes C L 2011 Cholesterol effects on vesicle pools in chromaffin cells revealed by carbon-fiber microelectrode amperometry *Anal. Bioanal. Chem.* **400** 2963
[93] Harrison N L, Skelly M J, Grosserode E K, Lowes D C, Zeric T, Phister S and Salling M C 2017 Effects of acute alcohol on excitability in the CNS *Neuropharmacology* **122** 36–45
[94] Woodbury D J and Miller C 1990 Nystatin-induced liposome fusion. a versatile approach to ion channel reconstitution into planar bilayers *Biophys. J.* **58** 833–9
[95] Woodbury D J 1999 Building a bilayer model of the neuromuscular synapse *Cell Biochem. Biophys.* **30** 303–29
[96] Ly H V and Longo M L 2004 The influence of short-chain alcohols on interfacial tension, mechanical properties, area/molecule, and permeability of fluid lipid bilayers *Biophys. J.* **87** 1013–33
[97] Wittenberg N J, Zheng L, Winograd N and Ewing A G 2008 Short-chain alcohols promote accelerated membrane distention in a dynamic liposome model of exocytosis *Langmuir* **24** 2637–42
[98] Chanturiya A, Leikina E, Zimmerberg J and Chernomordik L V 1999 Short-chain alcohols promote an early stage of membrane hemifusion *Biophys. J.* **77** 2035–45
[99] Janoff A S, Pringle M J and Miller K W 1981 Correlation of general anesthetic potency with solubility in membranes *Biochim. Biophys. Acta Biomembr.* **649** 125–8

Part III

Physiological systems of exocytosis

Chapter 11

Exocytosis in yeast: major players and mechanisms

Natasha Buwa, Leonora Martínez-Núñez and Mary Munson

Membrane trafficking is a crucial biological phenomenon known to impact various aspects of physiology, including cell polarization, growth, signaling, and migration. It is mediated by the transport of vesicles and their cargo to predetermined sites of delivery, which can either be within the cell (intracellular transport) or outside it (exocytosis). Vesicle transport during exocytosis involves the budding of vesicles by the donor organelle, motor-driven vesicle transport, and vesicle tethering and fusion with the plasma membrane (PM). The core components of the exocytic pathway are conserved from yeasts to mammals, and include the exocyst complex, soluble N-ethylmaleimide-sensitive factor attachment protein receptor (SNARE) fusion machinery, small GTPases, the cytoskeleton, and associated motors. This chapter summarizes the major components of the yeast exocytic machinery, with a special focus on the exocyst complex, SNAREs, and their various regulatory mechanisms.

11.1 Membrane trafficking in the yeast physiology

An enclosure consisting of a thin, two-dimensional phospholipid bilayer membrane separates all cells from the external microenvironment and compartmentalizes biochemical processes. Cellular compartmentalization is a distinguishing feature of eukaryotic cells, and membrane-driven partitioning of cellular contents gives rise to distinct organelles. Regulated transport to and from these compartments, as well as uptake (endocytosis) and export (exocytosis) of extracellular materials is mediated, in large part, by membrane trafficking. The importance of these processes for physiological functions, such as cell division and growth, signaling, cell polarization, and migration, as well as neuronal transmission and specialized functions such as phagocytosis has long been appreciated. The membrane trafficking system comprises the endoplasmic reticulum, the Golgi apparatus, and the endo-lysosomal

doi:10.1088/978-0-7503-3771-7ch11

system, including early, late, and recycling endosomes and lysosomes and the PM [1–4]. Membrane trafficking between these organelles is mediated by rapid, dynamic, and tightly regulated transport of vesicles and their cargo to a predetermined destination inside cells [5, 6]. Key events during vesicle transport include: cargo (proteins and lipids) sorting, the budding of vesicles by the donor organelle, motor-driven vesicle transport, and vesicle tethering and fusion with the acceptor organelle. Vesicle coats such as coat protein complex II (COPII) mediate vesicle budding and cargo sorting, while SNARE proteins mediate vesicle targeting and fusion [7].

The core membrane trafficking machinery is conserved across eukaryotes (from yeast to mammals) [8, 9], with differences observed in specialized processes such as the synapse. An impressive amount of our current knowledge in the membrane trafficking field comes from the use of the budding yeast *Saccharomyces cerevisiae*. *S. cerevisiae* cells are ovoid and proliferate by budding daughter cells; they are genetically and biochemically tractable, and numerous tools, reagents, and mutants are readily available. *S. cerevisiae* cells divide by forming a bud; a daughter cell buds off from the surface of the mother cell. The bud first appears during the G1 phase of the cell cycle and undergoes tip growth followed by spherical growth (uniform growth throughout the bud) until it separates from the mother cell after mitosis. The cell cycle stage can be identified based on the bud size relative to the mother cell. *S. cerevisiae* can propagate as either diploid or haploid cells. *S. cerevisiae* cells spatially and temporally regulate exocytosis in sync with cell growth and the cell cycle. Further, as with most eukaryotic cells, *S. cerevisiae* cells are asymmetric in nature and have a defined cell polarity. This is achieved by a concerted effect of the exocytic machinery, which includes the exocyst complex, SNARE fusion machinery, small GTPases, the cytoskeleton and associated motors, and polarized cargo distribution—making it an excellent model system for studying exocytosis.

Polarized exocytosis in yeast follows several well-defined steps (summarized in figure 11.1); it is initiated by the establishment of a polarity axis that depends on the position of the bud scar from the previous cell cycle [10]. This leads to polarization of the actin cables, which act as tracks for cargo-containing secretory vesicles that are to be transported from a donor compartment to the PM. This transport is rapid and is facilitated by the type V myosin motor protein Myo2 [11]. Once vesicles arrive at the PM, they are recognized by the multisubunit tethering complex (MTC) called the exocyst. The secretory vesicles and the PM are associated with SNARE proteins, namely, the vesicle SNAREs (v-SNAREs) and the target SNAREs (t-SNAREs). Vesicle recognition leads to activation and assembly of the SNARE complex between the v- and t-SNAREs; this is followed by SNARE-mediated fusion of the vesicle and PMs and cargo delivery at the sites of fusion [12]. Each of these components is essential for exocytosis, and interactions between them are necessary to efficiently coordinate and facilitate the exocytic process. This review summarizes the major exocytic components, including the exocyst, the cytoskeleton, and the SNARE fusion machinery, together with many additional regulatory mechanisms in yeast exocytosis, and makes relevant comparisons to exocytosis in other experimental systems.

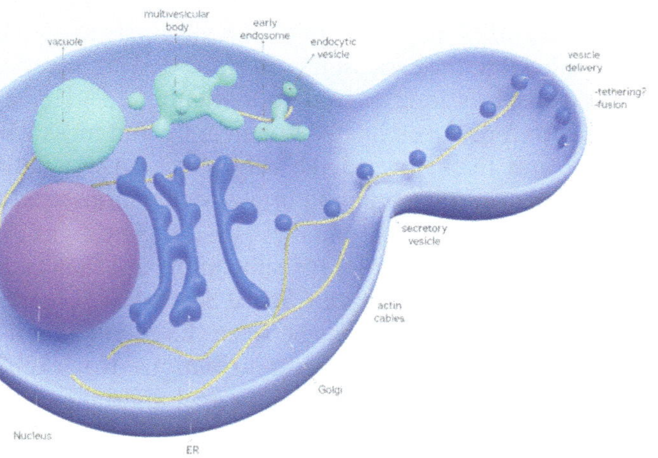

Figure 11.1. Overview of the secretory pathway in yeast: exocytosis and endocytosis. Membrane trafficking, a process primarily mediated by endocytosis and exocytosis, is crucial for physiological functions such as cell division and growth, signaling, and cell polarity. The core membrane trafficking machinery is conserved from yeast to mammals and moves cargoes between the endoplasmic reticulum, Golgi apparatus, endosomes, vacuole/lysosome, and the PM. Membrane trafficking between these organelles is mediated by the tightly regulated vesicle transport of protein and lipid cargoes. The uptake of external materials is mediated by endocytosis (marked in green) and is not the focus of this chapter. The secretion of cargo is mediated by exocytosis (marked in blue) and starts with the definition of a polarity axis and the polarization of actin cables. Actin cables, along with myosin motor proteins, act as tracks for cargo-containing secretory vesicles that are to be transported from a donor compartment to the PM in the bud. Vesicles arriving at the PM are recognized by the exocyst complex. The basic membrane fusion machinery consists of SNARE proteins that are associated with both secretory vesicles and the target PM; vesicle fusion with the PM results in the delivery of cargo.

11.2 The role and regulation of the exocyst in vesicle tethering during exocytosis

The exocyst is an evolutionarily conserved MTC that is thought to mediate the tethering of secretory vesicles to help drive their fusion to the PM. Although tethering complexes have been shown to be essential for membrane fusion and exocytosis *in vivo*, *in vitro* fusion does not require tethers [13]. The tethering of vesicles to bring about membrane fusion may occur using passive mechanisms, in which tethers hold vesicles in place near destination membranes. Alternatively, in an active mechanism, tethers can dynamically engage with, and even regulate the assembly and function of the SNARE fusion machinery [14]. The exocyst plays roles in exocytosis, endocytosis, cell polarization, cell division, migration, and autophagy [15–17]. The exocyst complex is composed of eight subunits: Sec3, Sec5, Sec6, Sec8, Sec10, Sec15, Exo70, and Exo84, which are present as a single copy and in equal stoichiometry in the yeast complex [18]. The components were originally identified in a yeast secretory pathway screen by the Schekman and Novick groups [19–22].

All exocyst subunits (except Sec3 in *S. cerevisiae*) are essential for exocytosis and cell growth [19, 23].

Structure and assembly: negative-stain electron microscopy (EM) and cryo-EM studies have revealed that the exocyst has an elongated architecture, with contiguous helical bundles packed as long curved rods (figure 11.2) [18, 24]. The rods pack against each other to form an elongated architecture that measures ~32 nm long, ~13 nm wide, and ~6 nm thick toward the central region. At its ends, the structure appears to have two arms (extending in opposite directions) and a tail, which could be more flexible than the α-helical bundles [18, 24]. The eight exocyst subunits are present as two four-subunit modules (or subcomplexes) that pack together to form the complex: Sec3-Sec5-Sec6-Sec8 and Sec10-Sec15-Exo70-Exo84. These modules are assembled by pairwise interactions between the eight subunits (Sec3-Sec5, Sec6-Sec-8, and Sec10-Sec15) with weaker contributions from other pairs [18]. The proteins of the multisubunit tethering family, including the exocyst, the conserved oligomeric Golgi (COG), the Golgi-associated retrograde protein (GARP), and Dsl, are composed of long helical bundles (hence, they are known collectively as complexes associated with tethering containing helical rods (CATCHR)) and are structurally similar [25–27]. All the subunits of the yeast exocyst (with the exception of small domains in Sec3 and Exo84) have been shown by cryo-EM to also have a similar fold, with an extended coiled-coil near the N-terminal end, known as the CorEx (core of exocyst) motif and an elongated helical bundle-containing structure towards their C-terminal ends [24]. The Sec3 coiled-coil domain is present near its center, and the N-terminal region is absent from the cryo-EM structure, likely owing to its flexible nature. The Sec6 and Sec15 C-terminal domains are positioned at opposite ends of the exocyst complex [24, 28], potentially allowing the complex to interact with vesicles and the PM simultaneously. A recent study that used subunit crosslinking and integrative models suggested that the Sec3 N-terminal region is localized near Sec6 and its SNARE-binding domains [29]. The CorEx motifs are likely a starting point for the assembly of the two exocyst modules, while the C-terminal domains interact with non-exocyst binding partners [28].

Exocyst complex assembly and stability both rely on its subunit interactions. Each protein in the complex is adjacent to at least four others, and the multivalent combination of the various interactions between the subunits is likely responsible for the stability of the complex [18, 30]. Notably, although mammalian exocyst connectivity is similar to that in yeast, no stable subcomplexes have been detected in yeast [18], whereas mammalian cells have two distinct tetrameric subcomplexes present in dynamic equilibrium with the octamer and monomers [31]. These tetrameric subcomplexes can interact with each other only if they are intact, their association with the PM is retained independently of the other, and disruption of any one subunit disrupts only its cognate subcomplex [31]. Further, the subunits of both subcomplexes localize at PM exocytic sites; tracking of individual secretory vesicles in yeast has revealed that the exocyst residence time at the PM is ~18 s, followed by release upon (and not before) vesicle fusion [32]. In contrast, the mammalian exocyst subunits arrive separately (with their cognate subcomplexes) at the PM at small intervals of ~80 ms and depart from the fusion site ~1.3 s post-fusion (except Sec3, which leaves the site ~0.4 s prior to fusion) [31]. Studies of yeast using single-molecule high-resolution fluorophore colocalization in combination with protein interactions from imaging complexes after translocation (PICT) revealed both the spatial positioning of individual subunits within the exocyst as well as the positioning of exocysts at membrane contact sites. These studies also indicated that ~14 (and not more than ~20) complexes are likely to be present in a ring at the vesicle–PM contact site [30], whereas in mammalian epithelial cells this number may be ~9 [31]. The organization of its subunits may allow the exocyst to position itself at the vesicle–PM interface with minimal interference during the fusion of the two membranes. Interestingly, some mammalian vesicle fusion systems, such as those found at neuronal synapses [33], exhibit kiss-and-run fusion behavior that likely regulates the submillisecond dynamics at these sites. The exocyst complex may determine whether vesicles undergo kiss-and-run or full fusion in mammalian cells and may therefore essentially choose the fate of fusion events [34].

Vesicle tethering/interactions between the PM and vesicles: the exocyst complex interacts with the SNARE machinery, the Sec1/Munc18 (SM) proteins, and the Rho and Rab GTPases to regulate membrane fusion [26, 35–41]. Mutations in exocyst genes disrupt exocytosis, leading to an accumulation of unfused secretory vesicles at sites of active exocytosis, such as yeast buds, indicating that vesicle transport is upstream of exocyst function and is unhindered in these cells [19, 22, 42]. Sec3 is likely the last subunit recruited to the exocyst, and therefore its loss has the least destabilizing effect on exocyst assembly [18]. Sec3, through its pleckstrin homology (PH) domain, interacts with phosphatidylinositol (4,5) bisphosphate (PI(4,5)P2) and localizes to the PM [38, 43]. The exocyst subunit Exo70 also binds PI(4,5)P2 through a polybasic domain [44] or via several other residues at the C-terminal region [45], and can lead to PI(4,5)P2 clustering [46]. The interactions of Sec3 and Exo70 with PI(4,5)P2 may be critical for recognition of exocytic sites on the PM, as well as the targeting of exocysts to these sites. The disruption of PI(4,5)P2 interactions caused by mutations of Sec3 or Exo70 renders the exocyst unable to bind to the PM [43, 45].

In addition to binding PI(4,5)P2 at the PM, Sec3 and Exo70 also interact with the small GTPases Rho1, Rho3, and Cdc42 to regulate exocytosis [47–50]. Further, Sec3 interacts with active, guanine triphosphate (GTP)-bound Cdc42, and Rho1 via the same PH domain, suggesting a likely spatiotemporal and context-dependent regulation of the exocyst. The mechanism responsible for the regulation of the exocyst by Rho GTPases is not well understood; however, it is likely that this occurs by the recruitment of exocyst subunits and/or by local activation (e.g. inducing conformational changes) of the assembled exocyst. Rho1 appears to regulate spatial exocyst recruitment, whereas Rho3 and Cdc42 induce its activation [49, 50]. Structural changes induced by activating Exo70 mutations alter the exocyst structure such that its abilities to tether secretory vesicles and bind the v-SNARE Snc2 *in vitro* are increased, while bypassing the dependence on Rho3/Cdc42 *in vivo* [41, 50]. Sec3 has also been reported to bind, through its PH domain, to the t-SNARE Sso2 to release its auto-inhibition (Yue *et al*, 2017). Furthermore, the binding of Sec3 to Sso2 has been shown to be independent of its binding to Cdc42 or PI(4,5)P2 [51], suggesting that the vesicle tethering and membrane fusion roles of Sec3 are distinct. The Sec6 subunit has also been shown to bind to the t-SNARE Sec9, plus both binary Sec9-Sso1 and ternary Sec9-Sso1-Snc2 complexes [36, 53]. Thus, the exocyst binds various players at the PM including PI(4,5)P2, Rho GTPases, and t-SNAREs, which are proposed to lead to vesicle tethering as well as the initiation of SNARE-complex assembly and membrane fusion (see below).

The tethering of vesicles by the exocyst is further suggested by studies that used ectopic targeting of the complex to non-canonical sites, such as mitochondria. Targeting of the exocyst subunit Sec3 to the mitochondrial membrane, by tagging it with the mitochondrial outer-membrane protein Tom20 in yeast, leads to exocyst relocalization and vesicle association at these non-canonical sites [24, 54]. Similarly, reconstituted mammalian exocyst subcomplexes containing Sec3 and Exo70 have been shown to independently bind PI(4,5)P2-containing membranes, and these interactions were sufficient for exocyst recruitment and tethering [55]. Indeed, the importance of a specific PM lipid signature has recently been highlighted by observations that the Exo70 subunit of the plant exocyst recognizes and binds multiple anionic phospholipids, such as PI(4)P and phosphatidic acid, which are unique to the plant PM [56]. Further, ergosterol and sphingolipids (abundantly found in post-Golgi exocytic vesicles) have been suggested to regulate exocytosis and membrane delivery [57]. A recent study also indicated that lipids, particularly sphingolipids, could regulate exocyst polarity by disrupting Sec3 localization [58]. Lipid contents with increased saturation and ergosterol also markedly affect exocytosis [59].

The exocyst also interacts with secretory vesicles through direct interactions between Sec6 and the v-SNARE Snc1/2 [60]. Indeed, it is likely that Snc2 provides additional sites for binding between Sec6 and the SNARE complex, as the ternary complex Sec9-Sso1-Snc2 has a higher affinity for Sec6 binding than the binary complex Sec9-Sso1 [36]. The exocyst also interacts with several SNARE regulatory proteins. Sec6 binds to the Sec1/Munc18 (SM) family protein Sec1 [23, 35]. Further, Sec6–Sec1 interaction is compatible with Sec6–exocyst assembly, suggesting that upon arrival at the PM, Sec6 may recruit Sec1 for SNARE-complex formation and membrane fusion [35]. A Sec1-interacting protein, Mso1, has also been reported to coprecipitate with the exocyst subunit Sec15 [61]. Together, these exocyst interactions probably provide added layers of stringency and directionality to vesicular transport [41, 60].

Other critical interaction partners of the exocyst are Sro7/77 and the type V myosin motor Myo2. Sro7/77 transiently interacts with the exocyst subunit Exo84 [62], and Sro7 functions as a Rab effector in vesicle tethering [63]. This Sro7 function may be mediated via a pathway parallel to exocyst-mediated tethering [64], as overexpression of Sro7 improves exocytosis and is able to rescue polarity defects caused by Sec3 mutants [62], and Sro7 binds Rab GTPase Sec4, which also binds to the exocyst subunit Sec15 [63]. Further, studies done using an *in vitro* assay with post-Golgi vesicles, purified exocyst, and Sro7 show increased vesicle–vesicle tethering in the presence of both the exocyst and Sro7, suggesting a cooperative function in vesicle tethering [41]. The Myo2 cargo-binding domain is structurally similar to exocyst subunits and has been suggested to 'integrate' with the exocyst to function in exocytosis [17, 65]. The exocyst has also been implicated in interactions with cytoskeletal components. In fission yeast, Sec3 regulates actin cytoskeleton recruitment to the sites of exocytosis and also mediates assembly of the actin network, endocytic actin patches, and actomyosin rings [66]. In mammalian cells, the exocyst subunit Exo70 directly interacts with Arp2/3, which is a branched actin nucleator [67]. This interaction has been shown to be essential for actin branching and lamellipodia formation during cell migration. The exocyst subunits Exo70 and Sec6 also interact with another actin nucleator, WAVE, in mammalian cells [68]. Thus, interactions between the exocyst and the cytoskeleton are mediated by distinct exocyst subunits and may regulate polarity determination.

Regulatory mechanisms—phosphorylation: exocyst-mediated exocytosis plays a crucial role in cell surface expansion leading to cell growth. However, cell cycle progression necessitates cell growth inhibition, which may occur through the inhibition of exocytosis. One of the mechanisms that regulates exocytosis during cell cycle progression is the phosphorylation of exocyst subunits, which can affect exocyst assembly and function. The exocyst subunits Sec3, Sec5, Exo70, and Exo84 have all been reported to be phosphorylated in several of their Ser/Thr residues [69–74]. Phosphorylation sites in exocyst subunits are mainly present in intrinsically disordered regions that are not visible in the cryo-EM structure [24] and also residues of Exo84 in the exocyst core [30]. Additionally, phosphorylation sites in some subunits are present near regions that interact with other subunits, or with non-exocyst partners such as SNAREs, Rabs, and Myo2 [29]. In general, phosphorylation of the subunits has been suggested to enhance assembly of the complex and hence exocytosis. Although subunit phosphorylation might increase exocyst function in mammalian cells, a recent report suggests a more intricate phospho-regulation of Exo70 [75]. Taken together, these studies strongly indicate context- and likely localization-specific regulation of exocyst subunit phosphorylation and exocyst function. Moreover, phosphorylation may regulate the *in vivo* stability of exocyst subunits; studies in yeast indicate that Sec3, which localizes to the PM, undergoes proteasomal degradation upon laser damage to the yeast cell wall/membrane. This in turn leads to exocyst relocalization and the inhibition of exocytosis [76].

Secretory vesicles tethered to the PM need to undergo fusion with the PM to release their contents. Exocytic membrane fusion is facilitated by SNARE-complex assembly, followed by SNARE-mediated membrane fusion. Evidence suggests that the exocyst complex may actively regulate fusion events by regulating the assembly of, and engaging, the SNARE fusion machinery. Figure 11.2 details the fundamental machinery including the exocyst complex and its various interactors in yeast exocytosis. See below for details.

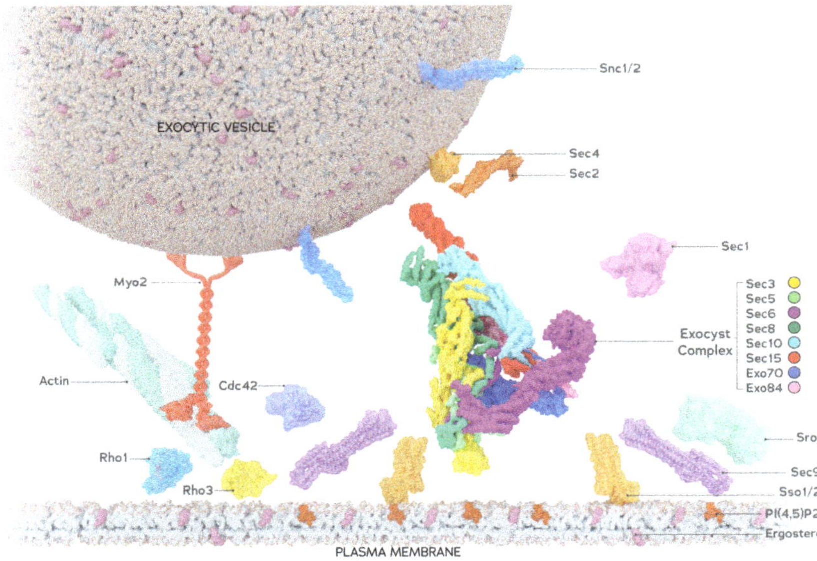

Figure 11.2. 3D architecture/schematic representation of the fundamental machinery of yeast exocytosis: the exocyst complex, SNARE fusion machinery, and actomyosin. Exocytosis, the final step in the secretory pathway, involves the transport and fusion of secretory vesicles arising from the Golgi apparatus with the PM. The secretory vesicles are transported on actin cables, which, along with associated motor proteins such as Myo2, serve as tracks for the transport of secretory vesicles. Rab GTPases such as Sec4 are essential for interaction with Myo2 in order to coordinate transport along the actin cables, as well as for exocyst recruitment. Rho GTPases are essential for actin dynamics and the establishment of polarity. The exocyst, a hetero-octameric complex composed of the subunits Sec3, Sec5, Sec6, Sec8, Sec10, Sec15, Exo70, and Exo84 (as indicated), tethers vesicles to the PM. The subunits pack side by side along their lengths as long curved rods and form an elongated structure as determined by negative-stain EM and cryo-EM studies [18, 24]. The C-terminal domains of Sec6 and Sec15 are positioned at opposite ends of the complex. On vesicles, Sec6 binds the v-SNARE Snc1/2, while Sec15 binds the Rab GTPase Sec4. At the PM, Sec3 and Exo70 can both interact with PI(4,5)P2 and with Rho GTPases, Sec3 can bind the t-SNARE Sso2, and Sec6 can also interact with the SM protein Sec1 and assembled SNARE complexes. Lipids such as PI(4,5)P2 and ergosterol are essential for proper recruitment of the exocyst and exocytosis. The exocyst regulates SNARE-complex assembly to facilitate membrane fusion. It is unclear whether the binding interactions happen simultaneously, or in a particular order, and what conformational changes occur during exocytosis. Most of the identified protein–protein interactions have not been mapped at high resolution. Figure not to scale.

11.3 The role and regulation of SNAREs in vesicle tethering and fusion during exocytosis

The SNARE machinery mediates the fusion of donor vesicle membranes with target organelle membranes. Membrane fusion is an energy-intensive process that needs to overcome the high energy barrier associated with the bilayers. SNARE proteins are used to overcome the activation energy of membrane fusion [77], and are highly regulated to prevent non-specific, potentially detrimental fusion events. SNAREs were identified in three independent approaches by the Rothman, Schekman, and Südhof labs, and much of the knowledge about yeast SNAREs comes from studies of mammalian neuronal (synaptic) SNAREs [78–80], which were found to be homologous to their yeast counterparts. The yeast exocytic SNAREs include the PM-associated Sso1 and Sso2 (homologs of mammalian syntaxin1), Sec9 (a homolog of mammalian SNAP-25), and vesicle-associated Snc1/2 (a homolog of mammalian VAMP1 or synaptobrevin) [81].

SNAREs are typically transmembrane proteins that contain one or two conserved, 60–70 residue, intrinsically disordered SNARE motif(s) that mediate interactions with other SNAREs by forming stable four-helix bundles [81–83]. The transmembrane domains are usually at the C-terminal ends, and several SNAREs are prenylated to bind membranes [84]. The SNARE-motif regions are located N-terminal to the transmembrane domains and form SNARE complexes linking the two membranes, termed SNAREpins [82]. SNARE-motif interactions and the formation of the parallel four-helix bundle are stabilized by 15 layers of hydrophobic residues packed together into the core [85]. The central layer of the core (also called the zero layer) contains arginine and glutamine residues that originate from v- and t-SNAREs, respectively (which are therefore also known as R- or Q-SNAREs), which form ionic interactions. This arrangement may stabilize the SNARE complex, as the hydrophobic layers shield the ionic interactions in the zero layer from the surrounding water and the resultant dielectric weakening, thereby enhancing electrostatic interactions [85, 86]. The asymmetric ionic layer, along with other asymmetric layers, may contribute to the proper alignment/registration of the four helices during complex assembly [86]. This ionic layer has also been shown to be required for efficient SNARE-complex dissociation and disassembly mediated by α-SNAP and N-ethylmaleimide sensitive factor (NSF) [87]. In several syntaxin-type SNAREs, such as in the yeast Sso1/2, an N-terminal three-helix-bundle regulatory domain called the Habc domain folds onto and closes the SNARE motif to prevent premature SNARE assembly [52, 88, 89]. Sec9 has two SNARE motifs and no Habc domain, plus a non-essential N-terminal domain [90]. Activation of the SNAREs involves conformational changes that open this Sso1/2 closed state, making the SNARE motif accessible for binding to other Q-SNAREs; SNARE-complex formation is then completed upon binding of the R-SNARE, which forms the four-helix bundle [85, 91, 92].

The binding of v- and t-SNAREs leads to zippering of the SNARE motifs from the N-terminal (further from the membrane) to the C-terminal end (closer to the membrane), driving the membranes into very close proximity [12]. Studies using optical tweezers have shown that neuronal SNARE-complex assembly occurs through sequential binary steps rather than a continuous process [85, 93]. The distinct steps involve the slow association of N-terminal domains and the stabilization of this intermediate due to repulsive forces produced by membranes in apposition, which is finally followed by rapid sequential zippering of the C-terminal and linker domains [85]. These steps are distinctly regulated by mechanical tension regulation arising from other players (discussed below) and from additional SNARE complexes. The transient N-terminal assembly state is formed at an intermembrane distance of 20 nm, and eventually stabilizes upon SNARE assembly at 8 nm [94]. EM studies have reported that the synaptic SNARE complex forms a 2.5 nm by

15 nm rod [95]. The zippering action and subsequent SNAREpin formation brings the membranes physically close together and provides the energy required to overcome the barrier for bilayer fusion. The activation energy of ~30kBT required for the spontaneous fusion of lipid vesicles is further reduced in the presence of the SNARE fusion machinery [96]. This makes the process more selective and physiologically relevant on the fusion timescale. Further, although only one SNARE complex may be sufficient for fusion [97], the ultrafast, submillisecond dynamics of the process, like those seen in neuronal synapses, result from the concerted action of at least three to six mechanically synchronized SNAREpins [98, 99], and also the fact that the SNAREs in such processes are primed in a partially zippered state [100].

SNARE-complex disassembly after membrane fusion is crucial in order to recycle SNAREs for further rounds of fusion and is mediated by ATP-hydrolyzing NSF and its partner α-SNAP (Sec18 and Sec17, respectively, in yeast). These factors bind the SNARE complex to form a 20S particle [101], which has α-SNAP binding antiparallel to the SNARE complex and NSF binding one end of the SNAP–SNARE complex. The resulting 20S particle is thus reported to measure ~22 nm in length; it is narrow (~6 nm) at one end and wide (~13.5 nm) at the other [95]. Cryo-EM studies have further shown four SNAP proteins to bind one SNARE complex and also bridge it with NSF; the bridge plays a crucial role in the disassembly dynamics [102]. NSF hydrolyzes ATP (through its D1 domain), undergoes conformational changes to tear the SNARE complex apart, and can mediate SNARE-complex disassembly in a single round of ATP hydrolysis [12, 102, 103]. It will be interesting to determine the mechanisms that prevent futile reassembly, or even misassembly, of the complex. In particular, does the syntaxin Habc domain fold back on the SNARE motif immediately after SNARE-complex disassembly? Following SNARE-complex disassembly, the v-SNAREs are recycled back to the donor membrane through endocytosis, while t-SNAREs are reorganized on the acceptor membranes. Figure 11.3 outlines the distinct steps in SNARE-mediated membrane fusion during exocytosis.

SNARE interactions are not specific, and SNARE proteins are usually found in cells in their inactive, non-complexed states [12, 104]. SNARE-complex assembly thus requires regulation via other proteins in cells, as the absence of regulators would cause the process to be slow, as well as potentially inaccurate. Some of the regulators of SNAREs include the SM proteins, tethering factors, and modifications of the SNARE proteins themselves.

Regulatory mechanisms—SM proteins: the SM family of proteins, whose members include Sec1 (yeast)/Munc18 (mammals), Vps45, Sly1, and Vps33, is as essential as the SNARE proteins for membrane fusion [19]. These different SM proteins function at different steps of the intracellular trafficking pathway; Sec1 and Munc18 are the primary exocytic SM proteins in yeast and mammals, respectively. One of the first SM protein structures was reported for Munc18 bound to syntaxin1 [105, 106], which stabilizes the auto-inhibited syntaxin1 conformation. SM proteins may act as clasps that bind the v- and t-SNAREs in zippering SNARE complexes and can organize them spatially and temporally [107, 108] The typical arch-shaped clasp

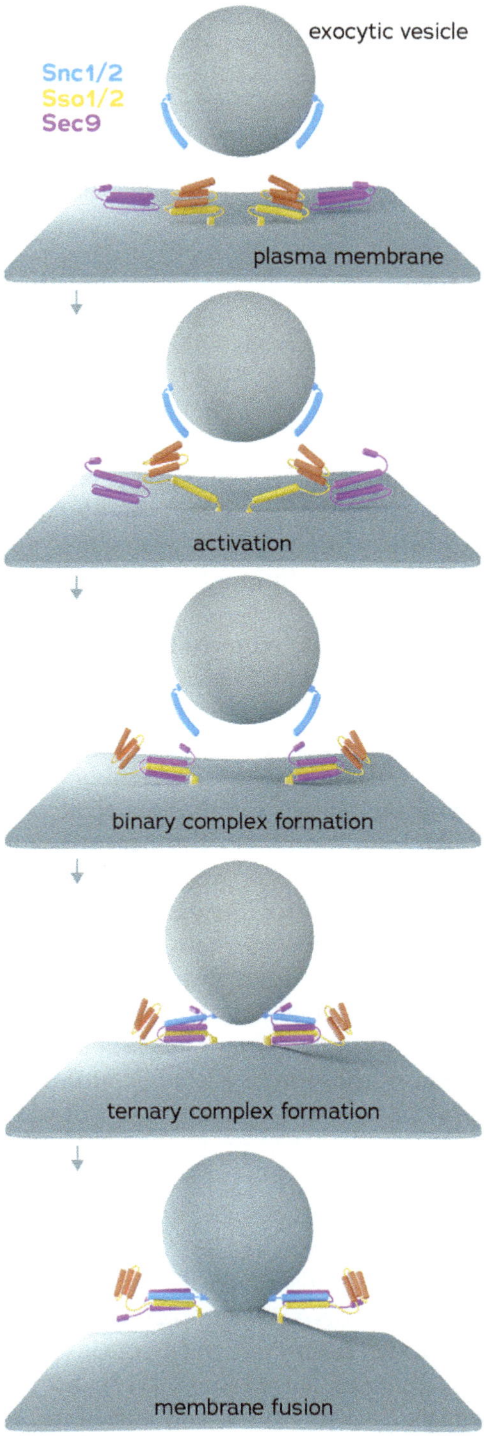

Figure 11.3. The steps of SNARE-mediated membrane fusion during exocytosis in yeast. The secretory vesicles and the PM have distinct SNARE proteins: the vesicle or v-SNAREs (Snc1/2) and the target or t-SNAREs (Sec9 and Sso1/2), respectively. SNARE assembly involves a complex formation of the v- and

t-SNAREs, followed by the fusion of the two membranes and eventual cargo delivery at the sites of fusion. SNARE activation through the exocyst and other regulators involves the release of the Sso1/2 inhibitory Habc domain from its SNARE motif, leading to conformational changes that make the SNARE motif accessible for binding to the other SNAREs. Activation is followed by formation of the binary Sec9-Sso1/2 complex and then the ternary Sec9-Sso1/2-Snc1/2 complex. SNARE-complex formation is completed upon N- to C-terminal zippering of the SNAREs. This zippering action and SNARE-complex formation physically bring the vesicle and target membranes into close proximity, eventually leading to membrane fusion and cargo release. Additional regulators have been omitted for clarity.

structure is contributed by a conserved ~600-residue sequence in SM proteins, particularly for Munc18; the syntaxin Q-SNARE is held in the clasp, and its Habc domain is folded on the SNARE motif [105]. The SNARE motif interacts with the Habc domain as well as Munc18, thus preventing interactions with other SNAREs. SNARE–SM interactions can occur through the N-peptide on syntaxins in addition to the Habc and SNARE motif [107]. Further, the SNARE–SM interactions can facilitate at least two conformations of the syntaxins, open and closed; while Munc18–syntaxin1 has a closed conformation, Vps45 (SM)-Tlg2 (syntaxin) has a more open conformation [106, 109].

Intriguingly, since SM proteins are required for vesicle fusion *in vivo*, it has been proposed that they may catalyze SNARE assembly by enhancing the accuracy and speed of the process [12, 77, 107]. Indeed, it has been shown that several SM proteins can catalyze the zippering of SNAREs [77, 110]. N-terminal cooperative binding of the SM–SNARE complex can induce SNARE assembly when SM binds both the v- and t-SNAREs. The SM protein clasp can thus act to align the different SNAREs and thereby induce SNAREpin formation. Furthermore, studies have shown that following NSF–α-SNAP-mediated disassembly of the syntaxin1–SNAP-25 heterodimers, Munc18 and another regulator, Munc13, are required for SNARE reassembly and fusion [111]. SM proteins can thus not only induce SNARE-complex assembly, but may also resist its premature disassembly. The SM protein binding of an individual syntaxin (Habc-mediated), as opposed to v- and t-SNAREs combined, may be the underlying distinction that determines whether the SM protein prevents or induces SNARE assembly. Accurate vesicle fusion thus requires SM and other regulatory proteins, implying that SM proteins play a crucial role in exocytosis.

The clasp mechanism with which SM proteins bind to closed SNAREs does not appear to be universal, as the yeast exocytic Sec1 protein binds to SNARE complexes and not to the individual SNAREs [112, 113]. Further, Sec1 binds to the complexed SNARE motif regions of the SNARE complex, and this interaction is independent of the Sso1/2 N-peptide or the Habc domain [113]. The interaction between Sec1 and the exocytic SNARE complex may instead be mediated via a groove between domains 1 and 2, opposite the syntaxin-binding cleft of Sec1 [114], which is also present in mammalian Munc18 [115]. Sec1 may also play a role following SNARE-complex assembly [116], and indeed, recombinant purified Sec1 increased the rate of SNARE-mediated liposome fusion in a dose-dependent manner [117]. The Sec1-binding protein Mso1, which also binds the Rab GTPase Sec4, has been shown to help target Sec1 at sites of secretion [118]. Mso1 has been shown to directly interact with the PM [119], and in concert with Sec4 [118] and the exocyst [61], Mso1 can facilitate SNARE-mediated fusion. As mentioned above, Sec1 also interacts with the exocyst complex, although the details of how all these regulators may cooperate to control SNARE assembly and membrane fusion remain to be elucidated.

Regulatory mechanisms—tethers: although *in vitro* fusion does not require tethers [13], tethering complexes have been shown to be essential for membrane fusion *in vivo*. The initial stages of membrane fusion involve the formation of a fusion pore, which can reseal or expand, causing the vesicle to retain its identity or integrate with the PM, respectively. These events cause either kiss-and-run or full fusion of vesicles with the PM. The function of the exocyst in active tethering and full fusion was exemplified in a recent study of mammalian cells [34]. This study showed that only in the presence of an exocyst can vesicles undergo full fusion with the membrane, which an artificial tether is incapable of achieving. Moreover, Exo70, one of the exocyst subunits that interact with the PM, is required for full fusion to occur [34]. It is therefore highly probable that the exocyst actively regulates fusion events by regulating the assembly of, and engaging, the SNARE fusion machinery.

Tethering factors also mediate SNARE-complex formation, as well as disassembly, to regulate fusion. The MTC family, including the exocyst, regulate SNARE/SM assembly and fusion [26, 108]. As described earlier, the exocyst subunit Sec3 binds the t-SNARE Sso2 to promote SNARE-complex assembly and vesicle fusion [51]. Exocyst Sec6 interacts with the SM protein Sec1 [23, 35]. Sec6 also interacts with Snc2 to regulate SNARE-complex localization and assembly [36, 60]. Apart from the exocyst, other MTCs of the CATCHR family, COG and Dsl1, also facilitate SNARE/SM assembly. The COG complex binds the SM protein Sly1, and disrupting this interaction impairs SNARE-complex formation for intra-Golgi and Golgi–endoplasmic reticulum (ER) retrograde transport [120], while the Dsl1 tethering complex binds Habc domains in SNAREs and position the SNAREs for assembly [121]. In an analogous fashion, structural studies have shown that Vps33 (an SM subunit of the non-CATCHR homotypic fusion and protein sorting complex (HOPS)) forms complexes with two SNARE-motif regions (one from each vacuolar membrane) to correctly position them for efficient SNARE assembly [110]. Intriguingly, HOPS and Vps33 also regulate Sec17 (α-SNAP)-mediated SNARE assembly–disassembly dynamics, which in turn regulates fusion [122, 123].

In neurons, the regulatory protein Munc13 contains an MUN domain that is structurally similar to the helical bundle domains of CATCHR MTCs [124], which can facilitate the transition from the inhibited syntaxin1-Munc18 complex to the SNARE complex [125]. Indeed, the MUN domain of Munc13 was recently shown to stabilize the Munc18–syntaxin1–VAMP2 complex, enhancing SNAP-25 binding and inducing SNARE-complex assembly [126–128]. Further, multiple Munc13 supramolecular assemblies can bridge the synaptic vesicle and PMs, a role similar to tethering by MTCs during exocytosis [129, 130]. These data underline the role of MTCs, in conjunction with SM proteins, in SNARE zippering and assembly and in membrane fusion.

Membrane fusion proteins likely function by deforming the outer leaflet of the bilayer, leading to a hemi-fusion structure termed a stalk [131]. Mechanical energy from SNARE zippering has been proposed to drive deformation of the membrane and facilitate the transition from the hemi-fusion state to fusion pore formation [132]. Studies using yeast vacuoles have shown that tethering factors are critical for overcoming the final energy barrier to fusion pore formation presented by SNARE complexes [133]. While SNAREs drive these vacuoles to a hemi-fused state, the binding of HOPS may increase the overall size of SNARE complexes and thus deform the hemi-fusion site, lowering the energy barrier for pore opening and resulting in complete fusion [133].

Regulatory mechanisms—phosphorylation: in addition to SM- and MTC-dependent regulation, post-translational modifications (including phosphorylation) also impact SNARE assembly and function, and can be either inhibitory or stimulatory, depending on the SNARE protein and trafficking stage [134]. Yeast t-SNARE Sso1 is phosphorylated by protein kinase A, resulting in inhibition of SNARE-complex assembly and exocytosis, while dephosphorylation of Sso1 promotes SNARE assembly [135]. Phosphorylation at the N-terminal region of Sso1 can promote its binding to Vsm1, a factor that inhibits SNARE assembly [136]. Phosphorylation of residues in the C-terminal region of the SNARE motif can impair SNARE zippering and fusion [137]. Protein kinase C (PKC)-dependent phosphorylation of v-SNARE VAMP8 at multiple threonine and serine residues in the SNARE motif suppresses fusion [137]. Remarkably, this phosphorylation motif is absent in neuronal VAMPs [137]. Phosphorylation of t-SNARE syntaxin3 leads to decreased binding to Munc18 and suppressed exocytosis in mast cells [138]. Phosphorylation of SNAP-25 by PKC was shown to abrogate binding to other SNAREs, as well as to change the subcellular localization of SNAP-25 [139]. Further, although phosphorylation impacts interactions with other SNAREs, this mechanism could potentially be used to prevent SNARE-complex misassembly and preserve cognate interactions. Phosphorylation of R-SNARE Ykt6 induces a conformational change resulting in a switch from closed cytosolic to open membrane-bound form, altered SNARE interactions, and a block in the secretory pathway [140]. Other potential phosphorylation sites in the N-terminal regions of Q-SNAREs can similarly regulate SNARE assembly [134].

Likewise, tyrosine phosphorylation of the SM protein Munc18 (Tyr473) abrogates fusion and synaptic transmission [141], while PKC-dependent serine phosphorylation enhances neurotransmitter release [142]. In adipocytes, however, insulin-triggered tyrosine phosphorylation of Munc18 stimulates GLUT4 vesicle exocytosis [143]. These Munc18 phosphorylation sites, located in the predicted SNARE-binding regions, can impact SNARE C-terminal zippering or the stability of the intermediate template complex [107].

Regulatory mechanisms—ubiquitination/proteasomal degradation: in addition to phosphorylation, both ubiquitination and palmitoylation have also been reported to affect SNARE protein stability and function. The yeast SNAREs Snc1 and Tlg1 are reported to be palmitoylated by palmitoyltransferase Swf1 at cysteine residues close to the cytoplasmic ends of their transmembrane domains [144]. Interestingly, this palmitoylation protects Tlg1 from mis-sorting and degradation, as unpalmitoylated Tlg1 has the propensity to be ubiquitinated by a ubiquitin ligase and enter the protein quality-control pathway [144]. In contrast, Snc1 polyubiquitination on lysines is required for its COPII-dependent endocytotic recycling post-fusion [145]. This ubiquitination occurs in the transmembrane domain and also decides the fate of Snc1: recycling versus degradation through the multivesicular endosome pathway [146]. Furthermore, ubiquitination and proteasomal degradation also regulate the availability of the synaptic SNARE SNAP-25 for SNARE-complex assembly and synaptic vesicle fusion [147]; this ubiquitination and degradation of SNAP-25 has been implicated in neurodegeneration. Recently, the Habc domain of syntaxin3 was reported to also bind ubiquitin in a non-covalent manner [148], although the way in which this binding might affect Habc-mediated SNARE inhibition or SNARE-complex assembly remains to be determined. These studies suggest a sustained, regulatory role for SNARE ubiquitination, through control of SNARE assembly and endocytotic recycling and degradation, in membrane fusion and exocytosis.

Regulatory mechanisms—membrane tension: the PM is a fluid, 2D structure that enables lateral movement of lipids in the bilayer. One of the properties of the PM that is suggested to regulate membrane trafficking in general, and exocytosis in particular, is membrane tension. Tension is a local, mechanical characteristic of the membrane, which describes the elastic stresses due to stretching and compression along its surface [149]. Lateral movement of lipids generally occurs from an area of low membrane tension to a high-tension area. A variety of biomechanical forces acting on the PM can modulate membrane tension, including (i) differences in hydrostatic pressure between cytosol and the external medium, (ii) interaction with underlying cytoskeletal components, (iii) stretching forces like those experienced during cell migration (front–rear membrane tension gradient [150–152]) and yeast cell growth [153], and (iv) membrane trafficking events, among which, endocytosis increases membrane tension and exocytosis decreases it [149, 154]. Hyper- and hypo-osmotic conditions regulate membrane tension in plant and mammalian cells, and hyperosmotic media have been reported to instantaneously and reversibly decrease the rate of complete fusion of secretory vesicles in mammalian cells, while increasing kiss-and-run events [155, 156]. Higher membrane tension can likely facilitate exocytosis and membrane fusion by increasing exposure of the lipid hydrophobic tails, thereby reducing the energy barrier needed to disrupt bilayer integrity. Further, SNARE proteins can be regulated by membrane tension. These proteins form a complex at the fusion site, generating an inward force that pulls the membranes together [77]. Intriguingly, a minimal machinery composed of SNARE proteins on model membranes shows that the fusion rates of vesicles are better on target membranes that are under high tension [157]. Finally, Rho GTPase-mediated membrane tension has also been suggested to facilitate exocytosis by enlarging the fusion pore [158]. It will be exciting to explore whether the exocyst and/or its subunits, especially Sec3 and Exo70 that bind the PM, are regulated by, or regulate, membrane tension to drive exocytosis.

11.4 Small Rab GTPases

Apart from the exocyst and SNAREs, small GTPases of the Ras superfamily play important roles in the regulation of exocytosis. Small GTPases cycle between GTP-bound (active) and guanosine diphosphate (GDP)-bound (inactive) forms and hence act as molecular switches. This cycling is catalyzed by two sets of proteins; guanine nucleotide exchange factors (GEFs) that promote GTP loading, and GTPase-activating proteins (GAPs), which enable GTP hydrolysis. Another family of proteins termed GDP-dissociation inhibitors (GDIs) prevent the dissociation of GDP, and thus regulate their activation. Activated small GTPase bind cognate effectors in order to regulate varied pathways, and some of the major effectors of Rho, Rab, and related GTPases are components of the cytoskeletal and secretory machineries [159].

On secretory vesicles, the exocyst subunit Sec15 binds the Rab family small GTPase Sec4 in its GTP-bound form [160] and the Sec4 GEF Sec2 [161]. Sec4 interacts with and recruits the exocyst through Sec15, which acts as a Sec4 effector [160]. Sec4 also interacts with the myosin Myo2 to regulate docking and fusion [65, 162]; and with the tomosyn homolog Sro7, which interacts with the exocyst as well as with t-SNAREs [62, 163, 164]. Rab GTPases thus recruit molecular motors, tethering factors, and components of the fusion machinery to regulate the exocytic pathway. Interestingly, Rab pathway cascades are such that an upstream Rab GTPase recruits GEF for a downstream Rab [165]. Thus, upstream Ypt31 Rab recruits GEF Sec2 to secretory vesicles for subsequent activation of Sec4 Rab [166], a cascade which eventually regulates exocytosis.

11.5 The cytoskeleton and small Rho GTPases: roles in exocytosis

Secretory vesicles use cytoskeletal networks such as actin and microtubules, along with their motors and Rho GTPases, for directional transport to their destination. The actin cytoskeleton in *S. cerevisiae* comprises actin filaments or cables, cortical actin, and a contractile actomyosin ring. Cortical actin is present as small, motile patches towards the cell cortex, whereas actin cables are long and present throughout the cytoplasm. The contractile actomyosin ring is largely formed when cells undergo mitotic division and stays through cytokinesis. Cortical actin and actin cables are localized in a polarized manner to sites of bud growth and cell division, while the actomyosin ring is present in the medial region for cell/bud constriction during cytokinesis [167]. Since actin cables span the length of the cell, secretory vesicles as well as organelles use these networks and cortical actin for transport to sites of polarized growth [168]. Polarization requires the establishment of a polarity axis, which is conferred by the cytoskeleton and associated proteins, such as Rho GTPases in eukaryotic cells [11, 40, 169]. In budding yeast, actin cables, along with the type V myosin Myo2 associated with these cables, serve as tracks to directionally transport secretory vesicles containing cargoes such as lipids, PM proteins, and cell wall components for bud tip growth and cell division. Binding to secretory vesicles activates Myo2, and around ten myosin motors are thought to associate with one secretory vesicle *en route* to growth sites [170]. Further, Myo2 is regulated by

Rab GTPases [162] and stays bound to vesicles until exocytosis completes [170]. As discussed earlier, Myo2 has a structural similarity to and interacts with the exocyst through Sec15, potentially yielding a secretory vesicle-tether–cytoskeleton attachment [17, 65]. In addition, most exocyst subunits require an intact actin cytoskeleton to reach sites of exocytosis at the PM [171]. In mammalian systems, actin and actin-binding proteins such as intersectin-IL and MyoVa (a mammalian homolog of Myo2) also directly interact with syntaxin and VAMP2 in the SNARE machinery [172, 173].

Interestingly, the actomyosin complex is recruited to fusion sites and reportedly coats secretory vesicles a few seconds *after* fusion pore opening, as observed in the salivary glands of live mice [174]. A few mechanisms have been proposed for actomyosin-driven maintenance of fusion pores in their open state and the prevention of kiss-and-run exocytosis: (i) actomyosin may provide the driving force for the extrusion or mechanical squeezing of cargo from vesicles through forces derived from its contractility, (ii) the actomyosin network may stabilize vesicles and prevent changes to their size or shape, (iii) actomyosin-generated mechanical forces may drive membrane remodeling by balancing membrane tension at the fusion pores [173, 175, 176]. Myo2-driven transport on actin cables also regulates the retrograde trafficking of Snc1 and Vps10 towards the *trans*-Golgi network [177].

Other studies further exemplify the combined roles of the cytoskeleton and the exocytic machinery in exocytosis. The exocyst subunit Sec3 appears to mediate the assembly of the actin network and its recruitment to sites of exocytosis in fission yeast [66]. In mammalian cells, the exocyst subunits Exo70 and Sec6 interact with the actin nucleators Arp2/3 and WAVE to mediate actin branching and lamellipodia formation [67, 68]. The cytoskeletal component formin Bni3 and Sec3 appear to be simultaneously regulated (by degradation) to redirect the routes and destinations of cargo delivery for cellular wound healing in budding yeast [76]. These data suggest, among other things, that (i) a preexisting polarized cytoskeleton as well as the exocytic machinery both need to be actively disrupted to halt cargo delivery, and (ii) although the cytoskeleton and exocytic machinery converge at one point, it is likely that their initial pathways are independent (as *both* need to be degraded to abrogate exocytosis). The second point is also supported by findings in fission yeast, which suggest that actin cables and the exocyst independently deliver secretory vesicles to sites of polarized growth [178]. Furthermore, in the moss *Physcomitrella patens*, the *Formin1F* gene encodes a very interesting fusion protein product that has an exocyst Sec10 protein at its N-terminus and a Class I formin at the C-terminus [179]. This protein dynamically associates with the exocyst through interactions with Sec6, and actin filaments influence this dynamic association.

Intriguingly, it has been observed in some mammalian systems that cortical actin can act as a physical barrier that prevents vesicle exocytosis and needs to be remodeled to allow vesicle fusion. Actin disruption stimulates insulin secretion in pancreatic β-cells; conversely, the stimulation of exocytosis in neuroendocrine cells induces local disruption of the cortical actin [173]. Actin–syntaxin4 interaction is also reported to diminish the availability of the SNARE and is therefore inhibitory to exocytosis [172]. Regulation of actin polymerization by small GTPases of the Rho and Rab family can further spatially and temporally control exocytosis.

While actin structures are the predominant cytoskeletal elements in exocytosis in *S. cerevisiae*, microtubules also contribute to exocytosis by regulating the polarization of mammalian cells [180]. The exocyst complex has been shown to associate with microtubules to regulate neurite growth and differentiation [181]. A microtubule-associated Rho GEF also interacts directly with Sec5 of the exocyst complex to regulate exocytosis [182]. Microtubule plus-end-associated motor kinesin-1 regulates the microtubule-based transport of exocytic vesicles to the cell periphery, along with the small GTPase Rab6 [183]. An interesting possibility exists in which actin and microtubules interact with each other and influence exocytosis in a controlled manner [184]. Indeed, the scaffolding protein IQGAP1, which binds microtubule plus-end-tracking proteins, has been shown to interact with exocyst subunits Sec3, Sec8, and Exo70 and regulate insulin and matrix metalloproteinase secretion [185–187]. Further, these interactions are mediated by the Rho GTPases Cdc42 and RhoA [187]. Similarly, studies in a moss phragmoplast system showed that the exocyst subunit Sec6 localizes to regions of microtubule overlap at sites of membrane delivery and with a plant SM homolog; this localization is crucial for vesicle fusion during cytokinesis [188].

The components of the actin cytoskeletal machinery are some of the major effectors of Rho GTPases [159]. The Rho GTPases Rho1, Rho3, and Cdc42 play a crucial role in the polarization of the actin cytoskeleton. These small GTPases in their GTP-bound form localize to sites of nascent bud formation and recruit downstream effectors, including the actin-binding machinery and actin nucleators such as formins [189]. In yeast, Rho3 has been implicated in exocytic functions at multiple tiers: regulating actin polarization, Myo2-dependent transport of exocytic vesicles, as well as Exo70-mediated docking and fusion with the PM [190]. As discussed earlier, Sec3 and Exo70, the PM-associated subunits of the exocyst complex, also interact with Rho GTPases [37, 38, 43]. Further, Cdc42-dependent actin polymerization also promotes the docking of secretory vesicles onto actin cables in a PI(4,5)P2-dependent manner [191]. Rho GTPases can not only mediate polarization of the exocyst at sites of active exocytosis, but also stimulate its activity [41, 50]. Additionally, Rho GTPases could regulate fusion via the SNARE fusion machinery [192]. A Rho-associated kinase (ROCK) in neurite cells phosphorylates syntaxin1, inhibiting SNARE-complex assembly and vesicle fusion [193]. The exocyst and SNAREs thus coordinate with actomyosin and their associated regulatory proteins to regulate polarized exocytosis.

11.6 Outstanding questions

The past few decades have seen an unprecedented growth in the field of membrane trafficking, and in particular exocytosis, yet major functional questions remain to be explored. What are the molecular details of the exocyst-dependent tethering of vesicles to the PM? What is the role of the exocyst–Myo2 interaction? What are the roles of PI(4,5)P2 and both Rab and Rho GTPases in these processes? How does the exocyst regulate SNARE assembly and membrane fusion? How and where is the exocyst complex assembled in cells, and is the assembly sequential or hierarchical? How are exocyst complexes activated as needed? What roles do post-translational modifications play in exocytosis? Given that multiple copies of exocyst subunits have been detected at a contact site, do multiple exocyst complexes cooperate to facilitate vesicle fusion with multiple SNARE complexes? Addressing these questions could not only add significantly to our knowledge of how exocytosis is regulated in physiological conditions, but also how it is dysregulated in pathological ones.

References

[1] Schekman R 1985 Protein localization and membrane traffic in yeast *Annu. Rev. Cell Biol.* **1** 115–43

[2] Hicke L and Schekman R 1990 Molecular machinery required for protein transport from the endoplasmic reticulum to the Golgi complex *BioEssays* **12** 253–8

[3] Lippincott-Schwartz J, Roberts T H and Hirschberg K 2000 Secretory protein trafficking and organelle dynamics in living cells *Annu. Rev. Cell Dev. Biol.* **16** 557–89

[4] Muth T R and Caplan M J 2003 Transport protein trafficking in polarized cells *Annu. Rev. Cell Dev. Biol.* **19** 333–66

[5] Pryer N K, Wuestehube L J and Schekman R 1992 Vesicle-mediated protein sorting *Annu. Rev. Biochem.* **61** 471–516

[6] Lee M C S, Miller E A, Goldberg J, Orci L and Schekman R 2004 Bi-directional protein transport between the ER and Golgi *Annu. Rev. Cell Dev. Biol.* **20** 87–123

[7] Bonifacino J S and Glick B S 2004 The mechanisms of vesicle budding and fusion *Cell* **116** 153–66

[8] Bennett M K and Scheller R H 1993 The molecular machinery for secretion is conserved from yeast to neurons *Proc. Natl Acad. Sci. USA* **90** 2559–63

[9] More K, Klinger C M, Barlow L D and Dacks J B 2020 Evolution and natural history of membrane trafficking in eukaryotes *Curr. Biol.* **30** R553–64

[10] Drubin D G 1991 Development of cell polarity in budding yeast *Cell* **65** 1093–6

[11] Pruyne D, Legesse-Miller A, Gao L, Dong Y and Bretscher A 2004 Mechanisms of polarized growth and organelle segregation in yeast *Annu. Rev. Cell Dev. Biol* **20** 559–91

[12] Yoon T Y and Munson M 2018 SNARE complex assembly and disassembly *Curr. Biol.* **28** R397–401

[13] Wickner W and Schekman R 2008 Membrane fusion *Nat. Struct. Mol. Biol.* **15** 658–64

[14] Whyte J R C and Munro S 2002 Vesicle tethering complexes in membrane traffic *J. Cell Sci.* **115** 2627–37

[15] Heider M R and Munson M 2012 Exorcising the exocyst complex *Traffic* **13** 898–907

[16] Martin-Urdiroz M, Deeks M J, Horton C G, Dawe H R and Jourdain I 2016 The exocyst complex in health and disease *Front. Cell Dev. Biol.* **4** 1–22

[17] Mei K and Guo W 2018 The exocyst complex *Curr. Biol.* **28** 922–5

[18] Heider M R, Gu M, Duffy C M, Mirza A M, Marcotte L L and Walls A C *et al* 2016 Subunit connectivity, assembly determinants and architecture of the yeast exocyst complex *Nat. Struct. Mol. Biol.* **23** 59–66

[19] Novick P, Field C and Schekman R 1980 Identification of 23 complementation groups required for post-translational events in the yeast secretory pathway *Cell* **21** 205–15

[20] TerBush D R and Novick P S 6 1995 Sec8, and Secl5 are components of a multisubunit complex which localizes to small bud tips in *Saccharomyces cerevisiae J. Cell Biol.* **130** 299–312
[21] TerBush D R, Maurice T, Roth D and Novick P 1996 The exocyst is a multiprotein complex required for exocytosis in *Saccharomyces cerevisiae EMBO J.* **15** 6483–94
[22] Guo W, Grant A and Novick P 1999 Exo84p is an exocyst protein essential for secretion *J. Biol. Chem.* **274** 23558–64
[23] Wiederkehr A, De Craene J O, Ferro-Novick S and Novick P 2004 Functional specialization within a vesicle tethering complex: bypass of a subset of exocyst deletion mutants by Sec1p or Sec4p *J. Cell Biol.* **167** 875–87
[24] Mei K, Li Y, Wang S, Shao G, Wang J and Ding Y *et al* 2018 Cryo-EM structure of the exocyst complex *Nat. Struct. Mol. Biol.* **25** 139–46
[25] Yu I M and Hughson F M 2010 Tethering factors as organizers of intracellular vesicular traffic *Annu. Rev. Cell Dev. Biol.* **26** 137–56
[26] Dubuke M L and Munson M 2016 The secret life of tethers: the role of tethering factors in SNARE complex regulation *Front. Cell Dev. Biol.* **4** 1–8
[27] Ungermann C and Kümmel D 2019 Structure of membrane tethers and their role in fusion *Traffic* **20** 479–90
[28] Lepore D M, Martínez-Núñez L and Munson M 2018 Exposing the elusive exocyst structure *Trends Biochem. Sci.* **43** 714–25
[29] Ganesan S J, Feyder M J, Chemmama I E, Fang F, Rout M P and Chait B T *et al* 2020 Integrative structure and function of the yeast exocyst complex *Protein Sci.* **29** 1486–501
[30] Picco A, Irastorza-Azcarate I, Specht T, Böke D, Pazos I and Rivier-Cordey A S *et al* 2017 The *in vivo* architecture of the exocyst provides structural basis for exocytosis *Cell* **168** 400–12
[31] Ahmed S M, Nishida-Fukuda H, Li Y, McDonald W H, Gradinaru C C and Macara I G 2018 Exocyst dynamics during vesicle tethering and fusion *Nat. Commun.* **9** 5140
[32] Donovan K W and Bretscher A 2015 Tracking individual secretory vesicles during exocytosis reveals an ordered and regulated process *J. Cell Biol.* **210** 181–9
[33] Aravanis A M, Pyle J L and Tsien R W 2003 Single synaptic vesicles fusing transiently and successively without loss of identity *Nature* **423** 643–7
[34] An S J, Rivera-Molina F, Anneken A, Xi Z, McNellis B and Polejaev V I *et al* 2021 An active tethering mechanism controls the fate of vesicles *Nat. Commun.* **12** 5434
[35] Morgera F, Sallah M R, Dubuke M L, Gandhi P, Brewer D N and Carr C M *et al* 2012 Regulation of exocytosis by the exocyst subunit Sec6 and the SM protein Sec1 *Mol. Biol. Cell.* **23** 337–46
[36] Dubuke M L, Maniatis S, Shaffer S A and Munson M 2015 The exocyst subunit Sec6 interacts with assembled exocytic SNARE complexes *J. Biol. Chem.* **290** 28245–56
[37] Wu H, Rossi G and Brennwald P 2008 The ghost in the machine: small GTPases as spatial regulators of exocytosis *Trends Cell Biol.* **18** 397–404
[38] Yamashita M, Kurokawa K, Sato Y, Yamagata A, Mimura H and Yoshikawa A *et al* 2010 Structural basis for the Rho-and phosphoinositide-dependent localization of the exocyst subunit Sec3 *Nat. Struct. Mol. Biol.* **17** 180–6
[39] Mukherjee D, Sen A and Aguilar R C 2014 Rhogtpase-binding proteins, the exocyst complex and polarized vesicle trafficking *Small GTPases* **5** e983870

[40] Zeng J, Feng S, Wu B and Guo W 2017 Polarized exocytosis *Cold Spring Harb. Perspect. Biol.* **9** 1–16

[41] Rossi G, Lepore D, Kenner L, Czuchra A B, Plooster M and Frost A *et al* 2020 Exocyst structural changes associated with activation of tethering downstream of Rho/Cdc42 GTPases *J. Cell Biol.* **219** 1–15

[42] Finger F P and Novick P 1997 Sec3p is involved in secretion and morphogenesis in *Saccharomyces cerevisiae Mol. Biol. Cell* **8** 647–62

[43] Zhang X, Orlando K, He B, Xi F, Zhang J and Zajac A *et al* 2008 Membrane association and functional regulation of Sec3 by phospholipids and Cdc42 *J. Cell Biol.* **180** 145–58

[44] Liu J, Zuo X, Yue P and Guo W 2007 Phosphatidylinositol 4,5-bisphosphate mediates the targeting of the exocyst to the plasma membrane for exocytosis in mammalian cells *Mol. Biol. Cell* **18** 4483–92

[45] He B, Xi F, Zhang X, Zhang J and Guo W 2007 Exo70 interacts with phospholipids and mediates the targeting of the exocyst to the plasma membrane *EMBO J.* **26** 4053–65

[46] Pleskot R, Cwiklik L, Jungwirth P, Žárský V and Potocký M 2015 Membrane targeting of the yeast exocyst complex *Biochim. Biophys. Acta—Biomembr* **1848** 1481–9

[47] Robinson N G G, Guo L, Imai J, Toh-e A, Matsui Y and Tamanoi F 1999 Rho3 of *Saccharomyces cerevisiae*, which regulates the actin cytoskeleton and exocytosis, is a GTPase which interacts with Myo2 and Exo70 *Mol. Cell. Biol.* **19** 3580–7

[48] Zhang X, Bi E, Novick P, Du L, Kozminski K G and Lipschutz J H *et al* 2001 Cdc42 interacts with the exocyst and regulates polarized secretion *J. Biol. Chem.* **276** 46745–50

[49] Guo W, Tamanoi F and Novick P 2001 Spatial regulation of the exocyst complex by Rho1 GTPase *Nat. Cell Biol.* **3** 353–60

[50] Wu H, Turner C, Gardner J, Temple B and Brennwald P 2010 The Exo70 subunit of the exocyst is an effector for both Cdc42 and Rho3 function in polarized exocytosis *Mol. Biol. Cell* **21** 430–42

[51] Yue P, Zhang Y, Mei K, Wang S, Lesigang J and Zhu Y *et al* 2017 Sec3 promotes the initial binary t-SNARE complex assembly and membrane fusion *Nat. Commun.* **8** 1–12

[52] Munson M, Chen X, Cocina A E, Schultz S M and Hughson F M 2000 Interactions within the yeast t-SNARE Sso1p that control SNARE complex assembly *Nat. Struct. Biol.* **7** 894–902

[53] Sivaram M V S, Saporita J A, Furgason M L M, Boettcher A J and Munson M 2005 Dimerization of the exocyst protein Sec6p and its interaction with the t-SNARE Sec9p *Biochemistry* **44** 6302–11

[54] Luo G, Zhang J and Guo W 2014 The role of Sec3p in secretory vesicle targeting and exocyst complex assembly *Mol. Biol. Cell* **25** 3813–22

[55] Maib H and Murray D H 2021 A mechanism for exocyst-mediated tethering via Arf6 and PIP5K1C driven phosphoinositide conversion *BioRxiv*:2021.10.14.464363

[56] Synek L, Pleskot R, Sekereš J, Serrano N, Vukašinović N and Ortmannová J *et al* 2021 Plasma membrane phospholipid signature recruits the plant exocyst complex via the EXO70A1 subunit *Proc. Natl Acad. Sci. USA* **118** e2105287118

[57] Klemm R W, Ejsing C S, Surma M A, Kaiser H J, Gerl M J and Sampaio J L *et al* 2009 Segregation of sphingolipids and sterols during formation of secretory vesicles at the trans-Golgi network *J. Cell Biol.* **185** 601–12

[58] Guo Q, Zhang T, Meng N, Duan Y, Meng Y and Sun D *et al* 2020 Sphingolipids are required for exocyst polarity and exocytic secretion in *Saccharomyces cerevisiae Cell Biosci.* **10** 1–11

[59] Degreif D, Cucu B, Budin I, Thiel G and Bertl A 2019 Lipid determinants of endocytosis and exocytosis in budding yeast *Biochim. Biophys. Acta—Mol. Cell Biol. Lipids* **1864** 1005–16

[60] Shen D, Yuan H, Hutagalung A, Verma A, Kümmel D and Wu X *et al* 2013 The synaptobrevin homologue Snc2p recruits the exocyst to secretory vesicles by binding to Sec6p *J. Cell Biol.* **202** 509–26

[61] Knop M, Miller K J, Mazza M, Feng D J, Weber M and Keränen S *et al* 2005 Molecular interactions position Mso1p, a novel PTB domain homologue, in the interface of the exocyst complex and the exocytic SNARE machinery in yeast *Mol. Biol. Cell* **16** 4543–56

[62] Zhang X, Wang P, Gangar A, Zhang J, Brennwald P and TerBush D *et al* 2005 Lethal giant larvae proteins interact with the exocyst complex and are involved in polarized exocytosis *J. Cell Biol.* **170** 273–83

[63] Rossi G, Watson K, Kennedy W and Brennwald P 2018 The tomosyn homologue, Sro7, is a direct effector of the Rab GTPase, Sec4, in post-Golgi vesicle tethering *Mol. Biol. Cell* **29** 1476–86

[64] Rossi G, Watson K, Demonch M, Temple B and Brennwald P 2015 *In vitro* reconstitution of Rab GTPase-dependent vesicle clustering by the yeast lethal giant larvae/tomosyn homolog, Sro7 *J. Biol. Chem.* **290** 612–24

[65] Jin Y, Sultana A, Gandhi P, Franklin E, Hamamoto S and Khan A R *et al* 2011 Myosin V transports secretory vesicles via a Rab GTPase cascade and interaction with the exocyst complex *Dev. Cell* **21** 1156–70

[66] Jourdain I, Dooley H C and Toda T 2012 Fission yeast Sec3 bridges the exocyst complex to the actin cytoskeleton *Traffic.* **13** 1481–95

[67] Liu J, Zhao Y, Sun Y, He B, Yang C and Svitkina T *et al* 2012 Exo70 stimulates the Arp2/3 complex for lamellipodia formation and directional cell migration *Curr. Biol* **22** 1510–5

[68] Biondini M, Sadou-Dubourgnoux A, Paul-Gilloteaux P, Zago G, Arslanhan M D and Waharte F *et al* 2016 Direct interaction between exocyst and Wave complexes promotes cell protrusions and motility *J. Cell Sci.* **129** 3756–69

[69] Guo W, Luo G, Luca F C and Zhang J 2013 Mitotic phosphorylation of Exo84 disrupts exocyst assembly and arrests cell growth *J. Cell Biol.* **202** 97–111

[70] Duan Y, Guo Q, Zhang T, Meng Y, Sun D and Luo G *et al* 2019 Cyclin-dependent kinase-mediated phosphorylation of the exocyst subunit Exo84 in late G1 phase suppresses exocytic secretion and cell growth in yeast *J. Biol. Chem.* **294** 11323–32

[71] Tay Y D, Leda M, Spanos C, Rappsilber J, Goryachev A B and Sawin K E 2019 Fission yeast NDR/LATS kinase Orb6 regulates exocytosis via phosphorylation of the exocyst complex *Cell Rep.* **26** 1654–67

[72] Chen X W, Leto D, Xiao J, Goss J, Wang Q and Shavit J A *et al* 2011 Exocyst function regulated by effector phosphorylation *Nat. Cell Biol.* **13** 580–8

[73] Uhm M, Bazuine M, Zhao P, Chiang S H, Xiong T and Karunanithi S *et al* 2017 Phosphorylation of the exocyst protein Exo84 by TBK1 promotes insulin-stimulated GLUT4 trafficking *Sci. Signal* **10** eaah5085

[74] Ren J and Guo W 2012 ERK1/2 regulate exocytosis through direct phosphorylation of the exocyst component Exo70 *Dev. Cell* **22** 967–78

[75] Mao L, Zhan Y, Wu B, Yu Q, Xu L and Hong X *et al* 2020 ULK1 phosphorylates Exo70 to suppress breast cancer metastasis *Nat. Commun.* **11** 117
[76] Kono K, Saeki Y, Yoshida S, Tanaka K and Pellman D 2012 Proteasomal degradation resolves competition between cell polarization and cellular wound healing *Cell.* **150** 151–64
[77] Südhof T C and Rothman J E 2009 Membrane fusion: grappling with SNARE and SM proteins *Science* **323** 474–7
[78] Rizo J and Südhof T C 2012 The membrane fusion enigma: SNAREs, Sec1/Munc18 proteins, and their accomplices guilty as charged? *Annu. Rev. Cell Dev. Biol* **28** 279–308
[79] Söllner T, Whiteheart S W, Brunner M, Erdjument-Bromage H, Geromanos S and Tempst P *et al* 1993 SNAP receptors implicated in vesicle targeting and fusion *Nature* **362** 318–24
[80] Kaiser C A and Schekman R 1990 Distinct sets of SEC genes govern transport vesicle formation and fusion early in the secretory pathway *Cell.* **61** 723–33
[81] Bock J B, Matern H T, Peden A A and Scheller R H 2001 A genomic perspective on membrane compartment organization *Nature* **409** 839–41
[82] Weber T, Zemelman B V, McNew J A, Westermann B, Gmachl M and Parlati F *et al* 1998 SNAREpins: minimal machinery for membrane fusion *Cell.* **92** 759–72
[83] Paumet F, Rahimian V and Rothman J E 2004 The specificity of SNARE-dependent fusion encoded in the SNARE motif *Proc. Natl. Acad. Sci. USA* **101** 3376–80
[84] McNew J A, Søgaard M, Lampen N M, Machida S, Ye R R and Lacomis L *et al* 1997 Ykt6p, a prenylated SNARE essential for endoplasmic reticulum-Golgi transport *J. Biol. Chem.* **272** 17776–83
[85] Gao Y, Zorman S, Gundersen G, Xi Z, Ma L and Sirinakis G *et al* 2012 Single reconstituted neuronal SNARE complexes zipper in three distinct stages *Science* **337** 1340–3
[86] Ossig R, Schmitt H D, De Groot B, Riedel D, Keränen S and Ronne H *et al* 2000 Exocytosis requires asymmetry in the central layer of the SNARE complex *EMBO J.* **19** 6000–10
[87] Scales S J, Yoo B Y and Scheller R H 2001 The ionic layer is required for efficient dissociation of the SNARE complex by α-SNAP and NSF *Proc. Natl Acad. Sci. USA* **98** 14262–7
[88] Dulubova I, Sugita S, Hill S, Hosaka M, Fernandez I and Südhof T C *et al* 1999 A conformational switch in syntaxin during exocytosis: role of munc18 *EMBO J.* **18** 4372–82
[89] Margittai M, Fasshauer D, Jahn R and Langen R 2003 The Habc domain and the SNARE core complex are connected by a highly flexible linker *Biochemistry* **42** 4009–14
[90] Brennwald P, Kearns B, Champion K, Keränen S, Bankaitis V and Novick P 1994 Sec9 is a SNAP-25-like component of a yeast SNARE complex that may be the effector of Sec4 function in exocytosis *Cell* **79** 245–58
[91] Nicholson K L, Munson M, Miller R B, Filip T J, Fairman R and Hughson F M 1998 Regulation of SNARE complex assembly by an N-terminal domain of the t-SNARE Sso1p *Nat. Struct. Biol.* **5** 793–802
[92] Poirier M A, Xiao W, Macosko J C, Chan C, Shin Y K and Bennett M K 1998 The synaptic SNARE complex is a parallel four-stranded helical bundle *Nat. Struct. Biol.* **5** 765–9
[93] Zhang Y 2017 Energetics, kinetics, and pathway of SNARE folding and assembly revealed by optical tweezers *Protein Sci.* **26** 1252–65

[94] Wang Y J, Li F, Rodriguez N, Lafosse X, Gourier C and Perez E *et al* 2016 Snapshot of sequential SNARE assembling states between membranes shows that N-terminal transient assembly initializes fusion *Proc. Natl Acad. Sci. USA* **113** 3533–8

[95] Hohl T M, Parlati F, Wimmer C, Rothman J E, Söllner T H and Engelhardt H 1998 Arrangement of subunits in 20 S particles consisting of NSF, SNAPs, and SNARE complexes *Mol. Cell.* **2** 539–48

[96] François-Martin C, Rothman J E and Pincet F 2017 Low energy cost for optimal speed and control of membrane fusion *Proc. Natl Acad. Sci. USA* **114** 1238–41

[97] Van Den Bogaart G, Holt M G, Bunt G, Riedel D, Wouters F S and Jahn R 2010 One SNARE complex is sufficient for membrane fusion *Nat. Struct. Mol. Biol.* **17** 358–64

[98] Li X, Radhakrishnan A, Grushin K, Kasula R, Chaudhuri A and Gomathinayagam S *et al* 2019 Symmetrical organization of proteins under docked synaptic vesicles *FEBS Lett.* **593** 144–53

[99] Manca F, Pincet F, Truskinovsky L, Rothman J E, Foret L and Caruel M 2019 SNARE machinery is optimized for ultrafast fusion *Proc. Natl Acad. Sci. USA* **116** 21328–9

[100] Yang Y, Shin J Y, Oh J M, Jung C H, Hwang Y and Kim S *et al* 2010 Dissection of SNARE-driven membrane fusion and neuroexocytosis by wedging small hydrophobic molecules into the SNARE zipper *Proc. Natl Acad. Sci. USA* **107** 22145–50

[101] Söllner T, Bennett M K, Whiteheart S W, Scheller R H and Rothman J E 1993 A protein assembly-disassembly pathway *in vitro* that may correspond to sequential steps of synaptic vesicle docking, activation, and fusion *Cell* **75** 409–18

[102] Zhou Q, Huang X, Sun S, Li X, Wang H W and Sui S F 2015 Cryo-EM structure of SNAP-SNARE assembly in 20S particle *Cell Res.* **25** 551–60

[103] Ryu J K, Min D, Rah S H, Kim S J, Park Y and Kim H *et al* 2015 Spring-loaded unraveling of a single SNARE complex by NSF in one round of ATP turnover *Science* **347** 1485–9

[104] Tsui M M K and Banfield D K 2000 Yeast Golgi SNARE interactions are promiscuous *J. Cell Sci.* **113** 145–52

[105] Misura K M S, Scheller R H and Weis W I 2000 Three-dimensional structure of the neuronal-Sec1-syntaxin 1a complex *Nature* **404** 355–62

[106] Burkhardt P, Hattendorf D A, Weis W I and Fasshauer D 2008 Munc18a controls SNARE assembly through its interaction with the syntaxin N-peptide *EMBO J.* **27** 923–33

[107] Zhang Y and Hughson F M 2021 Chaperoning SNARE folding and assembly *Annu. Rev. Biochem.* **90** 581–603

[108] Baker R W and Hughson F M 2016 Chaperoning SNARE assembly and disassembly *Nat. Rev. Mol. Cell Biol.* **17** 465–79

[109] Eisemann T J, Allen F, Lau K, Shimamura G R, Jeffrey P D and Hughson F M 2020 The sec1/munc18 protein vps45 holds the qa-snare tlg2 in an open conformation *Elife* **9** 1–18

[110] Baker R W, Jeffrey P D, Zick M, Phillips B P, Wickner W T and Hughson F M 2015 A direct role for the Sec1/Munc18-family protein Vps33 as a template for SNARE assembly *Science* **349** 1111–4

[111] Ma C, Su L, Seven A B, Xu Y and Rizo J 2013 Reconstitution of the vital functions of Munc18 and Munc13 in neurotransmitter release *Science* **339** 421–5

[112] Carr C M, Grote E, Munson M, Hughson F M and Novick P J 1999 Sec1p binds to SNARE complexes and concentrates at sites of secretion *J. Cell Biol.* **146** 333–44

[113] Togneri J, Cheng Y S, Munson M, Hughson F M and Carr C M 2006 Specific SNARE complex binding mode of the Secl/Munc-18 protein, Seclp *Proc. Natl Acad. Sci. USA* **103** 17730–5
[114] Hashizume K, Cheng Y S, Hutton J L, Chiu C H and Carr C M 2009 Yeast Seclp functions before and after vesicle docking *Mol. Biol. Cell.* **20** 4673–85
[115] Weber-Boyvat M, Chernov K G, Aro N, Wohlfahrt G, Olkkonen V M and Jäntti J 2016 The Secl/Muncl8 protein groove plays a conserved role in interaction with Sec9p/SNAP-25 *Traffic* **17** 131–53
[116] Grote E, Carr C M and Novick P J 2000 Ordering the final events in yeast exocytosis *J. Cell Biol.* **16** 439–52
[117] Scott B L, Van Komen J S, Irshad H, Liu S, Wilson K A and McNew J A 2004 Secl p directly stimulates SNARE-mediated membrane fusion *in vitro J. Cell Biol.* **167** 75–85
[118] Weber-Boyvat M, Aro N, Chernov K G, Nyman T and Jäntti J 2011 . Seclp and Msolp C-terminal tails cooperate with the SNAREs and Sec4p in polarized exocytosis *Mol. Biol. Cell.* **22** 230–44
[119] Weber-Boyvat M, Zhao H, Aro N, Yuan Q, Chernov K and Peränen J *et al* 2013 A conserved regulatory mode in exocytic membrane fusion revealed by Msolp membrane interactions *Mol. Biol. Cell.* **24** 331–41
[120] Laufman O, Kedan A, Hong W and Lev S 2009 Direct interaction between the COG complex and the SM protein, Slyl, is required for Golgi SNARE pairing *EMBO J.* **28** 2006–17
[121] Travis S M, DAmico K, Yu I M, McMahon C, Hamid S and Ramirez-Arellano G *et al* 2020 Structural basis for the binding of SNAREs to the multisubunit tethering complex Dsll *J. Biol. Chem.* **295** 10125–35
[122] Schwartz M L, Nickerson D P, Lobingier B T, Plemel R L, Duan M and Angers C G *et al* 2017 Sec17 (α-SNAP) and an SM-tethering complex regulate the outcome of SNARE zippering *in vitro* and *in vivo Elife* **6** e27396
[123] Song H, Orr A, Duan M, Merz A J and Wickner W 2017 Sec17/sec18 act twice, enhancing membrane fusion and then disassembling cis-SNARE complexes *Elife* **6** e26646
[124] Pei J, Ma C, Rizo J and Grishin N V 2009 Remote homology between Muncl3 MUN domain and vesicle tethering complexes *J. Mol. Biol.* **391** 509–17
[125] Ma C, Li W, Xu Y and Rizo J 2011 Muncl3 mediates the transition from the closed syntaxin-Muncl8 complex to the SNARE complex *Nat. Struct. Mol. Biol.* **18** 542–9
[126] Wang X, Gong J, Zhu L, Wang S, Yang X and Xu Y *et al* 2020 Muncl3 activates the Muncl8-1/syntaxin-1 complex and enables Muncl8-1 to prime SNARE assembly *EMBO J.* **39** e103631
[127] Shu T, Jin H, Rothman J E and Zhang Y 2020 Muncl3-1 MUN domain and Muncl8-1 cooperatively chaperone SNARE assembly through a tetrameric complex *Proc. Natl Acad. Sci. USA* **117** 1036–41
[128] Kalyana Sundaram R V, Jin H, Li F, Shu T, Coleman J and Yang J *et al* 2021 Muncl3 binds and recruits SNAP25 to chaperone SNARE complex assembly *FEBS Lett.* **595** 297–309
[129] Sakamoto H, Ariyoshi T, Kimpara N, Sugao K, Taiko I and Takikawa K *et al* 2018 Synaptic weight set by Muncl3-1 supramolecular assemblies *Nat. Neurosci.* **21** 41–9
[130] Quade B, Camacho M, Zhao X, Orlando M, Trimbuch T and Xu J *et al* 2019 Membrane bridging by muncl3-1 is crucial for neurotransmitter release *Elife* **8** e42806

[131] Chernomordik L V and Kozlov M M 2008 Mechanics of membrane fusion *Nat. Struct. Mol. Biol.* **15** 675–83
[132] Risselada H J and Mayer A 2020 SNAREs, tethers and SM proteins: how to overcome the final barriers to membrane fusion? *Biochem.* J **477** 243–58
[133] Agostino M D, Risselada H J, Lürick A, Ungermann C and Mayer A 2017 A tethering complex drives the terminal stage of SNARE-dependent membrane fusion *Nature* **551** 634–8
[134] Laidlaw K M E, Livingstone R, Al-Tobi M, Bryant N J and Gould G W 2017 SNARE phosphorylation: a control mechanism for insulin-stimulated glucose transport and other regulated exocytic events *Biochem. Soc. Trans.* **45** 1271–7
[135] Marash M and Gerst J E 2001 T-SNARE dephosphorylation promotes SNARE assembly and exocytosis in yeast *EMBO J.* **20** 411–21
[136] Marash M and Gerst J E 2003 Phosphorylation of the autoinhibitory domain of the Sso t-SNAREs promotes binding of the Vsm1 SNARE regulator in yeast *Mol. Biol. Cell.* **14** 3114–25
[137] Malmersjö S, Di Palma S, Diao J, Lai Y, Pfuetzner R A and Wang A L *et al* 2016 Phosphorylation of residues inside the SNARE complex suppresses secretory vesicle fusion *EMBO J.* **35** 1810–21
[138] Tadokoro S, Shibata T, Inoh Y, Amano T, Nakanishi M and Hirashima N *et al* 2016 Phosphorylation of syntaxin-3 at Thr 14 negatively regulates exocytosis in RBL-2H3 mast cells *Cell Biol. Int.* **40** 589–96
[139] Snyder D A, Kelly M L and Woodbury D J 2006 SNARE complex regulation by phosphorylation *Cell Biochem. Biophys.* **45** 111–23
[140] McGrath K, Agarwal S, Tonelli M, Dergai M, Gaeta A L and Shum A K *et al* 2021 A conformational switch driven by phosphorylation regulates the activity of the evolutionarily conserved SNARE Ykt6 *Proc. Natl Acad. Sci. USA* **118** e2016730118
[141] Meijer M, Dörr B, Lammertse H C, Blithikioti C, Weering J R and Toonen R F *et al* 2018 Tyrosine phosphorylation of Munc18-1 inhibits synaptic transmission by preventing SNARE assembly *EMBO J.* **37** 300–20
[142] Genç Ö, Kochubey O, Toonen R F, Verhage M and Schneggenburger R 2014 Munc18-1 is a dynamically regulated PKC target during short-term enhancement of transmitter release *Elife* **3** e01715
[143] Jewell J L, Oh E, Ramalingam L, Kalwat M A, Tagliabracci V S and Tackett L *et al* 2011 Munc18c phosphorylation by the insulin receptor links cell signaling directly to SNARE exocytosis *J. Cell Biol.* **193** 185–99
[144] Valdez-Taubas J and Pelham H 2005 Swf1-dependent palmitoylation of the SNARE Tlg1 prevents its ubiquitination and degradation *EMBO J.* **24** 2524–32
[145] Xu P, Hankins H M, MacDonald C, Erlinger S J, Frazier M N and Diab N S *et al* 2017 COPI mediates recycling of an exocytic SNARE by recognition of a ubiquitin sorting signal *Elife* **6** e28342
[146] Ma M and Burd C G 2019 Retrograde trafficking and quality control of yeast synaptobrevin, Snc1, are conferred by its transmembrane domain *Mol. Biol. Cell.* **30** 1729–42
[147] Sharma M, Burré J and Südhof T C 2012 Proteasome inhibition alleviates SNARE-dependent neurodegeneration *Sci. Transl. Med.* **147** 147ra113
[148] Giovannone A J, Reales E, Bhattaram P, Nackeeran S, Monahan A B and Syed R *et al* 2020 The Habc domain of syntaxin 3 is a ubiquitin binding domain *Sci. Rep.* **10** 21350

[149] Kozlov M M and Chernomordik L V 2015 Membrane tension and membrane fusion *Curr. Opin. Struct. Biol.* **33** 61–7

[150] Veale K J, Offenhäuser C, Whittaker S P, Estrella R P and Murray R Z 2010 Recycling endosome membrane incorporation into the leading edge regulates lamellipodia formation and macrophage migration *Traffic* **11** 1370–9

[151] Lieber A D, Schweitzer Y, Kozlov M M and Keren K 2015 Front-to-rear membrane tension gradient in rapidly moving cells *Biophys. J.* **108** 1599–603

[152] Hetmanski J H R, de Belly H, Busnelli I, Waring T, Nair R V and Sokleva V *et al* 2019 Membrane tension orchestrates rear retraction in matrix-directed cell migration *Dev. Cell* **51** 460–75

[153] Chang F 2017 Forces that shape fission yeast cells *Mol. Biol. Cell.* **28** 1819–24

[154] Wang G and Galli T 2018 Reciprocal link between cell biomechanics and exocytosis *Traffic* **19** 741–9

[155] Wen P J, Grenklo S, Arpino G, Tan X, Liao H S and Heureaux J *et al* 2016 Actin dynamics provides membrane tension to merge fusing vesicles into the plasma membrane *Nat. Commun.* **7** 12604

[156] Wang G, Nola S, Bovio S, Bun P, Coppey-Moisan M and Lafont F *et al* 2018 Biomechanical control of lysosomal secretion via the VAMP7 hub: a tug-of-war between VARP and LRRK1 *iScience* **4** 127–43

[157] Kliesch T T, Dietz J, Turco L, Halder P, Polo E and Tarantola M *et al* 2017 Membrane tension increases fusion efficiency of model membranes in the presence of SNAREs *Sci. Rep.* **7** 12070

[158] Bretou M, Jouannot O, Fanget I, Pierobon P, Larochette N and Gestraud P *et al* 2014 Cdc42 controls the dilation of the exocytotic fusion pore by regulating membrane tension *Mol. Biol. Cell.* **25** 3195–209

[159] Anitei M and Hoflack B 2012 Bridging membrane and cytoskeleton dynamics in the secretory and endocytic pathways *Nat. Cell Biol.* **14** 11–9

[160] Guo W, Roth D, Walch-Solimena C and Novick P 1999 The exocyst is an effector for Sec4P, targeting secretory vesicles to sites of exocytosis *EMBO J.* **15** 1071–80

[161] Medkova M, France Y E, Coleman J and Novick P 2006 The rab exchange factor Sec2p reversibly associates with the exocyst *Mol. Biol. Cell.* **17** 2757–69

[162] Lipatova Z, Tokarev A A, Jin Y, Mulholland J, Weisman L S and Segev N 2008 Direct interaction between a myosin v motor and the Rab GTPases Ypt31/32 is required for polarized secretion *Mol. Biol. Cell.* **19** 4177–87

[163] Lehman K, Rossi G, Adamo J E and Brennwald P 1999 Yeast homologues of tomosyn and lethal giant larvae function in exocytosis and are associated with the plasma membrane SNARE, Sec9 *J. Cell Biol.* **146** 125–40

[164] Grosshans B L, Andreeva A, Gangar A, Niessen S, Yates J R and Brennwald P *et al* 2006 The yeast lgl family member Sro7p is an effector of the secretory Rab GTPase Sec4p *J. Cell Biol.* **172** 55–66

[165] Hutagalung A H and Novick P J 2011 Role of Rab GTPases in membrane traffic and cell physiology *Physiol. Rev.* **91** 119–49

[166] Pfeffer S R 2017 Rab GTPases: master regulators that establish the secretory and endocytic pathways *Mol. Biol. Cell.* **28** 712–5

[167] Akram Z, Ahmed I, Mack H, Kaur R, Silva R C and Castilho B A *et al* 2020 Yeast as a model to understand actin-mediated cellular functions in mammals-illustrated with four actin cytoskeleton proteins *Cells* **9** 1–42

[168] Pruyne D W, Schott D H and Bretscher A 1998 Tropomyosin-containing actin cables direct the Myo2p-dependent polarized delivery of secretory vesicles in budding yeast *J. Cell Biol.* **143** 1931–45

[169] Brennwald P and Rossi G 2007 Spatial regulation of exocytosis and cell polarity: yeast as a model for animal cells *FEBS Lett.* **581** 2119–24

[170] Donovan K W and Bretscher A 2012 Myosin-V is activated by binding secretory cargo and released in coordination with rab/exocyst function *Dev. Cell* **23** 769–81

[171] Boyd C, Hughes T, Pypaert M and Novick P 2004 Vesicles carry most exocyst subunits to exocytic sites marked by the remaining two subunits, Sec3p and Exo70p *J. Cell Biol.* **167** 889–901

[172] Jewell J L, Luo W, Oh E, Wang Z and Thurmond D C 2008 Filamentous actin regulates insulin exocytosis through direct interaction with syntaxin 4 *J. Biol. Chem.* **283** 10716–26

[173] Porat-Shliom N, Milberg O, Masedunskas A and Weigert R 2013 Multiple roles for the actin cytoskeleton during regulated exocytosis *Cell. Mol. Life Sci.* **70** 2099–121

[174] Masedunskas A, Sramkova M, Parente L, Sales K U, Amornphimoltham P and Bugge T H *et al* 2011 Role for the actomyosin complex in regulated exocytosis revealed by intravital microscopy *Proc. Natl Acad. Sci. USA* **108** 13552–7

[175] Nightingale T D, Cutler D F and Cramer L P 2012 Actin coats and rings promote regulated exocytosis *Trends Cell Biol.* **22** 329–37

[176] Ebrahim S, Liu J and Weigert R 2018 The actomyosin cytoskeleton drives micron-scale membrane remodeling *in vivo* via the generation of mechanical forces to balance membrane tension gradients *BioEssays* **40** 1–7

[177] Nguyen V, Smothers J, Ballhorn P, Kottapalli S, Ly A and Villarreal J *et al* 2021 Myosin V-mediated transport of Snc1 and Vps10 toward the trans-Golgi network *Eur. J. Cell Biol.* **100** 151143

[178] Bendezú F O and Martin S G 2011 Actin cables and the exocyst form two independent morphogenesis pathways in the fission yeast *Mol. Biol. Cell.* **22** 44–53

[179] van Gisbergen P A C, Wu S Z, Chang M, Pattavina K A, Bartlett M E and Bezanilla M 2018 An ancient Sec 10-formin fusion provides insights into actin-mediated regulation of exocytosis *J. Cell Biol.* **217** 945–57

[180] Hammer J A and Sellers J R 2012 Walking to work: roles for class v myosins as cargo transporters *Nat. Rev. Mol. Cell Biol.* **13** 13–26

[181] Vega I E and Hsu S C 2001 The Exocyst complex associates with microtubules to mediate vesicle targeting and neurite outgrowth *J. Neurosci.* **21** 3839–48

[182] Pathak R, Delorme-Walker V D, Howell M C, Anselmo A N, White M A and Bokoch G M *et al* 2012 The microtubule-associated rho activating factor gef-h1 interacts with exocyst complex to regulate vesicle traffic *Dev. Cell* **23** 397–411

[183] Grigoriev I, Splinter D, Keijzer N, Wulf P S, Demmers J and Ohtsuka T *et al* 2007 Rab6 regulates transport and targeting of exocytotic carriers *Dev. Cell* **13** 305–14

[184] Müller M T, Schempp R, Lutz A, Felder T, Felder E and Miklavc P 2019 Interaction of microtubules and actin during the post-fusion phase of exocytosis *Sci. Rep.* **9** 1–15

[185] Rittmeyer E N, Daniel S, Hsu S C and Osman M A 2008 A dual role for IQGAP1 in regulating exocytosis *J. Cell Sci.* **121** 391–403

[186] Sakurai-Yageta M, Recchi C, Le Dez G, Sibarita J B, Daviet L and Camonis J *et al* 2008 The interaction of IQGAP1 with the exocyst complex is required for tumor cell invasion downstream of Cdc42 and RhoA *J. Cell Biol.* **181** 985–98

[187] Akhmanova A and Noordstra I 2017 Linking cortical microtubule attachment and exocytosis *F1000Research* **6** 1–12

[188] Tang H, de Keijzer J, Overdijk E J R, Sweep E, Steentjes M and Vermeer J E M *et al* 2019 Exocyst subunit Sec6 is positioned by microtubule overlaps in the moss phragmoplast prior to cell plate membrane arrival *J. Cell Sci.* **132** jcs222430

[189] Dong Y, Pruyne D and Bretscher A 2003 Formin-dependent actin assembly is regulated by distinct modes of Rho signaling in yeast *J. Cell Biol.* **161** 1081–92

[190] Adamo J E, Rossi G and Brennwald P 1999 The Rho GTPase Rho3 has a direct role in exocytosis that is distinct from its role in actin polarity *Mol. Biol. Cell.* **10** 4121–33

[191] Wen P J, Osborne S L, Zanin M, Low P C, Wang H T A and Schoenwaelder S M *et al* 2011 Phosphatidylinositol(4,5)bisphosphate coordinates actin-mediated mobilization and translocation of secretory vesicles to the plasma membrane of chromaffin cells *Nat. Commun.* **2** 491

[192] Croisé P, Eestay-Aahumada C, Gasman S and Ory S 2014 Rho GTPases, phosphoinositides, and actin: a tripartite framework for efficient vesicular trafficking *Small GTPases* **5** e29469

[193] Sakisaka T, Baba T, Tanaka S, Izumi G, Yasumi M and Takai Y 2004 Regulation of SNAREs by tomosyn and ROCK: implication in extension and retraction of neurites *J. Cell Biol.* **166** 17–25

Chapter 12

Stimulus–secretion coupling in the adrenal medulla

Arun Anantharam and Ronald W Holz

The adrenal medulla is a key effector of the sympathetic nervous system in the periphery. It responds to sympathetic stimulation by releasing hormones—including the catecholamines, epinephrine and norepinephrine—into the circulation. This chapter describes important features of the ongoing communication between the sympathetic nerves that innervate the adrenal medulla and the consequent secretory responses of chromaffin cells. The emphasis of the chapter is less on the molecular features of communication than its physiological underpinnings. The adrenal medulla has been a mainstay of research efforts since the early part of the twentieth century, yet there remain glaring gaps in our knowledge of stimulus–secretion coupling in this system. To the extent possible, these knowledge gaps are discussed and potential areas for future study, highlighted.

12.1 Anatomy and innervation

The 'fight or flight' response encapsulates a range of physiological changes that are evolutionarily designed to prepare an individual to respond to a stressor (Goldstein and Kopin 2007, Goldstein 2010). This stressor may take the form of an external, physical threat to one's safety, an emotionally charged event, or metabolic disturbances. Though the precise response to each stressor may vary, all share some basic characteristics of sympathetic activation (Cannon 1940): the heart rate and strength of contraction increase as blood is directed toward the skeletal muscles and the brain and away from non-exigent activities such as digestion (McCorry 2007); respiration is facilitated, and the pupils dilate to improve vision and assess the threat; furthermore, the body mobilizes energy stores to supplement this demanding state through gluconeogenesis in the liver (Tank and Lee Wong 2015). In the periphery, a key effector of the autonomic response to stress and the 'fight or flight' reaction is the adrenal medulla (Cannon 1940).

doi:10.1088/978-0-7503-3771-7ch12

The adrenal glands are situated above the kidneys in the perirenal space (Megha *et al* 2022). 'Medulla' refers to the central region of the adrenal gland, which is surrounded by tissue of distinct mesodermal origin, the adrenal cortex. In contrast to cortical cells, chromaffin cells, which constitute the principal secretory units of the medulla, share, at least in part, a common lineage with sympathetic neurons (Saito *et al* 2012, Furlan *et al* 2017). The secretion produced by the medulla is dependent on input from preganglionic, sympathetic fibers which pass into the gland via the splanchnic nerves (De Robertis and Ferreira 1957, Grynszpan-Winograd 1974). Most preganglionic neurons exit the spinal cord and form cholinergic synapses with noradrenergic neurons, either in the paravertebral sympathetic chain ganglia or other ganglia (e.g. celiac ganglia, mesenteric ganglia, etc.) (Wolf *et al* 2016). The splanchnic efferents that innervate the adrenal medulla, on the other hand, exit the spinal cord, pass through the sympathetic chain ganglia, and then form cholinergic synapses on chromaffin cells (Wolf *et al* 2016). The splanchnic efferents are referred to as 'preganglionic,' whereas the chromaffin cells can be thought of as being 'postganglionic.' The adrenal medulla is an endocrine tissue; catecholamines and peptides are delivered into the interstitial space and then into the circulation through transcytosis across the vascular endothelium.

The close morphological relationship between splanchnic neurons and chromaffin cells is similar in form to the relationship between pre- and postsynaptic cells within the brain (Grynszpan-Winograd 1974). In adrenal slices imaged by electron microscopy, abundant examples of vesicle-laden nerve terminals are visible, and almost every chromaffin cell has its own synaptic contact (figure 12.1). These contacts display features found in synapses throughout the central and peripheral nervous systems, including typical pre- and postsynaptic densities, striking accumulations of small synaptic vesicles, both round and flat, plus variable accumulations of somewhat larger dense-cored synaptic vesicles, and abundant mitochondria. They house

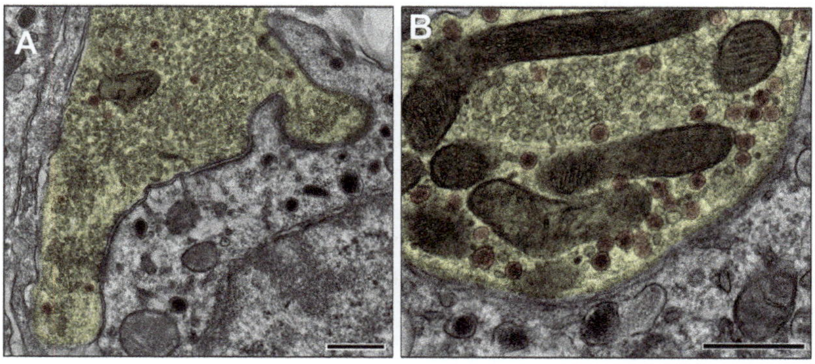

Figure 12.1. Electron micrographs of splanchnic nerve processes in the adrenal medullae of aldehyde-perfused C57BL/6J mice: (A and B) vesicle-filled nerve terminals (highlighted in yellow), appose chromaffin cells (uncolored) across typically 25 nm-wide synaptic clefts, bordered intermittently by pre-and postsynaptic electron densities. The majority of synaptic vesicles are small and electron lucent, but also present in the nerves are variable numbers of larger dense-cored vesicles (DCVs; highlighted in orange), which are demonstrably more abundant in panel (B). Bar, 0.5 μm. Original data/figure kindly provided by Dr John Heuser (NIH).

electron-lucent, pleomorphic vesicles, in addition to some dense-core vesicles (DCVs). DCVs appear to comprise only a small proportion of the total number of vesicles in these splanchnic synapses, and their numbers vary dramatically from one synapse to the next (compare the DCVs in figure 12.1(A) vs. those in figure 12.1(B)). Likewise, the number of mitochondria varies widely from one to the next, as again can be seen by comparing figures 12.1(A) and (B). Careful vesicle-size measurements made using these images indicate a mean of 50 nm for the electron-lucent vesicles, versus a mean of 80 nm for the DCVs—values that are consistent with previously published electron microscopic studies (De Robertis and Ferreira 1957, Grynszpan-Winograd 1974).

Early attempts to describe peripheral autonomic innervation by Hillarp and others led to the generalization that '...each axon has a certain extension, and innervates in its course a certain number of cells, which react as a functional unit, the neuro-effector unit' (Hillarp 1949). Thus, a single nerve fiber likely contacts multiple chromaffin cells; Iijima estimated this number to be 25 (Iijima *et al* 1992), although other estimates have been lower (Kajiwara *et al* 1997). Within the gland, chromaffin cells are organized into readily perceptible 'clusters,' consisting of perhaps 100 cells (Iijima *et al* 1992, Wolf *et al* 2016, Ruiz *et al* 2022). The organization of chromaffin cells into clusters suggests that they might operate in manner analogous to the 'motor unit' of skeletal muscle (Wolf *et al* 2016). Indeed, multi-channel, *in vivo* recordings of chromaffin cell activity in the intact rat adrenal show frequent coupled activity in cells within distances of less than 100 microns—within the size distribution of a cell cluster (Ruiz *et al* 2022). However, networked cellular activity is not solely confined to clusters. Temporal synchronization of chromaffin cell activity, sometimes separated by several hundred microns, is also observed (Ruiz *et al* 2022).

A fundamental property of the adrenal medulla as an effector of the sympathetic stress response is the mutability of its secretory output. Therefore, knowledge of splanchnic discharge rates at rest and during stress is important. In humans, the splanchnic discharge rate has been estimated to be approximately 1 Hz at rest and between 10–20 Hz during full activation (de Diego *et al* 2008). Animal models have also been used to measure changes in discharge rates caused by specific stressors of the sympathetic nervous system (Niijima 1975). For example, in rabbits subjected to IV insulin injections—a stimulus which strongly stimulates epinephrine secretion from the adrenal medulla because of hypoglycemic stress—the firing of splanchnic fibers may increase to 40 Hz from a basal rate of approximately 10 Hz. In rats, the basal firing frequency of splanchnic nerves varies from approximately 2–5 Hz (Morrison and Cao 2000). Direct injection of glutamate into the rostral ventrolateral medulla (i.e. a brain region commanding sympathetic outflow (Strack *et al* 1989, Jansen *et al* 1995)) causes an increase in splanchnic discharge to a value of approximately 10 Hz. Interestingly, these discharge rates appear to depend on whether splanchnic nerves are targeting epinephrine-secreting or norepinephrine-secreting chromaffin cells (Morrison and Cao 2000). Chromaffin cells are thought to mostly produce and secrete either one catecholamine or the other, but usually not both (Cooper *et al* 1992, 1994). One confounding feature of all these experiments is that it is necessary to anaesthetize animals to measure splanchnic activity.

Anesthesia may blunt the 'true' discharge frequency or otherwise distort the rate at which splanchnic action potentials fire.

12.2 Cholinergic transmission in the medulla: the role of nicotinic receptors

The finding that acetylcholine (ACh)—released by splanchnic nerves onto chromaffin cells in the adrenal gland—causes the subsequent release of catecholamines was first made by Feldberg and colleagues (Feldberg *et al* 1934). In this study, it was shown that electrical stimulation of the splanchnic nerve and injections of ACh into the abdominal aortas of anaesthetized cats produced a similar vasopressor effect (Feldberg *et al* 1934). That this effect wholly depended on substances whose release was stimulated by ACh, and not by ACh itself, was demonstrated by the lack of a change in blood pressure when the adrenal gland was removed (Feldberg *et al* 1934).

With the advent of electrophysiological techniques, the electrical properties of dispersed chromaffin cells could be directly assayed (Fenwick *et al* 1982a). These early studies showed that there is a close temporal relationship between ACh application and the appearance of action potentials (Brandt *et al* 1976, Fenwick *et al* 1982a). Later, Parkington and colleagues showed that direct *in situ* stimulation of the splanchnic nerves is similarly associated with the appearance of chromaffin cell action potentials (Holman *et al* 1994). The frequency of these action potentials increases with the frequency of nerve stimulation. Indeed, repetitive stimulation of the guinea pig splanchnic at frequencies of up to 3 Hz produced action potentials in chromaffin cells with no evidence of fatigue or desensitization (Holman *et al* 1994). However, at 30 Hz, and even intermittently at 10 Hz, some stimuli did fail to evoke action potentials (Holman *et al* 1994). Action potentials were absent when splanchnic stimulation was performed in the presence of hexamethonium—a non-depolarizing nicotinic acetylcholine receptor blocker—demonstrating the necessity of nicotinic receptors in the process of action potential generation (Holman *et al* 1994).

Chromaffin cells are thought to express multiple α and β subunits of nicotinic receptors. Messenger RNAs for α3, α5, α7, and β4 have all been identified in the bovine medulla, and the expression of combinations of these subunits in heterologous systems produces currents that are, in some instances, similar to those recorded in the primary cell (Criado 2018). In the human medulla, a greater diversity of subunit subtype messenger RNAs (mRNAs) has been identified (Albillos and McIntosh 2018). But even here, it is the α3, α7, and β4 subunits that appear to be expressed at the highest levels (Albillos and McIntosh 2018). As discussed later in this chapter, the expression of other receptor subunits, including the α9 subunit, may be upregulated in response to stressful conditions (Colomer *et al* 2010).

The predominant chromaffin cell receptor in bovine tissue is thought to be composed of two α3 and three β4 subunits, where channel opening is triggered by binding of two ACh molecules at the orthostatic site (Criado 2018). The nicotinic receptor conducts Na^+, K^+, and Ca^{2+}, although its permeability to Ca^{2+} is probably quite low (Zhou and Neher 1993). ACh-triggered membrane depolarization activates fast, voltage-gated Na^+ (Na_v) channels to initiate the action potential (Fenwick *et al* 1982c). In the guinea pig, mouse, and cow, these Na_vs show a high sensitivity to tetrodotoxin (TTX) (Fenwick *et al* 1982c, Holman *et al* 1994, Vandael *et al* 2015). As a result, action potentials are strongly suppressed when TTX is included in the extracellular solution. Chromaffin cells also express a variety of plasma membrane Ca^{2+} channels (Fenwick *et al* 1982b; Garcia *et al* 2006). These channels open during the action potential and provide an entry pathway for the Ca^{2+} necessary for exocytosis. Curiously, the expression of specific channel subtypes relative to others varies tremendously by organism. In bovine cells, for example, L-type channels carry only 15%–20% of the total Ca^{2+} current, and over 50% of the Ca^{2+} current is carried by P/Q channels. However, in mouse and rat cells, L-type channels carry approximately 50% of the total Ca^{2+} current, and only 15% is carried by P/Q (Garcia *et al* 2006). Presumably, certain characteristics of these channels are important for the fine control of exocytosis in different species.

12.3 Cholinergic transmission in the medulla: the role of muscarinic receptors

Nicotine and ACh have similar effects on the medulla with respect to stimulating secretion. This suggests that the principal action of synaptically released ACh is on nicotinic receptors. However, as reported by Feldberg and then Douglas, muscarine, perfused into the cat adrenal, can also liberate catecholamines (Feldberg *et al* 1934, Douglas and Poisner 1965). We now appreciate that the efficacy of muscarinic agonists as secretagogues is not absolute. In fact, the coupling of muscarinic receptor activation to secretion is highly species-dependent (Olivos and Artalejo 2008). Cells harvested from the bovine adrenal medulla secrete little catecholamine in response to muscarinic stimulation (Fisher *et al* 1981, Cheek and Burgoyne 1985, O'Sullivan and Burgoyne 1989). On the other hand, muscarinic stimulation of cat (Douglas and Poisner 1965) and rat (Guo and Wakade 1994, Guo *et al* 1996) adrenals causes copious release of catecholamines.

Transcripts for five subtypes of muscarinic receptor have been identified in chromaffin cells in various species (M1–M5) (Inoue *et al* 2018). The relative importance of the receptors for Ca^{2+} mobilization and/or secretion depends on the organism in question. Muscarinic stimulation of bovine cells stimulates phospholipase C and increases intracellular Ca^{2+} via a pathway involving IP3 (Fisher *et al* 1981, Olivos and Artalejo 2008). However, no appreciable secretion is observed. Evidently, the amount of Ca^{2+} mobilized in this way is insufficient to elicit exocytosis (Cheek and Burgoyne 1985). On the other hand, stimulation of muscarinic receptors in rat chromaffin cells causes mobilization of Ca^{2+} from caffeine-sensitive intracellular stores to a level that, on its own, enables exocytosis to occur (Guo *et al* 1996). Exocytosis does not require external Ca^{2+}, as evinced by the fact that secretion still occurs when it is absent from the bathing medium (Guo *et al* 1996).

Muscarinic receptor activation has also been linked to changes in membrane potential, resulting in membrane depolarization and even the firing of action potentials (Holman *et al* 1998, Olivos and Artalejo 2008, Inoue *et al* 2012). For example, in a guinea pig adrenal preparation, stimulation of the splanchnic input at 10 Hz causes a slow depolarization that yields action potentials, even in the presence of hexamethonium (Holman *et al* 1998). How does the depolarization occur? Inoue and colleagues have shown that, in dissociated guinea pig chromaffin cells, muscarinic receptors activate non-specific cation channels and inhibit TWIK-related acid sensitive K^+ (TASK) channels (Inoue *et al* 2018). TASK channels have also been shown to be internalized in response to muscarinic receptor stimulation (Matsuoka *et al* 2020) These downstream effects, in which, on the one hand, Na^+ influx is facilitated and on the other, K^+ efflux is inhibited, may underlie the membrane-depolarizing action of muscarine (Olivos and Artalejo 2008).

In the rat adrenal, the situation is more complicated. As previously stated, the evidence is quite clear that for dissociated cells in culture, the exposure of cells to muscarine causes secretion, principally by stimulating the mobilization of Ca^{2+} from intracellular stores (Guo *et al* 1996). However, when internal Ca^{2+} stores are experimentally depleted, muscarine is still capable of eliciting catecholamine release, implicating an alternate pathway for fusion involving external Ca^{2+} (Guo *et al* 1996). The extent to which these results reflect the physiological situation remains unclear. In the 'intact' rat adrenal preparation, secretion caused by splanchnic stimulation is dramatically reduced by ~ 70% in the presence of hexamethonium (Wakade and Wakade 1983). Thus, *in situ*, it appears that chromaffin cell secretion primarily depends on a pathway coupled to nicotinic receptor activation. Interestingly, if ACh is *injected* into the adrenal, hexamethonium has only a minor inhibitory effect on catecholamine output (Wakade and Wakade 1983). The simplest explanation for this observation is that in the intact gland, postsynaptic nicotinic receptors are closely apposed to ACh release sites, and muscarinic receptors are consigned to extrasynaptic locations. This might explain why muscarinic receptors are more readily activated by injected ACh, which likely spreads widely to extrasynaptic regions, than by endogenously released ACh (Wakade and Wakade 1983).

In the end, muscarinic receptors may primarily function to complement the effects of nicotinic receptor activation (Olivos and Artalejo 2008). This is definitely the case in the bovine adrenal medulla, but also may be true in other systems, including the rat (NB: the chicken adrenal appears to depend purely on transmission via muscarinic receptors for chromaffin cell secretion (Knight and Baker 1986)). Quite possibly, under physiological situations associated with strong splanchnic stimulation, muscarinic receptors augment secretion that is coupled to nicotinic receptor activation by increasing intracellular Ca^{2+}, depolarizing the membrane potential, and modulating the release apparatus via signaling proteins such as protein kinase C (PKC) (TerBush *et al* 1988, TerBush and Holz 1990). Overall, the relative importance of muscarinic receptor stimulation in the secretory response is complicated and remains incompletely understood, even today.

12.4 Peptidergic transmission in the medulla: the role of PACAP

We now appreciate that splanchnic neurons house a variety of peptide neurotransmitters which may be co-released with ACh onto chromaffin cells. These include opioid peptides, neuropeptide Y, and pituitary adenylate cyclase-activating polypeptide (PACAP) (Guerineau 2019). The consequences of peptidergic stimulation of chromaffin cells remain poorly understood. In some cases, peptides released by splanchnic processes appear to modulate ongoing chromaffin cell secretion, yet in others, they directly trigger chromaffin cell secretion (Guerineau 2019). This section will focus exclusively on the role of PACAP in stimulating release from chromaffin cells.

PACAP was first described as causing an increase in cyclic adenosine monophosphate (cAMP) in the hypothalamus (Miyata *et al* 1989). Once PACAP became available as a pharmacological agent, it was applied to adrenal medullae as well as dissociated chromaffin cells and was shown to provoke a long-lasting, non-desensitizing release of catecholamines (Watanabe *et al* 1995, Babinski *et al* 1996, Przywara *et al* 1996). Subsequent studies established the necessity of PACAP for sympathetic adjustments to insulin-induced hypoglycemia—a process regulated by adrenomedullary hormones—but evidence that PACAP caused chromaffin cell secretion *in situ* was lacking (Hamelink *et al* 2002, Smith and Eiden, 2012, Eiden *et al* 2018). This evidence was eventually provided by Smith and colleagues, who showed that high-frequency splanchnic stimulation of an adrenal slice fails to elicit catecholamine release when a PAC1 receptor antagonist, PACAP (6–38), is included in the slice-bathing solution (Kuri *et al* 2009, Smith and Eiden 2012).

The notion that that some other 'non-cholinergic substance' released by splanchnic neurons might be responsible for maintaining the tonic release of catecholamines during high rates of neural activity was first articulated by Wakade (Wakade 1988). This was based on experiments showing that catecholamine secretion rapidly desensitized when the splanchnic was stimulated at 10 Hz. Release recovered if the stimulation frequency were subsequently decreased to 3 Hz. The release stimulated by 3 Hz stimulation was insensitive to blockage by both hexamethonium and atropine. The sustained nature of PACAP-dependent catecholamine release is certainly compatible with Wakade's prescient statement.

In comparison to ACh—which causes Na^+ entry through nicotinic receptors to depolarize the membrane potential, initiate action potentials, and eventually a Ca^{2+} influx to trigger catecholamine release—the molecular mechanisms by which PACAP acts to stimulate secretion are poorly understood. The identity of at least part of the PACAP signaling pathway is encoded in its name, pituitary adenylate cyclase-activating polypeptide. Indeed, there is strong evidence that, in multiple cell types, stimulation of the PAC1 receptor leads to activation of adenylate cyclase and cAMP production, leading to an increase in Ca^{2+} (Przywara *et al* 1996, Dickson and Finlayson 2009, Blechman and Levkowitz 2013). However, the PAC1 receptor is also known to couple to $G\alpha_q$ and phospholipase C β (PLCβ), which makes it unclear exactly which Gα and phospholipase C couple receptor activation to exocytosis (Blechman and Levkowitz 2013, Eiden *et al* 2018).

In the pancreatic beta cell, stimulation of Gα_s activates the Epac-Rap-PLCε signaling 'module,' causes Ca^{2+} mobilization from internal stores, and ultimately causes insulin granule exocytosis (Dzhura *et al* 2010, Dzhura *et al* 2011). In chromaffin cells, the direct pharmacological activation of Epac with 8-pCPT-2-O-Me-cAMP—a cAMP analog—causes sustained secretion in a manner analogous to PACAP (Kuri *et al* 2009). This effect is inhibited by $NiCl_2$ which also, incidentally, inhibits PACAP-stimulated secretion (Stroth *et al* 2013). In support of a role for the Epac-Rap-PLCε signaling module in chromaffin cells, evidence (unpublished) from our lab does point to a critical role for PLCε in the pathway that couples PAC1 receptor activation to secretion (figure 12.2). Little PACAP-stimulated increase in Ca^{2+} or exocytosis is evident in mice that lack expression of PLCε.

One important difference between the aforementioned beta cell signaling pathway and the chromaffin cell pathway is that Ca^{2+} entry, rather than mobilization from internal stores, is thought to underlie the secretory response to PACAP in chromaffin cells (Hill *et al* 2011), although this view has not been universally shared (Tanaka *et al* 1996). Smith, Eiden, and colleagues have suggested that PACAP induces Ca^{2+} entry through a low-voltage-activated (LVA), T-type channel that is sensitive to blockage by $NiCl_2$ (Kuri *et al* 2009, Hill *et al* 2011). $NiCl_2$ is proposed to block T-type channels by binding to a site within the permeation path for Ca^{2+} and/or binding to a site on the S3-S4 extracellular domain and stabilizing the channel in the closed state (Lee *et al* 1999, Kang *et al* 2006). However, a role for the

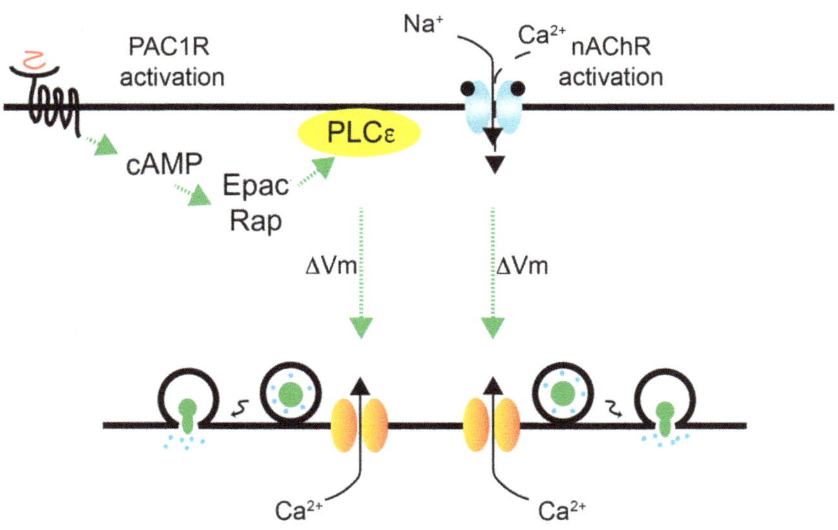

Figure 12.2. Pathways for fusion in the chromaffin cell. *Binding of ACh to nicotinic receptors (nAChRs)* causes cation influx, strong depolarization, the firing of action potentials, and a Ca^{2+} influx through voltage-gated channels that triggers exocytosis. Note that the important contribution of muscarinic receptors is not considered. *Binding of PACAP to PAC1 receptors* causes the activation of adenylate cyclase and cAMP production. Cyclic AMP binds to Epac, which stimulates Rap and PLCε activity. PLCε, likely through a pathway that involves PKC, causes depolarization that opens voltage-gated Ca^{2+} channels. Ca^{2+} enters the cell and exocytosis ensues.

T-type channel has been questioned because T-type currents are barely detectable in the mouse chromaffin cell in which these experiments were performed (Garcia *et al* 2006). Carbone and colleagues showed that T-type currents do emerge in rat chromaffin cells after a multi-day incubation with pharmacological activators of the Epac signaling pathway (Novara *et al* 2004). However, this type of experiment has little relevance to those assessing the effects of acute PACAP stimulation. Therefore, the immediate secretory response to PACAP stimulation may instead involve a Ca^{2+} channel that is more abundantly expressed in the medulla.

12.5 Spontaneous versus evoked activity in the medulla

Recent observations of the electrical properties of chromaffin cells suggest they are more than simple 'relay elements' whose activation depends purely on splanchnic stimulation. Instead, chromaffin cells display intrinsic pacemaking activity that can be modulated in several ways, including pH, membrane depolarization, blockers, and even TTX (Carbone *et al* 2014, Lingle *et al* 2018, Milman *et al* 2021). This activity has been observed in numerous studies performed in cultured cells as well as in adrenal slices. Pacemaking spikes occur at a frequency of approximately 1 Hz in mouse chromaffin cells (Lingle *et al* 2018) and at a much lower frequency in bovine cells (Brandt *et al* 1976). On occasion, spikes transition to slow-wave bursts that ostensibly arise from synchronized depolarizing activity (Lingle *et al* 2018). The mechanisms underlying intrinsic electrical activity, be it spikes or bursts, are not yet resolved. Several reasonable explanations have been proposed based on the biophysical properties of channels known to be active during the inter-spike intervals, including voltage-gated L-type (Cav1.3) and big potassium (BK) channels (Lingle *et al* 2018). Intriguingly, Guerineau and colleagues recently identified a background Na^+ conductance which may be important in setting the resting membrane potential to the relatively depolarized value of between −40 to −55 mV typical of chromaffin cells (Milman *et al* 2021). It is easy to conceive how a persistently active Na^+ conductance, operating at, or near the resting potential, could depolarize the membrane potential towards the threshold for the activation of the Na_v and K_v channels and initiate spiking/bursting phenotypes in the chromaffin cell.

Looming over the notion of intrinsic electrical activity of chromaffin cells is the question of its actual impact on the secretory response. A single action potential in a dissociated mouse chromaffin cell has been estimated to release between one and two granules (Moser and Neher 1997); in adrenal slices, the number might approach seven granules (Moser and Neher 1997). One presumes that the increased levels of intracellular Ca^{2+} which accompany the transition from action potential spikes to bursts would have a much greater impact on granule fusion and catecholamine release.

Pacemaking activity may be a phenomenon that is dependent upon the experimental preparation. Indeed, spontaneous secretory events detected in mouse adrenal slices by cyclic voltammetry are almost completely inhibited by hexamethonium. This indicates that they are stimulated by acetylcholine release from cholinergic nerve terminals present in the preparation and not by spontaneous activity (Petrovic *et al* 2010).

Recent experiments continuously monitored the electrical activity of individual adrenomedullary chromaffin cells in living anesthetized rats using multielectrode arrays to record extracellular action potentials that are generated by chromaffin cells (Ruiz *et al* 2022). Under basal conditions, chromaffin cells fired action potentials with frequencies between 0.2 and 4 Hz. Severing of the splanchnic nerves resulted in almost complete loss of basal activity. This finding supports the view that basal spiking in the intact adrenal medulla with its normal innervation and blood supply largely reflects basal splanchnic nerve stimulation that is regulated by the central nervous system (CNS) and not by activity intrinsic to chromaffin cells. This technique enabled the investigation of medullary function under the control of the CNS at the cellular level under normal and stressed conditions. It revealed networks of neuronally coupled cells with greatly increased action potential frequencies upon hypoglycemic stress induced by insulin shock.

What may be the function *in vivo* of the variety of voltage-sensitive Na^+, Ca^{2+}, and K^+ channels if they are not involved in pacemaking activity? Action potentials are only the first step in the events leading to exocytosis. Catecholamine secretion (measured by amperometry) can be closely coupled to the action potentials, but can also occur asynchronously hundreds of milliseconds after the action potential when the membrane is partially repolarized (Zhou and Misler 1995, Elhamdani *et al* 1998). The rich complement of channels in chromaffin cells may regulate ion flow during and after the action potential, thereby influencing the rise of cytosolic Ca^{2+} and the extent and timing of exocytosis.

12.6 Activity-dependent remodeling of the adrenal medulla

It has been appreciated for many years that the activity-dependent remodeling of synapses is a fundamental feature of neurons in the CNS. The remodeling can be structural as well as functional. Therefore, it is not particularly surprising that the adrenal medulla is also capable of undergoing remodeling. Invariably, the 'goal' of the resultant remodeling is to regulate catecholamine release. Several examples of this have recently emerged in the literature. For example, hypothermia is known to cause a strong sympathetic response and elevated blood epinephrine (Goldstein and Kopin 2007). In animals subjected to cold stress (ambient temperatures of 4 °C for five days), gap junction protein expression in the medulla was upregulated, resulting in increased electrical coupling between chromaffin cells (Colomer *et al* 2008). The increased electrical coupling enhances epinephrine secretion by enabling the rapid propagation of action potentials from one cell to the next (Colomer *et al* 2008). A separate study found α9 nicotinic receptors to be upregulated in the adrenal medulla in response to cold stress. The α9 receptors have a higher permeability to Ca^{2+} than the typical α3β4 receptor and may therefore be favored in conditions in which a greater Ca^{2+}-triggered release is desired (Colomer *et al* 2010). In contrast to cold stress, which modifies the secretory apparatus to produce more secretion, intermittent hypoxic stress downregulates nicotinic receptor expression and consequently reduces catecholamine secretion (Souvannakitti *et al* 2010).

It has been proposed that during 'stressed' states there is an increased synaptic release of PACAP (Eiden *et al* 2018). Accordingly, the perfusion of rat adrenal slices with PACAP for a period of 10 min was shown to enhance the electrical coupling between chromaffin cells (Hill *et al* 2012). The mechanism responsible for the upregulation of gap junction proteins was shown to involve PKC, which notably is also an important signaling metabolite in the PACAP-stimulated signaling pathway for fusion (Kuri *et al* 2009, Hill *et al* 2012).

Functional remodeling can also occur at the level of the synapse. One context in which this happens was recently described by Whim and colleagues (Wang *et al* 2016). The counter-regulatory response describes the series of metabolic changes that underlie an organism's ability to avoid the devastating consequences of hypoglycemia (Holz and Anantharam 2016, Ma *et al* 2018, Wang *et al* 2016). A key feature of the response is elevated sympathetic activity, which, by increasing epinephrine secretion, stimulates hepatic glucose production and the release of gluconeogenic substrates (Wang *et al* 2016). It now appears that fasting stressors promote epinephrine secretion in part by enhancing the probability of presynaptic vesicle exocytosis. Thus, when action potentials invade splanchnic boutons, more neurotransmitter is released into the synapse and a greater amount of epinephrine appears in the circulation. The change in vesicle release probability relies on paracrine signaling between chromaffin cells and splanchnic neurons. Neuropeptide Y (NPY) secreted by (postsynaptic) chromaffin cells diffuses to and binds Y5 NPY receptors on the splanchnic boutons. The activated Y5 receptors then stimulate intracellular events that strengthen synaptic vesicle exocytosis (Wang *et al* 2016). Figure 12.3 summarizes many of the concepts described in this section.

12.7 Concluding perspectives

The chromaffin cell has enjoyed 'favored-model' status for investigations of exocytosis for many decades. It is on account of this system that the basic features of exocytosis, including its Ca^{2+} dependence, its regulation by phospholipids and proteins, and the properties of its fusion pores, are now so well understood (reviewed in (Anantharam and Kreutzberger 2019)). Similar features have subsequently been identified in other secretory cell types, from neurons in the brain to beta cells in the pancreas, signaling the standing of the chromaffin cell as a secretory lodestar. Without diminishing the value of what has been done, it is important to recognize that much of our knowledge comes from work in dissociated cells, in which important regulatory elements, including innervation, blood supply, paracrine, autocrine, and gap junction signaling, are all absent. Thus, while a great deal is known about the intracellular aspects of membrane trafficking and fusion within a chromaffin cell, our knowledge of stimulus–secretion coupling *in situ* is limited. Fortunately, new, powerful techniques are available that can readdress the classical questions that originally motivated research into the function of the adrenal medulla.

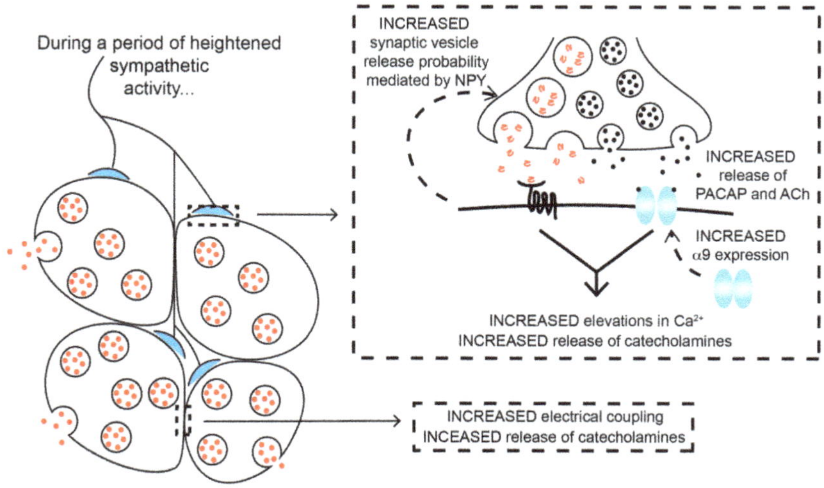

Figure 12.3. Pathways for activity-dependent remodeling of the adrenal medulla. On the left side of diagram, adrenal chromaffin cells are organized in 'clusters,' in which one nerve fiber innervates multiple cells within that cluster (Iijima *et al* 1992). Chromaffin cells release catecholamines (denoted by red dots) in response to presynaptic input from splanchnic nerves. In response to heightened sympathetic input (i.e. stress), remodeling occurs at the synapses as well as at the level of chromaffin cells. At the synapses, increased activity gives rise to a greater release of transmitters, including PACAP (red squiggles; (Kuri *et al* 2009, Stroth *et al* 2013)) and ACh (black dots). Stress also increases the expression of α9 receptors on chromaffin cells (Colomer *et al* 2010). These changes cause sustained increases in the Ca^{2+} within, and greater release of catecholamines from, chromaffin cells. The release of NPY from chromaffin cells enhances vesicle release probability at terminals by binding to and activating presynaptic Y5 receptors in a paracrine manner (Wang *et al* 2016). This, in turn, further enhances the release of presynaptic neurotransmitters. Heightened sympathetic activity also increases the electrical coupling between chromaffin cells (Colomer *et al* 2008). At least some part of this change is likely to be driven by a PACAP-dependent signaling cascade (Hill *et al* 2012).

Future studies may integrate information about the physiological or metabolic state of an animal with detailed analyses of autonomic synapse function in adrenal slices (Wang *et al* 2016), or, in *ex vivo*, semi-intact adrenal preparations (Wolf *et al* 2016). Importantly, the function of individual chromaffin cells in intact adrenal medullae can now also be studied *in vivo* in anesthetized rats using multielectrode arrays and sophisticated signal-processing techniques that have been developed for studies of electrical activity in the CNS (Ruiz *et al* 2022). It is anticipated that in addition to electrical activity, single-cell catecholamine secretion and perhaps single-granule exocytosis will also be measured using multielectrode array cyclic voltammetry in living animals. On the horizon is the possibility of simultaneously investigating the function of tens to hundreds of individual adrenal chromaffin cells *in situ* under basal and stressed conditions in unanesthetized, freely behaving animals using implantable devices (Sunwoo *et al* 2019). Thus, one can look forward to a deeper understanding of the function of the adrenal medulla that incorporates the biochemical and biophysical insights gained *in vitro* with mechanistic studies in living animals.

Acknowledgements

This chapter comes with the requisite disclaimer that only some of the many critical contributions to the field are discussed herein. The exclusion of important work is unintended. The writing of this chapter was made possible by support provided to A A by NIH grant R01NS122534. The authors thank Dr. John Heuser (NIH) for providing the electron microscopy images shown in figure 12.1.

References

Albillos A and McIntosh J M 2018 Human nicotinic receptors in chromaffin cells: characterization and pharmacology *Pflugers Archiv.* **470** 21–7

Anantharam A and Kreutzberger A J B 2019 Unraveling the mechanisms of calcium-dependent secretion *J. Gen. Physiol.* **151** 417–34

Babinski K *et al* 1996 Pituitary adenylate-cyclase activating polypeptide (PACAP) evokes long-lasting secretion and de novo biosynthesis of bovine adrenal medullary neuropeptides *Neuropeptides* **30** 572–82

Blechman J and Levkowitz G 2013 Alternative splicing of the pituitary adenylate cyclase-activating polypeptide receptor PAC1: mechanisms of fine tuning of brain activity *Front. Endocrinol.* **4** 55

Brandt B L, Hagiwara S, Kidokoro Y and Miyazaki S 1976 Action potentials in the rat chromaffin cell and effects of acetylcholine *J. Physiol.* **263** 417–39

Cannon W B 1940 The adrenal medulla *Bull. NY Acad. Med.* **16** 3–13 https://www.ncbi.nlm.nih.gov/pmc/articles/PMC1911485/

Carbone E, Calorio C and Vandael D H 2014 T-type channel-mediated neurotransmitter release *Pflugers Archiv* **466** 677–87

Cheek T R and Burgoyne R D 1985 Effect of activation of muscarinic receptors on intracellular free calcium and secretion in bovine adrenal chromaffin cells *Biochim. Biophys. Acta* **846** 167–73

Colomer C *et al* 2008 Functional remodeling of gap junction-mediated electrical communication between adrenal chromaffin cells in stressed rats *J. Neurosci.* **28** 6616–26

Colomer C, Olivos-Ore L A, Vincent A, McIntosh J M, Artalejo A R and Guerineau N C 2010 Functional characterization of alpha9-containing cholinergic nicotinic receptors in the rat adrenal medulla: implication in stress-induced functional plasticity *J. Neurosci.* **30** 6732–42

Cooper B R, Jankowski J A, Leszczyszyn D J, Wightman R M and Jorgenson J W 1992 Quantitative determination of catecholamines in individual bovine adrenomedullary cells by reversed-phase microcolumn liquid chromatography with electrochemical detection *Anal. Chem.* **64** 691–4

Cooper B R, Wightman R M and Jorgenson J W 1994 Quantitation of epinephrine and norepinephrine secretion from individual adrenal medullary cells by microcolumn high-performance liquid chromatography *J. Chromatogr. B Biomed. Appl.* **653** 25–34

Criado M 2018 Acetylcholine nicotinic receptor subtypes in chromaffin cells *Pflugers Archiv.* **470** 13–20

de Diego A M, Gandia L and Garcia A G 2008 A physiological view of the central and peripheral mechanisms that regulate the release of catecholamines at the adrenal medulla *Acta Physiol. (Oxf.)* **192** 287–301

De Robertis E and Ferreira. A V 1957 Submicroscopic changes of the nerve endings in the adrenal medulla after stimulation of the splanchnic nerve *J. Biophys. Biochem. Cytol* **3** 611–4

Dickson L and Finlayson K 2009 VPAC and PAC receptors: from ligands to function *Pharmacol. Ther.* **121** 294–316

Douglas W W and Poisner A M 1965 Preferential release of adrenaline from the adrenal medulla by muscarine and pilocarpine *Nature* **208** 1102–3

Dzhura I *et al* 2010 Epac2-dependent mobilization of intracellular Ca(2+) by glucagon-like peptide-1 receptor agonist exendin-4 is disrupted in beta-cells of phospholipase C-epsilon knockout mice *J. Physiol.* **588** 4871–89

Dzhura I *et al* 2011 Phospholipase C-epsilon links Epac2 activation to the potentiation of glucose-stimulated insulin secretion from mouse islets of Langerhans *Islets* **3** 121–8

Eiden L E, Emery A C, Zhang L and Smith C B 2018 PACAP signaling in stress: insights from the chromaffin cell *Pflugers Arch.* **470** 79–88

Elhamdani A, Zhou Z and Artalejo C R 1998 Timing of dense-core vesicle exocytosis depends on the facilitation L-type Ca channel in adrenal chromaffin cells *J. Neurosci.* **18** 6230–40

Feldberg W, Minz B and Tsudzimura H 1934 The mechanism of the nervous discharge of adrenaline *J. Physiol.* **81** 286–304

Fenwick E M, Marty A and Neher E 1982a A patch-clamp study of bovine chromaffin cells and of their sensitivity to acetylcholine *J. Physiol.* **331** 577–97

Fenwick E M, Marty A and Neher E 1982b Sodium and calcium channels in bovine chromaffin cells *J. Physiol.(Lond)* **331** 599–635

Fenwick E M, Marty A and Neher E 1982c Sodium and calcium channels in bovine chromaffin cells *J. Physiol.* **331** 599–635

Fisher S K, Holz R W and Agranoff B W 1981 Muscarinic receptors in chromaffin cell cultures mediate enhanced phospholipid labeling but not catecholamine secretion *J. Neurochem.* **37** 491–7

Furlan A *et al* 2017 Multipotent peripheral glial cells generate neuroendocrine cells of the adrenal medulla *Science* **357** eaal3753

Garcia A G, Garcia-De-Diego A M, Gandia L, Borges R and Garcia-Sancho J 2006 Calcium signaling and exocytosis in adrenal chromaffin cells *Physiol. Rev.* **86** 1093–131

Goldstein D S 2010 Adrenal responses to stress *Cell Mol. Neurobiol* **30** 1433–40

Goldstein D S and Kopin I J 2007 Evolution of concepts of stress *Stress* **10** 109–20

Grynszpan-Winograd O 1974 Adrenaline and noradrenaline cells in the adrenal medulla of the hamster: a morphological study of their innervation *J. Neurocytol.* **3** 341–61

Guerineau N C 2019 Cholinergic and peptidergic neurotransmission in the adrenal medulla: a dynamic control of stimulus-secretion coupling *IUBMB Life* **72** 553–67

Guo X, Przywara D A, Wakade T D and Wakade A R 1996 Exocytosis coupled to mobilization of intracellular calcium by muscarine and caffeine in rat chromaffin cells *J. Neurochem.* **67** 155–62

Guo X and Wakade A R 1994 Differential secretion of catecholamines in response to peptidergic and cholinergic transmitters in rat adrenals *J. Physiol.* **475** 539–45

Hamelink C, Tjurmina O, Damadzic R, Young W S, Weihe E, Lee H W and Eiden L E 2002 Pituitary adenylate cyclase-activating polypeptide is a sympathoadrenal neurotransmitter involved in catecholamine regulation and glucohomeostasis *Proc. Natl Acad. Sci. USA* **99** 461–6

Hill J, Chan S A, Kuri B and Smith C 2011 Pituitary adenylate cyclase-activating peptide (PACAP) recruits low voltage-activated T-type calcium influx under acute sympathetic stimulation in mouse adrenal chromaffin cells *J. Biol. Chem.* **286** 42459–69

Hill J, Lee S K, Samasilp P and Smith C 2012 Pituitary adenylate cyclase-activating peptide enhances electrical coupling in the mouse adrenal medulla *Am. J. Physiol. Cell Physiol.* **303** C257–266

Hillarp N A 1949 The functional organization of the peripheral autonomic innervation *Acta Physiol. Scand.* **17** 120–9

Holman M E, Coleman H A, Tonta M A and Parkington H C 1994 Synaptic transmission from splanchnic nerves to the adrenal medulla of guinea-pigs *J. Physiol.* **478** 115–24

Holman M E, Tonta M A, Coleman H A and Parkington H C 1998 Muscarinic receptor activation in guinea-pig chromaffin cells causes decreased membrane conductance and depolarization *J. Auton. Nerv. Syst.* **68** 140–4

Holz R W and Anantharam A 2016 Food deprivation induces presynaptic plasticity in the autonomic nervous system *Proc. Natl Acad. Sci. USA* **113** 5766–7

Iijima T, Matsumoto G and Kidokoro Y 1992 Synaptic activation of rat adrenal medulla examined with a large photodiode array in combination with a voltage-sensitive dye *Neuroscience* **51** 211–9

Inoue M, Harada K, Matsuoka H, Nakamura J and Warashina A 2012 Mechanisms and roles of muscarinic activation in guinea-pig adrenal medullary cells *Am. J. Physiol. Cell Physiol* **303** C635–644

Inoue M, Matsuoka H, Harada K and Kao L S 2018 Muscarinic receptors in adrenal chromaffin cells: physiological role and regulation of ion channels *Pflugers Archiv.* **470** 29–38

Jansen A S, Nguyen X V, Karpitskiy V, Mettenleiter T C and Loewy A D 1995 Central command neurons of the sympathetic nervous system: basis of the fight-or-flight response *Science* **270** 644–6

Kajiwara R, Sand O, Kidokoro Y, Barish M E and Iijima T 1997 Functional organization of chromaffin cells and cholinergic synaptic transmission in rat adrenal medulla *Jpn. J. Physiol.* **47** 449–64

Kang H W, Park J Y, Jeong S W, Kim J A, Moon H J, Perez-Reyes E and Lee J H 2006 A molecular determinant of nickel inhibition in Cav3.2 T-type calcium channels *J. Biol. Chem.* **281** 4823–30

Knight D E and Baker P F 1986 Observations on the muscarinic activation of catecholamine secretion in the chicken adrenal *Neuroscience* **19** 357–66

Kuri B A, Chan S A and Smith C B 2009 PACAP regulates immediate catecholamine release from adrenal chromaffin cells in an activity-dependent manner through a protein kinase C-dependent pathway *J. Neurochem.* **110** 1214–25

Lee J H, Gomora J C, Cribbs L L and Perez-Reyes E 1999 Nickel block of three cloned T-type calcium channels: low concentrations selectively block alpha1H *Biophys. J.* **77** 3034–42

Lingle C J, Martinez-Espinosa P L, Guarina L and Carbone E 2018 Roles of Na(+), Ca(2+), and K(+) channels in the generation of repetitive firing and rhythmic bursting in adrenal chromaffin cells *Pflugers Archiv.* **470** 39–52

Ma Y, Wang Q, Joe D, Wang M and Whim M D 2018 Recurrent hypoglycemia inhibits the counterregulatory response by suppressing adrenal activity *J. Clin. Invest.* **128** 3866–71

Matsuoka H, Harada K, Mashima K and Inoue M 2020 Muscarinic receptor stimulation induces TASK1 channel endocytosis through a PKC-Pyk2-Src pathway in PC12 cells *Cell Signal* **65** 109434

McCorry L K 2007 Physiology of the autonomic nervous system *Am. J. Pharm. Educ.* **71** 78

Megha R, Wehrle C J, Kashyap S and Leslie S W 2022 *Anatomy, Abdomen and Pelvis, Adrenal Glands (Suprarenal Glands)* (Treasure Island, FL: In StatPearls) https://www.ncbi.nlm.nih.gov/books/NBK482264/

Milman A, Venteo S, Bossu J L, Fontanaud P, Monteil A, Lory P and Guerineau N C 2021 A sodium background conductance controls the spiking pattern of mouse adrenal chromaffin cells in situ *J. Physiol.* **599** 1855–83

Miyata A, Arimura A, Dahl R R, Minamino N, Uehara A, Jiang L, Culler M D and Coy. D H 1989 Isolation of a novel 38 residue-hypothalamic polypeptide which stimulates adenylate cyclase in pituitary cells *Biochem. Biophys. Res. Commun.* **164** 567–74

Morrison S F and Cao W H 2000 Different adrenal sympathetic preganglionic neurons regulate epinephrine and norepinephrine secretion *Am. J. Physiol.* **279** R1763–1775

Moser T and Neher E 1997 Rapid exocytosis in single chromaffin cells recorded from mouse adrenal slices *J. Neurosci.* **17** 2314–23

Niijima A 1975 An electrophysiological study on the regulatory mechanism of blood sugar level in the rabbit *Brain Res.* **87** 195–9

Novara M, Baldelli P, Cavallari D, Carabelli V, Giancippoli A and Carbone E 2004 Exposure to cAMP and beta-adrenergic stimulation recruits Ca(V)3 T-type channels in rat chromaffin cells through Epac cAMP-receptor proteins *J. Physiol.* **558** 433–49

O'Sullivan A J and Burgoyne R D 1989 A comparison of bradykinin, angiotensin II and muscarinic stimulation of cultured bovine adrenal chromaffin cells *Biosci. Rep.* **9** 243–52

Olivos L and Artalejo A R 2008 Muscarinic excitation-secretion coupling in chromaffin cells *Acta Physiol. (Oxf.)* **192** 213–20

Petrovic J, Walsh P L, Thornley K T, Miller C E and Wightman R M 2010 Real-time monitoring of chemical transmission in slices of the murine adrenal gland *Endocrinology* **151** 1773–83

Przywara D A, Guo X, Angelilli M L, Wakade T D and Wakade A R 1996 A non-cholinergic transmitter, pituitary adenylate cyclase-activating polypeptide, utilizes a novel mechanism to evoke catecholamine secretion in rat adreanl chromaffin cells *J. Biol. Chem.* **271** 10545–50

Ruiz J R L, Ernst S A, Holz R W and Stuenkel E L 2022 Basal and stress-induced network activity in the adrenal medulla *in vivo Front. Endocrinol.* **13** 875865

Saito D, Takase Y, Murai H and Takahashi Y 2012 The dorsal aorta initiates a molecular cascade that instructs sympatho-adrenal specification *Science* **336** 1578–81

Smith C B and Eiden L E 2012 Is PACAP the major neurotransmitter for stress transduction at the adrenomedullary synapse? *J. Mol. Neurosci.* **48** 403–12

Souvannakitti D, Kuri B, Yuan G, Pawar A, Kumar G K, Smith C, Fox A P and Prabhakar N R 2010 Neonatal intermittent hypoxia impairs neuronal nicotinic receptor expression and function in adrenal chromaffin cells *Am. J. Physiol. Cell Physiol.* **299** C381–388

Strack A M, Sawyer W B, Platt K B and Loewy A D 1989 CNS cell groups regulating the sympathetic outflow to adrenal gland as revealed by transneuronal cell body labeling with pseudorabies virus *Brain Res.* **491** 274–96

Stroth N, Kuri B A, Mustafa T, Chan S A, Smith C B and Eiden L E 2013 PACAP controls adrenomedullary catecholamine secretion and expression of catecholamine biosynthetic enzymes at high splanchnic nerve firing rates characteristic of stress transduction in male mice *Endocrinology* **154** 330–9

Sunwoo S H *et al* 2019 Chronic and acute stress monitoring by electrophysiological signals from adrenal gland *Proc. Natl Acad. Sci. USA* **116** 1146–51

Tanaka K, Shibuya I, Nagamoto T, Yamashita H and Kanno T 1996 Pituitary adenylate cyclase-activating polypeptide causes rapid Ca^{2+} release from intracellular stores and long lasting Ca^{2+} influx mediated by Na^{+} influx-dependent membrane depolarization in bovine adrenal chromaffin cells *Endocrinology* **137** 956–66

Tank A W and Lee Wong. D 2015 Peripheral and central effects of circulating catecholamines *Compr. Physiol.* **5** 1–15

TerBush D R, Bittner M A and Holz R W 1988 Ca^{2+} influx causes rapid translocation of protein kinase C to membranes. Studies of the effects of secretagogues in adrenal chromaffin cells *J. Biol. Chem.* **263** 18873–9

TerBush D R and Holz R W 1990 Activation of protein kinase C is not required for exocytosis from bovine adrenal chromaffin cells: the effects of protein kinase C(19–31), Ca/CaM kinase II(291–317), and staurosporin *J. Biol. Chem.* **265** 21179–84

Vandael D H, Ottaviani M M, Legros C, Lefort C, Guerineau N C, Allio A, Carabelli V and Carbone E 2015 Reduced availability of voltage-gated sodium channels by depolarization or blockade by tetrodotoxin boosts burst firing and catecholamine release in mouse chromaffin cells *J. Physiol.* **593** 905–27

Wakade A R 1988 Noncholinergic transmitter(s) maintains secretion of catecholamines from rat adrenal medulla for several hours of continuous stimulation of splanchnic neurons *J. Neurochem.* **50** 1302–8

Wakade A R and Wakade T D 1983 Contribution of nicotinic and muscarinic receptors in the secretion of catecholamines evoked by endogenous and exogenous acetylcholine *Neuroscience* **10** 973–8

Wang M, Wang Q and Whim M D 2016 Fasting induces a form of autonomic synaptic plasticity that prevents hypoglycemia *Proc. Natl Acad. Sci. USA* **113** E3029–3038

Watanabe T, Shimamoto N, Takahashi A and Fujino M 1995 PACAP stimulates catecholamine release from adrenal medulla: a novel noncholinergic secretagogue *Am. J. Physiol.* **269** E903–909

Wolf K, Zarkua G, Chan S A, Sridhar A and Smith C 2016 Spatial and activity-dependent catecholamine release in rat adrenal medulla under native neuronal stimulation *Physiol. Rep* **4** e12898

Zhou Z and Misler S 1995 Action potential-induced quantal secretion of catecholamines from rat adrenal chromaffin cells *J. Biol. Chem.* **270** 3498–505

Zhou Z and Neher E 1993 Calcium permeability of nicotinic acetylcholine receptor channels in bovine adrenal chromaffin cells *Pflugers Archiv.* **425** 511–7

IOP Publishing

Exocytosis: From Molecules to Cells

Arun Anantharam and Jefferson Knight

Chapter 13

Synaptic vesicle dynamics at the calyx of Held and other central synapses

Takafumi Miki, Yuki Hashimotodani and Takeshi Sakaba

The calyx of Held synapse has been used as a model synapse with which to study the presynaptic mechanisms of transmitter release. The biophysics of the vesicle pool scheme has been well established at the calyx of Held synapse. Molecular mechanisms have been investigated to explain how synaptic vesicles are dynamically regulated. The correspondence between morphology, biophysics, and molecular biology needs to be resolved. The concepts developed at the calyx synapse are useful for our understanding other central synapses, but further investigation is required to establish a general model of transmitter release which explains the diversity and plasticity of different types of synapse.

The calyx of Held synapse in the auditory brainstem has been used as a model synapse of synaptic transmission since the 1990s [1]. A large presynaptic terminal wraps around the soma of the postsynaptic neuron, and this unique structure allows one to directly record from pre- and postsynaptic compartments (figure 13.1(a)). Synaptic responses are very large (on the order of several nA under a voltage clamp) and a single action potential (AP) evokes the fusion of 200–300 vesicles [2]. This synapse is designed for high-fidelity transmission of auditory signals [3]. Experimentally, the synapse has been widely used to understand the presynaptic mechanism of transmitter release using electrophysiology. Furthermore, morphological studies using electron microscopy have provided quantitative information about the number of docked vesicles and active zones, as well as the topology of the Ca^{2+} channels in the active zones [4, 5]. Later, this information was combined with genetics, allowing the molecular mechanism of transmission to be dissected [6]. Here, we focus on a biophysical analysis of transmitter release at the calyx of Held and other central synapses.

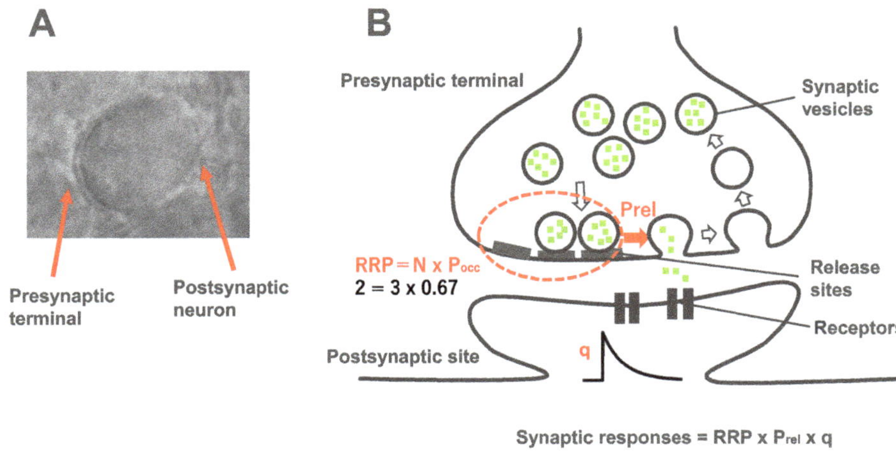

Figure 13.1. The readily releasable pool (RRP) of synaptic vesicles.(a) Photo image of the rat calyx of Held synapse (around postnatal day 8–11). A single terminal wraps the soma of the postsynaptic neuron. (b) In the presynaptic terminal, synaptic vesicles are recruited to the release sites. Docked vesicles at release sites are considered to be the vesicles within the RRP (readily releasable pool). The RRP is determined by the number of release sites (N) and the probability of the site occupancy (P_{occ}). In this example, there are three release sites and Pocc is 0.67, so that two sites are occupied by synaptic vesicles. The docked vesicles are fused immediately upon Ca^{2+} influx with P_{rel}. Released transmitters act on postsynaptic receptors and elicit postsynaptic responses. Single-vesicle fusion elicits postsynaptic responses, q. The vesicle membrane is retrieved via endocytosis. Vesicles are recycled for further rounds of fusion.

13.1 Regulation of release probability at the calyx of Held

Katz and colleagues have proposed that synaptic strength is determined by the product of N (the number of release sites), P_r (release probability), and q (quantal size, reference [7]). Later, P_r was split into P_{occ} (probability of vesicle occupancy at the release site) and P_{rel} (release probability per vesicle, reference [8]). The so-called readily releasable pool (RRP) of synaptic vesicles is the product of N and P_{occ}. This means that not all the release sites have docked vesicles [9]. The RRP is thought to correspond to morphologically docked vesicles at the active zone (figure 13.1(b)), which release neurotransmitters rapidly upon Ca^{2+} influx. In this article, we define the RRP as the vesicles which release transmitters in response to the strongest possible physiological stimulus (action potential, AP). It is widely accepted that P_{rel} is regulated by the local Ca^{2+} concentration at the release site [10]. In addition to the RRP, a recycling pool of vesicles, which feeds the vesicles to the RRP, has mainly been studied using electrophysiology, live imaging, and electron microscopy [11]. In the presynaptic terminal, some vesicles are not used during neuronal activity under physiological conditions. These vesicles belong to the resting pool, which comprises the majority of the vesicles in the presynaptic terminal [12, 13]. Table 13.1 summarizes the parameter that can be obtained by each method. Table 13.2 summarizes the obtained parameters from various synapse types.

Table 13.1. Various methods used to estimate N, RRP, P_{rel}, P_{occ}, and P_r. N: The number of functional release sites. RRP: The readily releasable pool of synaptic vesicles. P_{rel}: Release probability per vesicle. P_{occ}: Probability of vesicle occupancy at the release site. P_r: Release probability.

Method	N	RRP (N*Pocc)	P_{rel}	P_{occ}	P_r(Pocc*P_{rel})	Ref.
Presynaptic depolarization		✓[a]	✓[ab]			[21, 24, 42]
Variance-mean analysis	✓[c]				✓[c]	[41]
Cumulative variance- mean analysis	✓[c]	✓[c]	✓[c]	✓[c]	✓[c]	[44]
SMN method		✓[d]	✓[d]			[14, 15]

[a]: may include synaptic vesicles, which cannot be released by APs (such as the SRP vesicles).
[b]: in combination with an AP stimulus
[c]: needs correction by intra- and inter-synaptic variability.
[d]: needs correction for incomplete depletion of vesicle pool.

Table 13.2. The RRP size per synapse, release probability (P_{rel}) and probability of vesicle occupancy at the release site (P_{occ}) at various synapse types. hMFBs: hippocampal mossy fiber boutons. BC: basket cell. PC: Purkinje cell. cMFB: cerebellar mossy fiber boutons. GC: granule cell. PF: parallel fiber. IN: cerebellar interneuron.

Synapse type	RRP (vesicles)	P_{rel}	P_{occ}	Ref.
calyx of Held	1000–2000	0.1–0.3	>>0.5	[14, 16, 19, 20, 23, 24, 72]
hMFB-CA3	50–100	0.1	?	[37]
BC-PC	100	0.2–0.3	?	[39]
cMFB-GC	5–20	0.6	?	[41–43]
PF-IN	2	0.6	0.3–0.5	[44, 45, 66]
IN-IN	1–4	0.8–0.9	0.5–0.7	[9]

At the calyx of Held synapse, the release probability (P_r and P_{rel}) has been measured by several methods. Schneggenburger *et al* [14] changed the external Ca^{2+} concentration to measure the Ca^{2+} dependence of synaptic strength. This analysis assumed that P_{rel} is maximized and approaches one at very high Ca^{2+} concentration (10 mM). By fitting the Ca^{2+} dependence to a Hill function, they estimated a P_{rel} value of 0.2–0.3 under physiological conditions (2 mM Ca^{2+}/1 mM Mg^{2+}). Schneggenburger *et al* also estimated P_{rel} by another method, often referred to as back extrapolation or more recently as the Schneggenburger–Meyer–Neher (SMN) method [15]. In this analysis, they assumed that high-frequency stimulation depletes the RRP rapidly. By comparing the number of vesicles released by a single AP with the RRP size, they estimated P_{rel}. Synaptic responses decline during high-frequency stimulation but do not vanish completely. Schneggenburger *et al* [14] assumed that the remaining responses are mediated by the release of vesicles which are newly recruited to the empty release sites during stimulation. In a plot of cumulative excitatory postsynaptic currents (EPSCs) amplitudes during an AP train, Schneggenburger *et al* estimated the RRP size by back extrapolating the vesicle recruitment to the onset of stimulation using a line fitted to the steady state after synaptic depression (the SMN method). They estimated P_{rel} and the RRP to be 0.2–0.3 and 600, respectively. The same conclusion was derived by Bollmann *et al* [16]. In the third type of analysis, variance-mean analysis [17, 18] has been used to estimate N and P_r (= P_{rel} x P_{occ}). The analysis assumes that the relationship between the variance and the mean of synaptic responses (obtained by repeatedly recording synapses using a given stimulation protocol) should follow a parabolic relationship. Meyer *et al* [19] and Scheuss *et al* [20] estimated a P_r value of 0.2–0.3. In calyx of Held synapses, P_r is similar to P_{rel}, suggesting that P_{occ} may be close to one (or at least ≫0.5). Overall, the three independent methods converge to similar values of P_{rel} and P_r.

In subsequent studies, a larger vesicle pool (the releasable pool (RP)) was estimated using strong presynaptic stimulation. Sun and Wu [21], Schneggenburger and Neher [22], and Sakaba and Neher [23] estimated the RP size of 2000–4000 vesicles by inducing a sustained increase of the internal Ca^{2+} concentration using flash photolysis of caged Ca^{2+} or by long-lasting presynaptic depolarization. Half of the RP was hardly used during an AP train [24]. APs mainly evoke vesicle fusion from the super-primed pool (SPP, 0–500 vesicles that have a release-time constant of 1–2 ms during a depolarizing pulse, references [25, 26], see also reference [27]) and the fast-releasing pool (FRP, 1000–1500 vesicles with a release-time constant of 2–3 ms). Strong stimulation additionally recruits the slowly-releasing pool (SRP, 1000–2000 vesicles with a release-time constant of 20–30 ms, reference [23]), which is not used under physiological conditions [24]. Our recent understanding is that the original estimation of P_{rel} made by Schneggenburger and colleagues is close to P_{rel} calculated as the sum of the SPP and FRP vesicles (see below).

Bollmann *et al* [16] and Schneggenburger and Neher [22] clarified the Ca^{2+} regulation of P_{rel} at the calyx of Held synapse. Using flash photolysis of caged Ca^{2+}, they estimated the intracellular Ca^{2+} sensitivity of transmitter release. Later, the intracellular Ca^{2+} sensitivity was shown to be regulated by synaptotagmin2 [6]. Once the relationship between the intracellular Ca^{2+} and the release kinetics is established, the local Ca^{2+} concentration at the transmitter release sites can be estimated from the release kinetics during an AP. The estimated local Ca^{2+} concentration (10–25 μM) at the release site was lower than previously thought (> 50 μM; reference [28]). By including both caged Ca^{2+} and the fast Ca^{2+} buffer BAPTA in the recording patch pipette, Bollmann *et al* [29] created a fast Ca^{2+} transient with an amplitude of 5–10 μM in the calyx terminal, which mimics an AP-evoked EPSC, as recorded in a 1–2 mM external Ca^{2+} concentration. Relatively low peak Ca^{2+} concentration is consistent with the model in which vesicle fusion is controlled by Ca^{2+} channels loosely coupled to release sites. The loose coupling model is also consistent with the observation that the slow Ca^{2+} buffer EGTA effectively blocks fast transmitter release at the calyx of Held synapse [2]. BAPTA and EGTA have a similar affinity for Ca^{2+}. Near Ca^{2+} channels, Ca^{2+} buffers with slow Ca^{2+}-binding rates (e.g. EGTA) cannot capture Ca^{2+} effectively upon Ca^{2+} influx, whereas those with fast binding rates (e.g. BAPTA) can buffer Ca^{2+} effectively [30]. Therefore, if slow Ca^{2+} buffers block transmitter release effectively, the coupling between Ca^{2+} channels and the release sites must be loose. The results of electron microscopy and modeling for a Ca^{2+} channel distribution are consistent with loose coupling [5, 31]. It has been shown that this coupling becomes somewhat tighter with postnatal development [5, 32], enabling fast transmitter release [3].

Active-zone scaffold proteins play an important role in coupling between Ca^{2+} channels and synaptic vesicles. Rab3-interacting molecules (RIMs) 1 and 2 are important regulators of coupling Ca^{2+} channels and transmitter release sites. In RIM1/2 knockout (KO) mice, Ca^{2+} current amplitudes and P_{rel} were much reduced at the calyx of Held synapse [33]. RIM binding proteins (RIM-BPs) are also supposed to be important regulators, but Ca^{2+} current amplitudes were not much reduced by KO. However, the coupling between Ca^{2+} channels and synaptic vesicles was impaired [34]. In KO mice of another scaffold protein, cytomatrix at the active zone-associated structural protein (CAST), the phenotype seems somewhat milder [35]. Ca^{2+} current amplitudes and the RRP size were reduced, but the AP-evoked EPSCs were not impaired. Active-zone scaffold proteins form a protein complex, and the loss of a single type of protein may not lead to a strong phenotype. The exact role of each protein and complex needs further analysis.

13.2 Regulation of release probability at other synapses

At the central synapses, P_r differs considerably between synapses, which is one of the major mechanisms of functional diversity among synapses. It is generally assumed that the extent of short-term plasticity (facilitation vs depression) critically depends on the initial value of P_r [36].

At hippocampal mossy fiber–CA3 synapses, Vyleta and Jonas [37] estimated a P_{rel} of 0.1 using the SMN method. They estimated that the RRP contains 50–100 vesicles. A low P_{rel} is consistent with a high sensitivity of transmitter release to EGTA [37] and a low local Ca^{2+} concentration estimated by caged Ca^{2+} experiments at the release site (around 5 μM; see [38]). A low P_{rel} is also consistent with short-term facilitation observed at this synapse type.

At the inhibitory synapse between cerebellar basket cells and Pukinje cells, P_{rel} was estimated to be 0.2–0.3 [39]. At inhibitory synapses, transmitter release is insensitive to EGTA, suggesting tight coupling between Ca^{2+} channels and synaptic vesicles [40]. Tight coupling allows rapid inhibition by interneurons. Nevertheless, the peak local Ca^{2+} concentration at the release site is estimated to be 20–30 μM [39], which is similar to that of the calyx of Held synapse.

At cerebellar mossy fiber synapses, where each presynaptic contact with the postsynaptic neuron is physically small, P_r is high (0.6), while the number of release sites (N) is small (estimated by fluctuation analysis, reference [41]). Two types of AP-evoked vesicle with higher (0.8) and lower (0.3) P_{rel} values were subsequently described [42, 43], which might correspond to SPP and FRP vesicles at the calyx, respectively. At cerebellar parallel fiber–interneuron synapses, P_{rel} is high (0.6) at a concentration of 3 mM external Ca^{2+} [44, 45] but is comparable to other synapses (0.39) at physiological Ca^{2+} concentrations (1.5 mM Ca^{2+}, reference [45]). Interestingly, this synapse type is facilitated by the initial low P_{occ} and

Ca^{2+}-dependent synaptic vesicle replenishment to the RRP. In general, it is unclear whether facilitation is mediated by rapid vesicle replenishment, activation of the facilitation sensor of the docked vesicles, or both (see below).

13.3 Synaptic vesicle pools at the calyx of Held

Simultaneous pre- and postsynaptic recordings allow one to dissect vesicle pools quantitatively. As described above, the calyx terminal has three pools, the SPP (super-primed pool), the FRP (fast-releasing pool), and the SRP (slowly-releasing pool, figure 13.2(a)). The SPP and the FRP are involved in AP-evoked release, whereas the SRP is not directly involved [24]. Therefore, the RRP is thought to be composed of the SPP and the FRP.

The initial analysis identified the FRP and the SRP. The different timescales of the FRP and the SRP may be caused by either different degrees of coupling between Ca^{2+} channels and vesicles (reference [46], positional priming) or by a difference in the willingness to fuse (reference [47], molecular priming). It remains to be determined which is the correct mechanism, but a complication may be caused by the fact that the release timescale of the SRP (20–30 ms) and the replenishment time constant of the SRP (100 ms) are close, so that it is difficult to

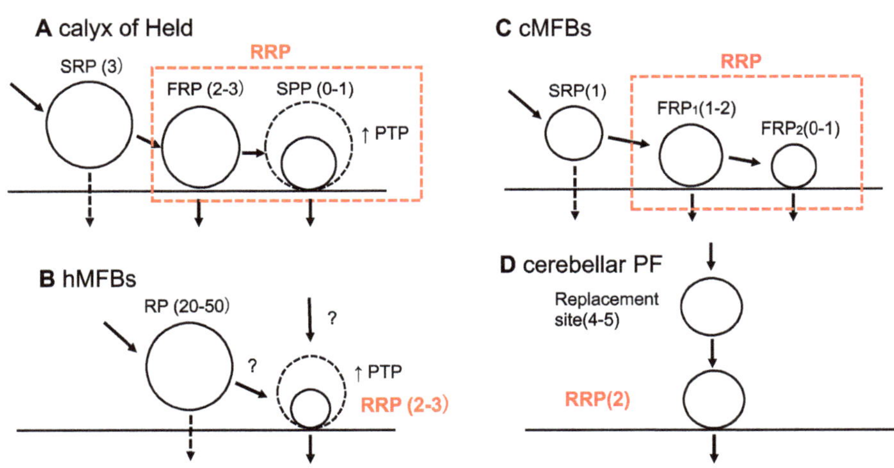

Figure 13.2. Vesicle pool schemes at several central synapses.(a) The vesicle pool scheme at the calyx of Held. Vesicles are recruited to the SRP and then become the FRP through molecular or positional priming. Some vesicles undergo further maturation to become vesicles of the SPP, which have a higher intracellular Ca^{2+} sensitivity for fusion. PTP (post-tetanic potentiation) enhances the SPP size. The numbers indicate the number of vesicles in each pool per active zone. (b) The vesicle pool scheme at the hippocampal mossy fiber bouton (hMFB). In response to the APs, the vesicles from the RRP are used. In response to a strong depolarizing pulse or flash photolysis of caged Ca^{2+} , the RP is recruited. The RRP is smaller than the RP, and it is unclear whether vesicles transit from the RP to the RRP. PTP enhances the RRP size. (c) The vesicle pool scheme at the cerebellar mossy fiber bouton (cMFB). Vesicles are rapidly recruited to the SRP and further matured to enter the FRP. The FRP has two populations (FRP_1 and FRP_2). (d) The vesicle pool scheme at the synapse between cerebellar parallel fibers and molecular-layer interneurons. Vesicles are first recruited to the replacement site and then transferred to the docking site. Vesicle recruitment to the docking site can be very rapid during stimulation (on the order of 10 ms).

separate the two processes. In other words, the SRP vesicles are the incompletely-formed vesicles, which reflect an intermediate process of the molecular and positional priming of synaptic vesicles. Following complete depletion of the FRP and the SRP, vesicle replenishment of the SRP is fast, whereas the replenishment of the FRP is slow [23, 48]. When only the FRP is depleted, conversion from the SRP to the FRP is rapid, allowing for sustained transmission during repetitive firing [49].

As described above, later analysis has revealed existence of the SPP. Initially, the FRP was found to be inhomogeneous; high-P_{rel} and low-P_{rel} vesicles contribute to the initial and subsequent phase of synaptic depression seen during an AP train, respectively [27]. Mueller *et al* also showed that the intracellular Ca^{2+} sensitivity differs by a factor of two using flash photolysis experiments. Later, a high-P_{rel} fraction was classified as a distinct pool, the SPP [25]. The size of the SPP varies among different calyx of Held synapses [26], which causes a large variation in the amplitudes of synaptic responses to a single AP and in short-term plasticity. The SPP size is dependent on the activation of diacylglycerol, PKC, and the C1 domain of Munc13; prolonged stimulation increases the SPP size and elicits post-tetanic potentiation of synaptic responses [26, 50].

We have not yet fully understood how vesicles transit between the three pools. Identification of the physical correlates of vesicle pools at the nanometer scale is necessary to solve this issue. The SRP is likely to be recruited from the recycling pool of synaptic vesicles. The FRP is likely to be recruited from the SRP, because the FRP, when depleted by strong stimulation, is refilled at the expense of the SRP—at least under some experimental conditions [49]. Conversion from the SRP to the FRP needs intact actin [49] and calmodulin may accelerate the process through Munc13–1 [51]. Conversion from the FRP to the SPP is slow in the resting condition and is accelerated during activity [49]. Alternatively, the SPP and FRP may be recruited from the SRP in parallel pathways. Neher and Brose [50] proposed that the SPP and the FRP represent tight and loose docking of synaptic vesicles at the release site, respectively. In this scheme, the SPP and FRP vesicles could release transmitters upon Ca^{2+} influx, but the FRP vesicles had to be matured to become SPP vesicles before fusion. A recent rapid-freeze electron microscopy (EM) technique gave rise to this idea [52, 53]. The KO of priming molecules such as Munc13s changes the proximity of synaptic vesicles to the plasma membrane at the nanometer scale. To understand the molecular states of vesicles and correlate them with functionally determined vesicle pools, one needs to resolve the dynamics of synaptic vesicles and of the release machinery by live imaging [54, 55].

13.4 Vesicle pools at other synapses

Vesicle pools were dissected at other synapses as well. In hippocampal culture preparation, hypertonic sucrose solution was used to deplete the RRP [56]. Cross-depletion experiments suggest that the RRP could be deleted by a high-frequency AP train and sucrose application [56]. Moulder *et al* [57] indicated that some vesicles were reluctant to release in response to APs [58]. Therefore, the vesicle pool determined by sucrose solution may be larger than the RRP defined in this article. Nevertheless, sucrose elicits release in Ca^{2+}-independent manner and bypasses Ca^{2+}-dependent processes [59], and it has been widely used to estimate the number of readily releasable vesicles. In CA3–CA1 synapses in the slice preparation, the RRP can be split into the pre-primed pool and the recruited pool [60], which contribute to the initial and subsequent phase of synaptic depression, respectively. This is somewhat similar to the concept of SPP vs FRP, but Gustaffson *et al* proposed that the distinction between the pre-primed pool and the non-primed pool is due to their location within active zones, as in Trommershausser *et al* [61], whereas the SPP and the FRP are proposed to reflect tight- and loose-release complexes, respectively [50].

At hippocampal mossy fiber synapses (figure 13.2(b)), Vyleta and Jonas [37] estimated the RRP of 50–100 vesicles using an AP train, whereas Hallermann *et al* [62] and Midorikawa and Sakaba [63] estimated the RP size to be 500–1000 vesicles using long depolarizing pulses. Hallermann *et al* [62] split the RP into two components, but the fast component was still several times larger than the RRP estimated by Vyleta and Jonas [37]. Flash photolysis experiments by Fukaya *et al* [38] suggest that intrinsic Ca^{2+} sensitivity for release is similar for all vesicles, which may speak for positional heterogeneity (different degrees of coupling between Ca^{2+} channels and vesicles), explaining the heterogeneity of P_{rel} within the RRP. Vyleta and Jonas [37] have proposed that endogenous Ca^{2+} buffers have the same properties as 0.3 mM BAPTA, a fast Ca^{2+} buffer. In unperturbed terminals, endogenous Ca^{2+} buffers may restrict the spread of Ca^{2+} signals [36, 64] and may primarily deplete vesicles relatively close to Ca^{2+} channels during APs. Alternatively, Vyleta and Jonas [37] may have mainly depleted a pool which is similar to the SPP at the calyx synapse and has higher intracellular Ca^{2+} sensitivity for release. Jonas and colleagues have observed that post-tetanic potentiation is mediated by the increased RRP [65], suggesting that the RRP size can be dynamically regulated.

At cerebellar mossy fiber synapses (figure 13.2(c)), the small RRP is rapidly depleted, and replenishment is very fast, with a time constant of tens of milliseconds. The small RRP was confirmed by experiments using long depolarizing pulses and was split into fast and slow components (with release-time constants of 0.4 and 5 ms, respectively; see reference [42]), presumably corresponding to FRP and SRP at the calyx of Held. The fast release component (FRP vesicle) matched the number of AP-evoked vesicles [44], which is consistent with the view that SRP vesicles do not contribute to AP-evoked release at the calyx [24]. During high-frequency AP trains, the initial phase of synaptic depression is determined by the high-P_{rel} vesicles (referred to as FRP_2), followed by release of low-P_{rel} vesicles (referred to as FRP_1), which are rapidly recruited [42]. FRP_2 and FRP_1 may correspond to the SPP and FRP, respectively, at the calyx of Held. This issue remains to be examined because it is unknown whether the SPP at the calyx of Held exists at the cerebellar mossy fiber synapse.

At cerebellar parallel fiber–interneuron synapses, a different scheme of vesicle pools has been proposed. Miki *et al* [44] utilized the stochastic properties of vesicular event numbers during high-frequency stimulation, providing separate estimations of N, P_{occ}, and P_{rel} at the cerebellar parallel fiber-molecular layer interneuron synapse (figure 13.2(d)). The RRP is small at rest, because P_{occ} is not maximal. Rapid synaptic vesicle replenishment increases P_{occ} and hence the RRP, causing synaptic facilitation. The results suggest that P_{occ} may determine the extent of facilitation vs depression, because P_{rel} stays high during the train. The rapid vesicle replenishment seems actin dependent [44], which is similar to the rapid conversion from the SRP to the FRP at the calyx synapse. Using synaptic latency analysis and modeling, Miki *et al* [66] have postulated a 'replacement site' upstream of the release site, which feeds the vesicles rapidly. Upon the arrival of an AP later in a train, some vesicles at the replacement sites can be recruited to the release site and then be released as a two-step release. If we assume that the vesicles for such two-step release are part of the RRP, they have apparently lower P_{rel}, producing slower release kinetics than the vesicles which are at the release sites before stimulation. Therefore, they may functionally behave like the FRP at the calyx synapses or FRP_1 at cerebellar mossy fiber synapses. However, the underlying mechanisms are different, because there is no direct release from the replacement site, whereas vesicles can release transmitter from the FRP or FRP_1. Alternatively, it could be that the RRP and the pool for replacement sites at the parallel fiber synapse are the same as the SPP (or tightly-docked vesicles) and FRP (or loosely-docked vesicles) at the calyx of Held, respectively. The transition from the FRP to the SPP is so fast and efficient that some FRP vesicles are quickly converted to the SPP and contribute to neurotransmission at cerebellar synapses during neuronal activity. In line with this idea, flash-and-freeze experiments with electron microscopy indicate ultrafast vesicle recruitment to release sites [53, 67]. At the calyx synapse, the conversion is not fast, and some FRP vesicles remain non-super-primed under resting conditions. It would be interesting to understand the correspondence between the vesicle pools in the calyx and those at the cerebellar synapses. At inhibitory synapses between basket and Purkinje cells, a single pool of the RRP has been observed to date [39]. This conclusion was drawn because flash photolysis and AP trains deplete similar numbers of vesicles. It is not yet known whether there is any sub-pool within the RRP at inhibitory synapses.

The identification of vesicle pools at the calyx of Held and other central synapses needs further molecular dissection. Accumulated experimental observations and a well-established vesicle pool scheme at the calyx of Held synapse may help us to understand how different synapses are tuned to their particular function.

13.5 Visualization of synaptic vesicle dynamics at the calyx of Held and other synapses

So far, the vesicle pool concept has mainly been developed using electrophysiology and modeling. Visualization of synaptic vesicle dynamics may allow physical identification of vesicle pools. Zenisek *et al* [68] pioneered the imaging of single-vesicle dynamics in dissociated goldfish retinal bipolar cells using total internal reflection fluorescence (TIRF) microscopy (figure 13.3(a)). TIRF microscopy has very good resolution in the *z*-axis (50–100 nm), which provides detailed vesicle

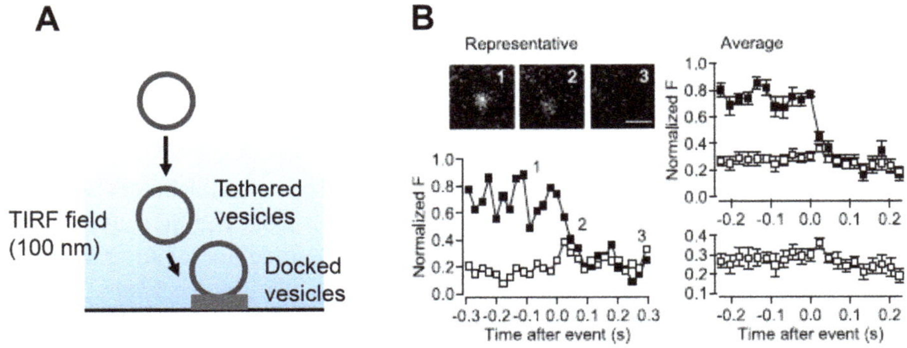

Figure 13.3. Visualization of synaptic vesicle dynamics with TIRF microscopy. (a) TIRF (total internal fluorescence) microscopy visualizes vesicles close to the plasma membrane (50–100 nm in the *z*-axis). Docked vesicles are considered to be within 30 nm of the plasma membrane. (b) By staining the vesicle membrane with FM1–43, one can see the fusion of synaptic vesicles, which is accompanied by a loss of dye in the center of a given vesicle and a transient increase of fluorescence in its surroundings, reflecting the diffusion of dye along the plasma membrane. In the left-hand panel, representative images and the timescales of the fluorescence changes in the center (closed circles) and surrounding region (open circles) are shown. In right-hand panel, average data are shown. Copyright 2020. Published under the PNAS license.

dynamics near the plasma membrane. On the other hand, the resolutions of the *x*- and *y*-axes are diffraction limited, which means that single vesicles (30–50 nm in diameter) appear as single dots. Sparse labeling of synaptic vesicles with FM1–43 allows one to look at the dynamics of synaptic vesicles before and during fusion. One can see the fusion of synaptic vesicles, which accompanies the loss of dyes in the center and a transient increase in the surrounding fluorescence, reflecting the diffusion of dyes along the plasma membrane (figure 13.3(b)). Zenisek *et al* have shown that some vesicles are already resident at rest and are rapidly fused at a discrete location, presumably a release site, whereas others are recruited to the release site only upon prolonged stimulation, reflecting synaptic vesicle recruitment to the release site. Newly recruited vesicles cannot be fused immediately, but need several hundreds of ms for fusion, presumably reflecting priming. Interestingly, some vesicles are fused outside the release site during prolonged stimulation. The timescales of the release events are a good match with those obtained via presynaptic capacitance measurements [69].

Following Zenisek *et al* [68], Midorikawa and Sakaba [70] applied TIRF microscopy to the dissociated calyx of Held presynaptic terminal. They showed that the RP, presumably reflecting both the FRP and the SRP, comprises those vesicles tethered at the plasma membrane before stimulation. Newly tethered vesicles are not immediately fusion competent, but need at least several seconds for fusion. In addition, synaptic vesicle recruitment to the RRP is likely mediated by the priming of vesicles which are not fusion competent at the onset of stimulation but become fusion competent with a time constant of hundreds of milliseconds. The overall results suggest that vesicle recruitment is not as rapid as that of bipolar cells. A similar analysis has been carried out at hippocampal mossy fiber terminals by Midorikawa and Sakaba [63] and results were essentially similar to those for the calyx of Held.

Miki *et al* [71] applied TIRF microscopy to cerebellar mossy fiber terminals (figure 13.3(b)). They found that the RRP corresponds to those vesicles already resident and ready for fusion upon Ca^{2+} influx. Following depletion of the RRP, vesicles which are within the TIRF field (100 nm) are fused in response to sustained depolarization or a train of action potentials, suggesting that newly replenished vesicles are already close to the membrane. At the same time, vesicles are recruited to the TIRF field more rapidly than they are to the calyx synapse. In addition, newly tethered vesicles can be fused with maturation times of several hundreds of milliseconds, which is much faster than that of the calyx of Held. Therefore, cerebellar mossy fiber terminals have more efficient vesicle recruitment and priming processes than those of the calyx of Held synapse.

TIRF allows live imaging of synaptic vesicles, but its spatial resolution along the *z*-axis is somewhat limited (50–100 nm). Recently, using rapid-freeze EM techniques, the detailed steps of synaptic vesicle dynamics have been dissected. By combining these with mouse genetics, Imig *et al* [52] indicated that morphological docking and synaptic vesicle priming are identical processes, in agreement with the idea that morphologically docked vesicles constitute the RRP. Before priming, synaptic vesicles are in the pre-docked state and are 30 nm away from the plasma membrane. Moreover, Chang *et al* [53] and Kusick *et al* [67] have shown that docking takes place on a very fast timescale (10 ms), which is similar to the transition from the replacement site to the docked site postulated by Miki *et al* [66].

13.6 Physiological relevance of vesicle pools at the calyx of Held and other central synapses

Vesicle pool dynamics are related to short-term plasticity. It is generally considered that synaptic depression is mediated by depletion of the RRP. Synaptic depression at the calyx of Held is mediated by depletion of the SPP and the FRP. The initial phase of depression is mediated by depletion of the SPP [26]. Afterwards, depletion of the FRP and rapid transition from the SRP to the FRP [25] determine the timescale of depression. Quantitative dissection of three pools during the train is not trivial because the decomposition is not always unique, but the contributions of the three pools to depression have been dissected in a relatively non-biased manner [72]. At other synapses, such as cerebellar mossy fiber synapses, the depletion of FRP_2 contributes to the initial phase of depression, and the depletion of FRP_1 and the rapid recruitment of vesicles to FRP_2 contribute to the steady state after depression [43]. Although other mechanisms, such as Ca^{2+} current inactivation and attenuation of the AP waveform [73], may potentially cause synaptic depression, it is generally thought that depletion of the SPP/FRP (RRP) is a dominating factor.

Short-term facilitation, which is elicited by paired or repetitive stimulation and lasts for millisecond to second timescales, has been considered to be driven by residual Ca^{2+} [10]. Because transmitter release depends on the third or fourth power of the Ca^{2+} concentration at the release site, a small increase in the Ca^{2+} concentration may cause massive facilitation when the initial P_{rel} is low. Currently, facilitation is considered to be caused not only by residual Ca^{2+}, but also by Ca^{2+} current facilitation, the activation of Ca^{2+} sensors for facilitation, enhanced vesicle recruitment, and the saturation of endogenous Ca^{2+} buffers [74, 75]. At the calyx of Held, facilitation is not always observed at a concentration of 2 mM of external Ca^{2+}, but is seen more pronouncedly at reduced concentrations of external Ca^{2+}. Felmy *et al* [76] showed that facilitation at the calyx synapse is mediated by three mechanisms: residual Ca^{2+}, Ca^{2+} current facilitation, and another factor, such as local Ca^{2+} buffer saturation. Felmy *et al* [76] excluded changes in the intracellular Ca^{2+} sensitivity for release as a possible mechanism of facilitation using Ca^{2+} uncaging experiments. Jackman *et al* [74] have shown that synaptotagmin7 has a dominant role in synaptic facilitation in several types of central synapse. This may mean that synaptotagmin7 is the Ca^{2+} sensor for facilitation. Synaptotagmin7 is also involved in asynchronous release and synaptic vesicle replenishment of the RRP [77, 78]. Miki *et al* [44] have shown that the facilitation observed at cerebellar parallel fiber synapses is mediated by enhanced recruitment of vesicles to the release site. In addition, Rozov *et al* [36] and Vyleta and Jonas [37] have postulated that a first stimulus causes local saturation of fast endogenous Ca^{2+} buffers at the release sites, which leads to an increased rise in local Ca^{2+} during a second stimulation. Such an increase causes facilitation of transmitter release.

Post-tetanic potentiation is caused by repetitive stimulation at high frequency and/or repetitive stimulation over a long period [79, 80]. At the calyx of Held, post-tetanic potentiation is mediated by an increase in the SPP [26, see also 81]. An increase in the SPP is caused by the activation of Ca^{2+}-sensitive PKC and Munc13/Munc18 [82–85]. At hippocampal mossy fiber synapses, post-tetanic potentiation causes an increase in the RRP estimated by the SMN method and an increase in the number of docked vesicles observed by electron microscopy [65]. In addition to vesicle pool dynamics, Borst and colleagues have suggested that Ca^{2+} current facilitation is likely to be responsible for post-tetanic potentiation at the calyx of Held [86].

Activity-dependent forms of long-term plasticity, such as long-term potentiation and depression, have not been found at the calyx of Held synapse. However, developmental changes have been thoroughly studied by von Gersdorff's, Takahashi's, and Wang and Young's groups. With development, synaptic responses become faster, and short-term depression is reduced. At the presynaptic side, action potential waveforms get faster, and the coupling between Ca^{2+} channels and vesicles becomes tighter. P_{rel} lowered by faster AP waveforms and a larger RRP size contributes to reduced synaptic depression [3, 5, 32, 87–89]. In addition, the release kinetics of the FRP and the SRP became considerably faster [90]. In mature animals at physiological external Ca^{2+} concentrations, the rat calyx of Held synapses around P32 show almost no depression or facilitation, which allows for fast and reliable signal transfer, contributing to high-frequency auditory signaling [89].

13.7 Summary

The calyx of Held synapse has been used as a model synapse with which to study the presynaptic mechanisms of transmitter release. The biophysics of the vesicle pool scheme has been well established at the calyx of Held synapse, and the dynamics of three pools (SRP, FRP, and SPP) are important for our understanding of short-term plasticity. In addition, molecular mechanisms have been investigated using a combination of electrophysiology, imaging, genetics, and electron microscopy to explain how synaptic vesicles are dynamically regulated before fusion events occur. Nevertheless, the correspondence between morphology (docked vesicles etc.), biophysics (functionally defined vesicle pools), and molecular biology (the molecular regulation of docking, priming, and fusion) needs to be resolved further. Recent advances in super-resolution imaging and electron microscopy will help us to understand molecular mechanisms at the nanometer scale in coming years. The concepts developed at the calyx synapse are useful for understanding other central synapses, but the establishment of a general model of transmitter release which explains the diversity and plasticity of different types of synapse needs further investigation. In addition, it remains to be seen how the dynamic regulation of synaptic vesicle pools contributes to neural circuits' function and behavior.

Acknowledgements

We thank Erwin Neher and Alain Marty for their valuable comments on the manuscript. This study was supported by JSPS KAKENHI (JP18K06472, JP21H02584 to TM, JP18H02530, JP19H04760, JP20KK0171, JP21H02598 to TS) and the JSPS Core-to-Core Program A (Advanced Research Networks) (JPJSCCA20220007).

References

[1] Forsythe I D 1994 Direct patch recording from identified presynaptic terminals mediating glutamatergic EPSCs in the rat CNS, *in vitro J. Physiol.* **479** 381–7

[2] Borst J G G and Sakmann B 1996 Calcium influx and transmitter release in a fast CNS synapse *Nature* **383** 431–4

[3] Taschenberger H, Leão R M, Rowland K C, Spirou G A and von Gersdorff H 2002 Optimizing synaptic architecture and efficiency for high-frequency transmission *Neuron* **36** 1127–43

[4] Sätzler K, Söhl L F, Bollmann J H, Borst J G, Frotscher M, Sakmann B and Lübke J H 2002 Three-dimensional reconstruction of a calyx of Held and its postsynaptic principal neuron in the medial nucleus of the trapezoid body *J. Neurosci.* **22** 10567–79

[5] Nakamura Y, Harada H, Kamasawa N, Matsui K, Rothman J S, Shigemoto R, Silver R A, DiGregorio D A and Takahashi T 2015 Nanoscale distribution of presynaptic Ca^{2+} channels and its impact on vesicular release during development *Neuron* **85** 145–58

[6] Sun J, Pang Z P, Qin D, Fahim A T, Adachi R and Südhof T C 2007 A dual-Ca^{2+}-sensor model for neurotransmitter release in a central synapse *Nature* **450** 676–82

[7] Katz B 1969 *The Release of Neural Transmitter Substances (The Sherrington Lectures)* (Liverpool: Liverpool University Press)

[8] Zucker R S 1973 Changes in the statistics of transmitter release during facilitation *J. Physiol.* **229** 787–810

[9] Pulido C, Trigo F F, Llano I and Marty A 2015 Vesicular release statistics and unitary postsynaptic current at single GABAergic synapses *Neuron* **85** 159–72

[10] Dodge F A Jr and Rahamimoff R 1967 Co-operative action a calcium ions in transmitter release at the neuromuscular junction *J. Physiol.* **193** 419–32

[11] Alabi A A and Tsien R W 2012 Synaptic vesicle pools and dynamics *Cold Spring Haeb. Perspect. Biol.* **4** a013680
[12] Denker A, Kröhnert K, Bückers J, Neher E and Rizzoli S O 2011 The reserve pool of synaptic vesicles acts as a buffer for proteins involved in synaptic vesicle recycling *Proc. Natl Acad. Sci. USA* **108** 17183–8
[13] Denker A, Bethani I, Kröhnert K, Körber C, Horstmann H, Wilhelm B G, Barysch S V, Kuner T, Neher E and Rizzoli S O 2011 A small pool of vesicles maintains synaptic activity *in vivo Proc. Natl Acad. Sci. USA* **108** 17177–82
[14] Schneggenburger R, Meyer A C and Neher E 1999 Released fraction and total size of a pool of immediately available transmitter quanta at a calyx synapse *Neuron* **23** 399–409
[15] Neher E 2015 Merits and limitations of vesicle pool models in view of heterogeneous populations of synaptic vesicles *Neuron* **87** 1131–42
[16] Bollmann J H, Sakmann B and Borst J G 2000 Calcium sensitivity of glutamate release in a calyx-type terminal *Science* **289** 953–7
[17] Clements J D and Silver R A 2000 Unveiling synaptic plasticity. A new graphical and analytical approach *Trends Neurosci.* **23** 105–13
[18] Schmidt H, Brachtendorf S, Arendt O, Hallermann S, Ishiyama S, Bornschein G, Gall D, Schiffmann S N, Heckmann M and Eilers J 2013 Nanodomain coupling at an excitatory cortical synapse *Curr. Biol.* **23** 244–9
[19] Meyer A C, Neher E and Schneggenburger R 2001 Estimation of quantal size and number of functional active zones at the calyx of Held synapse by nonstationary EPSC variance analysis *J. Neurosci.* **21** 7889–900
[20] Scheuss V, Schneggenburger R and Neher E 2002 Separation of presynaptic and postsynaptic contributions to depression by covariance analysis of successive EPSCs at the calyx of Held synapse *J. Neurosci.* **22** 728–39
[21] Sun J Y and Wu L G 2001 Fast kinetics of exocytosis revealed by simultaneous measurements of presynaptic capacitance and postsynaptic currents at a central synapse *Neuron* **30** 171–82
[22] Schneggenburger R and Neher E 2000 Intracellular calcium dependence of transmitter release rates at a fast central synapse *Nature* **406** 889–93
[23] Sakaba T and Neher E 2001 Calmodulin mediates rapid recruitment of fast-releasing synaptic vesicles at a calyx-type synapse *Neuron* **32** 1119–31
[24] Sakaba T 2006 Roles of the fast-releasing and the slowly releasing vesicles in synaptic transmission at the calyx of Held *J. Neurosci.* **26** 5863–71
[25] Lee J S, Ho W K, Neher E and Lee S H 2013 Superpriming of synaptic vesicles after their recruitment to the readily releasable pool *Proc. Natl Acad. Sci. USA* **110** 15079–84
[26] Taschenberger H, Woehler A and Neher E 2016 Superpriming of synaptic vesicles as a common basis for intersynapse variability and modulation of synaptic strength *Proc. Natl Acad. Sci. USA* **113** E4548–57
[27] Müller M, Goutman J D, Kochubey O and Schneggenburger R 2010 Interaction between facilitation and depression at a large CNS synapse reveals mechanisms of short-term plasticity *J. Neurosci.* **30** 2007–16
[28] Zucker R S 1996 Exocytosis: a molecular and physiological perspective *Neuron* **17** 1049–55
[29] Bollmann J H and Sakmann B 2005 Control of synaptic strength and timing by the release-site Ca^{2+} signal *Nat. Neurosci.* **8** 426–34

[30] Neher E 1998 Usefulness and limitations of linear approximations to the understanding of Ca^{++} signals *Cell Calcium* **24** 345–57
[31] Keller D, Babai N, Kochubey O, Han Y, Markram H, Schürmann F and Schneggenburger R 2015 An exclusion zone for Ca^{2+} channels around docked vesicles explains release control by multiple channels at a CNS synapse *PLoS Comput. Biol.* **11** e1004253
[32] Fedchyshyn M J and Wang L Y 2005 Developmental transformation of the release modality at the calyx of Held synapse *J. Neurosci.* **25** 4131–40
[33] Han Y, Kaeser P S, Südhof T C and Schneggenburger R 2011 RIM determines Ca^{2+} channel density and vesicle docking at the presynaptic active zone *Neuron* **69** 304–16
[34] Acuna C, Liu X, Gonzalez A and Sudhof T C 2015 RIM-BPs mediate tight coupling of action potentials to Ca(2+)-triggered neurotransmitter release *Neuron* **87** 1234–47
[35] Dong W *et al* 2018 CAST/ELKS proteins control voltage-gated Ca^{2+} channel density and synaptic release probability at a mammalian central synapse *Cell Rep.* **24** 284–93
[36] Rozov A, Burnashev N, Sakmann B and Neher E 2001 Transmitter release modulation by intracellular Ca^{2+} buffers in facilitating and depressing nerve terminals of pyramidal cells in layer 2/3 of the rat neocortex indicates a target cell-specific difference in presynaptic calcium dynamics *J. Physiol.* **531** 807–26
[37] Vyleta N P and Jonas P 2014 Loose coupling between Ca^{2+} channels and release sensors at a plastic hippocampal synapse *Science* **343** 665–70
[38] Fukaya R, Maglione M, Sigrist S and Sakaba T 2021 Rapid Ca^{2+} channel accumulation contributes to cAMP-mediated increase in transmission at hippocampal mossy fiber synapses *Proc. Natl Acad. Sci. USA* **118** e2016754118
[39] Sakaba T 2008 Two Ca(2+)-dependent steps controlling synaptic vesicle fusion and replenishment at the cerebellar basket cell terminal *Neuron* **57** 406–19
[40] Bucurenciu I, Kulik A, Schwaller B, Frotscher M and Jonas P 2008 Nanodomain coupling between Ca^{2+} channels and Ca^{2+} sensors promotes fast and efficient transmitter release at a cortical GABAergic synapse *Neuron* **57** 536–45
[41] Saviane C and Silver R A 2006 Fast vesicle reloading and a large pool sustain high bandwidth transmission at a central synapse *Nature* **439** 983–7
[42] Ritzau-Jost A, Delvendahl I, Rings A, Byczkowicz N, Harada H, Shigemoto R, Hirrlinger J, Eilers J and Hallermann S 2014 Ultrafast action potentials mediate kilohertz signaling at a central synapse *Neuron* **84** 152–63
[43] Ritzau-Jost A, Jablonski L, Viotti J, Lipstein N, Eilers J and Hallermann S 2018 Apparent calcium dependence of vesicle recruitment *J. Physiol.* **596** 4693–707
[44] Miki T, Malagon G, Pulido C, Llano I, Neher E and Marty A 2016 Actin- and myosin-dependent vesicle loading of presynaptic docking sites prior to exocytosis *Neuron* **91** 808–23
[45] Malagon G, Miki T, Tran V, Gomez L C and Marty A 2020 Incomplete vesicular docking limits synaptic strength under high release probability conditions *Elife* **9** e52137
[46] Wadel K, Neher E and Sakaba T 2007 The coupling between synaptic vesicles and Ca^{2+} channels determines fast neurotransmitter release *Neuron* **53** 563–75
[47] Wölfel M, Lou X and Schneggenburger R 2007 A mechanism intrinsic to the vesicle fusion machinery determines fast and slow transmitter release at a large CNS synapse *J. Neurosci.* **27** 3198–210
[48] Wu L G and Borst J G G 1999 The reduced release probability of releasable vesicles during recovery from short-term synaptic depression *Neuron* **23** 821–32

[49] Lee J S, Ho W-K and Lee S-H 2012 Actin-dependent rapid recruitment of reluctant synaptic vesicles into a fast-releasing vesicle pool *Proc. Natl Acad. Sci. USA* **109** E765–74
[50] Neher E and Brose N 2018 Dynamically primed synaptic vesicle states: key to understand synaptic short-term plasticity *Neuron* **100** 1283–91
[51] Lipstein N, Sakaba T, Cooper B H, Lin K H, Strenzke N, Ashery U, Rhee J S, Taschenberger H, Neher E and Brose N 2013 Dynamic control of synaptic vesicle replenishment and short-term plasticity by Ca^{2+}-calmodulin-Munc13-1 signaling *Neuron* **79** 82–96
[52] Imig C, Min S W, Krinner S, Arancillo M, Rosenmund C, Südhof T C, Rhee J, Brose N and Cooper B H 2014 The morphological and molecular nature of synaptic vesicle priming at presynaptic active zones *Neuron* **84** 416–31
[53] Chang S, Trimbuch T and Rosenmund C 2018 Synaptotagmin-1 drives synchronous Ca^{2+}-triggered fusion by C2B-domain-mediated synaptic-vesicle-membrane attachment *Nat. Neurosci.* **21** 33–40
[54] An S J and Almers W 2004 Tracking SNARE complex formation in live endocrine cells *Science* **306** 1042–6
[55] Degtyar V, Hafez I M, Bray C and Zucker R S 2013 Dance of the SNAREs: assembly and rearrangements detected with FRET at neuronal synapses *J. Neurosci.* **33** 5507–23
[56] Rosenmund C and Stevens C F 1996 Definition of the readily-releasable pool of vesicles at hippocampal synapses *Neuron* **16** 1197–207
[57] Moulder K L, Jiang X, Taylor A A, Olney J W and Mennerick S 2007 Vesicle pool heterogeneity at hippocampal glutamate and GABA synapses *J. Neurosci.* **27** 9846–54
[58] Wang S S H, Held R G, Wong M Y, Liu C, Karakhanyan A and Kaeser P S 2016 Fusion competent synaptic vesicles persist upon active zone disruption and loss of vesicle docking *Neuron* **91** 777–91
[59] Orlando M, Schmitz D, Rosenmund C and Herman M A 2019 Calcium-independent exo-endocytosis coupling at small central synapses *Cell Rep.* **12** 3767–74 e3
[60] Gustafsson B, Ma R and Hanse E 2019 The small and dynamic pre-primed pool at the release site; a useful concept to understand release probability and short-term synaptic plasticity? *Front. Synap. Neurosci.* **11** 7
[61] Trommershäuser J, Schneggenburger R, Zippelius A and Neher E 2003 Heterogeneous presynaptic release probabilities: functional relevance for short-term plasticity *Biophys. J.* **84** 1563–79
[62] Hallermann Sm Pawlu C, Jonas P and Heckmann M 2003 A large pool of releasable vesicles in a cortical glutamatergic synapse *Proc. Natl Acad. Sci. USA* **100** 8975–80
[63] Midorikawa M and Sakaba T 2017 Kinetics of releasable synaptic vesicles and their plastic changes at hippocampal mossy fiber synapses *Neuron* **96** 1033–40
[64] Blatow M, Caputi A, Burnashev N, Monyer H and Rozov A 2003 Ca^{2+} buffer saturation underlies paired pulse facilitation in calbindin-D28k-containing terminals *Neuron* **38** 79–88
[65] Vandael D, Borges-Merjane C, Zhang X and Jonas P 2020 Short-term plasticity at hippocampal mossy fiber synapses is induced by natural activity patterns and associated with vesicle pool engram formation *Neuron* **107** 509–21
[66] Miki T, Nakamura Y, Malagon G, Neher E and Marty A 2018 Two-component latency distributions indicate two-step vesicular release at simple glutamatergic synapses *Nat. Commun.* **9** 3943

[67] Kusick G F, Chin M, Raychaudhuri S, Lippmann K, Adula K P, Hujber E J, Vu T, Davis M W, Jorgensen E M and Watanabe S 2020 Synaptic vesicles transiently dock to refill release sites *Nat. Neurosci.* **23** 1329–38

[68] Zenisek D, Steyer J A and Almers W 2000 Transport, capture and exocytosis of single synaptic vesicles at active zones *Nature* **406** 849–54

[69] Mennerick S and Matthews G 1996 Ultrafast exocytosis elicited by calcium current in synaptic terminals of retinal bipolar neurons *Neuron* **17** 1241–9

[70] Midorikawa M and Sakaba T 2015 Imaging exocytosis of single synaptic vesicles at a fast CNS presynaptic terminal *Neuron* **88** 492–8

[71] Miki T, Midorikawa M and Sakaba T 2020 Direct imaging of rapid tethering of synaptic vesicles accompanying exocytosis at a fast central synapse *Proc. Natl Acad. Sci. USA* **117** 14493–502

[72] Neher E and Taschenberger H 2021 Non-negative matrix factorization as a tool to distinguish between synaptic vesicles in different functional states *Neuroscience* **458** 182–202

[73] Kawaguchi S Y and Sakaba T 2015 Control of inhibitory synaptic outputs by low excitability of axon terminals revealed by direct recording *Neuron* **85** 1273–88

[74] Jackman S L, Turecek J, Belinsky J E and Regehr W G 2016 The calcium sensor synaptotagmin 7 is required for synaptic facilitation *Nature* **529** 88–91

[75] Jackman S and Regehr W G 2017 The mechanisms and functions of synaptic facilitation *Neuron* **94** 447–64

[76] Felmy F, Neher E and Schneggenburger R 2003 Probing the intracellular calcium sensitivity of transmitter release during synaptic facilitation *Neuron* **37** 801–11

[77] Liu H, Bai H, Hui E, Yang L, Evans C S, Wang Z, Kwon S E and Chapman E R 2014 Synaptotagmin 7 functions as a Ca^{2+}-sensor for synaptic vesicle replenishment *Elife* **3** e01524

[78] Luo F and Südhof T C 2017 Synaptotagmin-7-mediated asynchronous release boosts high-fidelity synchronous transmission at a central synapse *Neuron* **94** 826–39

[79] Korogod N, Lou X and Schneggenburger R 2005 Presynaptic Ca^{2+} requirements and developmental regulation of posttetanic potentiation at the calyx of Held *J. Neurosci.* **25** 5127–37

[80] Habets R L and Borst J G 2005 Post-tetanic potentiation in the rat calyx of Held synapse *J. Physiol.* **564** 173–87

[81] Habets R L and Borst J G 2007 Dynamics of the readily releasable pool during post-tetanic potentiation in the rat calyx of Held synapse *J. Physiol.* **581** 467–78

[82] Korogod N, Lou X and Schneggenburger R 2007 Posttetanic potentiation critically depends on an enhanced Ca(2+) sensitivity of vesicle fusion mediated by presynaptic PKC *Proc. Natl Acad. Sci. USA* **104** 15923–8

[83] Lou X, Korogod N, Brose N and Schneggenburger R 2008 Phorbol esters modulate spontaneous and Ca^{2+}-evoked transmitter release via acting on both Munc13 and protein kinase C *J. Neurosci.* **28** 8257–67

[84] Fioravante D, Chu Y, Myoga M H, Leitges M and Regehr W G 2011 Calcium-dependent isoforms of protein kinase C mediate posttetanic potentiation at the calyx of Held *Neuron* **70** 1005–19

[85] Genc O, Kochubey O, Toonen R F, Verhage M and Schneggenburger R 2014 Munc18-1 is a dynamically regulated PKC target during short-term enhancement of transmitter release *Elife* **3** e01715

[86] Habets R L and Borst J G 2006 An increase in calcium influx contributes to post-tetanic potentiation at the rat calyx of Held synapse *J. Neurophysiol.* **96** 2868–76

[87] Taschenberger H and von Gersdorff H 2000 Fine-tuning an auditory synapse for speed and fidelity: developmental changes in presynaptic waveform, EPSC kinetics, and synaptic plasticity *J. Neurosci.* **20** 9162–73

[88] Iwasaki S and Takahashi T 2001 Developmental regulation of transmitter release at the calyx of Held in rat auditory brainstem *J. Physiol.* **534** 861–71

[89] Lorteije J A, Rusu S I, Kushmerick C and Borst J G 2009 Reliability and precision of the mouse calyx of Held synapse *J. Neurosci.* **29** 13770–84

[90] Chen Z, Das B, Nakamura Y, DiGregorio D A and Young S M Jr. 2015 Ca^{2+} channel to synaptic vesicle distance accounts for the readily releasable pool kinetics at a functionally mature auditory synapse *J. Neurosci.* **35** 2083–100

IOP Publishing

Exocytosis: From Molecules to Cells

Arun Anantharam and Jefferson Knight

Chapter 14

Mechanisms of exocytosis in mammalian fertilization

Lisa M Mehlmann

Exocytosis plays an important role in mammalian fertilization. In the sperm, exocytosis of the large secretory vesicle called the acrosome is required for the release of enzymes needed to permit the sperm to traverse the outermost coat of the egg, the zona pellucida (ZP). In the egg, exocytosis of vesicles located in the cortical region (cortical granules) occurs shortly after sperm–egg fusion. The contents of the vesicles that are released modify the ZP such that it is no longer permeable to sperm; thus, this exocytic event is an important polyspermy prevention mechanism. The molecular mechanisms of exocytosis in both sperm and egg are similar to those utilized by other secretory cells and involve SNAREs and their interacting proteins. This chapter reviews current knowledge about the molecular events required for exocytosis in mammalian sperm and egg cells.

14.1 Overview of fertilization

Fertilization is a complex process that occurs in sexually reproducing organisms in order to combine genetic material from two different individuals into a new, genetically distinct organism. The haploid egg and sperm DNA combine during this process to become a diploid cell that subsequently divides and ultimately forms a complete organism. The process of exocytosis in both the egg and the sperm is critical for achieving successful fertilization. This chapter outlines the major exocytic events that occur in sperm and eggs and highlights the main proteins that are known to play key roles in these processes.

Before the sperm contacts the cell membrane of the egg, it undergoes the exocytosis of a large vesicle, the acrosome, located in its head, anterior to the tightly condensed sperm DNA. The acrosome contains enzymes needed to penetrate the egg's outer acellular coat, the ZP [1]. The 'acrosome reaction' (AR), or exocytosis of this large vesicle, occurs when the sperm is in the vicinity of the egg

doi:10.1088/978-0-7503-3771-7ch14

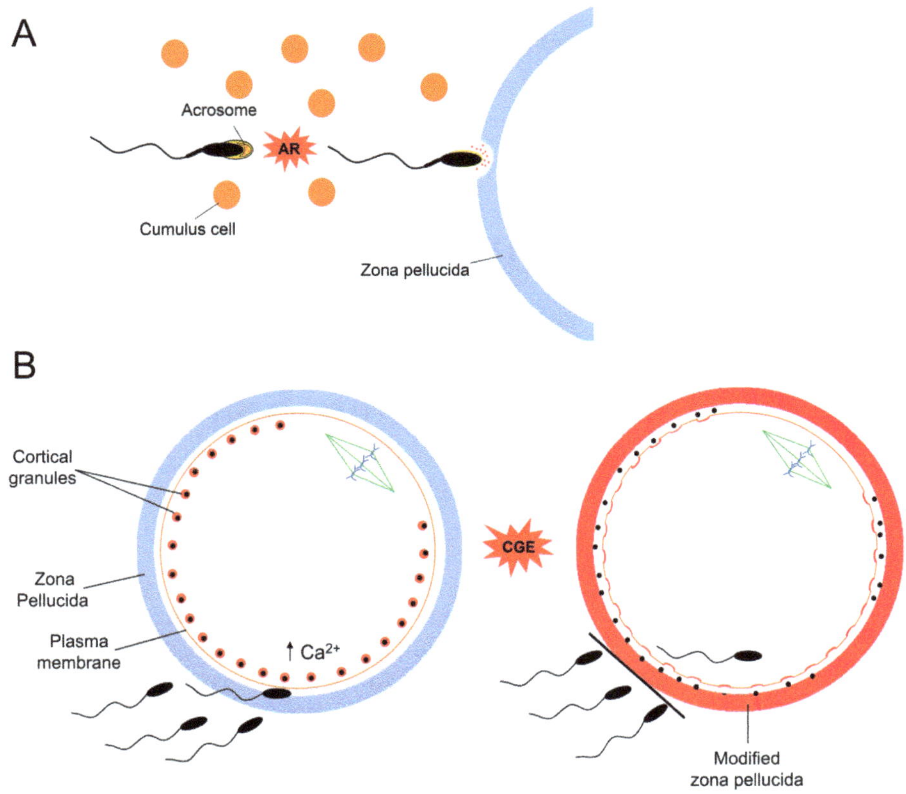

Figure 14.1. Overview of exocytosis at fertilization. (a) As the sperm reaches the egg, it traverses the somatic, cumulus cell layers that surround the egg, and undergoes acrosomal exocytosis (AR). The contents of the acrosome contain enzymes that are required to digest the egg's outer coat, the ZP. (b) Once a sperm penetrates the ZP and fuses with the egg's plasma membrane, it sets off an intracellular signaling cascade involving the release of intracellular Ca^{2+}, which triggers cortical granule exocytosis (CGE). The contents of these cortical vesicles modify the ZP so that only one sperm can enter.

(figure 14.1(a)). It was once thought that mammalian sperm undergo the AR only after binding to the ZP, as sperm can fertilize in the absence of the surrounding somatic cells, acrosome-intact sperm can bind to eggs, and components of the ZP trigger exocytosis [2–4]. However, the use of fluorescent sperm in mice has clearly shown that acrosomal exocytosis begins when the sperm are in the oviduct, prior to reaching the egg [5, 6]. The precise sites of the onset of the AR in other mammalian species are not as clear; however, the presence of somatic cells increases the percentage of acrosome-reacted sperm and improves fertilization rates in the bovine [7], suggesting that somatic cells have a role in inducing the AR.

It is critical that only one sperm enters the egg. The entry of more than one sperm is termed polyspermy, and polyspermic eggs are not viable. Eggs have therefore evolved mechanisms to prevent more than one sperm from entering. Aquatic species, including marine invertebrates and frogs, that encounter numerous sperm from external fertilization, utilize two lines of defense against polyspermy. The 'fast block' involves the depolarization of the plasma membrane, resulting in a rapid, transient electrical block [8–10]. The 'slow block' follows the fast block and involves the release of cortical granules [11]. Cortical granules are enzyme-filled vesicles located in the outermost part of the egg, and their exocytosis is stimulated by the release of Ca^{2+} at fertilization. The contents of the cortical granules modify the outermost coat of the egg to create a barrier that prevents more than one sperm from entering the egg. While aquatic organisms utilize both fast and slow blocks to polyspermy, the mechanism utilized by mammalian eggs is the slower, mechanical block which occurs shortly after sperm–egg fusion (figure 14.1(b)). In mammals, the contents of the cortical granules cleave proteins that comprise the egg's extracellular coat, the 'ZP,' such that it is no longer receptive to sperm [12]. Despite this polyspermy prevention mechanism, polyspermy accounts for ~15% of spontaneous abortions [13].

14.2 Triggers of acrosomal exocytosis in mammalian sperm

Mature sperm are unable to fertilize eggs immediately after ejaculation. Rather, they must be 'capacitated' in the female reproductive tract prior to fertilization. Capacitation is a complex process that includes many modifications to sperm, including increased membrane fluidity due to the removal of cholesterol from the plasma membrane [14, 15], an increase in cytoplasmic alkalinization [16–18], increased tyrosine phosphorylation due to the activation of protein kinase A (PKA; [19–22]), and changes in membrane potential that lead to an influx of extracellular Ca^{2+} [23].

Many cell signaling events occur during the process of sperm capacitation. The sperm consists of three primary segments: the head, the neck, and the principal piece of the tail. The entire sperm contains numerous ion channels as well as Ca^{2+} storage sites in the head and neck that regulate intracellular pH ($[pH]_i$) and Ca^{2+} concentration. These include voltage-gated H^+ channels, Na^+/H^+ exchangers, HCO_3^- transporters, K^+ channels, Na^+ channels, Na^+/K^+ ATPases, Ca^{2+} ATPases, Na^+/Ca^{2+} exchangers, voltage-gated Ca^{2+} channels, and intracellular Ca^{2+} store receptors (see [17] for review). The particular channels expressed in sperm vary by species, but the overall changes that occur during capacitation are similar.

There are several proteins in the female reproductive tract that serve to remove cholesterol from the sperm membrane during capacitation. This process is necessary for normal fertility [24]. *In vitro*, capacitation can be accomplished by adding serum albumin to the capacitation medium, but capacitation also occurs when other substances such as β-cyclodextrins are substituted [22]. Once the sperm is exposed to the relatively high HCO_3^- concentrations present in the female reproductive tract, HCO_3^- (along with Ca^{2+}) activates a soluble adenylate cyclase (sAC or ADCY10), which stimulates the production of cyclic adenosine monophosphate (cAMP) and in turn activates PKA. PKA phosphorylates numerous proteins in the sperm, including ion channels that regulate $[pH]_i$ and membrane potential [23]. An increase in $[pH]_i$, as well as membrane hyperpolarization, stimulate Ca^{2+} influx through the Ca^{2+} channel CatSper, the function of which is critical for sperm capacitation and hyperactivation [25–27]. Hyperactivation involves a change in the sperm swimming pattern from a wavelike to a whiplike motion and is required for normal fertilization, as it permits the transport of sperm through the thick oviductal secretions and cumulus cell layers, as well as penetration of the sperm through the ZP [28].

All the changes that occur during capacitation prime the sperm to respond to activators of the AR. The physiological activators are progesterone and ZP3. Progesterone is a steroid hormone synthesized by the corpus luteum and cumulus cells that surround the egg after the induction of oocyte maturation, and ZP3 is one of three (or four, depending on the species) glycoproteins that comprise the egg's ZP [29]. Both inducers act through membrane receptors in the sperm that ultimately cause an influx of extracellular Ca^{2+} [30–32]. Ca^{2+} from both the extracellular environment, as well as from internal stores, is required for the AR [33]; see below.

The molecular events of acrosomal exocytosis are similar to those of other secretory systems, such as neurons and neuroendocrine cells. However, the AR is a specialized exocytic event because the AR is an all-or-nothing response that occurs once during the lifetime of a sperm. The acrosome is a single, large vesicle that fuses in hundreds of places along the plasma membrane and there is no membrane recycling after the AR occurs [33]. On the contrary, the outer acrosomal membrane and plasma membrane overlying it vesiculate and are removed [34]. The contents of the acrosome include enzymes needed to digest the ZP surrounding the egg [35], and this process is tightly regulated so that it does not occur prematurely. Soluble *N*-ethylmaleimide sensitive factor (NSF)-attachment protein receptors (SNAREs) have been identified in sperm from all mammalian species studied [33, 36]; together with other proteins such as Rabs, they mediate acrosomal exocytosis. The major signal that initiates the onset of the AR is Ca^{2+}.

Immediately upon stimulation of the sperm with progesterone or ZP3, Ca^{2+} channels in the plasma membrane open to permit the influx of extracellular Ca^{2+}. Cytosolic Ca^{2+} increases within minutes after stimulation with ZP3 or progesterone [37–39]. Ca^{2+} has downstream effects that lead to a sustained release of Ca^{2+} from intracellular stores and, ultimately, to the AR (see below).

Most of the proteins known to regulate exocytosis in other cell types have been identified in mammalian sperm. These include NSF [40–43], α-SNAP [40, 42], Rab3A [41, 44], Rab27 [45], SNAP25 and SNAP23 [46, 47], synaptotagmins I, IV, VI, and VIII [36, 48–50], Munc18–1 [51], Munc18–2 [50], Munc13 [52], and complexins I and II [50, 53, 54]. The particular protein isoforms responsible for any given event vary somewhat from species to species, but all are needed in some way to achieve the AR.

Much of what is known about the key regulators of the AR has been obtained via studies using sperm that have been permeabilized by the bacterial pore toxin streptolysin O (SLO). This toxin aggregates within the lipid bilayer of cells and forms stable pores without disrupting the integrity of the cell membrane as a whole [55] (such disruption occurs when non-ionic detergents are used to permeabilize cells) and leaves cell signaling molecules intact and functional [56]. Under some experimental conditions, SLO permeabilizes sperm without compromising motility [57] or acrosomal integrity [58]. The application of this toxin to capacitated sperm permits the influx of extracellular Ca^{2+} through the pores, bypassing the first physiological step in the AR, which is normally entry of Ca^{2+} through Ca^{2+} channels in response to the binding of progesterone or ZP3. The generation of these pores permits the introduction of substances that would otherwise be unable to enter the sperm, such as antibodies and recombinant proteins; thus, SLO has been a useful tool for teasing out each step of the AR.

14.3 Early stages of the AR: Rabs, Ca^{2+}, and cAMP

Rabs are small GTPases that act as molecular switches by alternating between active (GTP-loaded) and inactive (GDP-loaded) states [59]. Their cycling is catalyzed by guanine nucleotide exchange factors (GEFs) and GTPase activating proteins (GAPs). Rabs localize to specific compartments in cells which, in the case of the AR, are secretory vesicles. Rabs recruit effector proteins to the secretory vesicles that regulate different steps of membrane trafficking. Rab3A and Rab27 have been identified in mammalian sperm [41, 45, 60], where they localize to the acrosomal region of the sperm head [45, 60]. Rab27 is largely loaded with GTP and associates with membranes in capacitated but not yet activated sperm, whereas Rab3A mainly becomes associated with membranes after sperm activation [45]. Antibodies against both proteins block the Ca^{2+}-induced AR [40, 45]. Recombinant Rab3A that has been activated by loading it with GTPγS stimulates the AR in permeabilized sperm [41, 44], and Rab3A that has been loaded with GDPβS inhibits it [41]. Importantly, both Rab3A and Rab27 engineered to be arginine-rich (so that they cross the plasma membrane of intact sperm) also trigger the AR when they have been prenylated and loaded with GTPγS, showing that experiments in which sperm are permeabilized with SLO can be extended to events that occur under more physiological conditions [61, 62].

The acrosome itself is the major Ca^{2+} storage site in the sperm [63]. The influx of extracellular Ca^{2+} that occurs shortly after sperm stimulation with progesterone is followed by the release of Ca^{2+} into the cytoplasm through inositol 1,4,5-trisphosphate (IP_3)-sensitive Ca^{2+} channels [63]. This release of Ca^{2+} is stimulated by activated Rab3A and is required for the AR [63]. The use of a photoactivatable Ca^{2+} chelator, NP-EGTA-AM, in permeabilized sperm has been a valuable tool for examining the sequential steps involved in the AR. This chelator diffuses into the sperm and the acrosome, where it binds up Ca^{2+} until it is inactivated by exposure to UV light, thereby restoring the Ca^{2+} store within the acrosome [40]. Using this chelator, it has been shown that antibodies against Rab3A block the AR when added prior to the release of Ca^{2+} from the acrosome but not when added subsequently, demonstrating that Rab3A activity is required before IP_3-mediated Ca^{2+} release can take place [40]. Likewise, anti-Rab27 antibodies prevent the Ca^{2+}-induced AR prior to the efflux of Ca^{2+} from the acrosome [45]. The initial Ca^{2+} influx that occurs when Ca^{2+} channels are activated after progesterone or ZP3 stimulation indirectly activates Rab27, and this activation recruits Rab3A GEF activity, which activates Rab3A [33, 45]. Rab3A, in turn, activates downstream effectors that lead to IP_3-induced Ca^{2+} release from the acrosome (figure 14.2).

Another signaling molecule that is required for the AR is the second messenger, cAMP [33, 64, 65], which is synthesized by the sAC (or ADCY10), which is

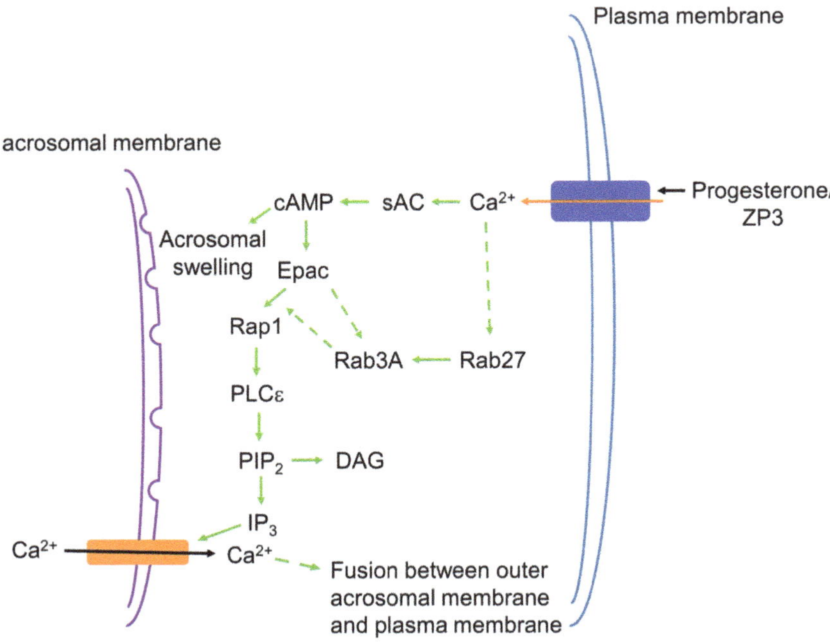

Figure 14.2. Steps leading to the efflux of Ca^{2+} from the acrosome prior to acrosomal exocytosis; sAC, soluble adenylate cyclase; cAMP, cyclic adenosine monophosphate; Epac, exchange factor directly activated by cAMP; Rap1, Ras-related protein 1; PLCε, phospholipase Cε; PIP_2, phosphatidylinositol 4,5-bisphosphosphate; DAG, diacylglycerol; IP_3, inositol 1,4,5-trisphosphate.

stimulated by the influx of extracellular Ca^{2+} caused by progesterone or ZP3 [33, 66–68]. Cyclic AMP acts through a PKA-independent mechanism, instead stimulating Epac (an exchange protein directly activated by cAMP; [64, 68]). Epac catalyzes the conversion of GDP into GTP on another small G-protein, Rap1 [68]. Rap1 activates phospholipase Cε, which cleaves the membrane phospholipid phosphatidylinositol 4,5-bisphosphosphate (PIP_2) to produce diacylglycerol (DAG) and IP_3. IP_3 binds to a receptor/Ca^{2+} channel located in the acrosomal membrane, releasing stored Ca^{2+} into the cytoplasm [64, 65], and DAG helps to maintain high levels of IP_3 [69]. It should be noted that there is evidence for the role of another phospholipase C, PLCδ4, in acrosomal exocytosis, as sperm from PLCδ4$^{-/-}$ mice are unable to mount a sustained release of Ca^{2+} in response to ZP3 or progesterone [39]. In this case, the activation of PLCδ4 could be due to the initial Ca^{2+} increase that occurs upon the activation of ZP3 or progesterone stimulation [70]. In any case, the efflux of Ca^{2+} from the acrosome is required for the subsequent stages of the AR. Cyclic AMP is required for the acrosomal swelling that occurs prior to the efflux of Ca^{2+} and brings the acrosomal membrane into close proximity with the plasma membrane; this swelling is a necessary intermediate step in the AR [34, 68, 71].

SNARE proteins mediate membrane fusion events in species from yeast to mammals [72], and sperm membrane fusion is no exception. As mentioned above, sperm contain many members of this protein family; the particular players vary somewhat in different mammalian species, but the general mechanisms by which they act are similar. Much information about the function of SNAREs in sperm has been discovered from the use of SLO-permeabilization, antibodies or recombinant proteins, and bacterial toxins. Several members of the SNARE protein family, including SNAP25, syntaxins, and VAMPs, are cleaved by botulinum and tetanus toxins but only when these proteins are in their monomeric forms rather than bound in *cis*- or *trans*-SNARE complexes [73]. Thus, it has been possible to tease out the sequential steps in acrosomal exocytosis by applying these toxins at different stages of exocytosis (i.e. before or after the efflux of Ca^{2+} from the acrosome).

Some of the earliest studies of SNARE proteins in mammalian sperm demonstrated that the chaperone proteins NSF and α-SNAP are required for both Ca^{2+}- and Rab3A-induced acrosomal exocytosis [40–42]. Antibodies against each of these proteins, as well as recombinant α-SNAP, block the AR in permeabilized sperm [42]. The addition of NSF/α-SNAP renders SNAREs sensitive to cleavage by tetanus toxin [42]. Thus, in capacitated, unstimulated sperm, SNAREs are in toxin-resistant *cis* complexes that must be disassembled by NSF/α-SNAP prior to achieving membrane fusion via *trans*-SNARE complexes. NSF is phosphorylated and kept inactive in sperm prior to the AR, but Rab3A activates the phosphatase PTP1B to dephosphorylate it ([43]; figure 14.3). Dephosphorylated, active NSF can then interact with α-SNAP, which stimulates the ATPase activity of NSF to promote the disassembly of *cis*-SNARE complexes. Thus, the activity of NSF/α-SNAP appears to be downstream of Rab3A activation (post-tethering) and upstream of Ca^{2+} efflux from the acrosome. This supports the hypothesis that SNARE proteins reside in *cis*-SNARE complexes prior to sperm activation (pre-fusion), that they dissociate into monomeric, toxin-sensitive forms after NSF/α-SNAP activation, and that they subsequently assemble into *trans*-SNARE complexes for membrane fusion.

Both SNAP25 and SNAP23 are expressed in human sperm and both appear to be required for the AR. This is based on the ability of bacterially expressed SNAP25, as well as antibodies against SNAP23, to block Ca^{2+}-induced exocytosis [46]. However, SNAP25 has been more extensively studied in human sperm, and is absolutely required for the AR because toxins that cleave it (botulinum neurotoxin A (BoNT/A) and botulinum neurotoxin E (BoNT/E)) inhibit the AR [46]. Neither BoNT/A nor BoNT/E cleave human SNAP23 [74, 75]. Therefore, if SNAP23 is

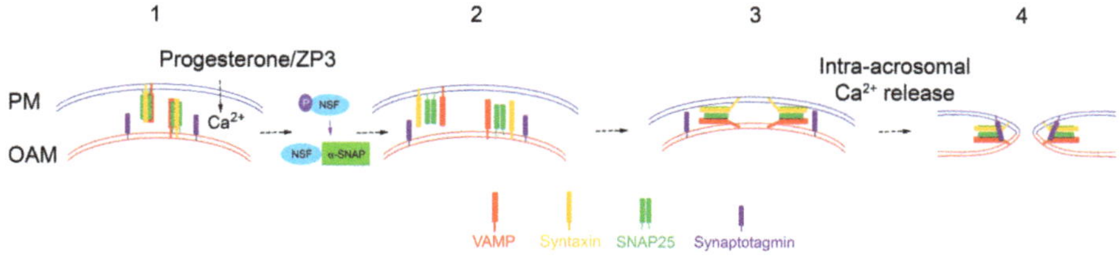

Figure 14.3. Schematic showing the basic core fusion machinery that operates during sperm acrosomal exocytosis: (1) influx of extracellular Ca^{2+} due to the stimulation of sperm by progesterone/ZP3; (2) disassembly of *cis*-SNARE complexes in response to the activation of α-SNAP, followed by (3) the formation of loose *trans*-SNARE complexes that then (4) form tight *trans*-SNARE complexes following the release of Ca^{2+} from the acrosomal store.

involved in the AR in humans, its activity is not enough to substitute for cleaved and inactive SNAP25, as has been observed in other cell types [74]. In contrast, SNAP23 appears to be the relevant SNARE in boar sperm AR [47, 50].

Syntaxins are t-SNAREs that are required to dock the acrosome to the plasma membrane [76]. Syntaxins 1A, 1B, 2, 3, 4, and 6 have all been identified in sperm. The isoforms that have been investigated for function (i.e. syntaxins 1, 2, 4, and 6) seem to have a role in mammalian sperm exocytosis [46, 47, 49]; the particular syntaxins expressed in sperm vary from species to species.

Vesicle-associated membrane proteins (VAMPs, or synaptobrevins) are v-SNAREs that are normally anchored in the vesicle membrane. VAMPs are expressed in mammalian sperm [50, 54]. In human sperm, antibodies against VAMP2 block the Ca^{2+}-induced AR [40, 46]. In capacitated boar sperm, VAMP3 is found in a ~75 kDa protein complex containing SNAP23 and syntaxin 1B [47, 50] and this trimeric complex is thought to mediate acrosomal docking to the plasma membrane prior to the AR. In contrast, in acrosome-reacted sperm, a 95 kDa protein complex containing SNAP23, syntaxin 3, and VAMP2 occurs, suggesting that different heterotrimeric SNARE complexes mediate docking and fusion. Further, this complex may need to interact with an additional protein, which has been identified as the regulatory protein, complexin [50].

14.4 Other regulatory proteins that control Ca^{2+}-induced acrosomal exocytosis: complexins, synaptotagmins, and Munc proteins

Complexins are small, soluble proteins that localize to the acrosomal region of mouse and human sperm [53, 54] and to membranes in capacitated, but not uncapacitated, boar sperm [50]. Complexins bind to the groove between the VAMP and syntaxin helices of the SNARE complex but not to individual SNARE proteins, stabilizing the fully assembled SNARE complex [77]. Both complexin 1 and complexin 2 are expressed in mouse sperm [54, 78] and mice deficient in complexin 1 do not readily undergo ZP-induced AR [54] and are subfertile [78]. In human sperm, antibodies against complexins, as well as the application of recombinant complexin to permeabilized sperm, inhibit the Ca^{2+}- and Rab3A-induced AR [53]. In sperm, complexin appears to act as a transient fusion clamp that is released by Ca^{2+} and synaptotagmin [33].

Synaptotagmins (Syts) have been studied most extensively in neurons. The 17 different isoforms have diverse functions, the best understood of which is to act as a Ca^{2+} sensor for vesicle exocytosis [79]. Types I, IV, VI, and VIII have been found in mammalian sperm [36, 48–50, 53]. Antibodies against Syt VI and Syt VIII inhibit exocytosis in permeabilized human [48] and mouse [49] sperm, respectively. In addition, the cytoplasmic domains of synaptotagmins block Ca^{2+}-induced AR [48, 49]. Syt VI is phosphorylated *in vitro* [48] and in capacitated, unstimulated sperm [80]. The introduction of phosphorylated Syt VI or phosphomimetic mutants into permeablized human sperm inhibits the Ca^{2+}-induced AR [53]. Syt VI is dephosphorylated by calcineurin after Ca^{2+} efflux from the acrosome [81], at which point it can displace complexin that is bound to the SNARE complex and promote the zippering of loose *trans*-SNARE complexes that form just before acrosome fusion with the plasma membrane.

The Munc18 and Munc13 proteins are also important for the Ca^{2+}-induced AR [51, 52]. Munc18–1 is expressed in human sperm and antibodies against it block the Ca^{2+}- and Rab3A-induced AR [51]. Munc18–1 functions at a step before the efflux of Ca^{2+} from the acrosome, when SNAREs transition between *cis* and loose *trans* complexes before Ca^{2+} stimulates fusion [51]. Munc13 is also expressed in the acrosomal region of human sperm and an antibody against Munc13 blocks the Ca^{2+}-induced AR [52]. Like Munc18, Munc13 is important at an early step in exocytosis, before Ca^{2+} efflux from the acrosome takes place [52].

In summary, the AR is a complex event that is tightly regulated so that a spermatozoan, which can only undergo one acrosome reaction, exocytoses the acrosome when it is in the correct time and place to do so. This process is similar to that performed for other secretory cell types; it involves vesicle tethering, mediated by Rab proteins; docking, mediated by loose *trans-SNARE* complexes; priming of the fusion machinery by NSF/α-SNAP; efflux of Ca^{2+} from the acrosome to stimulate synaptotagmin, which displaces complexin; expansion of the fusion pore; and finally, membrane fusion.

14.5 Mammalian oocytes undergo constitutive exocytosis prior to oocyte maturation

Immature mouse oocytes (herein referred to as 'oocytes,' as opposed to mature, metaphase II stage 'eggs') are arrested at the diplotene stage of meiotic prophase from around the time of birth to the time just before ovulation. Meiotically arrested oocytes are surrounded by somatic cells that comprise the ovarian follicle. After puberty, follicles are recruited to begin growth and this growth continues under the influence of follicle-stimulating hormone (FSH), which is released by the anterior pituitary. As the oocyte reaches its full size, the follicle cells surrounding it acquire luteinizing hormone (LH) receptors that permit the follicle to respond to the surge of LH that occurs just before ovulation. Oocytes remain arrested in prophase until the LH surge, at which time they are stimulated to mature to the metaphase II stage, in which they undergo a secondary arrest that is relieved upon fertilization (see [82, 83] for reviews).

Meiotically arrested oocytes synthesize many RNAs and proteins that are needed to support early embryonic development after fertilization. Oocytes also actively secrete proteins including growth differentiation factor 9 (GDF9) and bone morphogenetic protein 15 (BMP15) [84]. These growth factors act on the follicle cells to stimulate their proliferation as well as the synthesis of cholesterol and metabolic byproducts that are transmitted to the oocyte and needed for oocyte viability [84, 85]. Little is known about the molecular mechanisms of constitutive secretion in oocytes, as there are no ultrastructural data showing growth factors within vesicles or exosomes. Measurements of secreted products in the follicular fluid of maturing oocytes have shown that exocytosis is a developmentally regulated process, in which some factors are secreted prior to the onset of maturation and some during maturation [86]. It should be noted that there is a disconnect between oocyte and follicle cells during maturation, such that gap junction connections are broken; the ensuing leakage through gap junctional remnants could possibly account for some of the proteins that are presumably 'secreted' due to their appearance in follicular fluid during oocyte maturation.

14.6 Release of Ca^{2+} from intracellular stores in the egg is the trigger for cortical granule exocytosis after fertilization

Soon after fertilization, the eggs in species as diverse as marine invertebrates, amphibians, and mammals undergo the exocytosis of cortical granules, a major polyspermy prevention mechanism. As in other secretory systems, Ca^{2+} is essential for this process. Ca^{2+} is stored in the endoplasmic reticulum (ER) and is released into the cytoplasm at fertilization shortly after sperm–egg fusion. In mammals, a sperm-specific phospholipase C, PLCζ, enters the egg soon after sperm–egg fusion and cleaves phosphatidylinositol 4,5-bisphosphate (PIP_2) in the egg membrane to produce DAG and inositol 1,4,5-trisphosphate (IP_3; [87]). IP_3 binds to a receptor on the ER, which is also a Ca^{2+} channel, causing the channel to open and stored Ca^{2+} to be released into the egg's cytoplasm. In mammalian eggs, much of the released Ca^{2+} is pumped back into the ER and re-released in a process of Ca^{2+}-induced Ca^{2+} release that leads to a series of repetitive Ca^{2+} oscillations that persist for several hours [88, 89]. Cortical granules are released during the first few oscillations [90]. Although the mechanistic details leading to the release of Ca^{2+} at fertilization differ in aquatic species, Ca^{2+} release stimulates cortical granule exocytosis in these species and the molecular machinery required for exocytosis is remarkably conserved from echinoderms to mammals [91].

14.7 SNARE proteins mediate constitutive exocytosis and CGE

Less is known about mechanisms of exocytosis in oocytes and eggs than in sperm. However, it is clear that SNARE proteins are required for both constitutive and regulated exocytosis [92] and that these processes are likely very similar to the exocytosis of the acrosomal vesicle in sperm. As in sperm acrosomal exocytosis, cortical granule exocytosis is a one-time event, in which exocytosis occurs without compensatory endocytosis and the separation of *trans*-SNARE complexes by NSF/α-SNAP in preparation for another round of exocytosis. The microinjection of antibodies against NSF/α-SNAP blocks CGE in activated mouse eggs, suggesting that these proteins are required for CGE [93]. This is in contrast to NSF in sea urchin eggs, which appears not to be required for CGE [94]. Because cortical granules are not renewed after exocytosis, NSF/α-SNAP in mouse eggs are likely to act in a pre-fusion step, as in sperm, and are required to disassemble preformed *cis*-SNARE complexes located on both cortical granule and plasma membranes prior to the formation of *trans*-SNARE complexes after Ca^{2+} release at fertilization (figure 14.4). However, it should be noted that the localization of SNAREs on both cortical granule and plasma membranes has not been confirmed using methods

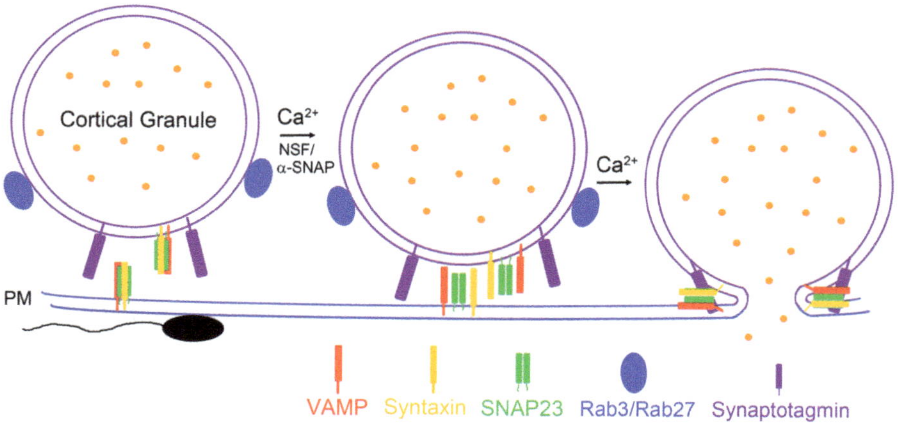

Figure 14.4. Model showing the core fusion machinery required for CGE. SNAREs reside in *cis*-SNARE complexes on both cortical granule and plasma membranes. In response to sperm–egg fusion, Ca^{2+} is released from the ER, stimulating the activity of NSF/α-SNAP, which disassembles SNARE complexes into monomeric forms. Finally, Ca^{2+} release is sensed by synaptotagmin, which stimulates the formation of *trans*-SNARE complexes that zip together to cause fusion between the CG and plasma membranes.

that would allow this distinction, so at present the hypothesis that these SNARE complexes are located on both membranes remains somewhat speculative.

SNAP23 is essential for both constitutive and regulated exocytosis in mouse oocytes and eggs, respectively [92]. SNAP23 protein is highly expressed in mouse and pig oocytes and eggs [92, 95]. SNAP23-depleted, fully-grown mouse oocytes are unable to maintain meiotic arrest and mature inappropriately within the ovarian follicle [92], demonstrating the importance of exocytosis prior to oocyte maturation. It is possible that constitutive exocytosis ensures the insertion of gap junction proteins into the oocyte plasma membrane, as patent gap junctions between the oocyte and cumulus cells are required for the maintenance of meiotic arrest [96], and the half-life of Cx37, the connexin that comprises the gap junctions, is very short [97]. Constitutive exocytosis is also likely to be necessary for the secretion of oocyte-derived factors that act on the cumulus cells to regulate follicle growth and differentiation and, in turn, cumulus cells produce factors that are required for oocyte growth and development [84]. In porcine eggs, SNAP23 appears to associate with syntaxin 2 and VAMP1 [95], although the function of these proteins in exocytosis has not been tested. The proteins that interact with SNAP23 and participate in mouse oocyte constitutive exocytosis have not been identified. However, the syntaxin 4 protein is expressed in mouse oocytes [98] and our lab has identified other syntaxins that may play a role in this process (L. Mehlmann, unpublished results). There have been no other published reports on the molecules that participate in constitutive exocytosis in oocytes.

A 1998 report concluded that SNAP25 is essential for cortical granule exocytosis in mouse eggs, based on the observation that microinjected BoNT/A reduced CGE by about half in activated mouse eggs [99]. The authors of that study reported a relatively weak band for SNAP25 in a western blot using a large amount of total protein. However, a subsequent study clearly showed that the SNAP25 protein is not expressed in mouse eggs, whereas SNAP23 is abundantly expressed [92]. Moreover, specific depletion of SNAP23 in mouse eggs completely inhibits CGE [92]. A possible explanation for the disparate conclusions from the two studies could be that in contrast to human SNAP23, BoNT/A cleaves mouse SNAP23, albeit to a lesser extent than SNAP25 [100]. SNAP25 and SNAP23 are 59% identical and 72% similar at the amino acid level [100] and it is conceivable that the SNAP25 antibody used by Ikebuchi *et al* [99] cross-reacted with SNAP23 enough to produce a faint signal on a western blot.

Sea urchin eggs contain a single syntaxin and VAMP which mediate CGE [91], whereas mammalian eggs contain multiple syntaxins and VAMPs. Specific syntaxin proteins that interact with SNAP23 in mouse eggs have not yet been identified, although, as mentioned above, syntaxin 4 is expressed [98]. Tetanus toxin, which cleaves VAMPs 1, 2, and 3, reduces CGE in mouse eggs but does not block it completely, suggesting that VAMP1 and/or VAMP3 (VAMP2 is not expressed) participate in CGE but that another, as-yet unidentified VAMP may also have a role [101]. In porcine eggs, the co-localization of SNAP23, syntaxin 2, and VAMP1 suggests that VAMP1 might be involved in CGE [95], but a definitive role for syntaxin 2 and VAMP1 in porcine CGE has not been established.

Rab proteins are also involved in CGE. Rab3 was first identified as being required for CGE in sea urchin eggs [91]. Mammals express four Rab3 isoforms [102], and Rab3A has been reported be localized in the cortical region of mouse eggs [103, 104]; but see [105]. Bello *et al* [103] showed that (1) Rab3A co-localizes with cortical granules; (2) the injection of activated Rab3A causes CGE; and (3) the injection of an antibody against Rab3A blocks CGE. These results strongly suggest a role for Rab3A in CGE. In addition, Rab3A is no longer detectable at the surface of eggs that have undergone CGE [103], suggesting that the protein is lost during the process. Another report, however, indicated that Rab3A is localized throughout the cytoplasm and has a role in spindle migration during oocyte maturation and in asymmetric cell division, rather than in CGE [105]. Rab27 co-localizes with cortical granules and knocking down Rab27 with RNA interference or blocking its function with specific antibodies inhibits CGE [105], suggesting a role for this Rab isoform. The Rab3/Rab27 effector protein, Rabphilin 3A, is also likely to be involved in CGE. The Rabphilin 3A protein is localized in the cortical region of mouse eggs and inhibiting its function by microinjecting N- or C-terminal fragments blocks CGE [106]. Whether Rabphilin acts as an effector for Rab3A or Rab27, or both, remains to be elucidated, as well as the specific roles for each of the two Rabs.

Synaptotagmin I has also been identified as having a role in mouse egg CGE [107]. Synaptotagmin I was identified in the cortex region of oocytes and eggs and knocking down this protein using RNA interference (RNAi) or morpholinos diminishes CGE [107]. However, synaptotagmin I knockdown also affects Ca^{2+} oscillations after artificial egg activation and seems to affect meiotic spindle stability [108]. It is not readily apparent how synaptotagmin regulates all of these functions; more studies are required to clarify the role(s) this protein plays in oocyte function. Other exocytic proteins that function in CGE have not yet been identified.

It is not yet known how oocytes and eggs use the same proteins to undergo the two separate events of constitutive and regulated exocytosis. It is likely that one or more regulatory proteins is synthesized during oocyte maturation that allows mature eggs to undergo regulated exocytosis, as immature oocytes are unable to undergo CGE even with a Ca^{2+} stimulus [109]. In support of this idea, the levels of the Ca^{2+} sensor synaptotagmin I increase during mouse oocyte maturation [108]. Although far less is known about the function of SNAREs and interacting proteins in oocytes and eggs than in sperm, the availability of new and powerful tools such as Trim-away for the acute and specific depletion of proteins [92, 110, 111] could prove instrumental in filling the knowledge gaps.

Acknowledgments

I thank Bruce White and Mark Pixley for critically reading the manuscript. This work was supported by grant HD096037 from the NIH.

References

[1] Abou-Haila A and Tulsiani D R 2000 Mammalian sperm acrosome: formation, contents, and function *Arch. Biochem. Biophys.* **379** 173–82

[2] Bleil J D and Wassarman P M 1983 Sperm-egg interactions in the mouse: sequence of events and induction of the acrosome reaction by a zona pellucida glycoprotein *Dev. Biol.* **95** 317–24

[3] Fazeli A, Hage W J, Cheng F P, Voorhout W F, Marks A, Bevers M M and Colenbrander B 1997 Acrosome-intact boar spermatozoa initiate binding to the homologous zona pellucida *in vitro Biol. Reprod.* **56** 430–8

[4] Florman H M, Jungnickel M K and Sutton K A 2008 Regulating the acrosome reaction *Int. J. Dev. Biol.* **52** 503–10

[5] Hirohashi N and Yanagimachi R 2018 Sperm acrosome reaction: its site and role in fertilization *Biol. Reprod.* **99** 127–33

[6] Jin M, Fujiwara E, Kakiuchi Y, Okabe M, Satouh Y, Baba S A, Chiba K and Hirohashi N 2011 Most fertilizing mouse spermatozoa begin their acrosome reaction before contact with the zona pellucida during *in vitro* fertilization *Proc. Natl Acad. Sci. USA* **108** 4892–6

[7] Fukui Y 1990 Effect of follicle cells on the acrosome reaction, fertilization, and developmental competence of bovine oocytes matured *in vitro Mol. Reprod. Dev.* **26** 40–6

[8] Cross N L and Elinson R P 1980 A fast block to polyspermy in frogs mediated by changes in the membrane potential *Dev. Biol.* **75** 187–98

[9] Jaffe L A 1976 Fast block to polyspermy in sea urchin eggs is electrically mediated *Nature* **261** 68–71

[10] Jaffe L A 2018 The fast block to polyspermy: new insight into a century-old problem *J. Gen. Physiol.* **150** 1233–4

[11] Wong J L and Wessel G M 2006 Defending the zygote: search for the ancestral animal block to polyspermy *Curr. Top. Dev. Biol.* **72** 1–151

[12] Fahrenkamp E, Algarra B and Jovine L 2020 Mammalian egg coat modifications and the block to polyspermy *Mol. Reprod. Dev.* **87** 326–40

[13] Boue J, Bou A and Lazar P 1975 Retrospective and prospective epidemiological studies of 1500 karyotyped spontaneous human abortions *Teratology* **12** 11–26

[14] Visconti P E, Ning X, Fornes M W, Alvarez J G, Stein P, Connors S A and Kopf G S 1999a Cholesterol efflux-mediated signal transduction in mammalian sperm: cholesterol release signals an increase in protein tyrosine phosphorylation during mouse sperm capacitation *Dev. Biol.* **214** 429–43

[15] Visconti P E, Moore G D, Bailey J L, Leclerc P, Connors S A, Pan D, Olds-Clarke P and Kopf G S 1995b Capacitation of mouse spermatozoa. II. protein tyrosine phosphorylation and capacitation are regulated by a cAMP-dependent pathway *Development* **121** 1139–50

[16] Nishigaki T, Jose O, Gonzalez-Cota A L, Romero F, Trevino C L and Darszon A 2014 Intracellular pH in sperm physiology *Biochem. Biophys. Res. Commun.* **450** 1149–58

[17] Puga Molina L C, Luque G M, Balestrini P A, Marin-Briggiler C I, Romarowski A and Buffone M G 2018 Molecular basis of human sperm capacitation *Front. Cell. Dev. Biol.* **6** 72

[18] Zeng Y, Oberdorf J A and Florman H M 1996 pH regulation in mouse sperm: identification of Na^+-, Cl^--, and HCO_3^--dependent and arylaminobenzoate-dependent regulatory mechanisms and characterization of their roles in sperm capacitation *Dev. Biol.* **173** 510–20

[19] Galantino-Homer H L, Visconti P E and Kopf G S 1997 Regulation of protein tyrosine phosphorylation during bovine sperm capacitation by a cyclic adenosine 3'5'-monophosphate-dependent pathway *Biol. Reprod.* **56** 707–19

[20] Leclerc P, de Lamirande E and Gagnon C 1996 Cyclic adenosine 3',5'monophosphate-dependent regulation of protein tyrosine phosphorylation in relation to human sperm capacitation and motility *Biol. Reprod.* **55** 684–92

[21] Visconti P E, Bailey J L, Moore G D, Pan D, Olds-Clarke P and Kopf G S 1995a Capacitation of mouse spermatozoa. I. correlation between the capacitation state and protein tyrosine phosphorylation *Development* **121** 1129–37

[22] Visconti P E, Galantino-Homer H, Ning X, Moore G D, Valenzuela J P, Jorgez C J, Alvarez J G and Kopf G S 1999b Cholesterol efflux-mediated signal transduction in mammalian sperm. β-cyclodextrins initiate transmembrane signaling leading to an increase in protein tyrosine phosphorylation and capacitation *J. Biol. Chem.* **274** 3235–42

[23] Orta G, de la Vega-Beltran J L, Martin-Hidalgo D, Santi C M, Visconti P E and Darszon A 2018 CatSper channels are regulated by protein kinase A *J. Biol. Chem.* **293** 16830–41

[24] Cross N L 1998 Role of cholesterol in sperm capacitation *Biol. Reprod.* **59** 7–11

[25] Lishko P V and Mannowetz N 2018 CatSper: a unique calcium channel of the sperm flagellum *Curr. Opin. Physiol.* **2** 109–13

[26] Rahban R and Nef S 2020 CatSper: the complex main gate of calcium entry in mammalian spermatozoa *Mol. Cell. Endocrinol.* **518** 110951

[27] Sun X H, Zhu Y Y, Wang L, Liu H L, Ling Y, Li Z L and Sun L B 2017 The catsper channel and its roles in male fertility: a systematic review *Reprod. Biol. Endocrinol.* **15** 65

[28] Suarez S S 2008 Control of hyperactivation in sperm *Hum. Reprod. Update* **14** 647–57

[29] Wassarman P M and Litscher E S 2018 The mouse egg's zona pellucida *Curr. Top. Dev. Biol.* **130** 331–56

[30] Jungnickel M K, Marrero H, Birnbaumer L, Lemos J R and Florman H M 2001 Trp2 regulates entry of Ca^{2+} into mouse sperm triggered by egg ZP3 *Nat. Cell Biol.* **3** 499–02

[31] Lishko P V, Botchkina I L and Kirichok Y 2011 Progesterone activates the principal Ca^{2+} channel of human sperm *Nature* **471** 387–91

[32] Sabeur K, Edwards D P and Meizel S 1996 Human sperm plasma membrane progesterone receptor(s) and the acrosome reaction *Biol. Reprod.* **54** 993–1001

[33] Tomes C N 2015 The proteins of exocytosis: lessons from the sperm model *Biochem. J.* **465** 359–70

[34] Zanetti N and Mayorga L S 2009 Acrosomal swelling and membrane docking are required for hybrid vesicle formation during the human sperm acrosome reaction *Biol. Reprod.* **81** 396–405

[35] Tulsiani D R, Abou-Haila A, Loeser C R and Pereira B M 1998 The biological and functional significance of the sperm acrosome and acrosomal enzymes in mammalian fertilization *Exp. Cell. Res.* **240** 151–64

[36] Ramalho-Santos J, Moreno R D, Sutovsky P, Chan A W, Hewitson L, Wessel G M, Simerly C R and Schatten G 2000 SNAREs in mammalian sperm: possible implications for fertilization *Dev. Biol.* **223** 54–69

[37] Florman H M, Arnoult C, Kazam I G, Li C and O'Toole C M 1998 A perspective on the control of mammalian fertilization by egg-activated ion channels in sperm: a tale of two channels *Biol. Reprod.* **59** 12–6

[38] Foresta C, Rossato M and Di Virgilio F 1993 Ion fluxes through the progesterone-activated channel of the sperm plasma membrane *Biochem. J.* **294** 279–83

[39] Fukami K, Yoshida M, Inoue T, Kurokawa M, Fissore R A, Yoshida N, Mikoshiba K and Takenawa T 2003 Phospholipase Cδ4 is required for Ca^{2+} mobilization essential for acrosome reaction in sperm *J. Cell Biol.* **161** 79–88

[40] De Blas G A, Roggero C M, Tomes C N and Mayorga L S 2005 Dynamics of SNARE assembly and disassembly during sperm acrosomal exocytosis *PLoS Biol.* **3** e323

[41] Michaut M, Tomes C N, De Blas G, Yunes R and Mayorga L S 2000 Calcium-triggered acrosomal exocytosis in human spermatozoa requires the coordinated activation of Rab3A and N-ethylmaleimide-sensitive factor *Proc. Natl. Acad. Sci. USA* **97** 9996–10001

[42] Tomes C N, De Blas G A, Michaut M A, Farre E V, Cherhitin O, Visconti P E and Mayorga L S 2005 α-SNAP and NSF are required in a priming step during the human sperm acrosome reaction *Mol. Hum. Reprod.* **11** 43–51

[43] Zarelli V E, Ruete M C, Roggero C M, Mayorga L S and Tomes C N 2009 PTP1B dephosphorylates N-ethylmaleimide-sensitive factor and elicits SNARE complex disassembly during human sperm exocytosis *J. Biol. Chem.* **284** 10491–503

[44] Yunes R, Michaut M, Tomes C and Mayorga L S 2000 Rab3A triggers the acrosome reaction in permeabilized human spermatozoa *Biol. Reprod.* **62** 1084–9

[45] Bustos M A, Lucchesi O, Ruete M C, Mayorga L S and Tomes C N 2012 Rab27 and Rab3 sequentially regulate human sperm dense-core granule exocytosis *Proc. Natl Acad. Sci. USA* **109** E2057–66

[46] Tomes C N, Michaut M, De Blas G, Visconti P, Matti U and Mayorga L S 2002 SNARE complex assembly is required for human sperm acrosome reaction *Dev. Biol.* **243** 326–38

[47] Tsai P S, Garcia-Gil N, van Haeften T and Gadella B M 2010 How pig sperm prepares to fertilize: stable acrosome docking to the plasma membrane *PLoS One* **5** e11204

[48] Michaut M, De Blas G, Tomes C N, Yunes R, Fukuda M and Mayorga L S 2001 Synaptotagmin VI participates in the acrosome reaction of human spermatozoa *Dev. Biol.* **235** 521–9

[49] Hutt D M, Baltz J M and Ngsee J K 2005 Synaptotagmin VI and VIII and syntaxin 2 are essential for the mouse sperm acrosome reaction *J. Biol. Chem.* **280** 20197–203

[50] Tsai P S, Brewis I A, van Maaren J and Gadella B M 2012 Involvement of complexin 2 in docking, locking and unlocking of different SNARE complexes during sperm capacitation and induced acrosomal exocytosis *PLoS One* **7** e32603

[51] Rodriguez F, Zanetti M N, Mayorga L S and Tomes C N 2012 Munc18-1 controls SNARE protein complex assembly during human sperm acrosomal exocytosis *J. Biol. Chem.* **287** 43825–39

[52] Bello O D, Zanetti M N, Mayorga L S and Michaut M A 2012 RIM, Munc13, and Rab3A interplay in acrosomal exocytosis *Exp. Cell. Res.* **318** 478–88

[53] Roggero C M, De Blas G A, Dai H, Tomes C N, Rizo J and Mayorga L S 2007 Complexin/synaptotagmin interplay controls acrosomal exocytosis *J. Biol. Chem.* **282** 26335–43

[54] Zhao L, Burkin H R, Shi X, Li L, Reim K and Miller D J 2007 Complexin I is required for mammalian sperm acrosomal exocytosis *Dev. Biol.* **309** 236–44

[55] Hugo F, Reichwein J, Arvand M, Kramer S and Bhakdi S 1986 Use of a monoclonal antibody to determine the mode of transmembrane pore formation by streptolysin O *Infect. Immun.* **54** 641–5

[56] Stephens L, Jackson T and Hawkins P T 1993 Synthesis of phosphatidylinositol 3,4,5-trisphosphate in permeabilized neutrophils regulated by receptors and G-proteins *J. Biol. Chem.* **268** 17162–72

[57] Johnson L R, Moss S B and Gerton G L 1999 Maintenance of motility in mouse sperm permeabilized with streptolysin O *Biol. Reprod.* **60** 683–90

[58] Diaz A, Dominguez L, Fornes M W, Burgos M H and Mayorga L S 1996 Acrosome content release in streptolysin O permeabilized mouse spermatozoa *Andrologia* **28** 21–6

[59] Homma Y, Hiragi S and Fukuda M 2021 Rab family of small GTPases: an updated view on their regulation and functions *FEBS J.* **288** 36–55

[60] Yunes R, Tomes C, Michaut M, De Blas G, Rodriguez F, Regazzi R and Mayorga L S 2002 Rab3A and calmodulin regulate acrosomal exocytosis by mechanisms that do not require a direct interaction *FEBS Lett.* **525** 126–30

[61] Lopez C I, Belmonte S A, De Blas G A and Mayorga L S 2007 Membrane-permeant Rab3A triggers acrosomal exocytosis in living human sperm *FASEB J.* **21** 4121–30

[62] Bustos M A, Lucchesi O, Ruete M C and Tomes C N 2018 Membrane-permeable Rab27A is a regulator of the acrosome reaction: role of geranylgeranylation and guanine nucleotides *Cell. Signal.* **44** 72–81

[63] De Blas G, Michaut M, Trevino C L, Tomes C N, Yunes R, Darszon A and Mayorga L S 2002 The intraacrosomal calcium pool plays a direct role in acrosomal exocytosis *J. Biol. Chem.* **277** 49326–31

[64] Branham M T, Mayorga L S and Tomes C N 2006 Calcium-induced acrosomal exocytosis requires cAMP acting through a protein kinase a-independent, Epac-mediated pathway *J. Biol. Chem.* **281** 8656–66

[65] Ruete M C, Lucchesi O, Bustos M A and Tomes C N 2014 Epac, Rap and Rab3 act in concert to mobilize calcium from sperm's acrosome during exocytosis *Cell. Commun. Signal.* **12** 43

[66] Jaiswal B S and Conti M 2003 Calcium regulation of the soluble adenylyl cyclase expressed in mammalian spermatozoa *Proc. Natl Acad. Sci. USA* **100** 10676–81

[67] Jose O, Hernandez-Hernandez O, Chirinos M, Gonzalez-Gonzalez M E, Larrea F, Almanza A, Felix R, Darszon A and Trevino C L 2010 Recombinant human ZP3-induced sperm acrosome reaction: evidence for the involvement of T- and L-type voltage-gated calcium channels *Biochem. Biophys. Res. Commun.* **395** 530–4

[68] Lucchesi O, Ruete M C, Bustos M A, Quevedo M F and Tomes C N 2016 The signaling module cAMP/Epac/Rap1/PLCepsilon/IP_3 mobilizes acrosomal calcium during sperm exocytosis *Biochim. Biophys. Acta* **1863** 544–61

[69] Lopez C I, Pelletan L E, Suhaiman L, De Blas G A, Vitale N, Mayorga L S and Belmonte S A 2012 Diacylglycerol stimulates acrosomal exocytosis by feeding into a PKC- and PLD1-dependent positive loop that continuously supplies phosphatidylinositol 4,5-bisphosphate *Biochim. Biophys. Acta* **1821** 1186–99
[70] Fukami K, Takenaka K, Nagano K and Takenawa T 2000 Growth factor-induced promoter activation of murine phospholipase C δ4 gene *Eur. J. Biochem.* **267** 28–36
[71] Sosa C M, Zanetti M N, Pocognoni C A and Mayorga L S 2016 Acrosomal swelling is triggered by cAMP downstream of the opening of store-operated calcium channels during acrosomal exocytosis in human sperm *Biol. Reprod.* **94** 57
[72] Grissom J H, Segarra V A and Chi R J 2020 New perspectives on SNARE function in the yeast minimal endomembrane system *Genes (Basel)* **11** 899
[73] Hayashi T, McMahon H, Yamasaki S, Binz T, Hata Y, Sudhof T C and Niemann H 1994 Synaptic vesicle membrane fusion complex: action of clostridial neurotoxins on assembly *EMBO J.* **13** 5051–61
[74] Sadoul K, Berger A, Niemann H, Weller U, Roche P A, Klip A, Trimble W S, Regazzi R, Catsicas S and Halban P A 1997 SNAP-23 is not cleaved by botulinum neurotoxin E and can replace SNAP-25 in the process of insulin secretion *J. Biol. Chem.* **272** 33023–7
[75] Sikorra S, Donald S, Elliott M, Schwede S, Coker S F, Kupinski A P, Tripathi V, Foster K, Beard M and Binz T 2020 Engineering an effective human SNAP-23 cleaving botulinum neurotoxin a variant *Toxins (Basel)* **12** 804
[76] Rodriguez F, Bustos M A, Zanetti M N, Ruete M C, Mayorga L S and Tomes C N 2011 α-SNAP prevents docking of the acrosome during sperm exocytosis because it sequesters monomeric syntaxin *PLoS One* **6** e21925
[77] Chen X, Tomchick D R, Kovrigin E, Arac D, Machius M, Sudhof T C and Rizo J 2002 Three-dimensional structure of the complexin/SNARE complex *Neuron* **33** 397–409
[78] Zhao L, Reim K and Miller D J 2008 Complexin-I-deficient sperm are subfertile due to a defect in zona pellucida penetration *Reproduction* **136** 323–34
[79] Wolfes A C and Dean C 2020 The diversity of synaptotagmin isoforms *Curr. Opin. Neurobiol.* **63** 198–209
[80] Roggero C M, Tomes C N, De Blas G A, Castillo J, Michaut M A, Fukuda M and Mayorga L S 2005 Protein kinase C-mediated phosphorylation of the two polybasic regions of synaptotagmin VI regulates their function in acrosomal exocytosis *Dev. Biol.* **285** 422–35
[81] Castillo Bennett J, Roggero C M, Mancifesta F E and Mayorga L S 2010 Calcineurin-mediated dephosphorylation of synaptotagmin VI is necessary for acrosomal exocytosis *J. Biol. Chem.* **285** 26269–78
[82] Orisaka M, Miyazaki Y, Shirafuji A, Tamamura C, Tsuyoshi H, Tsang B K and Yoshida Y 2021 The role of pituitary gonadotropins and intraovarian regulators in follicle development: a mini-review *Reprod. Med. Biol.* **20** 169–75
[83] Richards J S and Ascoli M 2018 Endocrine, paracrine, and autocrine signaling pathways that regulate ovulation *Trends Endocrinol. Metab.* **29** 313–25
[84] Matzuk M M, Burns K H, Viveiros M M and Eppig J J 2002 Intercellular communication in the mammalian ovary: oocytes carry the conversation *Science* **296** 2178–80
[85] Su Y Q, Sugiura K, Wigglesworth K, O'Brien M J, Affourtit J P, Pangas S A, Matzuk M M and Eppig J J 2008 Oocyte regulation of metabolic cooperativity between mouse cumulus cells and oocytes: BMP15 and GDF9 control cholesterol biosynthesis in cumulus cells *Development* **135** 111–21

[86] Cakmak H, Franciosi F, Zamah A M, Cedars M I and Conti M 2016 Dynamic secretion during meiotic reentry integrates the function of the oocyte and cumulus cells *Proc. Natl Acad. Sci. USA* **113** 2424–9
[87] Swann K 2020 The soluble sperm factor that activates the egg: PLCζ and beyond *Reproduction* **160** V9–V11
[88] Kline D and Kline J T 1992 Thapsigargin activates a calcium influx pathway in the unfertilized mouse egg and suppresses repetitive calcium transients in the fertilized egg *J. Biol. Chem.* **267** 17624–30
[89] Stein P, Savy V, Williams A M and Williams C J 2020 Modulators of calcium signalling at fertilization *Open. Biol.* **10** 200118
[90] Ozil J P, Markoulaki S, Toth S, Matson S, Banrezes B, Knott J G, Schultz R M, Huneau D and Ducibella T 2005 Egg activation events are regulated by the duration of a sustained [Ca^{2+}]cyt signal in the mouse *Dev. Biol.* **282** 39–54
[91] Conner S, Leaf D and Wessel G 1997 Members of the SNARE hypothesis are associated with cortical granule exocytosis in the sea urchin egg *Mol. Reprod. Dev.* **48** 106–18
[92] Mehlmann L M, Uliasz T F and Lowther K M 2019 SNAP23 is required for constitutive and regulated exocytosis in mouse oocytes *Biol. Reprod.* **101** 338–46
[93] de Paola M, Bello O D and Michaut M A 2015 Cortical granule exocytosis is mediated by alpha-SNAP and N-ethilmaleimide sensitive factor in mouse oocytes *PLoS One* **10** e0135679
[94] Whalley T, Timmers K, Coorssen J, Bezrukov L, Kingsley D H and Zimmerberg J 2004 Membrane fusion of secretory vesicles of the sea urchin egg in the absence of NSF *J. Cell Sci.* **117** 2345–56
[95] Tsai P S, van Haeften T and Gadella B M 2011 Preparation of the cortical reaction: maturation-dependent migration of SNARE proteins, clathrin, and complexin to the porcine oocyte's surface blocks membrane traffic until fertilization *Biol. Reprod.* **84** 327–35
[96] Norris R P, Freudzon M, Mehlmann L M, Cowan A E, Simon A M, Paul D L, Lampe P D and Jaffe L A 2008 Luteinizing hormone causes MAP kinase-dependent phosphorylation and closure of connexin 43 gap junctions in mouse ovarian follicles: one of two paths to meiotic resumption *Development* **135** 3229–38
[97] Laird D W 2006 Life cycle of connexins in health and disease *Biochem. J.* **394** 527–43
[98] Iwahashi K, Kuji N, Fujiwara T, Tanaka H, Takahashi J, Inagaki N, Komatsu S, Yamamoto A, Yoshimura Y and Akagawa K 2003 Expression of the exocytotic protein syntaxin in mouse oocytes *Reproduction* **126** 73–81
[99] Ikebuchi Y, Masumoto N, Matsuoka T, Yokoi T, Tahara M, Tasaka K, Miyake A and Murata Y 1998 SNAP-25 is essential for cortical granule exocytosis in mouse eggs *Am. J. Physiol.* **274** C1496–500
[100] Vaidyanathan V V, Yoshino K, Jahnz M, Dorries C, Bade S, Nauenburg S, Niemann H and Binz T 1999 Proteolysis of SNAP-25 isoforms by botulinum neurotoxin types A, C, and E: domains and amino acid residues controlling the formation of enzyme-substrate complexes and cleavage *J. Neurochem.* **72** 327–37
[101] de Paola M, Garrido F, Zanetti M N and Michaut M A 2021 VAMPs sensitive to tetanus toxin are required for cortical granule exocytosis in mouse oocytes *Exp. Cell. Res.* **405** 112629
[102] Stenmark H and Olkkonen V M 2001 The Rab GTPase family *Genome. Biol.* **2** reviews3007.1

[103] Bello O D, Cappa A I, de Paola M, Zanetti M N, Fukuda M, Fissore R A, Mayorga L S and Michaut M A 2016 Rab3A, a possible marker of cortical granules, participates in cortical granule exocytosis in mouse eggs *Exp. Cell. Res.* **347** 42–51

[104] Masumoto N, Ikebuchi Y, Tahara M, Yokoi T, Tasaka K, Miyake A and Murata Y 1998 Expression of Rab3A in the cortical region in mouse metaphase II eggs *J. Exp. Zool.* **280** 91–6

[105] Wang H H, Cui Q, Zhang T, Wang Z B, Ouyang Y C, Shen W, Ma J Y, Schatten H and Sun Q Y 2016 Rab3A, Rab27A, and Rab35 regulate different events during mouse oocyte meiotic maturation and activation *Histochem. Cell. Biol.* **145** 647–57

[106] Masumoto N, Sasaki T, Tahara M, Mammoto A, Ikebuchi Y, Tasaka K, Tokunaga M, Takai Y and Miyake A 1996 Involvement of rabphilin-3A in cortical granule exocytosis in mouse eggs *J. Cell Biol.* **135** 1741–7

[107] Zhu X L *et al* 2019 Synaptotagmin 1 regulates cortical granule exocytosis during mouse oocyte activation *Zygote* **28** 97–102

[108] Zhu X L *et al* 2012 Synaptotagmin1 is required for spindle stability and metaphase-to-anaphase transition in mouse oocytes *Cell. Cycle.* **11** 818–26

[109] Ducibella T, Duffy P, Reindollar R and Su B 1990 Changes in the distribution of mouse oocyte cortical granules and ability to undergo the cortical reaction during gonadotropin-stimulated meiotic maturation and aging *in vivo Biol. Reprod.* **43** 870–6

[110] Clift D, McEwan W A, Labzin L I, Konieczny V, Mogessie B, James L C and Schuh M 2017 A method for the acute and rapid degradation of endogenous proteins *Cell* **171** 1692–706

[111] Clift D, So C, McEwan W A, James L C and Schuh M 2018 Acute and rapid degradation of endogenous proteins by Trim-Away *Nat. Protoc.* **13** 2149–75

IOP Publishing

Exocytosis: From Molecules to Cells

Arun Anantharam and Jefferson Knight

Chapter 15

Molecular regulation of multivesicular endosome fusion and exosome secretion

Alan W Weisgerber, Anarkali Mahmood, Ann M Wehman and Michelle K Knowles

Exosomes are vesicles secreted by cells when multivesicular endosomes (MVEs) fuse with the plasma membrane. Exosome biogenesis begins with the formation of intraluminal vesicles (ILVs) within endosomes. These MVEs mature and are trafficked to the plasma membrane, where they dock and eventually fuse to release the intraluminal vesicles as exosomes. Post-fusion, exosomes can remain attached to the cell membrane, where they can affect cells locally, or they can be released into interstitial tissue or body fluids to interact with more distant cells. In this review, we focus on the release of exosomes and the imaging and isolation methods used to measure the trafficking, docking, and fusion of MVEs; we also discuss the proteins found to regulate each of these steps. In addition, we highlight important open questions and technical hurdles, as identifying further regulators of exosome secretion has the potential to aid in the treatment of disease.

15.1 Introduction

Exosomes are small vesicles that are secreted by cells, can signal to cells, and can be taken up by cells. Exosome biogenesis begins with the inward budding of an endosomal membrane to form ILVs (figure 15.1). These ILVs contain cargo sorted from the cytoplasm and the endosomal membrane and carry signatures of the cell, such as messenger RNAs (mRNAs), long noncoding RNAs (lncRNAs), microRNAs (miRNAs), lipids, cytokines, transmembrane receptors, and growth and transcription factors [1, 2]. MVEs are known to fuse with lysosomes for ILV cargo degradation [3] (figure 15.1, green arrow), but fusion with the plasma membrane and the release of exosomes from the cell were first proposed in 1974 [4]

doi:10.1088/978-0-7503-3771-7ch15

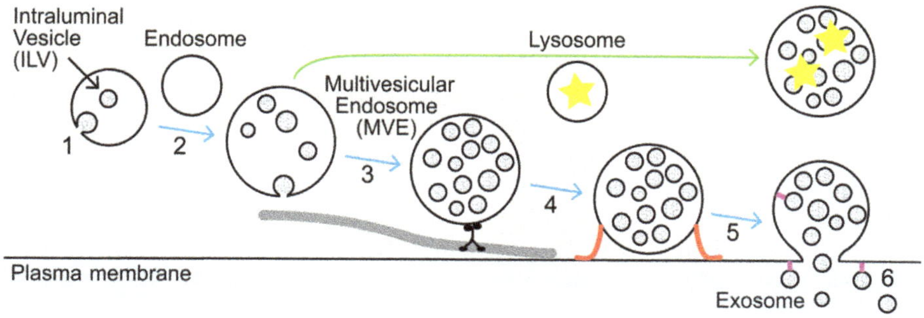

Figure 15.1. Stages of exosome secretion. Exosome secretion begins with (1) an endosome that invaginates to form ILVs and becomes an MVE. As MVEs mature, they can (2) fuse with other endosomes. MVEs can also fuse with lysosomes (green arrow) for cargo degradation by lysosomal enzymes (yellow stars) or they can (3) be trafficked to the plasma membrane to (4) dock and (5) fuse with the plasma membrane (blue arrows). Fusion releases ILVs as exosomes outside the cell, but exosomes can remain attached to the plasma membrane (purple), limiting them to local interactions. (6) Once exosomes diffuse away from the source cell into the environment, they can be found in body fluids or interact with neighboring or distant cells.

and then demonstrated in 1983 [5, 6]. To release exosomes, MVEs are directed to the plasma membrane, where they dock and fuse (figure 15.1, blue arrows). The ILVs are then secreted and the term 'exosome' is used to describe the now extracellular vesicles. After MVE fusion, exosomes can remain associated with the source cell, suggesting there is an additional step for environmental release dependent on molecules that adhere or tether exosomes to the cell surface (figure 15.1, purple). Exosomes can thereby be regulated to interact locally or diffuse away to interact with distant cells. Thus, exosome release can convey cellular material for signaling or uptake, which can affect health and cellular function [3, 7–12].

Exosomes constitute one subset of extracellular vesicles (EVs). EVs can also bud from the plasma membrane, and these microvesicles have a much larger size range (~50 nm–1 μm in diameter) than the ILV-derived exosomes (~30–100 nm in diameter). However, the considerable overlap in EV sizes complicates the isolation or identification of exosomes from other EV subpopulations once they have been secreted. The term exosome initially referred to EVs collected via the centrifugation of extracellular media from cultured cells [13], and the history of extracellular vesicles is reviewed here [14]. Recently, there has been a shift to call EVs isolated by size small EVs (sEVs), while acknowledging that this population contains a mix of <150 nm EVs including both ILV-derived exosomes and plasma-membrane-derived microvesicles (text box 1) [15]. However, much of the work on exosome secretion has been determined from the isolation of sEVs. Therefore, we will use the terms EV or sEV for bulk isolated vesicles and reserve the term exosome for cases in which MVEs, their fusion, or ILV release were specifically measured.

Extracellular vesicle (EV) terminology

- Exosomes: originate as intraluminal vesicles (ILVs) released upon fusion of the multivesicular endosome (MVE) membrane with the plasma membrane. Exosomes are typically 30–100 nm in diameter and secretion is often measured directly using single MVE imaging assays.
- Small extracellular vesicles (sEVs): isolated EVs <150 nm in diameter, including exosomes and small vesicles formed by plasma membrane budding. EVs are collected in bulk and the small EVs are isolated and measured by various methods.

15.2 Exosomes in physiology

The physiological roles of exosomes in immune signaling were first discovered in the 1990s [16]. Antigens are presented on sEVs expelled from immune cells [17, 18], which also contain endosome-associated major histocompatibility complex (MHC) class II complexes [16]. Lipids that modulate immunity [1] and miRNAs that affect inflammation are also transported in sEVs [19]. EV release can also protect cells by providing decoy membranes for bacterial toxins [20]. Indeed, sEV secretion increases as cellular stress increases [13, 21–24], suggesting that cells may respond to stress by releasing what they cannot degrade, prioritizing their survival by outsourcing degradation. Beyond immunity, EVs also facilitate communication between adipose cells and skeletal muscle upon exercise to induce lipolysis [25], and molecules transported by sEVs can act as trophic support for other cells [26]. Thus, exosomes and other sEVs play important roles in immune response and metabolism.

Exosome secretion also plays fundamental roles, such as facilitating cell migration during leukocyte chemotaxis and epithelial migration [27, 28]. Fibronectin and extracellular matrix (ECM) proteins are sorted onto exosomes that allow for polarized cell migration, and secreted exosomes provide attachment points for migrating cells [29]. Motile cells display bursts of exosome secretion just prior to adhesion formation, and impairment of the exosome secretion process leads to short-lived protrusions during motility [29]. Purified sEVs can promote adhesion, leading to increased cell motility when they are adhered to a surface [30, 31]. Thus, EVs play diverse physiological roles.

15.3 Exosomes in disease states

Exosomes have risen to fame for their role in disease states, including cancer, neurodegeneration, and viral spread. The isolation of EVs from bodily fluids is therefore emerging as a promising area of clinical research that uses sEVs as biomarkers for early detection and prevention [7, 12, 32]. EV secretion is upregulated in many types of cancer cell [33–37], where molecules, such as lncRNAs and miRNAs [2], are passed by cancer cells [11, 38–43]. EVs secreted by tumor cells are functional and can upregulate tumorigenic properties in otherwise healthy cells by, for example, priming a metastatic niche; blocking exosome release in tumor cells suppresses their tumorigenic properties [39, 44–47]. The ability of EVs to be targeted at and taken up by other cells has created considerable interest in their use as drug delivery vehicles [48]. Hence, exosomes and other EVs play a crucial role in the progression and spread of cancer [49–51] and clinical trials have demonstrated their therapeutic potential [52].

Exosome secretion is also thought to be a mechanism by which neurodegenerative diseases are spread. EVs containing misfolded protein aggregates have been detected in the blood or cerebral spinal fluid of patients with Parkinson's disease, Alzheimer's disease, Creutzfeldt–Jakob disease, and amyotrophic lateral sclerosis (see the full review in [53]). These EVs can transfer misfolded proteins [54–56], and several *in vivo* studies have demonstrated that exosomes play a role in the progression of several neurodegenerative diseases [57]. In the nervous system, sEVs can also transport proteins and regulatory RNAs that aid in homeostasis, plasticity, and neuroprotection [58]. Hence, exosomes and other EVs play key roles in neurodegenerative disease progression [53, 59], yet can also be protective in neurons [58].

MVE secretion is also involved in viral propagation [60], and viral particles are commonly found in sEV preparations [61, 62]. Recent work on murine β-coronaviruses, a model system for human pathogens such as SARS-CoV2, demonstrated that the release of coronavirus particles occurs via a lysosomal pathway that progresses through an MVE state [63]. Vaccinia viruses and hepatitis E viruses also exploit the MVE pathway to drive the egress of viral particles [60, 64]. Thus, the study of MVE fusion is likely to identify druggable targets that can disrupt viral spread.

15.4 The diversity of multivesicular endosomes, intraluminal vesicles, and exosomes

Since the initial observations that MVEs can fuse with the plasma membrane to release exosomes [5, 6], the way in which MVEs deviate from lysosomal degradation remains an open question. The MVE secretory and degradative pathways appear interconnected, as cells secrete more EVs when lysosomal degradation is blocked [65–67]. MVEs are part of a larger family of lysosome-related organelles (LROs), which includes late endosomes, Weibel–Palade bodies, melanosomes, and lysosomes [68]. These LROs can fuse with the plasma membrane to secrete contents, including EVs. As LROs are part of an endolysosomal continuum, the cargo in these organelles overlaps. For example, CD63, a tetraspanin that is enriched in ILVs [69, 70] and is often used as an exosome marker, was previously used as a lysosome acidification marker [71] and a lysosome fusion marker [67]. CD63 probes partially overlap with other late endosomal/lysosomal markers [72, 73], and CD63 has previously been called LAMP3. Based on this evidence, CD63 has been accepted as the preferred marker for ILVs and exosomes, even though it is found on the plasma membrane and other endolysosomal membranes.

MVEs and ILVs range in size and number depending on the cell type, organism, and condition. In HeLa cells, there are approximately 200 CD63-enriched organelles per cell and MVEs are 400–600 nm in diameter, as measured by electron microscopy (EM) [74]. The ILVs in HeLa cell MVEs are 30–70 nm in diameter, but stimulation with epidermal growth factor slightly increases ILV size (30–100 nm) and inhibition of the endosomal sorting complex required for transport (ESCRT) proteins, known regulators of ILV biogenesis, decreases the ILV diameter to less than 50 nm [75]. Using structured illumination microscopy, MVEs in MDA-MD-231 cancer cells were measured to be 1 μm in diameter. The MVE size was dependent on Munc13-4 to drive MVE maturation, and loss of the protein reduced the size of MVEs to 0.5 μm [39]. In *C. elegans* epithelial cells, MVEs are 400 nm on average, and most ILVs fall within a 40–80 nm range, based on EM analyses [76]. Thus, the expected size of exosomes can change under different conditions, which is likely to alter their cargo and physical properties.

15.5 Exosome analysis methods

Two main strategies exist for measuring exosome secretion, each with distinct advantages and disadvantages. One approach measures sEVs isolated from bulk EV collections as an approximation for exosomes, while the other directly visualizes MVE fusion with the plasma membrane in individual cells. New assays are also being developed to measure exosome secretion [77–79].

15.5.1 Bulk collection of sEVs

Bulk collection methods use differential ultracentrifugation, sucrose gradient centrifugation, field flow fractionation, and other separation methods to concentrate EVs from cell culture media or body fluids and to isolate sEVs from larger (>150 nm) EVs [80]. Commercial kits for EV extraction are also commonly used, although these often sacrifice purity for speed and ease of use [81–84]. Differential ultracentrifugation methods are typically purer but have the largest loss of sEVs compared to the starting sample [82]. With all these approaches, the sEV fraction contains a mix of exosomes and small vesicles that originate from plasma membrane budding, and it remains a challenge to isolate exosomes. Further separation using the immunoprecipitation of exosome-enriched surface markers, such as tetraspanins, has been performed to enrich exosomes (CD63+, CD81+) and plasma-membrane-derived sEVs (CD9+) [26, 85]. However, exosomal tetraspanins are also present on the plasma membrane, hindering complete isolation of exosomes from small plasma-membrane-derived vesicles. Once isolated, sEVs are characterized by Western blot, mass spectrometry, or immuno-EM to verify that exosome markers are present. These bulk collection methods can be broadly applied to identify regulators of exosome secretion that limit any of the upstream steps, including ILV formation, MVE trafficking, docking, fusion, or tethering.

15.5.2 Direct imaging of MVE fusion events

For direct imaging of MVE fusion, two components are needed—a pH-sensitive fluorophore and a protein that is enriched on ILVs. MVEs have an acidic pH, which quenches fluorophores such as ecliptic pHluorin, pHuji, and even enhanced green fluorescent protein (EGFP) when they are trapped within the MVE (figure 15.2(a)). Upon fusion, MVE-localized fluorophores are exposed to the neutral pH of the extracellular space, rapidly unquench, and fluoresce. Surface fluorescence is imaged over time using total internal reflection fluorescence (TIRF) microscopy, and an increase in signal indicates a fusion event (figures 15.2(a)–(c)). This method is

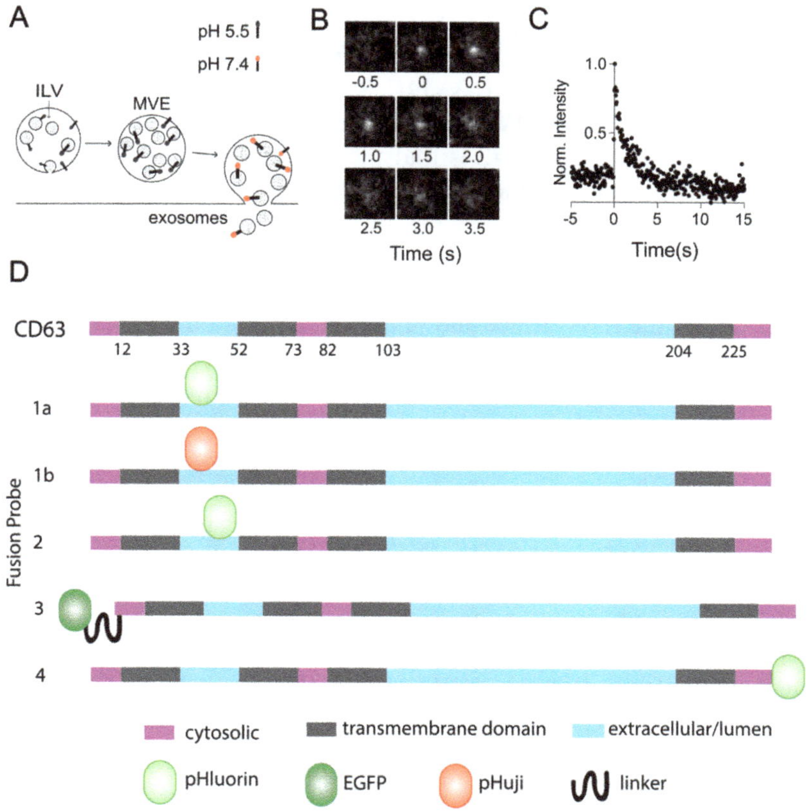

Figure 15.2. Assay and constructs used to measure single MVE fusion events. (a) The low pH inside the endosome lumen quenches pH-sensitive fluorescent proteins on an extracellular loop of CD63 until fusion occurs. (b) TIRF microscopy of one fusion event in a transiently transfected A549 cell shows a bright flash when endosomes fuse, followed by a gradual loss of intensity as CD63-pHuji (probe #1b) leaves the fusion site. (c) The intensity profile of this fusion event, aligned such that 0 s represents the moment of fusion. Using the intensity profile, the kinetics of release can be evaluated to determine whether exosomes stay at the fusion site or are released into the extracellular space. (d) CD63 protein schematic; numbers represent the start of each new domain. Fusion probes: #1a-b, which place pHluorin or pHuji after amino acid 36 in the first extracellular loop [74]; #2, which places pHluorin after aa 42 in the first extracellular loop [29] (a stabilized pHluorin variant (M153R) and a variant that adds mScarlet to the cytosolic C-terminus to also visualize MVEs prior to fusion exist for #2 [72]); #3 has EGFP on the cytosolic N-terminus of CD63 with a 22-amino-acid linker [67]; #4 has pHluorin at the cytosolic C-terminus of CD63 [74]. Several CD63 constructs are available on Addgene and the data in B–C has not been previously published.

analogous to the well-established secretion assays in neuroendocrine cells [86–88] and assays for constitutive fusion [89, 90]. It has the advantage that it distinguishes exosome release from plasma membrane budding, as the burst of signal associated with fusion arises from a pH change. As the same pH is maintained throughout plasma membrane budding, the EVs resulting from such budding do not cause a change in signal intensity. The imaging of individual fusion events also enables the kinetics of fusion and exosome release to be measured.

The second component of an MVE fusion imaging probe is a protein that is localized to ILVs. CD63 is a four-pass transmembrane protein enriched on ILVs [69, 70], but not localized solely to ILVs [91]. Several labs have focused on CD63 as a probe of choice for visualizing MVE fusion (figure 15.2(d)) [29, 39, 72, 74, 79, 92–94]. A key feature of probes #1 and #2 is that the pH-sensitive fluorescent proteins are located within an extracellular loop, which places the fluorophore inside the acidic MVE lumen. Indeed, 70%–90% of CD63-pHluorin probe #1 is contained within acidic vesicles near the surface, as measured by dequenching of the pHluorin fluorophore with NH_4Cl [74]. Similarly, immuno-EM studies suggest that two-thirds of CD63 resides on the ILV surface and one-third remains on the limiting endosomal membrane of MVEs [95]. CD63 with an N- or C-terminal label has also been used to measure fusion (probes #3–4), although these probes localize the pH-sensitive fluorophore to the neutral cytosol and acidic lumen of ILVs, leading to higher background fluorescence before fusion and a reduced increase in signal upon fusion [67]. Additional CD63 probes exist, including a stabilized pHluorin [72].

The choice of fusion probe can make a significant difference in the measurement of MVE fusion frequency (figure 15.2(d)). One study showed little MVE fusion in HeLa cells, with 0% of CD63+ vesicles fusing when probe #3 was used [67], whereas a similar study using probe #1 in HeLa cells measured numerous fusion events (~2–3 events/minute) [74]. Signal enhancement occurs when using probe #1, where pHluorin is contained within the acidic lumen of the MVE, thus reducing the background signal relative to the cytosolic EGFP present in probe #3. This makes fusion events easier to identify, but visualizing MVEs prior to fusion is challenging because pHluorin is completely quenched at low pH. The authors could not calculate the percent of CD63 + MVEs that fused using probe #1 [74], but the use of a doubly tagged CD63—a construct that contains both pHluorin, as shown in probe #2, and a C-terminal mScarlet—allowed for both the enhanced fusion signal and tracking of all possible CD63+ MVEs using dual color imaging [72]. These data suggest that probe-to-probe variations are present in the quantification of exosome secretion events.

The direct imaging method presents several challenges. First, data collection is time-consuming, as fusion events happen ~0–3 times per minute and are not well synchronized to a stimulus, in contrast to neuroendocrine cells, in which many fusion events occur in docked vesicles immediately after Ca^{2+} influx [96]. Second, data analysis is time intensive and subjective, but methods developed in ImageJ facilitate analysis and objectivity [97]. Third, overexpression of CD63 can complicate measurements, since CD63 is involved with the formation of ILVs [91] and the sorting of exosomal cargo proteins [98]. However, sEVs collected from cells expressing probe #1 showed no change in the number of CD81+ sEVs nor the number or morphology of MVEs [74], suggesting that the expression of probe #1 can be well tolerated.

15.6 Mechanisms by which MVE secretion releases exosomes

To better understand the regulation and role of exosomes in healthy and diseased states, numerous studies have focused on the way in which ILVs are formed within MVEs, MVE maturation, MVE trafficking, docking, fusion, and exosome release from the plasma membrane [79]. Several stages of the exosome secretion process are enhanced by Ca^{2+}, even though exosome secretion occurs constitutively from all cell types. This process is broken down into six stages below and these stages are depicted in figure 15.1.

15.6.1 Stage 1: ILV biogenesis and cargo sorting

Exosomes originate as ILVs that form within endolysosomes through an invagination process (figure 15.1, #1). Cargo sorting into ILVs can take place via canonical ESCRT-dependent mechanisms [99], a non-canonical ALG-2-interacting protein X (ALIX)–Syntenin–ESCRT-dependent mechanism [100], ESCRT-independent mechanisms that utilize the ceramide pathway [101], and/or CD63-dependent mechanisms [102]. As ILV biogenesis and cargo sorting have been reviewed in detail previously [103–105], we focus this review on the later portions of the MVE–exosome life cycle and only discuss a few proteins linked to ILV biogenesis (table 15.1). However, it is worth noting that in human melanoma cells, a subset of ILVs has been shown to fuse with the limiting membrane in a process called retrofusion [95]. Thus, ILVs have three potential fates: returning to the limiting endosomal membrane, becoming exosomes, or being degraded in lysosomes [95], making it more complicated to relate ILV numbers, cargoes, and sizes directly to exosomes without *a priori* knowledge of which subsets of ILVs or MVEs contribute to exosome release.

Table 15.1. Effectors of the maturation, trafficking, docking, and fusion of MVEs. CHO = Chinese hamster ovary cells, NRK = Normal rat kidney epithelial cells.

Protein	Cell Type	Method	Stage	References
RAL-1	*C. elegans* epidermis	EM	Biogenesis and trafficking	[110]
RalA/B	4T1 cells	EM	Biogenesis and trafficking	[110]
Rab7	*C. elegans* epidermis, HeLa	RNA interference (RNAi) screening	Biogenesis and maturation	[110]
Munc13-4	MDA-MB-231, A549, Panc-1	Confocal/TIRF	Maturation	[39]
Rab11	K562, *C. elegans* epidermis	Confocal, RNAi screening	Maturation and Docking	[39, 118, 119] [110]
Ykt6	HeLa, *Drosophila* epithelium	Confocal	Maturation	[123]
Rab35	Oligodendrites (Oli-neu), *C. elegans* epidermis	Confocal/TIRF, RNAi screening	Trafficking and Docking	[134, 110]
Rab27A/B	HeLa, *C. elegans* epidermis	TIRF, RNAi screening	Trafficking and Docking	[85, 110]

Cortactin	SCC61	TIRF	Trafficking and Docking	[93, 135]
Granuphilin/ Slp4a	HeLa	sEV collection	Ca^{2+} sensor for docking	[85]
Syt7	CHO, NRK, many cell types	Confocal/TIRF EM	Ca^{2+} sensor for fusion	[29, 67, 94, 156]
Syx4	HeLa	MVE fusion assay (TIRF)	Fusion	[74]
SYX-5	*C. elegans* epidermis	EM/light microscopy	Fusion	[110]
SNAP23	HeLa, tumor cells (SW480, HeLa, HepG2, A549)	MVE fusion assay (TIRF), sEV collection, Phosphoproteomics	Fusion	[74, 147]
VAMP7	K562	Confocal	Fusion	[146]
NSF	K562	sEV collection	Fusion and soluble N-ethylmaleimide sensitive factor attachment protein (SNAP) receptor (SNARE) recycling	[146]
Tetherin	HeLa	EM	Post-fusion tethering	[117]
RAB-1	*C. elegans* epidermis	RNAi screening	Not determined	[110]
RAB-2	*C. elegans* epidermis	RNAi screening	Not determined	[110]
Rab2b	HeLa	Short hairpin RNA (shRNA) screening	Not determined	[85]
Rab5a	HeLa	shRNA screening	Not determined	[85]
RAB-8	*C. elegans* epidermis	RNAi screening	Not determined	[110]
Rab9a	HeLa	shRNA screening	Not determined	[85]
RAB-39	*C. elegans* epidermis	RNAi screening	Not determined	[110]

Rab GTPases are often used as identifiers for different membrane organelles due to their dynamic and transient association with maturing membranes [106, 107]. Rab7, a marker of late endosomes, has been shown to have conflicting roles in exosome secretion in different systems. Hyperactivation of Rab7 led to a reduction in ILV formation in Vps34-deficient mouse embryonic fibroblasts, suggesting that Rab7-GTP on the endosome membrane could disrupt ILV formation [108]. It is possible that this response is not directly due to Rab7-GTP, but occurs due to proteins that interact with Rab7 or the activity of Vps34, a phosphatidylinositol (PI) kinase [108]. Other studies have shown that the loss of Rab7 leads to an increased number of ILVs in HeLa cells [109]. Rab7 is necessary for endosome maturation that supports lysosome fusion [106]. When Rab7 is depleted, lysosomal degradation is halted, leaving MVEs large and filled with ILVs [109]. However, Rab7 did not localize to MVEs that fused with the plasma membrane [67] and Rab7 silencing did not affect CD63+ exosome secretion from HeLa cells [85]. However, RAB-7 was identified in a screen for alae defects in *C. elegans* epidermis, a process requiring exosome secretion [110], and Rab7a is one of the top 100 proteins identified in sEVs [111], suggesting a potential role at some stage during exosome secretion. Together, these data emphasize that different pools of late endosomes or MVEs exist; however, they leave open the question of whether Rab7 plays a role in endosome maturation prior to ILV formation for secretion of a subtype of exosomes or whether Rab7 is a specific marker for immature MVBs destined for lysosomal degradation.

The Ras-related small GTPase RAL-1 facilitates ILV biogenesis, MVE trafficking or docking, and the secretion of exosomes in *C. elegans* [110]. RAL-1-deficient animals form fewer MVEs, which are enlarged and have a reduction in ILV number, suggesting a role for RAL-1 in ILV biogenesis. MVEs also remain further from the plasma membrane, as measured by EM, suggesting a role in trafficking or docking. RAL-1 localizes on the surface of MVEs and a constitutively active RAL-1 colocalizes with the syntaxin SYX-5 on the plasma membrane, suggesting a role in fusion. Surprisingly, proteins from the exocyst complex, a known RAL-1-effector involved in secretion, do not localize on the surface of MVEs, and exocyst proteins are not essential to MVE secretion [110]. The mammalian RAL-1 homologs RalA and RalB are both necessary for sEV secretion in 4T1 mammary cells [112, 113]. RalA and RalB depletion in breast cancer tumors reduced lung metastasis in mice, further suggesting that sEV secretion is necessary for metastasis [112]. Thus, RAL-1 orthologs appear to act at several important stages during exosome secretion.

15.6.2 Stage 2: MVE maturation

ILV biogenesis, MVE maturation (figure 15.1, #2), and MVE trafficking (figure 15.1, #3) to the plasma membrane are likely to occur simultaneously, although they are listed as separate stages here. MVE maturation primarily includes the acidification of the lumen, MVE fusion with other endosomes, and changes in proteins associated with the surface of MVEs. Proteins that regulate maturation are listed in table 15.1.

15.6.3 MVE acidification

The maturation of MVEs typically involves acidification by the major proton pump of the cell, the H+-vacuolar ATPase (V-ATPase) [114]. However, the role of V-ATPases in exosome secretion is a complicated issue. In the *C. elegans* epithelium, only certain V-ATPase subunits are necessary for MVE secretion. Mutations or RNAi-mediated depletion of *vha* subunits specifically block exosome secretion and lead to the accumulation of enlarged MVEs [110, 115], suggesting that maturation is affected, although the authors also propose a role for the V-ATPase in MVE fusion.

In contrast, knockdown or inactivation of the V-ATPase in HeLa cells led to the enhanced release of exosomes [116] and the accumulation of cholesterol in ILVs [117]. Similarly enhanced exosome secretion after V-ATPase inhibition was observed in H4 neuroglioma [56] and SH-SY5Y cells [66]. This suggests that the V-ATPase may be affecting exosome secretion via acidification, cholesterol content changes, or other unidentified mechanisms.

Rab11, Munc13-4 and Ca^{2+}-dependent maturation for fusion

Rab11 is a small GTPase mainly studied for its role in endosomal recycling [106], but it is also implicated in MVE maturation for exosome secretion, together with the Ca^{2+} sensor Munc13-4. The amount of Rab11 increases on MVEs and MVEs became enlarged in the presence of Ca^{2+} using an ionophore in K562 cells [118]. The overexpression of a dominant negative Rab11 was able to reduce the amount of sEVs isolated from K562 cells [119], but Rab11a silencing was not sufficient to affect sEV secretion by HeLa cells [85], suggesting Rab11 redundancy or that the role of Rab11 may vary with cell type. GTP-bound Rab11+ endosomes fuse with MVEs to enlarge MVEs, which has been hypothesized to prepare the MVE for Ca^{2+}-triggered exocytosis [39]. A reduction in Munc13-4 in MDA-MB-231 cells reduced the size of MVEs, as measured by super-resolution microscopy. Munc13-4 binds membranes upon Ca^{2+} influx and is thought to form a bridge between MVEs and Rab11+ endosomes, facilitating the growth of MVEs. Munc13-4 has a similar role in RBL-2H3 cells, where it facilitates the fusion of Rab11+ vesicles with secretory granules in a Ca^{2+}-dependent fashion [120]. Without this Rab11- and Munc13-4-driven endosome–endosome fusion step, Ca^{2+}-stimulated sEV secretion is reduced.

Under basal conditions, Munc13-4 is required for the release of CD63+ sEVs, but not CD9+ sEVs in MDA-MB-231 cancer cells [39]. In the presence of Ca^{2+} ionophores, knockdown of Munc13-4 reduced the secretion of both CD63+ and CD9+ sEVs, as measured by a bulk assay of collected sEVs. Mutation of the Ca^{2+} binding sites impeded the ability of Munc-13-4 to rescue Ca^{2+}-stimulated sEV secretion. Intriguingly, the subcellular localization of EV cargo proteins is altered upon Ca^{2+} stimulation; CD9 is typically found on the plasma membrane under basal conditions but localizes to ILVs upon Ca^{2+} influx [39]. This suggests that Munc13-4 is required for exosome secretion and that CD9 likely shifts from microvesicles to exosomes upon Ca^{2+} stimulation. Overall, Munc13-4 and Rab11 grow MVEs and prime them for exosome secretion.

The R-SNARE Ykt6 may be involved in MVE maturation

Ykt6, a soluble N-ethylmaleimide sensitive factor attachment protein (SNAP) receptor (SNARE) protein involved in endoplasmic reticulum (ER)–Golgi transport and autophagosome–lysosome fusion [121] also plays an important role in sEV secretion. Ykt6 has been shown to be a marker of aggressive cancer in non-small cell lung cancer patients, in whom the upregulation of Ykt6 in tumors was correlated with increased sEV secretion [122]. Reducing the expression of Ykt6 led to an 80% reduction in sEV secretion in model cell lines (A549) [122]. Ykt6 also localizes to ILVs and sEVs, and the reduction of Ykt6 reduced sEV secretion in *Drosophila* and

HEK293 cells [123]. Ykt6 knockdown also caused CD63 to be retained within cells in lysosomes rather than to be secreted on sEVs. In HEK293 cells, the secretion of CD81+ sEVs was also reduced after Ykt6 siRNA knockdown [123], suggesting that Ykt6 acts similarly on the secretion of both CD81+ and CD63+ EVs. These data suggest that Ykt6 is required for MVEs to follow the secretory pathway and avoid degradation. Given the fusogenic roles of SNAREs, this could work by regulating MVE–endosome fusion, but further experiments are needed to identify when and where Ykt6 is involved in exosome secretion.

15.6.4 Stage 3: MVE trafficking

MVEs need to be directed toward the cell surface in preparation for exosome secretion (figure 15.1, #3). Like lysosomes and late endosomes, MVEs are trafficked outwards via associated motor proteins along actin filaments or microtubules [124–126]. In polarized cells, different MVEs can be trafficked to different sides of the cell [127], but the mechanism by which this occurs is not clear. EVs collected from the apical side of polarized MDCK cells were enriched in CD63, whereas EVs collected from the basolateral side were rich in CD9 and CD81 [127]. Additionally, ESCRT proteins were required for apical sEV secretion, while the ceramide pathway was required for basolateral sEV secretion [127]. Thus, cells can direct exosome secretions with distinct cargoes to different parts of the cell; future work is needed to identify the molecular mechanisms of trafficking regulation.

LRO-associated Rab27 is one of the best studied Rab GTPases in exosome secretion and is implicated in several stages from trafficking to docking to fusion (table 15.1). Rab27a and Rab27b were originally identified in an shRNA fluorescence-activated cell sorting (FACS)-based screen of 59 human Rabs that identified five Rabs required for the secretion of CD63-containing EVs (table 15.1) [85]. RAB-27 was also identified alongside eight other Rabs in an RNAi screen identifying 73 proteins required for exosome-dependent cuticle formation in worms [110]. The depletion of Rab27a in cancer cells decreased directional cell motility and sEV secretion was reduced by two to three times, as measured by a bulk centrifugation assay [29]. MVEs were enlarged after silencing Rab27a in HeLa cells, while a perinuclear accumulation of MVEs was noted after Rab27b silencing, suggesting disparate roles for the Rab27 paralogs in maturation vs. MVE trafficking [85]. Rab27a is required to recruit myosin Va motor proteins as well as actin regulators during melanosome trafficking [128], raising the possibility that Rab27b could similarly regulate motor protein-dependent trafficking of MVEs to the plasma membrane. Rab27a is also required for the trafficking of late endosomes that carry phosphatidylinositol 4-kinase type 2α [129], which is essential for the production of the plasma membrane lipids phosphatidylinositol 4-phosphate and PI(4,5)P_2. As PI (4,5)P_2 is involved in clustering SNARE proteins for membrane fusion [130], Rab27a could promote MVE trafficking through actin-based motors as well as membrane fusion through lipid–SNARE protein interactions on the MVE or plasma membrane.

15.6.5 Stage 4: MVE docking

Once MVEs are trafficked to the plasma membrane, membrane docking can occur (figure 15.1, #4). Stable docking is indicated by an attachment to the plasma membrane; many assays rely on both the proximity to the plasma membrane and vesicle immobility to determine stable docking [85, 131]. MVE docking has been measured using time-lapse, TIRF microscopy (by observing MVEs that are close to the membrane and immobile) and using EM (by observing MVEs that are close to the plasma membrane). Several proteins have been identified as regulators of MVE docking and these are depicted in figure 15.3(a) and listed in table 15.1.

Rab11, 27, and 35 are necessary for docking

Several Rab GTPases have been implicated in docking MVEs to the plasma membrane, including Rab11, Rab27, and Rab35. Rab11 colocalizes with immobile MVEs in K562 cells [118], and the expression of a constitutively active Rab11 increased the colocalization of Rab11 with CD63+ vesicles in A549 cells [39]. The mobility of GFP-CD63-labeled MVEs was increased after Rab27a and Rab27b knockdown, compared to that of control HeLa cells [85]. Rab27a knockdown also led to an approximate fourfold reduction in the fraction of immobilized CD63-labeled MVEs at the plasma membrane of SSC61 cells [93], suggesting that fewer MVEs were docked. This is consistent with data regarding granule secretion in

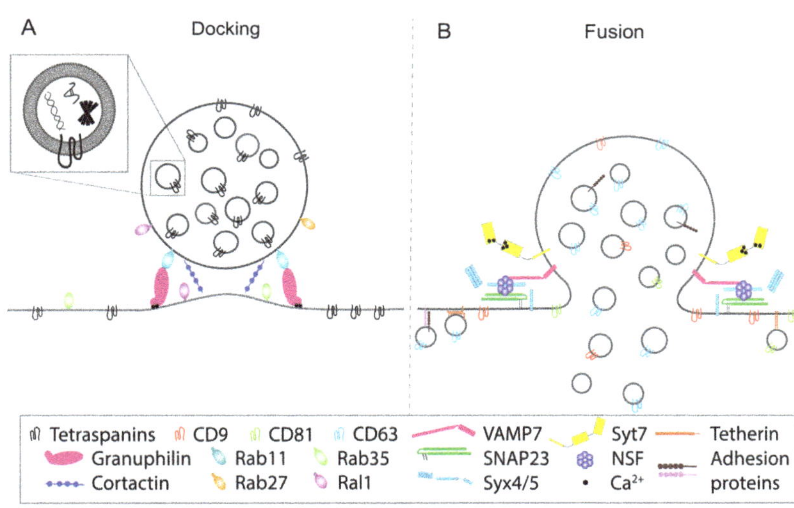

Figure 15.3. Protein regulators of MVE docking, fusion, and tethering. (a) Ral1, Rab11, Rab27, and Rab35 assist in trafficking and docking MVEs at the plasma membrane. Granuphilin and cortactin associate with proteins not included here to aid in docking MVEs. The inset zoom of an ILV shows the cargo (DNA, RNA, proteins, and fibrils) and the lipid bilayer shell. Tetraspanins decorate the ILV, MVE, and plasma membrane, and CD63 is enriched on the MVE and ILV membranes. (b) The fusion of MVEs is facilitated by SNARE proteins and the Ca^{2+} sensor Syt7. Long-lived exosome attachment has been observed for both CD81+ and CD63+ exosomes, but not CD9+ exosomes [74]. This likely occurs via tethering and/or adhesion proteins and may depend on the cell type.

neuroendocrine cells where Rab27 is involved with granule docking (in conjunction with granuphilin/Slp4) to bridge docked vesicles to the plasma membrane [132, 133]. In oligodendrocytes, GTP-bound Rab35 localizes to the cell surface; when Rab35 is knocked down, the density of MVEs near the cell surface decreases, as imaged using TIRF microscopy. Additionally, sEV secretion is impaired when Rab35 is inhibited [134]. Together, these data suggest that MVEs cannot dock properly without Rab11, Rab27, and/or Rab35 (figure 15.3(a)).

Cortactin regulates docking in a similar way to Rab27a

Cortactin, an actin-binding protein that facilitates actin filament branching, is a regulator of late endosomal trafficking and is involved in the docking of MVEs (figure 15.3(a)). Knockdown of cortactin in SCC61 carcinoma cells reduced the number of docked MVEs by about sixfold and promoted sEV release [93]. In EM studies, cortactin knockdown cells were noted to accumulate enlarged late endosomes/lysosomes [93, 135], and these cells contained MVEs with ILVs [93], suggesting that biogenesis is not affected. Cortactin works with Rab27a and Coronin1b to coordinate docking [94] and requires Arp2/3 and F-actin binding to facilitate exosome secretion [93]. Interestingly, cortactin overexpression can compensate for the loss of Rab27a [93], suggesting the two proteins play a similar role in docking during exosome secretion.

Ca^{2+} enhances MVE docking

Ca^{2+} and Ca^{2+} binding proteins are also involved with MVE docking (figure 15.3(a)). The use of a Ca^{2+} ionophore caused a reduction in the motion of CD63-labeled compartments near the plasma membrane [67], but the proteins that are involved with Ca^{2+}-dependent docking are as yet unknown. Granuphilin, which is also known as Synaptotagmin-like protein 4a (Slp4a), is a potential candidate. Granuphilin is a known effector of Rab27; knockdown of granuphilin reduced the secretion of CD63+ sEVs and led to the accumulation of CD63+ compartments within the cell, which was similar to the knockdown of the docking protein Rab27a [85]. Granuphilin has two C2 domains that bind to anionic lipids [136]. Granuphilin also interacts with Munc18s, SNARE accessory proteins that bind syntaxins [137, 138]. Granuphilin is involved with the Ca^{2+}-triggered exocytosis of Weibel–Palade bodies in endothelial cells [139], and the loss of granuphilin in insulin-secreting cells (mouse pancreatic β and MIN6 cells) also causes a loss of granule docking [140, 141]. However, more studies are needed to determine when and where granuphilin is involved in exosome secretion and whether Ca^{2+} alters granuphilin localization.

15.6.6 Stage 5: MVE fusion

Using bulk and direct imaging methods, the secretion of exosomes has been studied in a wide variety of cells, suggesting that practically all eukaryotic cells are capable of fusing MVEs with the plasma membrane for exosome secretion (figure 15.1, #5). The proteins that regulate MVE fusion are depicted in figure 15.3(b) and table 15.1.

SNARE regulation of MVE membrane fusion

SNARE proteins are essential for many membrane fusion processes (figure 15.3(b)). Four α-helical SNARE motifs tethered in opposing membranes assemble into a bundle to bring the membranes into close proximity for fusion [124, 125]. SNARE proteins on a vesicle membrane associate with SNARE proteins on a target membrane to provide energy for membrane fusion [142, 143]. For MVE fusion, the target plasma membrane SNAREs are likely to include syntaxins and SNAPs, while the MVE membrane SNAREs include vesicle-associated membrane proteins (VAMPs). The SNARE accessory protein and chaperone ATPase N-ethylmaleimide-sensitive factor (NSF) recycles these SNARE complexes and ensures the fidelity of SNARE complex formation [144, 145]. Indeed, NSF is required for sEV secretion in K562 cells [146], which is consistent with the view that SNAREs play a role in this process.

Knockdown of the plasma membrane SNAREs syntaxin4 (Syx4) or its binding partner SNAP23 reduces exosome secretion, as assessed by imaging MVE fusion events in HeLa cells [74]. The phosphorylation of SNAP23 is specifically needed for membrane fusion, as shown in single membrane fusion events [74] and in a bulk sEV collection assay [147]. When SNAP23 phosphorylation was blocked using an alanine mutant (S110A), the secretion of CD63/flotillin/syntenin-1-containing sEVs was reduced [74], while the phosphorylation of SNAP23 at Ser95 resulted in an increase in sEV secretion [147]. Interestingly, blocking SNAP23 phosphorylation with an alanine mutant (S110A) did not alter the secretion of CD81- and CD9-containing sEVs [74], raising the possibility that CD63-containing MVEs use Syx4/SNAP23, while different MVEs use different SNARE complexes to regulate membrane fusion.

At least one other plasma membrane SNARE acts during MVE fusion. In the *C. elegans* epidermis, the syntaxin SYX-5 has been shown to be essential in order for fusion to secrete exosomes [110]. In mutants that lack SYX-5, MVEs accumulate near the plasma membrane and exosome-dependent cuticle structures do not form, suggesting that the fusion step is reduced in the absence of SYX-5. Overall, the plasma membrane SNAREs are a large family of proteins and other Syx and SNAP proteins could facilitate fusion; however, this could depend on the cell type or the MVE cargo, and more studies are needed.

To complete the trimeric SNARE complex for membrane fusion, a vesicle bound SNARE protein is required. MVEs have been shown to contain VAMP7 on their surface in confocal imaging studies of K562 cells, where it partially overlaps with Rab11+ structures [146]. Overexpression of the N-terminal domain of VAMP7, a known inhibitor of SNARE complex formation [148], led to enlarged MVEs near the cell surface and a reduction in sEV secretion [146], suggesting inhibition of membrane fusion. VAMP7 also decorates late endosomes and lysosomes in Madin-Darby canine kidney (MDCK) cells, where overexpression of the N-terminal domain of VAMP7 similarly inhibited sEV secretion and epithelial cell migration [28]. VAMP7 interacts with a wide range of Q-SNAREs, such as Syx1a/SNAP25 and Syx4/SNAP23, as well as the Ca^{2+} sensor Syt7 [149]. These data suggest that VAMP7 is a vesicle-associated SNARE protein involved with MVE fusion to the plasma membrane.

Intracellular Ca^{2+} regulates MVE fusion

MVE fusion does not require stimulation, but increases in intracellular Ca^{2+} concentration also increase MVE fusion and sEV secretion [39, 67], suggesting that Ca^{2+} is an important regulator of MVE fusion. Ca^{2+} ionophores increase the number of fusion events in single MVE fusion assays [39, 67], although different fusion-event frequencies were observed in a variety of cell lines (CHO, NRK, NIH3T3, WM239, HeLa, MDCK, murine embryonic primary fibroblasts) using probe #3 (figure 15.2(d)) [67]. For example, HeLa cells lost 0% of their CD63+ MVEs until a Ca^{2+} ionophore was added, at which point CD63+ MVE loss increased to 4%. In contrast, primary fibroblasts lost 24% of their observable MVEs in the presence of an ionophore [67]. Similarly, more sEVs are secreted in the presence of Ca^{2+} ionophores in different cancer-cell lines, as measured by bulk collection methods [39]. Ca^{2+}-triggered secretion of sEVs has also been observed for α-synuclein-containing exosomes in SH-SY5Y cells [55], and neuronal depolarization leads to increased intracellular Ca^{2+} and increased sEV secretion in cultured cortical neurons [150]. Ca^{2+} also triggers lysosomal fusion in epithelial cells and fibroblasts [151]. These data demonstrate that the fusion of MVEs in order to release exosomes can be stimulated by intracellular Ca^{2+} in a variety of cell types, suggesting that a universal Ca^{2+} sensing mechanism exists.

One Ca^{2+} sensor has been identified in the MVE fusion process: synaptotagamin-7 (Syt7), as shown in figure 15.3(b) (yellow). Syt7 is a transmembrane protein which has two Ca^{2+} binding domains (figure 15.3(b), black dots) that associate with membranes and SNARE proteins to facilitate regulated exocytosis in neuroendocrine cells [152]. Syt7 binds the SNARE proteins Syx1a [153] and SNAP23 [154] in a Ca^{2+}-dependent manner [155] and is a Ca^{2+} sensor for lysosomal exocytosis in fibroblasts [156]. The palmitoylation of three cysteines near the transmembrane domain helps to target Syt7 to CD63+ compartments [157]; fusogenic MVEs contain Syt7 [67]. Indeed, the loss of Syt7 reduces sEV release in SCC61 cells [94]. Syt7 has also been shown to recruit insulin-containing vesicles to the plasma membrane after glucose stimulation [158], raising the possibility that Syt7 could recruit MVEs to the plasma membrane in addition to facilitating Ca^{2+}-dependent membrane fusion.

Although MVE fusion can be stimulated, MVEs do not appear poised to fuse in the same way as other secretory vesicles. In Chinese hamster ovary (CHO) cells, Ca^{2+} levels rise to a maximum within 10 s after Ca^{2+} ionophore treatment, but the first MVE fusion event is not observed for 35–100 s [67]. MVEs also continue to fuse over the next 7 min after Ca^{2+} stimulation [67]. This is in stark contrast to assays of dense core vesicles (DCV), where the release is tightly regulated and triggered by Ca^{2+}. DCV membrane fusion starts to occur 7 ms after stimulation (in chromaffin cells) and ends after approximately 2 s (in pancreatic β cells) for the docked, readily releasable pool [96], as measured by capacitance or excitatory postsynaptic current (EPSC) methods. The difference is more dramatic in the case of synaptic vesicle fusion, and synaptic vesicle release occurs from 0.1 ms to 1 s after stimulation [96]. PC12 cells, a neuroendocrine cell line, are slower in both synaptic vesicle (30 ms) and DCV fusion (8–20 s) due to a lack of docked vesicles molecularly prepared to fuse [96].

However, most CD63+ MVEs were observed to be docked prior to fusion, based on their proximity to the surface in a TIRF microscopy assay and their immobility (using probe #3) [67]. The delayed release after Ca^{2+} influx, while appearing docked, suggests that most docked MVEs are not ready to fuse on a molecular level. Interestingly, MVEs newly trafficked to the membrane were capable of membrane fusion, although most fusion occurred with docked MVEs [67]. Together, these data suggest that MVE fusion depends on Ca^{2+}, where Ca^{2+} first initiates a series of events possibly including MVE maturation, transport, docking, and finally fusion with the plasma membrane.

GPCR regulation of fusion
Exosome and sEV secretion can also be stimulated by histamine, which induces GPCR activation, G-protein signaling, kinase activation, and the subsequent phosphorylation of proteins [74]. Histamine significantly increased MVE fusion in HeLa and HUVEC cells [74]. GPCR activation with histamine led to delayed Ca^{2+} entry into cells (approximately 10s); however, chelating extracellular Ca^{2+} did not alter MVE fusion, suggesting that Ca^{2+} influx is not responsible for the histamine-induced increase in MVE fusion [74]. Therefore, histamine is likely acting through molecules downstream of the activation of the GPCR-linked small G-proteins. In a phosphoproteomics screen, two SNARE proteins (SNAP23 and Syntaxin binding protein 5 (STXBP5)) were phosphorylated after histamine activation, [74], raising the possibility that GPCR activation promoted SNARE-mediated MVE fusion. SNAP23 phosphorylation was affected by histamine treatment and this was shown to directly affect fusion [74], as described above. STXBP5 binds binary Syx/SNAP complexes such as Syx4/SNAP23 [159] and Syx1/SNAP25 [160–163] and negatively regulates SNARE complex formation, but the role of STXBP5 in MVE fusion has not yet been directly studied.

15.6.7 Stage 6: exosome secretion

The final stage of the exosome life cycle is the environmental release of exosomes from the secreting cell (figure 15.1, #6). EVs have been observed far from the site of release [26, 164], where they are taken up by endocytosis and a variety of other mechanisms [165]. For example, in zebrafish, EVs are secreted from the yolk syncytial layer into the blood, where the EVs then regulate the formation of a different organ, the caudal vein plexus [26]. Similarly, the oligimerization of alpha synuclein in cell culture has been induced by EVs collected from the cerebrospinal fluid (CSF) of patients with Parkinson's disease [166], suggesting that EVs can spread from the source of secretion into the CSF. This widespread distribution of EVs allows for intercellular communication, trophic support, and the progression of disease states.

Although some exosomes diffuse away after MVEs fuse with the plasma membrane, EM studies reveal that some exosomes stay in close proximity to the cell surface [16, 167] and cluster with one another [74, 75]. This is likely due to cell or matrix adhesion molecules found on the exosome surface that limit diffusion, including CD9, CD63, CD81, integrins, thrombospondins, basigin, and fibronectins [111, 168, 169], as well as tethering proteins such as tetherin [117]. Long-lasting exosome attachment to the cell surface is also measurable based on the loss of CD63 fluorescence from nascent MVE fusion sites (figures 15.2(b) and (c)). The signal from an MVE fusion event lasts about 100 s in HeLa cells (figure 15.2(d), probe #1) [74], making the fluorescence decay two orders of magnitude slower than soluble cargo release (i.e. neuropeptide Y (NPY) from a granule, <1.0 s) or the lateral diffusion of a transmembrane protein from a fusing vesicle into the plasma membrane (i.e. VAMP2, ~2 s). Intriguingly, while CD81+ exosomes disperse with similar kinetics to those of CD63+ exosomes in single MVE fusion assays, CD9+ exosomes diffuse away an order of magnitude faster than CD63+ exosomes [74], suggesting that the prolonged fluorescence observed for CD63 and CD81 is due to specific interactions with the cell surface, rather than technical limitations of exosome imaging using TIRF microscopy. In figure 15.3(b), CD9+ exosomes are shown to lack extra molecules that would adhere them to the surface; however, CD81+ and CD63+ vesicles appear with tethers or adhesion molecules. Together, these data demonstrate that the release of exosomes into tissues, body fluids, or culture media is also regulated by adhesion or tethering, making it important to determine whether CD9+ exosomes use different adhesion proteins or tethers than CD63+ and CD81+ exosomes to regulate their release. Exosome adherence may provide a mechanism by which cells can control whether interactions take place locally with tethered exosomes or over longer distances when exosomes are fully released.

Tetherin is a protein known to regulate exosome and viral particle attachment to the source cell (figure 15.3(b), orange). Tetherin has a transmembrane domain on one end and a glycosylphosphatidylinositol (GPI) linkage on the other, and forms homodimers that bridge two membranes. In EM studies of HeLa cells treated with the V-ATPase inhibitor Bafilomycin A1, secreted EVs appeared tethered to the cell surface at a concentration of ~6 EVs/μm^2 [117]. Bafilomycin treatment affects cholesterol trafficking [116, 117] and increases exosome release [117]. Tetherin inhibits the release of exosomes from the cell surface by about four times [117] and also affects viral particle release [170]. Tetherin's GPI linkage is essential for its tethering capabilities; tetherin mutants lacking the GPI linkage are unable to maintain high densities of exosomes on the cell surface [117]. Tetherin's role in attaching EVs to the cell surface was also confirmed in Epstein–Barr virus (EBV)-transformed B-cell lines [117]. However, the long-lived dynamics of CD63-pHluorin at MVE fusion sites was unchanged in tetherin knockdown cells, suggesting that other interactions attach CD63+ exosomes to the plasma membrane or extracellular matrix [74]. Thus, downregulating tetherin is likely to be only one of many mechanisms that control the spread of exosomes [117].

15.7 Conclusions

Substantial progress on the molecules involved with MVE trafficking, maturation, docking, and fusion has been made in recent years due to the growing knowledge that exosomes are regulators of a broad range of disease states. The secretion of exosomes is regulated at many points within the exosome life cycle; both early and late perturbations similarly disrupt the quantity of exosomes in the environment. Combining bulk assays with direct imaging of MVE fusion and/or quantitative EM studies is essential to identify when and where proteins act during the exosome secretion process.

Several gaps in our knowledge of the exosome lifecycle exist and future studies that address the molecular regulation of secretion are needed. Specifically, it is a priority to discover why some MVEs follow a secretory pathway, whereas others are on a path to degradation, in order to understand how cells decide to release exosomes. Once MVEs fail to become lysosomes, major open questions are: how many exosome secretion pathways exist and what molecules regulate them? There are at least four ILV biogenetic pathways, and current models suggest at least two exosome secretion pathways. Polarized epithelial cells secrete exosomes with different cargoes at apical and basolateral membranes [127]. Similarly, CD81+ or CD63+ exosomes remain tethered to the plasma membrane much longer than CD9+ exosomes [74]. Not all exosomes remain tethered, however; EVs can be fully released to affect cells far from the secretion site [26, 164]. More studies of the differential regulation of exosome secretion and the fate of exosomes based on exosome content are needed to discover how this occurs. Finally, the identification of druggable signaling pathways (i.e. G protein-coupled receptors (GPCRs)) that alter exosome secretion will be highly relevant to the diagnosis and treatment of diseases, as exosome secretion is often altered in disease states.

Acknowledgements

This work was funded by the NSF Division of Molecular and Cellular Biosciences, grant # 2122289.

References

[1] Boilard E 2018 Extracellular vesicles and their content in bioactive lipid mediators: more than a sack of microRNA *J. Lipid. Res.* **59** 2037–46

[2] Li Y, Yin Z, Fan J, Zhang S and Yang W 2019 The roles of exosomal miRNAs and lncRNAs in lung diseases *Signal Transduct. Target Ther.* **4** 47

[3] van Niel G, D'Angelo G and Raposo G 2018 Shedding light on the cell biology of extracellular vesicles *Nat. Rev. Mol. Cell Biol.* **19** 213

[4] Nunez E A, Wallis J and Gershon M D 1974 Secretory processes in follicular cells of the bat thyroid. 3. The occurrence of extracellular vesicles and colloid droplets during arousal from hibernation *Am. J. Anat.* **141** 179–201

[5] Harding C, Heuser J and Stahl P 1983 Receptor-mediated endocytosis of transferrin and recycling of the transferrin receptor in rat reticulocytes *J. Cell Biol.* **97** 329–39

[6] Pan B T and Johnstone R M 1983 Fate of the transferrin receptor during maturation of sheep reticulocytes *in vitro*: selective externalization of the receptor *Cell* **33** 967–78

[7] Clancy J and D'Souza-Schorey C 2018 Extracellular vesicles in cancer: purpose and promise *Cancer J.* **24** 65–9

[8] Colombo M, Raposo G and Théry C 2014 Biogenesis, secretion, and intercellular interactions of Exosomes and other extracellular vesicles *Annu. Rev. Cell Dev. Biol.* **30** 255–89

[9] Kowal J, Tkach M and Théry C 2014 Biogenesis and secretion of exosomes *Curr. Opin. Cell Biol.* **29** 116–25

[10] Meldolesi J 2018 Exosomes and ectosomes in intercellular communication *Curr. Biol.* **28** R435–R44

[11] Sun W, J-d L, Jiang H and Duan D 2018 Tumor exosomes: a double-edged sword in cancer therapy *Acta Pharmacol. Sin.* **39** 534

[12] Tkach M and Théry C 2016 Communication by extracellular vesicles: where we are and where we need to go *Cell* **164** 1226–32

[13] Johnstone R M, Adam M, Hammond J R, Orr L and Turbide C 1987 Vesicle formation during reticulocyte maturation. Association of plasma membrane activities with released vesicles (exosomes) *J. Biol. Chem.* **262** 9412–20

[14] Couch Y *et al* 2021 A brief history of nearly EV-erything – the rise and rise of extracellular vesicles *J. Extra. Ves.* **10** e12144

[15] Witwer K W and Théry C 2019 Extracellular vesicles or exosomes? On primacy, precision, and popularity influencing a choice of nomenclature *J. Extra. Ves.* **8** 1648167

[16] Raposo G, Nijman H W, Stoorvogel W, Liejendekker R, Harding C V, Melief C J and Geuze H J 1996 B lymphocytes secrete antigen-presenting vesicles *J. Exp. Med.* **183** 1161–72

[17] Bobrie A, Colombo M, Raposo G and Théry C 2011 Exosome secretion: molecular mechanisms and roles in immune responses *Traffic* **12** 1659–68

[18] Raposo G and Stahl P D 2019 Extracellular vesicles: a new communication paradigm? *Nat. Rev. Mol. Cell Biol.* **20** 509–10

[19] Alexander M, Hu R, Runtsch M C, Kagele D A, Mosbruger T L, Tolmachova T, Seabra M C, Round J L, Ward D M and O'Connell R M 2015 Exosome-delivered microRNAs modulate the inflammatory response to endotoxin *Nat. Commun.* **6** 7321

[20] Keller M D, Ching K L, Liang F X, Dhabaria A, Tam K, Ueberheide B M, Unutmaz D, Torres V J and Cadwell K 2020 Decoy exosomes provide protection against bacterial toxins *Nature* **579** 260–4

[21] Xiao X, Yu S, Li S, Wu J, Ma R, Cao H, Zhu Y and Feng J 2014 Exosomes: decreased sensitivity of lung cancer A549 cells to cisplatin *PLoS One* **9** e89534

[22] King H W, Michael M Z and Gleadle J M 2012 Hypoxic enhancement of exosome release by breast cancer cells *BMC Cancer* **12** 421

[23] Kanemoto S, Nitani R, Murakami T, Kaneko M, Asada R, Matsuhisa K, Saito A and Imaizumi K 2016 Multivesicular body formation enhancement and exosome release during endoplasmic reticulum stress *Biochem. Biophys. Res. Commun.* **480** 166–72

[24] Lehmann B D, Paine M S, Brooks A M, McCubrey J A, Renegar R H, Wang R and Terrian D M 2008 Senescence-associated exosome release from human prostate cancer cells *Cancer Res.* **68** 7864–71

[25] Vechetti I J Jr *et al* 2021 Mechanical overload-induced muscle-derived extracellular vesicles promote adipose tissue lipolysis *FASEB J.* **35** e21644

[26] Verweij F J *et al* 2019 Live tracking of inter-organ communication by endogenous exosomes *in vivo* *Dev. Cell* **48** 573–89.e4

[27] Colvin R A *et al* 2010 Synaptotagmin-mediated vesicle fusion regulates cell migration *Nat. Immunol.* **11** 495–502

[28] Proux-Gillardeaux V, Raposo G, Irinopoulou T and Galli T 2007 Expression of the longin domain of TI-VAMP impairs lysosomal secretion and epithelial cell migration *Biol. Cell* **99** 261–71

[29] Sung B H, Ketova T, Hoshino D, Zijlstra A and Weaver A M 2015 Directional cell movement through tissues is controlled by exosome secretion *Nat. Commun.* **6** 7164

[30] Koumangoye R B, Sakwe A M, Goodwin J S, Patel T and Ochieng J 2011 Detachment of breast tumor cells induces rapid secretion of exosomes which subsequently mediate cellular adhesion and spreading *PLoS One* **6** e24234

[31] Mu W, Rana S and Zöller M 2013 Host matrix modulation by tumor exosomes promotes motility and invasiveness *Neoplasia* **15** 875–87

[32] Munson P and Shukla A 2015 Exosomes: potential in cancer diagnosis and therapy *Medicines* **2** 310–27

[33] Giulietti M, Santoni M, Cimadamore A, Carrozza F, Piva F, Cheng L, Lopez-Beltran A, Scarpelli M, Battelli N and Montironi R 2018 Exploring small extracellular vesicles for precision medicine in prostate cancer *Fron. Oncol.* **8** 221

[34] Liu Y, Wang H and Wang J 2018 Exosomes as a novel pathway for regulating development and diseases of the skin *Biomed. Rep.* **8** 207–14

[35] Lowry M C, Gallagher W M and O'Driscoll L 2015 The role of Exosomes in breast cancer *Clin. Chem.* **61** 1457–65

[36] Shen J, Zhu X, Fei J, Shi P, Yu S and Zhou J 2018 Advances of exosome in the development of ovarian cancer and its diagnostic and therapeutic prospect *OncoTargets and Therapy* **11** 2831–41

[37] Vanni I, Alama A, Grossi F, Bello M and Coco S 2017 Exosomes: a new horizon in lung cancer *Drug Discov. Today* **22** 927–36

[38] Guo W, Gao Y, Li N, Shao F, Wang C, Wang P, Yang Z, Li R and He J 2017 Exosomes: new players in cancer *Oncol. Rep.* **38** 665–75

[39] Messenger S W, Woo S S, Sun Z and Martin T F J 2018 A Ca^{2+}-stimulated exosome release pathway in cancer cells is regulated by Munc13-4 *J. Cell Bio.* **217** 2877–90

[40] Zhang C, Ji Q, Yang Y, Li Q and Wang Z 2018 Exosome: function and role in cancer metastasis and drug resistance *Technol. Cancer Res. Treat.* **17** 1533033818763450

[41] Kharaziha P, Ceder S, Li Q and Panaretakis T 2012 Tumor cell-derived exosomes: a message in a bottle *Biochim. Biophys. Acta* **1826** 103–11

[42] Fu H, Yang H, Zhang X and Xu W 2016 The emerging roles of exosomes in tumor–stroma interaction *J. Cancer Res. Clin. Onco.* **142** 1897–907

[43] Kalluri R 2016 The biology and function of exosomes in cancer *J. Clin. Invest.* **126** 1208–15

[44] Bebelman M P, Smit M J, Pegtel M D and Baglio R S 2018 Biogenesis and function of extracellular vesicles in cancer *Pharmacol. Ther.* **188** 1–11

[45] Rashed M H *et al* 2017 Exosomal miR-940 maintains SRC-mediated oncogenic activity in cancer cells: a possible role for exosomal disposal of tumor suppressor miRNAs *Oncotarget* **8** 20145–64

[46] Peinado H, Lavotshkin S and Lyden D 2011 The secreted factors responsible for pre-metastatic niche formation: Old sayings and new thoughts *Seminars Cancer Biol.* **21** 139–46

[47] Chowdhury R, Webber J P, Gurney M, Mason M D, Tabi Z and Clayton A 2014 Cancer exosomes trigger mesenchymal stem cell differentiation into pro-angiogenic and pro-invasive myofibroblasts *Oncotarget* **6** 715–31

[48] Murphy D E, de Jong O G, Brouwer M, Wood M J, Lavieu G, Schiffelers R M and Vader P 2019 Extracellular vesicle-based therapeutics: natural versus engineered targeting and trafficking *Exp. Mol. Med.* **51** 1–12
[49] Nakamura K *et al* 2016 Exosomes promote ovarian cancer cell invasion through transfer of CD44 to peritoneal mesothelial cells *Mol. Cancer Res.* **15** 78–92
[50] Harris D A, Patel S H, Gucek M, Hendrix A, Westbroek W and Taraska J W 2015 Exosomes released from breast cancer carcinomas stimulate cell movement *PLoS One* **10** e0117495
[51] Weidle U H, Birzele F, Kollmorgen G and Ruger R 2017 The multiple roles of exosomes in metastasis *Cancer Genom. Proteom.* **14** 1–15
[52] Wiklander O P B, Brennan M, Lötvall J, Breakefield X O and El Andaloussi S 2019 Advances in therapeutic applications of extracellular vesicles *Sci. Transl. Med.* **11** eaav8521
[53] Howitt J and Hill A F 2016 Exosomes in the pathology of neurodegenerative diseases *J. Biol. Chem.* **291** 26589–97
[54] Lee H J, Patel S and Lee S J 2005 Intravesicular localization and exocytosis of alpha-synuclein and its aggregates *J. Neurosci.* **25** 6016–24
[55] Emmanouilidou E, Melachroinou K, Roumeliotis T, Garbis S D, Ntzouni M, Margaritis L H, Stefanis L and Vekrellis K 2010 Cell-produced alpha-synuclein is secreted in a calcium-dependent manner by exosomes and impacts neuronal survival *J. Neurosci.* **30** 6838–51
[56] Danzer K M, Kranich L R, Ruf W P, Cagsal-Getkin O, Winslow A R, Zhu L, Vanderburg C R and McLean P J 2012 Exosomal cell-to-cell transmission of alpha synuclein oligomers *Mol. Neurodegen.* **7** 42
[57] Asai H, Ikezu S, Tsunoda S, Medalla M, Luebke J, Haydar T, Wolozin B, Butovsky O, Kugler S and Ikezu T 2015 Depletion of microglia and inhibition of exosome synthesis halt tau propagation *Nat. Neurosci.* **18** 1584–93
[58] Krämer-Albers E M and Hill A F 2016 Extracellular vesicles: interneural shuttles of complex messages *Curr. Opin. Neurobiol.* **39** 101–7
[59] D'Anca M, Fenoglio C, Serpente M, Arosio B, Cesari M, Scarpini E A and Galimberti D 2019 Exosome determinants of physiological aging and age-related neurodegenerative diseases *Front. Aging Neurosci.* **11** 232
[60] Huttunen M, Samolej J, Evans R J, Yakimovich A, White I J, Kriston-Vizi J, Martin-Serrano J, Sundquist W I, Frickel E M and Mercer J 2021 Vaccinia virus hijacks ESCRT-mediated multivesicular body formation for virus egress *Life Sci. Alliance* **4** e202000910
[61] DeMarino C, Barclay R A, Pleet M L, Pinto D O, Branscome H, Paul S, Lepene B, El-Hage N and Kashanchi F 2019 Purification of high yield extracellular vesicle preparations away from virus *J. Vis. Exp.* **151** e59876
[62] Reiter K, Aguilar P P, Wetter V, Steppert P, Tover A and Jungbauer A 2019 Separation of virus-like particles and extracellular vesicles by flow-through and heparin affinity chromatography *J. Chromatogr.* A **1588** 77–84
[63] Ghosh S *et al* 2020 β-Coronaviruses use lysosomes for egress instead of the biosynthetic secretory pathway *Cell* **183** 1520–35
[64] Nagashima S, Jirintai S, Takahashi M, Kobayashi T, Tanggis, Nishizawa T, Kouki T, Yashiro T and Okamoto H 2014 Hepatitis E virus egress depends on the exosomal pathway, with secretory exosomes derived from multivesicular bodies *J. Gen. Virol.* **95** 2166–75

[65] Vingtdeux V *et al* 2007 Alkalizing drugs induce accumulation of amyloid precursor protein by-products in luminal vesicles of multivesicular bodies *J. Biol. Chem.* **282** 18197–205
[66] Alvarez-Erviti L, Seow Y, Schapira A H, Gardiner C, Sargent I L, Wood M J and Cooper J M 2011 Lysosomal dysfunction increases exosome-mediated alpha-synuclein release and transmission *Neurobiol. Dis.* **42** 360–7
[67] Jaiswal J K, Andrews N W and Simon S M 2002 Membrane proximal lysosomes are the major vesicles responsible for calcium-dependent exocytosis in nonsecretory cells *J. Cell Biol.* **159** 625–35
[68] Delevoye C, Marks M S and Raposo G 2019 Lysosome-related organelles as functional adaptations of the endolysosomal system *Curr. Opin. Cell Biol.* **59** 147–58
[69] Mathivanan S, Fahner C J, Reid G E and Simpson R J 2012 ExoCarta 2012: database of exosomal proteins, RNA and lipids *Nucleic Acids Res.* **40** D1241–4
[70] Momen-Heravi F *et al* 2013 Current methods for the isolation of extracellular vesicles *Biol. Chem.* **394** 1253–62
[71] Rost B R *et al* 2015 Optogenetic acidification of synaptic vesicles and lysosomes *Nat. Neurosci.* **18** 1845–52
[72] Sung B H, von Lersner A, Guerrero J, Krystofiak E S, Inman D, Pelletier R, Zijlstra A, Ponik S M and Weaver A M 2020 A live cell reporter of exosome secretion and uptake reveals pathfinding behavior of migrating cells *Nat. Commun.* **11** 2092
[73] Mathieu M *et al* 2021 Specificities of exosome versus small ectosome secretion revealed by live intracellular tracking of CD63 and CD9 *Nat. Commun.* **12** 4389
[74] Verweij F J *et al* 2018 Quantifying exosome secretion from single cells reveals a modulatory role for GPCR signaling *J. Cell Biol.* **217** 1129–42
[75] Edgar J R, Eden E R and Futter C E 2014 Hrs- and CD63-dependent competing mechanisms make different sized endosomal intraluminal vesicles *Traffic* **15** 197–211
[76] Wehman A M, Poggioli C, Schweinsberg P, Grant B D and Nance J 2011 The P4-ATPase TAT-5 inhibits the budding of extracellular vesicles in *C. elegans* embryos *Curr. Biol.* **21** 1951–9
[77] Cashikar A G and Hanson P I 2019 A cell-based assay for CD63-containing extracellular vesicles *PLoS One* **14** e0220007
[78] Hikita T, Miyata M, Watanabe R and Oneyama C 2018 Sensitive and rapid quantification of exosomes by fusing luciferase to exosome marker proteins *Sci. Rep.* **8** 14035
[79] Verweij F J *et al* 2021 The power of imaging to understand extracellular vesicle biology *in vivo Nat. Methods* **18** 1013–26
[80] Royo F, Théry C, Falcón-Pérez J M, Nieuwland R and Witwer K W 2020 Methods for separation and characterization of extracellular vesicles: results of a worldwide survey performed by the ISEV rigor and standardization subcommittee *Cells* **9** 1955
[81] Li P, Kaslan M, Lee S H, Yao J and Gao Z 2017 Progress in exosome isolation techniques *Theranostics* **7** 789–804
[82] Tang Y T *et al* 2017 Comparison of isolation methods of exosomes and exosomal RNA from cell culture medium and serum *Int. J. Mol. Med.* **40** 834–44
[83] Patel G K, Khan M A, Zubair H, Srivastava S K, Khushman M, Singh S and Singh A P 2019 Comparative analysis of exosome isolation methods using culture supernatant for optimum yield, purity and downstream applications *Sci. Rep.* **9** 5335
[84] Shtam T *et al* 2020 Evaluation of immune and chemical precipitation methods for plasma exosome isolation *PLoS One* **15** e0242732

[85] Ostrowski M *et al* 2010 Rab27a and Rab27b control different steps of the exosome secretion pathway *Nat. Cell Biol.* **12** 19–30
[86] Steyer J A and Almers W 1999 Tracking single secretory granules in live chromaffin cells by evanescent-field fluorescence microscopy *Biophys. J.* **76** 2262–71
[87] Allersma M W, Wang L, Axelrod D and Holz R W 2004 Visualization of regulated exocytosis with a granule-membrane probe using total internal reflection microscopy *Molecular Biology of the Cell* **15** 4658–68
[88] Barg S, Knowles M K, Chen X, Midorikawa M and Almers W 2010 Syntaxin clusters assemble reversibly at sites of secretory granules in live cells *Proc. Natl Acad. Sci. USA* **107** 20804–9
[89] Toomre D, Steyer J A, Keller P, Almers W and Simons K 2000 Fusion of constitutive membrane traffic with the cell surface observed by evanescent wave microscopy *J. Cell Bio.* **149** 33–40
[90] Schmoranzer J, Goulian M, Axelrod D and Simon S M 2000 Imaging constitutive exocytosis with total internal reflection fluorescence microscopy *J. Cell Biol.* **149** 23–32
[91] Hurwitz S N, Conlon M M, Rider M A, Brownstein N C and Meckes D G Jr. 2016 Nanoparticle analysis sheds budding insights into genetic drivers of extracellular vesicle biogenesis *J. Extr. Ves.* **5** 31295
[92] Sung B H and Weaver A M 2017 Exosome secretion promotes chemotaxis of cancer cells *Cell Adh. Migr.* **11** 187–95
[93] Sinha S, Hoshino D, Hong N H, Kirkbride K C, Grega-Larson N E, Seiki M, Tyska M J and Weaver A M 2016 Cortactin promotes exosome secretion by controlling branched actin dynamics *J. Cell Biol.* **214** 197–213
[94] Hoshino D, Kirkbride K C, Costello K, Clark E S, Sinha S, Grega-Larson N, Tyska M J and Weaver A M 2013 Exosome secretion is enhanced by invadopodia and drives invasive behavior *Cell Rep.* **5** 1159–68
[95] Perrin P, Janssen L, Janssen H, van den Broek B, Voortman L M, van Elsland D, Berlin I and Neefjes J 2021 Retrofusion of intralumenal MVB membranes parallels viral infection and coexists with exosome release *Curr. Bio.* **31** 3884–93
[96] Martin T F 2003 Tuning exocytosis for speed: fast and slow modes *Biochim. Biophys. Acta* **1641** 157–65
[97] Bebelman M P, Bun P, Huveneers S, van Niel G, Pegtel D M and Verweij F J 2020 Real-time imaging of multivesicular body-plasma membrane fusion to quantify exosome release from single cells *Nat. Protoc.* **15** 102–21
[98] Hurwitz S N, Nkosi D, Conlon M M, York S B, Liu X, Tremblay D C and Meckes D G Jr. 2017 CD63 regulates Epstein-Barr virus LMP1 exosomal packaging, enhancement of vesicle production, and noncanonical NF-κB Signaling *J. Virol.* **91** e02251
[99] Hurley J H 2015 ESCRTs are everywhere *EMBO J.* **34** 2398–407
[100] Baietti M F *et al* 2012 Syndecan-syntenin-ALIX regulates the biogenesis of exosomes *Nat. Cell Biol.* **14** 677–85
[101] Stuffers S, Sem Wegner C, Stenmark H and Brech A 2009 Multivesicular endosome biogenesis in the absence of ESCRTs *Traffic* **10** 925–37
[102] van Niel G, Charrin S, Simoes S, Romao M, Rochin L, Saftig P, Marks M S, Rubinstein E and Raposo G 2011 The tetraspanin CD63 regulates ESCRT-independent and -dependent endosomal sorting during melanogenesis *Dev. Cell* **21** 708–21

[103] Hessvik N P and Llorente A 2018 Current knowledge on exosome biogenesis and release *Cell. Mol. Life Sci.* **75** 193–208

[104] Piper R C and Katzmann D J 2007 Biogenesis and function of multivesicular bodies *Annu. Rev. Cell Dev. Biol.* **23** 519–47

[105] Kalluri R and LeBleu V S 2020 The biology, function, and biomedical applications of exosomes *Science* **367**

[106] Langemeyer L, Fröhlich F and Ungermann C 2018 Rab GTPase function in endosome and Lysosome biogenesis *Trends Cell Bio.* **28** 957–70

[107] Wandinger-Ness A and Zerial M 2014 Rab proteins and the compartmentalization of the endosomal system *Cold Spring Harb. Perspect. Biol.* **6** a022616

[108] Jaber N *et al* 2016 Vps34 regulates Rab7 and late endocytic trafficking through recruitment of the GTPase-activating protein armus *J. Cell Sci.* **129** 4424–35

[109] Vanlandingham P A and Ceresa B P 2009 Rab7 regulates late endocytic trafficking downstream of multivesicular body biogenesis and cargo sequestration *J. Biol. Chem.* **284** 12110–24

[110] Hyenne V *et al* 2015 RAL-1 controls multivesicular body biogenesis and exosome secretion *J. Cell Biol.* **211** 27–37

[111] Keerthikumar S *et al* 2016 ExoCarta: a web-based compendium of exosomal cargo *J. Mol. Biol.* **428** 688–92

[112] Ghoroghi S *et al* 2021 Ral GTPases promote breast cancer metastasis by controlling biogenesis and organ targeting of exosomes *eLife* **10** e61539

[113] Gurung S, Perocheau D, Touramanidou L and Baruteau J 2021 The exosome journey: from biogenesis to uptake and intracellular signalling *Cell Commun. Signal* **19** 47

[114] Huotari J and Helenius A 2011 Endosome maturation *EMBO J.* **30** 3481–500

[115] Liégeois S, Benedetto A, Garnier J M, Schwab Y and Labouesse M 2006 The V0-ATPase mediates apical secretion of exosomes containing Hedgehog-related proteins in *Caenorhabditis elegans J. Cell Biol.* **173** 949–61

[116] Kozik P, Hodson N A, Sahlender D A, Simecek N, Soromani C, Wu J, Collinson L M and Robinson M S 2013 A human genome-wide screen for regulators of clathrin-coated vesicle formation reveals an unexpected role for the V-ATPase *Nat. Cell Biol.* **15** 50–60

[117] Edgar J R, Manna P T, Nishimura S, Banting G and Robinson M S 2016 Tetherin is an exosomal tether *eLife* **5** e17180

[118] Savina A, Fader C M, Damiani M T and Colombo M I 2005 Rab11 promotes docking and fusion of multivesicular bodies in a calcium-dependent manner *Traffic* **6** 131–43

[119] Savina A, Vidal M and Colombo M I 2002 The exosome pathway in K562 cells is regulated by Rab11 *J. Cell Sci.* **115** 2505–15

[120] Woo S S, James D J and Martin T F 2017 Munc13-4 functions as a Ca^{2+} sensor for homotypic secretory granule fusion to generate endosomal exocytic vacuoles *Mol. Biol. Cell* **28** 792–808

[121] Kriegenburg F, Bas L, Gao J, Ungermann C and Kraft C 2019 The multi-functional SNARE protein Ykt6 in autophagosomal fusion processes *Cell Cycle* **18** 639–51

[122] Ruiz-Martinez M, Navarro A, Marrades R M, Viñolas N, Santasusagna S, Muñoz C, Ramírez J, Molins L and Monzo M 2016 YKT6 expression, exosome release, and survival in non-small cell lung cancer *Oncotarget* **7** 51515–24

[123] Gross J C, Chaudhary V, Bartscherer K and Boutros M 2012 Active Wnt proteins are secreted on exosomes *Nat. Cell Biol.* **14** 1036–45

[124] Granger E, McNee G, Allan V and Woodman P 2014 The role of the cytoskeleton and molecular motors in endosomal dynamics *Semin. Cell Dev. Biol.* **31** 20–9

[125] Mittelbrunn M, Vicente Manzanares M and Sánchez-Madrid F 2015 Organizing polarized delivery of exosomes at synapses *Traffic* **16** 327–37

[126] Morel E, Parton R G and Gruenberg J 2009 Annexin A2-dependent polymerization of actin mediates endosome biogenesis *Dev. Cell* **16** 445–57

[127] Matsui T, Osaki F, Hiragi S, Sakamaki Y and Fukuda M 2021 ALIX and ceramide differentially control polarized small extracellular vesicle release from epithelial cells *EMBO Rep.* **22** e51475

[128] Alzahofi N *et al* 2020 Rab27a co-ordinates actin-dependent transport by controlling organelle-associated motors and track assembly proteins *Nat. Commun.* **11** 3495

[129] Gerber P P *et al* 2015 Rab27a controls HIV-1 assembly by regulating plasma membrane levels of phosphatidylinositol 4,5-bisphosphate *J. Cell Biol.* **209** 435–52

[130] Murray D H and Tamm L K 2009 Clustering of syntaxin-1A in model membranes is modulated by phosphatidylinositol 4,5-bisphosphate and cholesterol *Biochemistry* **48** 4617–25

[131] Gandasi N R and Barg S 2014 Contact-induced clustering of syntaxin and munc18 docks secretory granules at the exocytosis site *Nat. Commun.* **5** 3914

[132] Fukuda M 2006 Rab27 and its effectors in secretory granule exocytosis: a novel docking machinery composed of a Rab27.effector complex *Biochem. Soc. Trans.* **34** 691–5

[133] Kasai K, Ohara-Imaizumi M, Takahashi N, Mizutani S, Zhao S, Kikuta T, Kasai H, Nagamatsu S, Gomi H and Izumi T 2005 Rab27a mediates the tight docking of insulin granules onto the plasma membrane during glucose stimulation *J. Clin. Invest.* **115** 388–96

[134] Hsu C *et al* 2010 Regulation of exosome secretion by Rab35 and its GTPase-activating proteins TBC1D10A-C *J. Cell Biol.* **189** 223–32

[135] Kirkbride K C, Hong N H, French C L, Clark E S, Jerome W G and Weaver A M 2012 Regulation of late endosomal/lysosomal maturation and trafficking by cortactin affects Golgi morphology *Cytoskeleton (Hoboken)* **69** 625–43

[136] Lyakhova T A and Knight J D 2014 The C2 domains of granuphilin are high-affinity sensors for plasma membrane lipids *Chem. Phys. Lipids* **182** 29–37

[137] Tsuboi T and Fukuda M 2006 The Slp4-a linker domain controls exocytosis through interaction with Munc18-1.syntaxin-1a complex *Mol. Biol. Cell* **17** 2101–12

[138] Fukuda M, Imai A, Nashida T and Shimomura H 2005 Slp4-a/granuphilin-a interacts with syntaxin-2/3 in a Munc18-2-dependent manner *J. Biol. Chem.* **280** 39175–84

[139] Bierings R, Hellen N, Kiskin N, Knipe L, Fonseca A V, Patel B, Meli A, Rose M, Hannah M J and Carter T 2012 The interplay between the Rab27A effectors Slp4-a and MyRIP controls hormone-evoked Weibel-Palade body exocytosis *Blood* **120** 2757–67

[140] Wang H, Ishizaki R, Kobayashi E, Fujiwara T, Akagawa K and Izumi T 2011 Loss of granuphilin and loss of syntaxin-1A cause differential effects on insulin granule docking and fusion *J. Biol. Chem.* **286** 32244–50

[141] Mizuno K, Fujita T, Gomi H and Izumi T 2016 Granuphilin exclusively mediates functional granule docking to the plasma membrane *Sci. Rep.* **6** 23909

[142] Chen Y A, Scales S J, Patel S M, Doung Y-C and Scheller R H 1999 SNARE complex formation is triggered by Ca^{2+} and drives membrane fusion *Cell* **97** 165–74

[143] Chen Y A and Scheller R H 2001 SNARE-mediated membrane fusion *Nat. Rev. Mol. Cell Biol.* **2** 98–106

[144] Zhao M and Brunger A T 2016 Recent advances in deciphering the structure and molecular mechanism of the AAA+ ATPase N-ethylmaleimide-sensitive factor (NSF) *J. Mol. Biol.* **428** 1912–26

[145] Ma C, Su L, Seven A B, Xu Y and Rizo J 2013 Reconstitution of the vital functions of Munc18 and Munc13 in neurotransmitter release *Science* **339** 421–5

[146] Fader C M, Sánchez D G, Mestre M B and Colombo M I 2009 TI-VAMP/VAMP7 and VAMP3/cellubrevin: two v-SNARE proteins involved in specific steps of the autophagy/multivesicular body pathways *Biochim. Biophys. Acta* **1793** 1901–16

[147] Wei Y *et al* 2017 Pyruvate kinase type M2 promotes tumour cell exosome release via phosphorylating synaptosome-associated protein 23 *Nat. Commun.* **8** 14041

[148] Martinez-Arca S, Alberts P, Zahraoui A, Louvard D and Galli T 2000 Role of tetanus neurotoxin insensitive vesicle-associated membrane protein (Ti-Vamp) in vesicular transport mediating neurite outgrowth *J. Cell Biol.* **149** 889–900

[149] Chaineau M, Danglot L and Galli T 2009 Multiple roles of the vesicular-SNARE TI-VAMP in post-Golgi and endosomal trafficking *FEBS Lett.* **583** 3817–26

[150] Fauré J *et al* 2006 Exosomes are released by cultured cortical neurones *Mol. Cell. Neurosci.* **31** 642–8

[151] Rodríguez A, Webster P, Ortego J and Andrews N W 1997 Lysosomes behave as Ca^{2+}-regulated exocytic vesicles in fibroblasts and epithelial cells *J. Cell Biol.* **137** 93–104

[152] MacDougall D D, Lin Z, Chon N L, Jackman S L, Lin H, Knight J D and Anantharam A 2018 The high-affinity calcium sensor synaptotagmin-7 serves multiple roles in regulated exocytosis *J. Gen. Physiol.* **150** 783–807

[153] Shao X, Li C, Fernandez I, Zhang X, Sudhof T C and Rizo J 1997 Synaptotagmin-syntaxin interaction: the C2 domain as a Ca^{2+}-dependent electrostatic switch *Neuron* **18** 133–42

[154] Chieregatti E, Chicka M C, Chapman E R and Baldini G 2004 SNAP-23 functions in docking/fusion of granules at low Ca^{2+} *Mol. Biol. Cell* **15** 1918–30

[155] Li C, Ullrich B, Zhang J Z, Anderson R G W, Brose N and Südhof T C 1995 Ca^{2+}-dependent and -independent activities of neural and non-neural synaptotagmins *Nature* **375** 594–9

[156] Martinez I, Chakrabarti S, Hellevik T, Morehead J, Fowler K and Andrews N W 2000 Synaptotagmin VII regulates Ca^{2+}-dependent exocytosis of lysosomes in fibroblasts *J. Cell Biol.* **148** 1141–49

[157] Flannery A R, Czibener C and Andrews N W 2010 Palmitoylation-dependent association with CD63 targets the Ca^{2+} sensor synaptotagmin VII to lysosomes *J. Cell Biol.* **191** 599–613

[158] Dolai S, Xie L, Zhu D, Liang T, Qin T, Xie H, Kang Y, Chapman E R and Gaisano H Y 2016 Synaptotagmin-7 functions to replenish insulin granules for exocytosis in human islet β-cells *Diabetes* **65** 1962–76

[159] Widberg C H, Bryant N J, Girotti M, Rea S and James D E 2003 Tomosyn interacts with the t-SNAREs syntaxin4 and SNAP23 and plays a role in insulin-stimulated GLUT4 translocation *J. Biol. Chem.* **278** 35093–101

[160] Gracheva E O, Burdina A O, Touroutine D, Berthelot-Grosjean M, Parekh H and Richmond J E 2007 Tomosyn negatively regulates both synaptic transmitter and neuropeptide release at the *C. elegans* neuromuscular junction *J. Physiol.* **585** 705–9

[161] Hatsuzawa K, Lang T, Fasshauer D, Bruns D and Jahn R 2003 The R-SNARE motif of tomosyn forms SNARE core complexes with syntaxin 1 and SNAP-25 and down-regulates exocytosis *J. Biol. Chem.* **278** 31159–66

[162] Sakisaka T *et al* 2008 Dual inhibition of SNARE complex formation by tomosyn ensures controlled neurotransmitter release *J. Cell Biol.* **183** 323–37

[163] Yizhar O, Matti U, Melamed R, Hagalili Y, Bruns D, Rettig J and Ashery U 2004 Tomosyn inhibits priming of large dense-core vesicles in a calcium-dependent manner *Proc. Natl Acad. Sci. USA* **101** 2578–83

[164] Hyenne V *et al* 2019 Studying the fate of tumor extracellular vesicles at high spatiotemporal resolution using the Zebrafish Embryo *Dev. Cell* **48** 554–72

[165] Mulcahy L A, Pink R C and Carter D R 2014 Routes and mechanisms of extracellular vesicle uptake *J. Extr. Ves.* **3** 24641

[166] Stuendl A, Kunadt M, Kruse N, Bartels C, Moebius W, Danzer K M, Mollenhauer B and Schneider A 2016 Induction of alpha-synuclein aggregate formation by CSF exosomes from patients with Parkinson's disease and dementia with Lewy bodies *Brain* **139** 481–94

[167] Simons M and Raposo G 2009 Exosomes – vesicular carriers for intercellular communication *Curr. Opin. Cell Biol.* **21** 575–81

[168] Kalra H *et al* 2012 Vesiclepedia: a compendium for extracellular vesicles with continuous community annotation *PLOS Biol.* **10** 1–5

[169] Kim D K *et al* 2013 EVpedia: an integrated database of high-throughput data for systemic analyses of extracellular vesicles *J. Extr. Ves.* **2** 20384

[170] Neil S J D, Zang T and Bieniasz P D 2008 Tetherin inhibits retrovirus release and is antagonized by HIV-1 Vpu *Nature* **451** 425–30

[162] Sakisaka T *et al* 2008 Dual inhibition of SNARE complex formation by tomosyn ensures controlled neurotransmitter release *J. Cell Biol.* **183** 323–37

[163] Yizhar O, Matti U, Melamed R, Hagalili Y, Bruns D, Rettig J and Ashery U 2004 Tomosyn inhibits priming of large dense-core vesicles in a calcium-dependent manner *Proc. Natl Acad. Sci. USA* **101** 2578–83

[164] Hyenne V *et al* 2019 Studying the fate of tumor extracellular vesicles at high spatiotemporal resolution using the Zebrafish Embryo *Dev. Cell* **48** 554–72

[165] Mulcahy L A, Pink R C and Carter D R 2014 Routes and mechanisms of extracellular vesicle uptake *J. Extr. Ves.* **3** 24641

[166] Stuendl A, Kunadt M, Kruse N, Bartels C, Moebius W, Danzer K M, Mollenhauer B and Schneider A 2016 Induction of alpha-synuclein aggregate formation by CSF exosomes from patients with Parkinson's disease and dementia with Lewy bodies *Brain* **139** 481–94

[167] Simons M and Raposo G 2009 Exosomes – vesicular carriers for intercellular communication *Curr. Opin. Cell Biol.* **21** 575–81

[168] Kalra H *et al* 2012 Vesiclepedia: a compendium for extracellular vesicles with continuous community annotation *PLOS Biol.* **10** 1–5

[169] Kim D K *et al* 2013 EVpedia: an integrated database of high-throughput data for systemic analyses of extracellular vesicles *J. Extr. Ves.* **2** 20384

[170] Neil S J D, Zang T and Bieniasz P D 2008 Tetherin inhibits retrovirus release and is antagonized by HIV-1 Vpu *Nature* **451** 425–30

[162] Sakisaka T *et al* 2008 Dual inhibition of SNARE complex formation by tomosyn ensures controlled neurotransmitter release *J. Cell Biol.* **183** 323–37

[163] Yizhar O, Matti U, Melamed R, Hagalili Y, Bruns D, Rettig J and Ashery U 2004 Tomosyn inhibits priming of large dense-core vesicles in a calcium-dependent manner *Proc. Natl Acad. Sci. USA* **101** 2578–83

[164] Hyenne V *et al* 2019 Studying the fate of tumor extracellular vesicles at high spatiotemporal resolution using the Zebrafish Embryo *Dev. Cell* **48** 554–72

[165] Mulcahy L A, Pink R C and Carter D R 2014 Routes and mechanisms of extracellular vesicle uptake *J. Extr. Ves.* **3** 24641

[166] Stuendl A, Kunadt M, Kruse N, Bartels C, Moebius W, Danzer K M, Mollenhauer B and Schneider A 2016 Induction of alpha-synuclein aggregate formation by CSF exosomes from patients with Parkinson's disease and dementia with Lewy bodies *Brain* **139** 481–94

[167] Simons M and Raposo G 2009 Exosomes – vesicular carriers for intercellular communication *Curr. Opin. Cell Biol.* **21** 575–81

[168] Kalra H *et al* 2012 Vesiclepedia: a compendium for extracellular vesicles with continuous community annotation *PLOS Biol.* **10** 1–5

[169] Kim D K *et al* 2013 EVpedia: an integrated database of high-throughput data for systemic analyses of extracellular vesicles *J. Extr. Ves.* **2** 20384

[170] Neil S J D, Zang T and Bieniasz P D 2008 Tetherin inhibits retrovirus release and is antagonized by HIV-1 Vpu *Nature* **451** 425–30

www.ingramcontent.com/pod-product-compliance
Ingram Content Group UK Ltd.
Pitfield, Milton Keynes, MK11 3LW, UK
UKHW051905100926
8556IPUK00005B/777

9 780750 3376